FLUVIAL FORMS AND PROCESSES

FLUVIAL FORMS AND PROCESSES

A New Perspective

David Knighton
Department of Geography, University of Sheffield, UK

Hodder Arnold

A MEMBER OF THE HODDER HEADLINE GROUP

First published in Great Britain in 1998 by
Arnold, a member of the Hodder Headline Group
338 Euston Road, London NW1 3BH
http://www.hoddereducation.com

Co-published in the United States of America by
Oxford University Press Inc.
198 Madison Avenue, New York, NY 10016

British Library Cataloguing in Publication Data
A catalogue entry for this book is available from the British Library

Library of Congress Cataloging-in-Publication Data
A catalog entry for this book is available from the Library of Congress

ISBN-10: 0 340 66313 8
ISBN-13: 978 0 340 66313 4

10 11 12

Production Editor: Liz Gooster
Production Controller: Sarah Kett
Cover Designer: Yvonne Booth

Composition by Saxon Graphics, Derby
Printed and bound in Malta.

For Celia, Karen and Tanya

CONTENTS

LIST OF SYMBOLS

a coefficient (specifically in width–discharge relation); channel initiation function

A cross-sectional area

A_b channel capacity (bankfull cross-sectional area)

$\mathrm{A_d}$ drainage basin area

$\bar{\mathrm{A}}_\mathrm{u}$ mean drainage area of streams of order u

b exponent (specifically in width–discharge relation)

B bifurcating link

B per cent silt–clay in channel banks

B_r braiding index

c coefficient (specifically in depth–discharge relation)

C cis-link

C concentration of solutes or solids; Chezy coefficient

CT cis–trans link

d mean depth

d_{max} maximum depth

D fractal dimension

D bed material size

$\bar{D}$ mean grain size

D_d drainage density

D_r rate of detachment by raindrop impact, delivery ratio

D_x grain diameter at which x per cent of bed material sample is finer

D_{50} median grain size

E gross erosion in catchment

f exponent in depth–discharge relation

ff Darcy–Weisbach friction factor

F Froude number

F fan area

g gravitational acceleration; coefficient in slope–discharge relation

h height above datum

H bed height, step height

I rainfall intensity, input

$\bar{I}_i$ mean interior link length

j exponent in suspended load–discharge relation

k coefficient in velocity–discharge relation; constant

K hydraulic conductivity
l_s mean length from channel heads to divides
L stream length; distance downstream; wavelength
L_i length of ith link
$\overline{L}_u$ mean length of streams of order u
L_Ω length of the highest-order stream
m exponent (specifically in velocity–discharge relation), mass
M magnitude, weighted per cent of silt–clay in channel perimeter
n Manning resistance coefficient; constant
N_u number of streams of order u
N(x) network width function
O output
p exponent in relation of ff to discharge
p_f, p_s proportions of total hillslope runoff in fast (p_f) and slow (p_s) components
P length of wetted perimeter
P_m mean annual precipitation
q discharge per unit width
q_{cr} critical discharge
q_h hillslope response function
q_n network response function
q_{sb} Q_{sb} per unit width
q_{ss} Q_{ss} per unit width
Q stream discharge
Q_b bankfull discharge
Q_d dominant discharge
Q_e most effective discharge
Q_{diss} dissolved load
Q_m mean annual discharge, mainstream discharge
Q_{ma} mean annual flood
Q_p peak discharge
Q_r discharge ratio (= Q_t/Q_m)
Q_s sediment load/discharge of bed-material load (= Q_{sb} + Q_{ss})
Q_{sb} bed-load discharge
Q_{ss} discharge of suspended fraction of bed-material load
Q_{susp} suspended sediment load
Q_t tributary discharge, threshold discharge
Q_2 discharge with a return period of 2 years (median annual flood)
$Q_{2.33}$ discharge with a return period of 2.33 years (mean annual flood)
r_c radius of curvature
R hydraulic radius
Re Reynolds number
R_A drainage area ratio
R_B bifurcation ratio
R_L stream length ratio
s channel gradient, energy gradient
s_r 'regime slope'

S source link
S storage; channel sinuosity
S_y mean annual sediment yield
t time
t_p time to peak discharge
T trans-link; tributary link
T recurrence interval
TB tributary bifurcating link
TF transient form ratio
TS tributary-source link
u stream order
u_* shear velocity
u_{*o} threshold shear velocity
v mean velocity, stream order
v_{cr} critical velocity
v_g velocity of groundwater flow
V peak velocity of flood wave
w width
w_b bed width
x distance; variable
x_c, X_c critical distances
y exponent in relation of Manning's n to discharge; channel form variable; water depth
z exponent in slope–discharge relation
α downstream rate of change of bed material size; coefficient
β exponent
γ specific weight of water ($= \rho g$)
δ downstream rate of change of channel gradient
δ_o thickness of laminar sublayer
Δ bedform height
η eddy viscosity; measure of grain packing
θ dimensionless shear stress; local path direction; slope angle
λ wavelength (specifically meander wavelength)
λ_* path wavelength
Λ_d dune wavelength
μ link magnitude; molecular viscosity
ν kinematic viscosity
ρ water density
ρ_s sediment density
σ sorting coefficient
τ shear stress
τ_o boundary shear stress
τ_{cr} critical shear stress
ϕ angle of repose and friction
Ψ bedform steepness ($= \Delta/\Lambda$)
ω stream power per unit width ($= \Omega/w$); maximum deviation angle
ω_{cr} threshold unit stream power
Ω stream power

ACKNOWLEDGEMENTS

The author and publishers would like to thank the following for permission to use copyright material in this book:

American Society of Civil Engineering Publications for our figure 5.21b, originally from Nanson, G.C. and Hickin, E.J. 1983: Channel migration and incision on the Beatton River. *Journal of Hydraulic Engineering* 109, pp 327–37, fig 2b; Arnold for our figure 4.12a, originally from Hoey, T.B. 1992: Temporal variations in bedload transport rates and sediment storage in gravel-bed rivers. *Progress in Physical Geography* 16, pp 319–38, fig 1b; Blackwell Publishers Ltd for our figure 3.3, originally from Burt, T.P. 1992: The hydrology of headwater catchments. In Calow, P. and Petts, G.E. (eds) *The Rivers Handbook: Hydrological and Ecological Principles*, pp 3–28, fig 1.7; our figure 3.8a, originally from Walling, D.E. and Webb, B. 1992: Water quality: 1. Physical characteristics. In Calow, P. and Petts, G.E. (eds) *The Rivers Handbook*, pp 48–72, fig 3.4; our figure 5.8, originally from Church, M. 1992: Channel morphology and typology. In Calow, P. and Petts, G.E. (eds) *The Rivers Handbook,* pp 126–43, fig 6.2; Blackwell Scientific Publications Ltd for our figure 2.17, originally from Bristow, C.S., Best, J.L. and Roy, A.G. 1993: Morphology and facies models of channel confluences. In Marzo, M. and Puigdefabregas, C. (eds) *Alluvial Sedimentation, Blackwell Scientific, International Association of Sedimentologists Special Publication* 17, pp 91–100, fig 1; our figure 6.1e, originally from Page, K.J. and Nanson, G.C. 1996: Stratigraphic architecture resulting from Late Quaternary evolution of the Riverine Plain, south-eastern Australia. *Sedimentology* 43, pp. 927–45, fig 9; Elsevier Science for our figure 3.8b, originally from Walling, D.E. 1983: The sediment delivery problem. *Journal of Hydrology* 65, pp 209–37, fig 2; our figure 4.7, originally from Webb, B.W. and Walling, D.E. 1982: The magnitude and frequency characteristics of fluvial transport in a Devon drainage basin and some geomorphological implications. *Catena* 9, pp 9–24, fig 6; our figure 4.8d, originally from Williams, G.P. 1989: Sediment concentration versus water discharge during single hydrologic events in rivers. *Journal of Hydrology* 111, pp 89–106, figs 3, 5, 6, 7; our figure 4.11, originally from Gomez, B. 1991: Bedload transport. *Earth-Science Reviews* 31, pp 89–132, fig 19; our figure 4.15, originally from Nanson, G.C. and Croke, J.C. 1992: A genetic classification of floodplains. *Geomorphology* 4, pp 459–86, figs 1, 2, 3 (parts); our figure 5.1, originally from

Rosgen, D.L. 1994: A classification of natural rivers. *Catena* 22, pp 169–99, fig 4; our figure 5.5, originally from Renwick, W.H. 1992: Equilibrium, disequilibrium and nonequilibrium landforms in the landscape. *Geomorphology* 5, pp 265–76, fig 1; our figure 5.12, originally from Yalin, M.S. 1992: *River Mechanics*, figs 2.1 and 3.7; our figure 6.11b, originally from Xu Jiongxin, 1996a: Underlying gravel layers in a large sand bed river and their influence on downstream-dam channel adjustment. *Geomorphology* 17, 351–9, fig 4; Geological Society of America for our figure 5.21c, originally from Nanson, G.C. and Hickin, E.J. 1986: A statistical examination of bank erosion and channel migration in western Canada. *Bulletin of the Geological Society of America* 97, pp 497–504, fig 3; Geological Society of London for our figure 5.16b, originally from Ferguson, R.I. 1993: Understanding braiding processes in gravel-bed rivers: progress and unsolved problems. In Best, J.L. and Bristow, C.S. (eds) *Braided Rivers, Special Publication of the Geological Society of London* 75, pp 73–87, fig 1; International Association of Hydrological Sciences for our figure 2.19b, originally from Naden, P.S. and Polarski, M. 1990: Derivation of river network variables from digitised data and their use in flood estimation. *Report to MAFF, Institute of Hydrology* 46, fig 3.5; our figure 2.19c, originally from Naden, P.S. 1993: A routing model for continental-scale hydrology. *International Association of Hydrological Sciences Publication* 214, pp 67–89, fig 4; our figure 3.8c, originally from Foster, I.D.L. and Walling, D.E. 1994: Using reservoir deposits to reconstruct changing sediment yields and sources in the catchment of the Old Mill Reservoir, south Devon, over the past 50 years. *Hydrological Sciences Journal* 39, pp 347–68, fig 4; New Zealand Hydrological Society for our figure 5.29b, originally from Wilson, D.D. 1985: Erosional and depositional trends in rivers of the Canterbury Plains, NZ. *Journal of Hydrology* (NZ) 24, 32–44, fig 8; Society of Economic Paleontologists and Mineralogists for our figure 5.16a, originally from Schumm, S.A. 1981: Evolution and response of the fluvial system, sedimentologic implications. *Society of Economic Paleontologists and Mineralogists Special Publication* 31, pp 19–29, fig 6; United States Geological Survey National Center for our figure 6.12, originally from Simon, A. 1994: Gradation processes and channel evolution in modified West Tennessee streams: process, response and form. *United States Geological Survey Professional Paper* 1470, pp 84, figs 28a, 28d and 28e; University of Chicago Press for our figure 4.8a, originally from Millman, J.D. 1983: World-wide delivery of river sediment to the oceans. *Journal of Geology* 91, pp 1–21, fig 1a; our figure 6.14, originally from Meade, R.H. 1982: Sources, sinks and storage of river sediment in the Atlantic drainage of the US. *Journal of Geology* 90, pp 235–52, fig 9a; John Wiley and Sons Ltd for our figure 2.10, originally from Dietrich, W.E. and Dunne, T. 1993: The channel head. In Beven, K. and Kirkby, M.J. (eds) *Channel Network Hydrology*, pp 175–219, fig 7.6; our figures 2.14 and 2.15e, originally from Willgoose, G.R., Bras, R.L. and Rodriguez-Iturbe, I. 1991a: Results from a new model of river basin evolution. *Earth Surface Processes and Landforms* 16, pp 237–54, fig 1 and pp 237–54, fig 4; our figure 2.15f, originally from Kashiwaya, K. 1987: Theoretical investigation of the time variation of drainage density. *Earth Surface*

Processes and Landforms 12, pp 39–46, fig 5; our figure 3.7, originally from Walling, D.E. and Webb, B. 1986: Solutes in river systems. In Trudgill, S.T. (ed.) *Solute Processes,* pp 251–327, figs 7.3b and 7.5c; our figure 3.8d, originally from Dearing, J.A. 1992: Sediment yields and sources in a Welsh upland lake-catchment during the past 800 years. *Earth Surface Processes and Landforms* 17, 1–22, fig 9; 4.8c, originally from Walling, D.E. and Webb, B.W. 1987b: Suspended load in gravel-bed rivers: UK experience. In Thorne, C.R., Bathurst, J.C. and Hey, R.D. (eds) *Sediment Transport in Gravel-bed Rivers,* pp 691–723, fig 22.5; our figure 4.12b, originally from Reid, I., Frostick, L.E. and Layman, J.T. 1985: The incidence and nature of bedload transport during flood flows in coarse-grained alluvial channels. *Earth Surface Processes and Landforms* 10, pp 33–44, figs 2b (part) and 3; our figure 5.14, originally from Sear, D.A. 1996: Sediment transport in pool–riffle sequences. *Earth Surface Processes and Landforms* 21, 241–62, fig 3; our figure 5.15, originally from Thompson, A. 1986: Secondary flows and the pool–riffle unit: a case study of the processes of meander development. *Earth Surface Processes and Landforms* 11, 631–41, fig 2; our figure 6.1b, originally from Hooke, J.M. and Redmond, C.E. 1989: Use of cartographic sources for analysing river channel change with examples from Britain. In Petts, G.E. (ed.) *Historical Change of Large Alluvial Rivers: Western Europe,* pp 79–93, fig 1; our figure 5.19c, originally from Thompson, A. 1986: Secondary flows and the pool–riffle unit: a case study of the processes of meander development. *Earth Surface Processes and Landforms* 11, 631–41, fig 4; our figure 5.29a, originally from Patrick, D.M., Smith, L.M. and Whitten, C.B. 1982: Methods of studying accelerated fluvial change. In Hey, R.D., Bathurst, J.C. and Thorne, C.R. (eds) *Gravel Bed Rivers,* pp 783–812, fig 28.2; our figure 6.3, originally from Rotnicki, A. 1991: Retrodiction of palaeodischarges of meandering and sinuous alluvial rivers and its palaeohydroclimatic implications. In Starkel, L., Gregory, K.J. and Thornes, J.B. (eds) *Temperate Palaeohydrology,* pp 431–71, fig 20.14; our figure 6.8, originally from Lewin, J. 1989: Floods in fluvial geomorphology. In Beven, K.J. and Carling, P.R. (eds) *Floods: Hydrological, Sedimentological and Geomorphological Implications,* pp 265–84, fig 16.6; our figure 6.16, originally from Sear, D.A., Newson, M.D. and Brookes, A. 1995: Sediment-related river maintenance: the role of fluvial geomorphology. *Earth Surface Processes and Landforms* 20, pp 629–47, figs 4 and 5.

1

INTRODUCTION

Overpopulation is the most serious problem facing the world today, with too many people making too many demands on a natural environment which is increasingly under stress. Rivers are a fundamental part of that environment. Their behaviour is of interest to a wide variety of concerns, ranging from water supply, navigation and power generation, to recreation and aesthetics. One of their main attractions to human populations is their unidirectional water flow, which provides a continuously renewable resource, a rapid removal system for unwanted substances, and a valuable source of energy. Such features are also of vital importance to the vast range of aquatic plants and animals inhabiting the fluvial ecosystem (Boon, 1992). In addition to their resource associations, rivers represent a potential threat to human populations and property through floods, drought, pollution and erosion. Whether as resource or hazard, rivers have political, social and economic as well as physical relevance. Most nations depend on sustainable river management.

Rivers are essentially agents of erosion and transportation, removing the water and sediment supplied to them from the land surface to the oceans. They provide the routeways that carry excess precipitation to the oceanic store, thereby completing the global hydrological cycle. Despite the fact that less than 0.005 per cent of continental water is stored in rivers at any one time, water flow is one of the most potent forces operating on the Earth's surface, in terms of both the total energy expended and the total amount of debris transported. Rivers transport on average about 19 000 million tonnes of material each year, 80 per cent in solid and 20 per cent in dissolved form (Meybeck, 1979; Milliman and Meade, 1983; Walling, 1987). In performing their erosional and transportational work, rivers have developed and continue to develop a wide range of network and channel forms. The character of those forms relative to the underlying processes at work in rivers is the principal concern of this book.

Rivers usually have well-defined spatial boundaries and can usefully be regarded as open systems in which energy and matter are exchanged with an external environment. The character and behaviour of the fluvial system

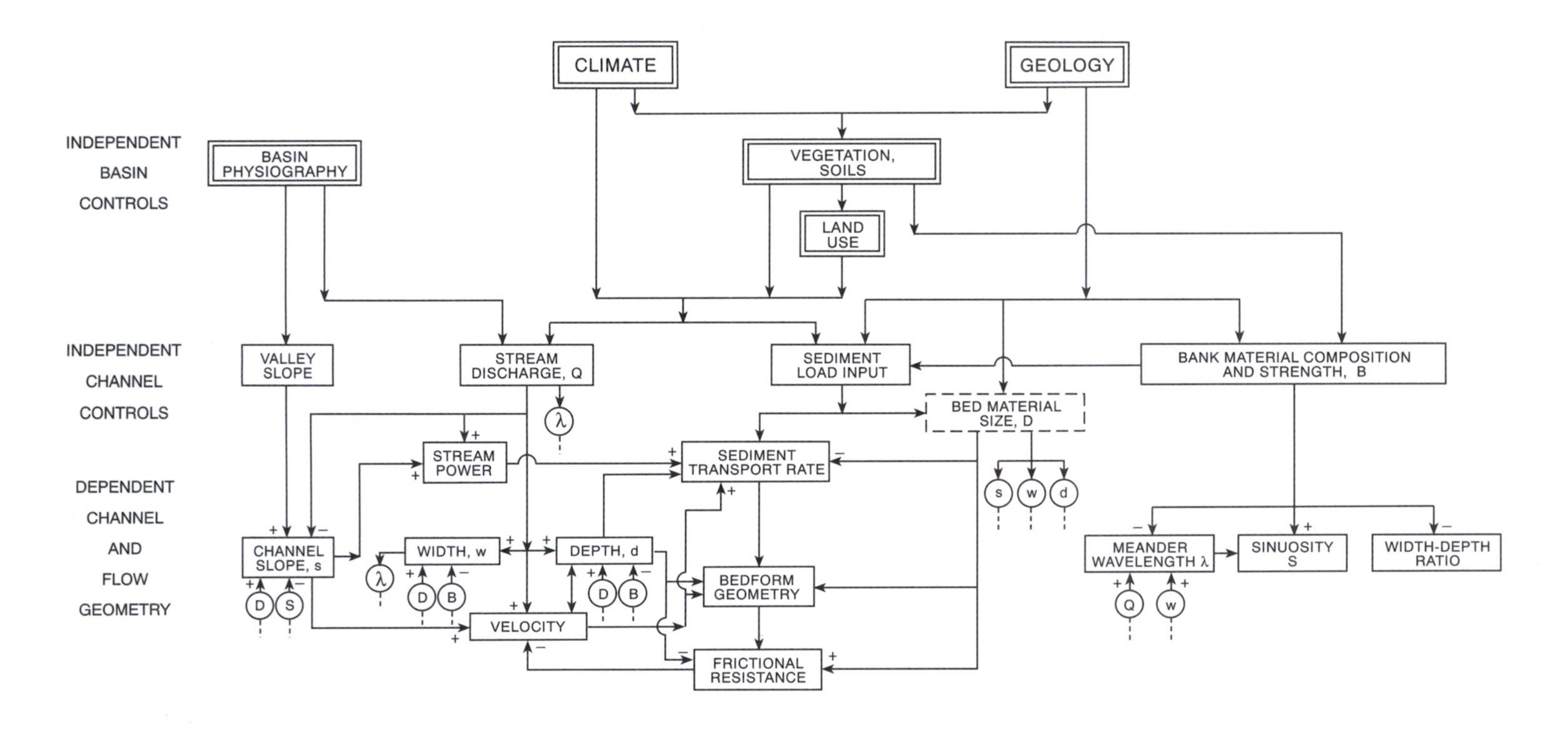

Figure 1.1 Interrelationships in the fluvial system. Relationships are indicated as direct (+) or inverse (–). Arrows indicate the direction of influence.

at any particular location reflect the integrated effect of a set of upstream controls, notably climate, geology, land use and basin physiography, which together determine the hydrologic regime and the quantity and type of sediment supplied (Fig. 1.1). Downstream controls such as the base-level are also important. Climate is of primary significance in that it provides the energy for the most important processes and, in combination with vegetation, directly influences basin hydrology and rates of erosion. Geology is less easily quantified but can have far-reaching effects at a variety of scales, particularly in constraining the nature and level of fluvial activity. In addition to these natural controls, human activities are becoming increasingly influential through river regulation and changing patterns of land use. There are already 36 000 dams worldwide and more than 200 large dams are completed each year (Gregory, 1995). Destruction of tropical forests is proceeding at a rate that could result in their virtual elimination within 40 years (Myers, 1989). Whether directly or indirectly, human interference with the physical environment commonly has an impact on rivers. They are, after all, the drains of the Earth's surface.

Rivers are dynamic entities whose characteristics vary over time and space with changes in environmental controls. Large shifts in climatic conditions over the last 20 000 years have significantly affected levels of fluvial activity in most parts of the world, although, given the sensitivity of alluvial channels, relatively modest climatic changes can trigger major episodes of fluvial adjustment (Knox, 1993, 1995). Consequently, global warming scenarios that predict an increased frequency of heavy rainfall and rising sea-level could induce widespread channel instability. Rivers in different environments are likely to be differentially affected. A major distinction is traditionally drawn between humid- and arid-zone rivers (Knighton and Nanson, 1997; Fig. 3.6, p. 79), with the latter being potentially more sensitive to changing rainfall patterns. Channels also become appreciably more responsive as the boundary sediment decreases in size from boulders to sand.

Channelled flow occurs over a large range of spatial scales, from small headwater streams to major rivers. Within a moderate-size drainage basin, a hierarchy of fluvial forms can be visualized: from the drainage network ($>10^5$ m) to the stream length (e.g. series of meanders), channel reach (e.g. single meander), channel unit (e.g. pool, riffle), subunit (e.g. point bar) and individual particle ($<10^{-1}$ m) scales. Perceptions of form–process relationships change, in proceeding from one scale to the next, even though ultimately all adjustment is the result of individual particle movements. All of these scales feature in this book.

The choice of an appropriate timescale has long been a source of debate among geomorphologists, as it influences our conception of equilibrium within streams, the relationship of cause and effect, and the significance attached to the magnitude–frequency characteristics of process action. In considering the conflict between short-term equilibrium and long-term evolution, Schumm and Lichty (1965) suggested a tripartite division into cyclic, graded and steady times, with corresponding periods of about 10^{4+}, 10^2 and

10^0 years. Progressive change over cyclic time is seen during the shorter span of graded time as a series of fluctuations about a mean state, underlying which are the day-to-day variations in streamflow when channel forms are essentially determined and independent. Over the intermediate or graded timescale and corresponding to that mean state, equilibrium channel forms may be expected to develop, adjusted to the average discharge of water and sediment delivered from the upstream catchment, and dependent on the valley characteristics inherited from the longer time period. Form–process relationships at this timescale are a primary consideration here.

Strong links are apparent between spatial and temporal scales of investigation. In many ways the approaches adopted by various groups differ in their choice of scale and in the significance given to change over time. At one end of the spectrum are mathematicians and physicists concerned with the development of rational theory with which to explain the detailed mechanics of flowing water at small spatial and temporal scales. The more pragmatic engineering approach has focused on stream behaviour over relatively short timespans of 10–100 years or less, with the sediment transport problem as a major concern, especially as it relates to the design of stable channels. To this end the Anglo-Indian school of engineers developed a set of semi-empirical equations, known collectively as 'regime theory', which were intended originally for use in the construction of stable irrigation canals.

Geomorphologists, at least in the first half of this century, have been more interested in long-term, large-scale landform development and in the reconstruction of events assumed to have led to the present forms, an approach owing much to the influence of W. M. Davis's cycle of erosion. Dissatisfaction with the levels of explanation achieved by this approach in the period after 1945 led to a greater concern with the action of contemporary processes and their relationship to form. With the rapid expansion in the use of statistical techniques, functional relations were sought between form and process variables. The emphasis thus shifted from the broad temporal and spatial scales of the denudation chronologist to the short time and small space scales more familiar to the engineer. Although a specific concern with river channel form and process has been the major outcome of these changes in approach, the historical element remains an important part of the geomorphological perspective (Schumm, 1977).

No one approach could be successful in describing and explaining all aspects of natural river systems. A major distinction can be drawn between *empirical* and *theoretical* approaches. The first chiefly involves the collection and analysis of field data in order to establish relationships between form variables, or between a form variable and factors summarizing some aspect of process. Paramount among the large range of relationships which could be used has been the power function:

$$y = \alpha x^{\beta} \tag{1.1}$$

where y is a variable dependent on x, and α and β are coefficients to be determined (Fig. 1.2). Nowhere is this more apparent than in the 'hydraulic geometry' approach pioneered by Leopold and Maddock (1953) as a basis for analysing stream response to changing discharge, both at particular cross-sections and in the downstream direction. The equation is flexible and easily linearized, but whether in its simple bivariate (equation 1.1) or extended multivariate form, its application has not always been adequately justified. Although techniques of analysis have become more sophisticated, the procedure remains largely inductive, with the attendant problem of making generalizations from empirical results usually founded on a restricted database. A strong statistical relationship sometimes gives the illusion that explanation has been achieved, and that the effects of underlying processes have in some way been captured, when often the independent variable is merely a surrogate for some aspect of process. Despite these shortcomings, the empirical approach has been dominant in geomorphology and has provided valuable insights into the workings of the fluvial system. Results of this form are an important preliminary step but, to be of lasting value, need to be embodied within a theoretical structure, even if only qualitatively.

The more deductive *theoretical* approach has as its main aim the formulation and testing of specific statements based on established principles, and commonly involves the construction of models of varying complexity. With the relative lack of established theory, geomorphologists have drawn on the experience gained in allied fields, notably hydraulic engineering, and have frequently argued by analogy between geomorphic and other systems (e.g. Leopold and Langbein, 1962). Characterized by complex interactions of many variables (Fig. 1.1), the fluvial system is eminently suited to the adoption of modelling strategies in which some degree of abstraction or simplification is introduced. Again, a distinction can be made between *deterministic* and *probabilistic* approaches.

Deterministic reasoning is based on the belief that physical laws control the behaviour of natural systems and that, once the laws are known, the behaviour can be predicted exactly or to a satisfactory level of accuracy for a given set of conditions. The basic equations used in modelling are: (a) the continuity equations for water and sediment; (b) the flow momentum equation; (c) a flow resistance equation; and (d) a sediment transport equation. Whereas (a) and (b) are well defined theoretically, being based respectively on the principle of mass conservation and Newton's Second Law of Motion, and (c) has a reasonably sound basis, many sediment transport equations include empirically derived coefficients. Deterministic modelling has been used in a wide range of contexts: from channel initiation (Smith and Bretherton, 1972) and drainage network development (Horton, 1945; Willgoose *et al.*, 1991a, b, c) at one end of the scale spectrum, through meander development (Blondeaux and Seminara, 1985) and the prediction of bed topography in meander bends (Bridge, 1977), to particle entrainment (Parker *et al.*, 1982; Andrews, 1983) at the other. It is possibly most useful when dealing with relatively small-scale problems. As situations become more complex, they are increasingly difficult to represent in terms of a set of closed equations, and consequently predictions become less reliable.

There appear to be two main arguments behind the adoption of an alternative, *probabilistic* strategy in model building. First, the natural world is so complex that a complete deterministic explanation can never be achieved even though each process may be deterministic, a view neatly summarized by Shreve (1975, p. 529): 'Geomorphic systems are descendants of antecedent states that are generally unknown, and they are invariably parts of larger systems from which they cannot be isolated . . . a probabilistic theory that takes account of the apparent randomness is evidently a necessity, because if our theories are to succeed, they must reflect the world as it is, not as we would like it to be.' The second argument goes one step further in not only recognizing an apparent randomness in physical systems, but claiming that inherent randomness is a fundamental property of such systems. Physical laws are regarded as not sufficient by themselves to determine the outcome of system interactions, however detailed are the observations. This view is a central theme in much of the theoretical work carried out by Leopold and Langbein, who intended that inherent randomness should be a basic principle governing behaviour in the fluvial system. Whatever view of randomness is taken, probabilistic methods have been widely used in the fluvial context, notably in drainage network analysis (Shreve, 1966, 1967) and network growth models (Dacey and Krumbein, 1976; Stark, 1991), and in the analysis of equilibrium channel geometry via minimum variance theory (Langbein, 1964a).

Inherent randomness is synonymous with deterministic complexity or *chaos* (Phillips, 1992). Chaos describes irregular behaviour in non-linear dynamical systems which, like the fluvial system, are often characterized by discontinuities or bifurcations during their evolution. Phillips (1992) argued that the potential for chaotic behaviour exists in many geomorphic systems, and that therefore non-linear dynamical systems theory could and should be applied to geomorphic problems. Thus far, applications of chaos theory have been limited, although the allied field of *fractal geometry* has been used to analyse drainage network structure (La Barbera and Rosso, 1989; Rinaldo *et al.*, 1992) and meander form (Snow, 1989; Stølum, 1996).

Very real problems exist in modelling a complex physical system whatever framework, deterministic or probabilistic, is adopted, and most progress is likely to be made with a mixed approach. Considering the inherent variability of natural streams, physical modelling provides an opportunity for scaling down space, accelerating change over time, and holding

Figure 1.2 (opposite) Power function relationships. (A) Relationship of bankfull discharge to drainage area, Upper Salmon River, Idaho (data from Emmett, 1975). (B) Relationship of channel width to bankfull discharge, British rivers (data from Nixon, 1959). (C) Relationship of meander wavelength to channel width: 1 = stream plate experiments (Gorycki, 1973); 2 = solution channel meanders in limestone; 3 = supraglacial stream meanders; 4 = river meanders; 5 = Gulf Stream meanders (Zeller, 1967).

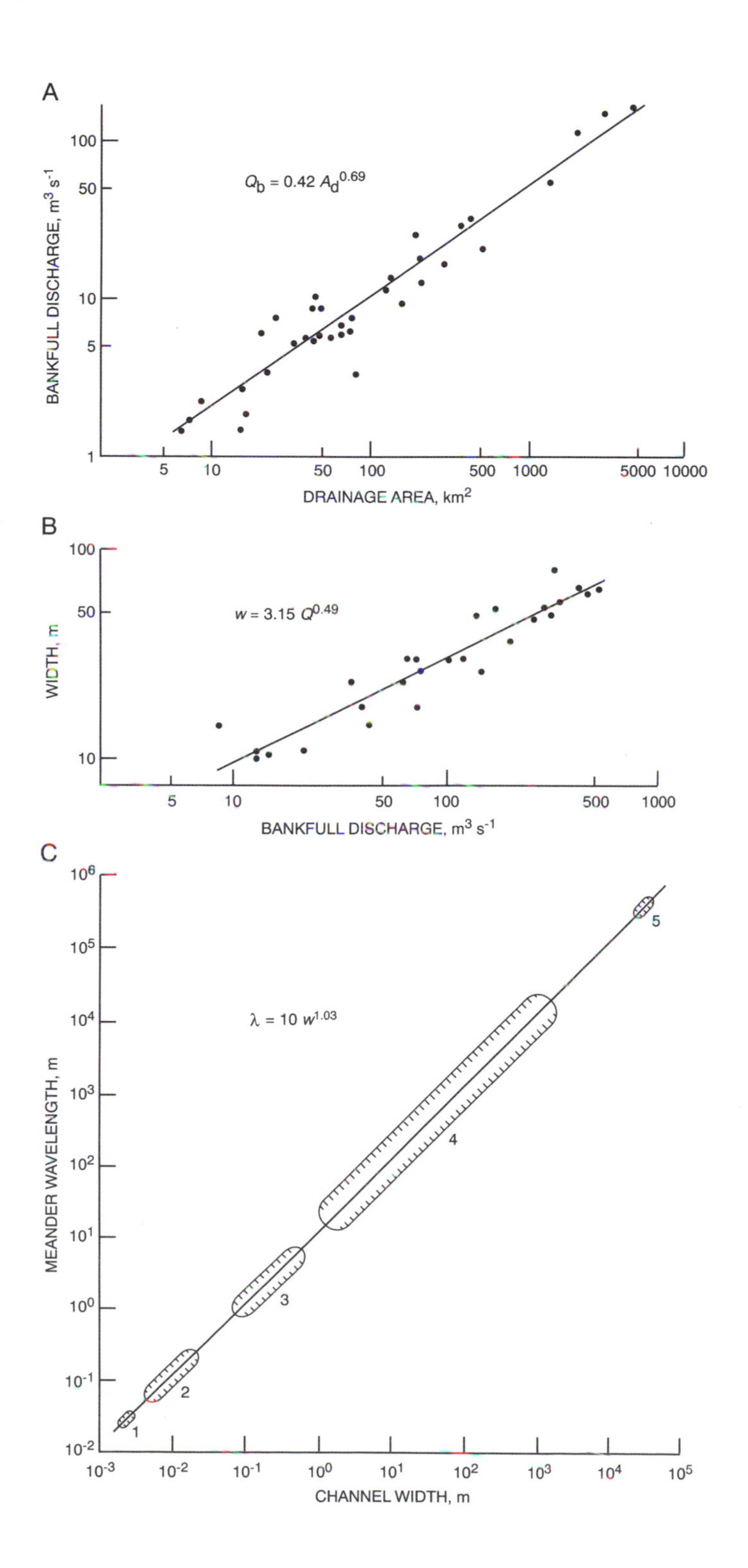

A
BANKFULL DISCHARGE, $m^3 s^{-1}$
$Q_b = 0.42 A_d^{0.69}$
DRAINAGE AREA, km^2
B
WIDTH, m
$w = 3.15 Q^{0.49}$
BANKFULL DISCHARGE, $m^3 s^{-1}$
C
MEANDER WAVELENGTH, m
$\lambda = 10 w^{1.03}$
CHANNEL WIDTH, m
1
2
3
4
5

certain conditions constant in order to identify detailed interactions. In this area the laboratory flume has occupied a pre-eminent position, even though it too has quite stringent limitations (Maddock, 1969). Hydraulic models have been used extensively to elucidate the sedimentary processes involved in the development of braiding (Leopold and Wolman, 1957; Ashmore, 1991a, b; Ashworth, 1996), complementing the explanations obtained from field-based and theoretical studies. Despite the use of various modelling strategies, geomorphology remains essentially a field science. With the ultimate objective of explaining the behaviour of natural streams, field observations provide the data necessary for empirical relationships and the testing of models, subject always to the constraints of an adequate sampling base and minimum disturbance to a system in which single effects are seldom the result of single causes.

2

DRAINAGE NETWORKS

Hillslopes and channels are the constituent elements of the drainage basin. Occurring in a hierarchy of sizes, the drainage basin is typically a well-defined topographic and hydrologic entity which is regarded as a fundamental spatial unit (Chorley, 1969). While the links between hillslopes and channels in the movement of water and material through the drainage basin should be recognized at the outset, the channel network itself is the initial focus here, firstly as regards its properties, and secondly as regards its possible mode of evolution. The first can be determined from the analysis of existing networks but the second can rarely be observed directly, and explanations of evolutionary tendencies often depend on the results of network analyses.

Network Analysis

Drainage network analysis has been used not only for the express purpose of identifying characteristics of network structure, but also as a basis for demonstrating the effects of environmental controls on the fluvial system, for suggesting how networks might evolve, and for indicating how basin output variables such as stream discharge are related to the net. The provision of an adequate database is therefore essential, involving such problems as consistent sampling and channel definition (Jarvis, 1977).

Because channel delineation in the field is time-consuming, most network analyses have been based on data derived from topographic maps which can be of variable accuracy and at best represent an 'average' stream network. The main issue is the unambiguous definition of fingertip tributaries and their headward limits, a problem complicated by short-term fluctuations in stream-head position (Gregory and Walling, 1968) and by the differences between perennial, intermittent and ephemeral streams. Except under favourable circumstances (Knighton *et al.*, 1992), aerial photographs provide only a partial answer to the drawbacks of cartographic sources, and are probably best used as a check on the integrity of map-derived networks.

The commonest practice is to take the blue-line network at an appropriate map scale (usually 1 : 25 000 in Britain) as the initial basis. Various methods for extending that network have been suggested, with the contour-crenulation method as the most popular, but their reliability can be highly variable, leading to inconsistency. Different methods, or even different operators using the same method, can produce large discrepancies, particularly in the length properties of headwater streams. Consistent and objective criteria for channel definition are needed if the results from different analyses are to be compared effectively.

Drainage Network Composition

Two themes dominate the early approaches: the importance of time, and geologic structure. The classification of streams by such terms as consequent, subsequent and obsequent, to which Davis (1899) attached evolutionary significance, reflects an early recognition of a hierarchical structure in drainage networks. Zernitz (1932) collated an elaborate set of terms and type examples intended to show the effects of geology and, to a lesser extent, topography on drainage patterns. Thus, for example, dendritic networks reflect a relative lack of geologic control, while trellis networks develop in areas with parallel belts of dipping strata having differential resistance to erosion. Applicable largely at a regional scale, such classifications are essentially qualitative, although a numerical methodology based on distinguishing characteristics has recently been suggested (Ichoku and Chorowicz, 1994).

Two related concepts introduced by Horton (1945), *stream order* and *drainage density*, laid the foundation for modern network analysis. Their development has tended to follow separate paths, with emphasis respectively on the internal composition or the overall geometry of networks.

Drainage network composition refers to the internal structure of networks, specifically their topologic and geometric properties. Horton's intention was to replace previous qualitative descriptions of drainage basins and their constituent networks with quantitative ones in which a hierarchical structure was explicitly recognized. The basis of the **Horton–Strahler approach** is a method of classifying segments of channel in terms of stream order. Assuming no triple junctions within an essentially dendritic network, the system of stream ordering (now most commonly by Strahler's (1952) modification of Horton's method) involves the following rules:

(a) fingertip tributaries originating at a source are designated order 1;
(b) the junction of two streams of order u forms a downstream channel segment of order $u + 1$;
(c) the junction of two streams of unequal order u and v, where $v > u$, creates a downstream segment having an order equal to that of the higher-order stream v (Fig. 2.1A).

The distinction between the streams of order u in (b) and (c), classified respectively as order-formative and order-excess by Jarvis (1976), underlines

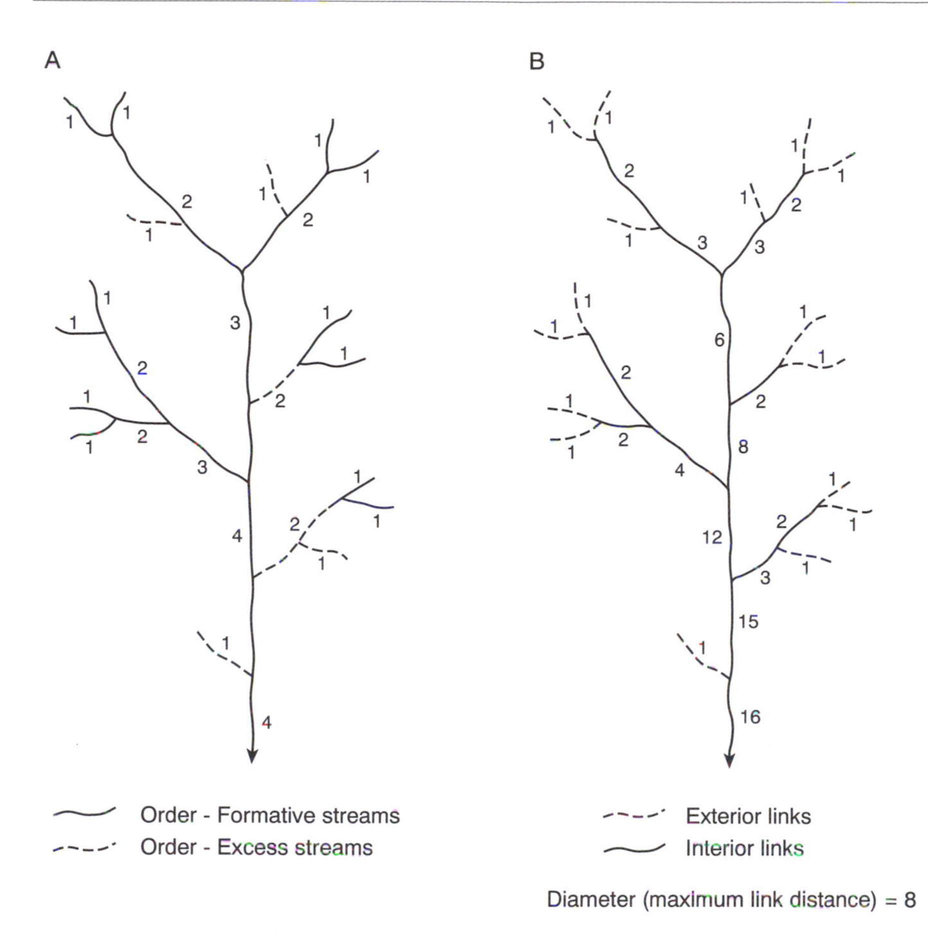

Figure 2.1 Systems of stream ordering. (A) Channel segments ordered by the Horton system as modified by Strahler (1952). (B) Channel links ordered by magnitude.

one serious drawback of the ordering system. Besides violating the associative law of algebra, rule (c) is out of accord with physical reality. Properties such as stream discharge can change when a lower-order tributary enters a higher-order stream, but the order of the main stream remains unaltered. This problem has partly been resolved by an alternative method which uses link magnitude as its basis (Fig. 2.1B).

The immediate outcomes of Horton's ordering scheme were the laws of stream numbers and stream lengths, to which a third, the law of drainage areas, was later added (Schumm, 1956). These so-called *laws of drainage network composition* (Table 2.1) approximate to geometric progressions of inverse (law of stream numbers) or direct (laws of stream lengths and drainage areas) form. Plots of log N_u, log $\overline{L}_u$ and log $\overline{A}_u$ against stream order (u) are therefore approximately linear (Fig. 2.2). Except in conditions of strong geologic control, the three ratios (R_B, R_L and R_A) tend to have rela-

Table 2.1 Laws of drainage network composition

	Ratio form	Functional form	Author
Law of stream numbers	$\frac{N_{u-1}}{N_u} \cong R_B$	$N_u \cong \alpha_1 e^{-\beta_1 u}$ where $\beta_1 = \ln R_B$	Horton (1945)
Law of stream lengths	$\frac{\bar{L}_u}{\bar{L}_{u-1}} \cong R_L$	$\bar{L}_u \cong \alpha_2 e^{\beta_2 u}$ where $\beta_2 = \ln R_L$	Horton (1945)
Law of drainage areas	$\frac{\bar{A}_u}{\bar{A}_{u-1}} \cong R_A$	$\bar{A}_u \cong \alpha_3 e^{\beta_3 u}$ where $\beta_3 = \ln R_A$	Schumm (1956)

N_u, $\bar{L}_u$, $\bar{A}_u$ are respectively the number, average length and average drainage area of streams of order u; R_B, bifurcation ratio; R_L, stream length ratio; R_A, drainage area ratio; α_1, α_2, α_3 are coefficients.

tively narrow ranges of 3–5, 1.5–3 and 3–6 respectively (Smart, 1972; Kirchner, 1993), suggesting either an underlying regularity in network structure or a lack of sensitivity on the part of the parameters.

Horton's work spawned a generation of essentially empirical analyses in which numerous regularities and relationships were recognized among network variables. A parallelism in the respective plots of mean stream length and mean drainage area against stream order suggested to Schumm (1956) a direct proportionality between length and area, leading to the concept of a *constant of channel maintenance*. It implies that, for a given set of environmental conditions, a minimum area is required for channel initiation, and as such represents an important principle in network evolution.

This theme of area–length association has been variously extended, notably by Hack (1957) in an empirical relationship of mainstream length (L) to drainage basin area (A_d):

$$L = 1.4 A_d^{0.6} \tag{2.1}$$

This appears to hold for a wide range of conditions (Ijjasz-Vasquez *et al.*, 1993). That the exponent is 0.6 rather than 0.5 suggests that drainage basins elongate with increasing size, and this has been interpreted as indicative of network development towards a structure which minimizes total energy expenditure (Ijjasz-Vasquez *et al.*, 1993). However, the exponent value may not be independent of basin size (Church and Mark, 1980), although it is generally agreed to be greater than 0.5. With that proviso, Rigon *et al.* (1996) showed that the relationship appears to hold for any point inside a basin.

The general applicability of the laws of network composition, demonstrated over a period of more than 20 years, implied a degree of organization in network structure not otherwise apparent. Each stream order has a

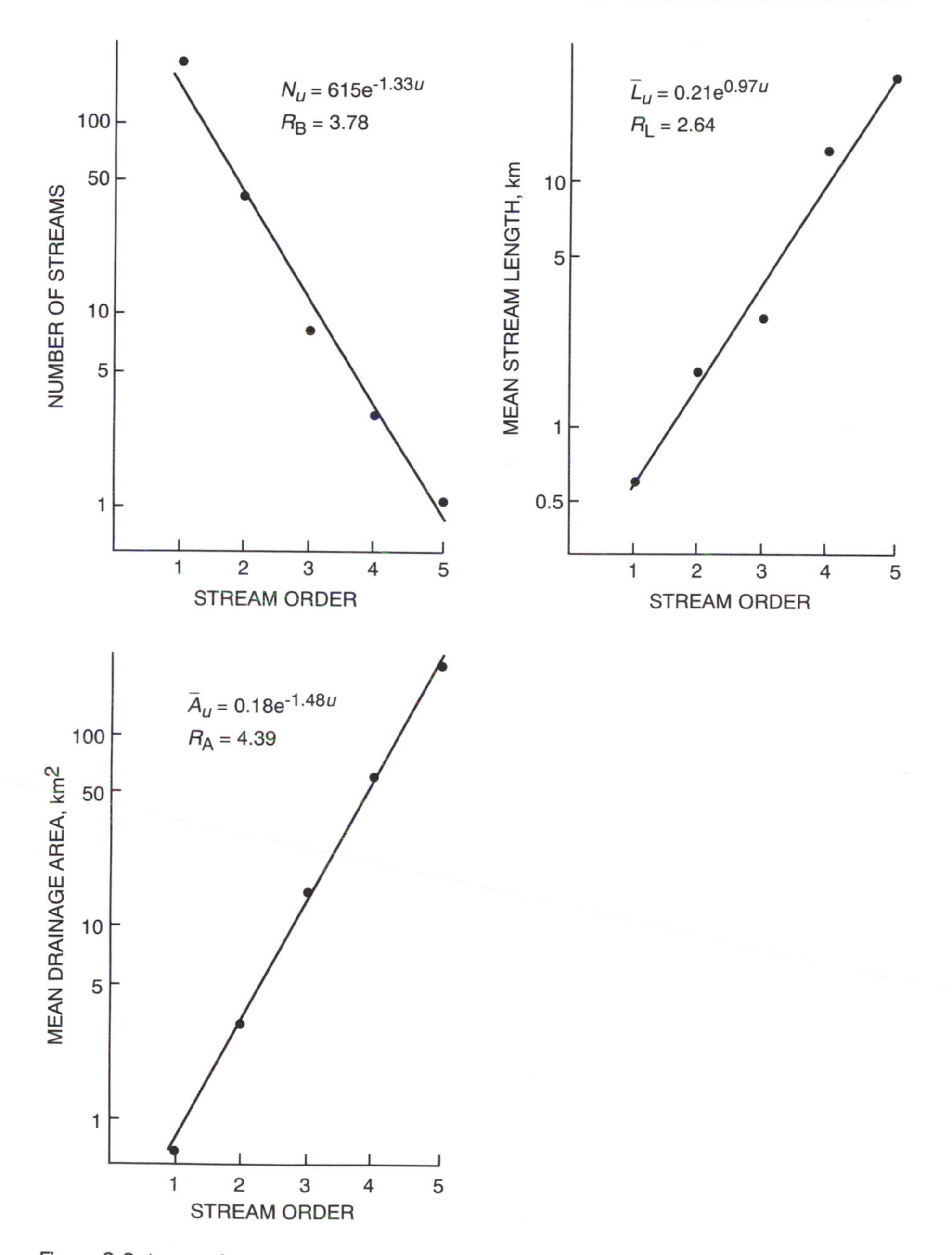

Figure 2.2 Laws of drainage network composition: Bollin-Dean network.

characteristic number of channels, average length and average drainage area. Horton regarded the laws as deterministic statements about natural networks, related through network development to the action of physical processes. However, they apply to any branching network, of which stream networks are merely one subset (Kirchner, 1993). Much of the regularity in the parameters, notably R_B and R_L, is a consequence of the ordering scheme itself, which is insensitive to variations in physical controls and which therefore fails to identify distinctive characteristics of drainage networks. Nevertheless, stream order remains useful as a crude index of basin size for descriptive purposes.

Repetitious application of Horton's methodology has been largely superseded by the broader-based **probabilistic–topologic approach** pioneered by Shreve (1966) and Smart (1968), which takes account of the randomness as well as the regularity in network structure. New models of network composition have been proposed, namely the random topology and random link length models, and new network variables defined, two developments which are closely related but logically independent of one another. To date, the main emphasis has been on the topologic and length properties of networks.

The Strahler stream segment has been replaced as the basic unit of network composition by the *link*, defined as an unbroken section of channel between successive nodes (sources, junctions or outlet). An *exterior link* extends from a source to the first junction downstream and therefore corresponds to a first-order stream, while an *interior link* connects two successive junctions or the last junction with the outlet. A network with M sources thus contains $2M - 1$ links, M being exterior and $M - 1$ interior. The most important topologic parameters are link *magnitude* and network *diameter* (Fig. 2.1B). The magnitude of a link is the number of sources upstream: thus an exterior link has a magnitude of 1, an interior link a magnitude equal to the sum of the magnitudes of the two links at the upstream end, and a network with M sources a magnitude of M. The additive property of magnitude overcomes the problem with stream ordering noted earlier. Diameter is the maximum link distance in a network and is a measure of the longitudinal extent of the network, with mainstream length as its geometric analogue.

The probabilistic–topologic approach is founded on two basic postulates:

(a) In the absence of strong geologic controls, natural channel networks are topologically random (Shreve, 1966); that is, all *topologically distinct channel networks* (TDCN) of a given magnitude are equally probable;

(b) The exterior and interior link lengths of drainage networks developed in similar environments have separate statistical distributions that are approximately independent of location within the basin (Shreve, 1967; Smart, 1968).

These two assumptions are reminiscent of Horton's laws in dealing respectively with topologic and length properties, but their implications are far wider. The random model is an attempt to establish a theoretical framework for the explanation and prediction of drainage network composition.

The number of TDCN for a given number of sources, N(M), is defined by

$$\mathrm{N}(M) = \frac{1}{2M-1}\binom{2M-1}{M} = \frac{(2M-2)!}{M!(M-1)!} \tag{2.2}$$

Thus the numbers of TDCN, or the ways in which the links can be interconnected, for magnitudes one to six are 1, 1, 2, 5, 14 and 42 respectively. Figure

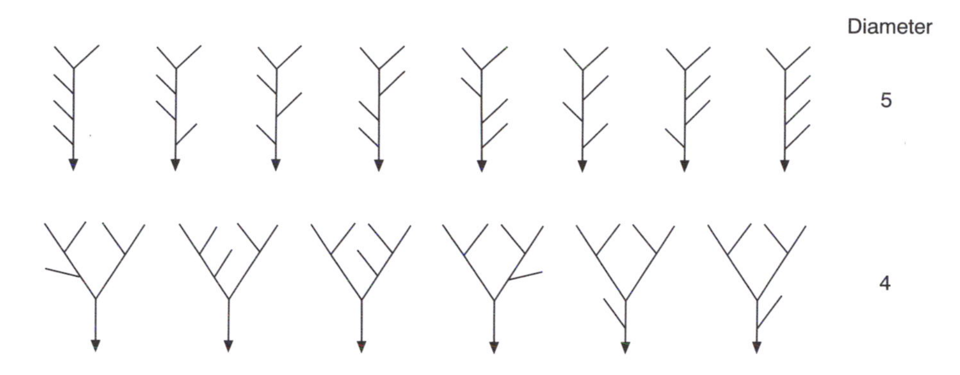

Figure 2.3 Schematic diagram of the N(5) = 14 topologically distinct channel networks (TDCN) of magnitude 5. The networks fall into two distinct diameter classes. In a topologically random population, each of these 14 networks should be equally probable.

2.3 illustrates the 14 possible TDCN for magnitude 5 networks. The only complete test of the random topology model (the first postulate) requires sampled networks of a given magnitude to be grouped into their respective TDCN classes, thereby forming an observed distribution which is then tested against the theoretical equi-frequency distribution. For $M > 5$, the sample sizes required to satisfy statistical tests are so large that a regrouping of TDCN becomes necessary. Direct tests have consequently been limited to networks of low magnitude. The need to aggregate TDCN represents a major drawback to the effective testing of the random topology model, since regrouping loses information.

The random topology model has successfully explained many of the previously observed relationships between network variables, notably Horton's law of stream numbers and the mainstream length–basin area relationship first reported by Hack (1957). It provides a theoretical explanation for the clustering of bifurcation ratios about 4, which begs the question: how can natural networks simultaneously satisfy Horton's seemingly deterministic law of stream numbers and the random topology model of equiprobable TDCN? One implies a sense of regularity in network structure and the other an absence of regularity. However, there is no real contradiction (Jarvis, 1977). Horton's law is not a law in the strict physical sense, but merely a statistical relationship describing the most probable state of network composition.

Early tests of the model tended to confirm its overall validity when applied at the network or TDCN level, even in areas with strong geologic control (Mock, 1976). However, small networks fit the model better than large ones, especially in basins of moderate to low relative relief, but even small networks can show systematic deviations (Abrahams, 1984a). Significant departures from topologic randomness have come to light, particularly when testing has been carried out at the more detailed link level, some of the most convincing evidence for regularities in network topology coming from studies of the

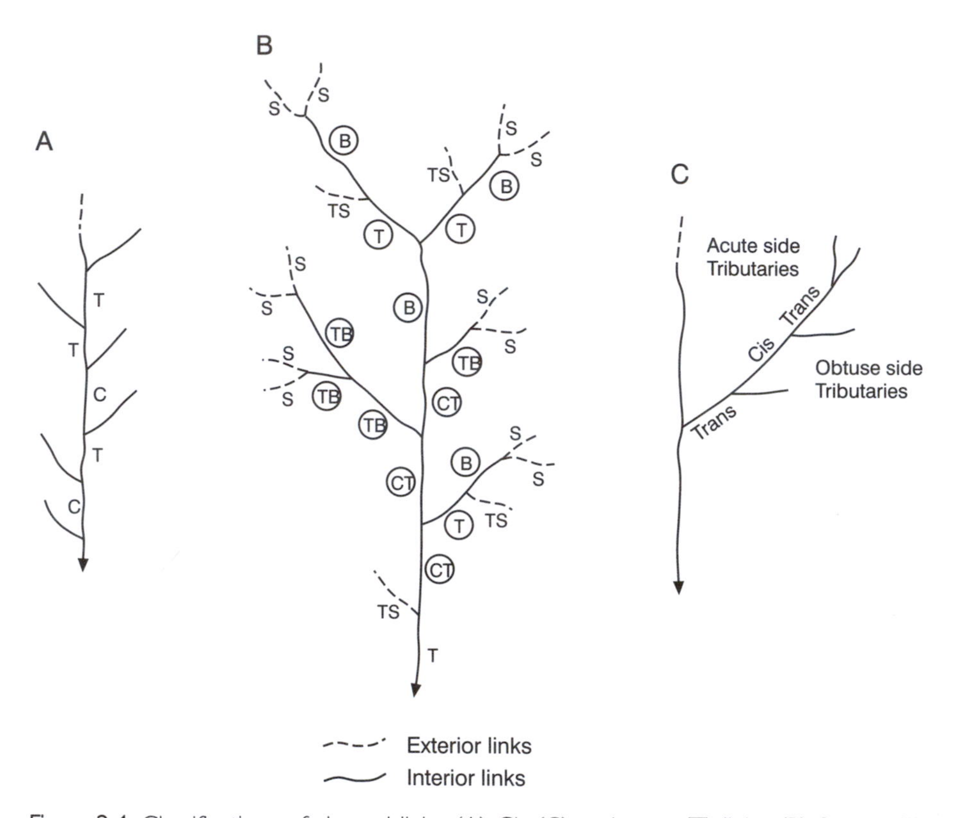

Figure 2.4 Classifications of channel links. (A) Cis (C) and trans (T) links. (B) Source (S), tributary-source (TS), bifurcating (B), tributary (T), cis–trans (CT) and tributary-bifurcating (TB) links as defined by Mock (1971). (C) Tributary arrangements (Flint, 1980).

arrangement of tributaries along main streams. Classifying interior links (Fig. 2.4A) as either *cis-links* (tributaries at either end of the link enter from the same side) or *trans-links* (tributaries at either end enter from opposite sides), James and Krumbein (1969) found that cis-links were less numerous than expected, a result confirmed in some (e.g. Smart, 1978) but not all cases (e.g. Dunkerley, 1977). Flint (1980) has pursued this theme and shown that, relative to the prediction of the random topology model, there is an excess of obtuse tributaries (Fig. 2.4C) upstream from main stream outlets, which can be attributed to the unequal availability of space for tributary development between the acute and obtuse sides. Valley sinuosity is also an influential factor because small tributaries tend to develop preferentially on the concave-downslope surfaces which characterize the outside of bends since such surfaces concentrate runoff (Abrahams, 1984b). Spatial requirements affect not only the arrangement of tributaries but also their size, with large tributaries being separated by clusters of small ones in a more regular way than might be expected to occur by chance (Jarvis and Sham, 1981). These several results do not constitute overall rejection of the random topology model, but rather illustrate how it can serve as a standard for comparison, enabling the roles of factors such as space-filling and basin relief to be better understood. Nevertheless, they do indicate the presence of systematic (non-random) tendencies in the branching pattern of drainage networks (Abrahams, 1987).

The random model has been concerned not only with topologic but also with length properties of drainage networks. Smart (1968, 1972) derived a random link length model in which the mean stream length of order u ($\bar{L}_u$) is expressed in terms of stream numbers:

$$\bar{L}_u = \bar{l}_i \prod_{a=2}^{u} (N_{a-1} - 1)/(2N_a - 1) \qquad u \geq 2 \tag{2.3}$$

where $\bar{l}_i$ is the mean interior link length and N_a is the number of streams of order a. The model explains the behaviour of ordered stream lengths and reveals that, as with the law of stream numbers, Horton's law of stream lengths is largely a result of applying the ordering scheme to a topologically random network.

The model rests on the assumption that interior link lengths are random variables independent of location in the basin or any topologic property such as stream order or link magnitude. While there appears to be no consistent tendency for link length to change downstream (Abrahams, 1984a), several tests cast doubt on the general validity of the assumption. In comparing the length distributions of cis- and trans-links, James and Krumbein (1969) found that their networks had a systematic deficiency of short cis-links. As regards his classification of interior links (Fig. 2.4B; Table 2.2), Mock (1976) noted a relation between link type and mean link length, with length increasing in the order: bifurcating (B), cis–trans (CT), tributary-bifurcating (TB) and tributary (T) links. This result has been confirmed elsewhere (Smart, 1978). Systematic deviations have also been found in the length

Table 2.2 Classification of link types (after Mock, 1971)

Type of link	Definition of link type having magnitude μ	
	Magnitude of upstream links	Magnitude of downstream link
Exterior		
Source (S) link	–	1
Tributary-source (TS) link	–	>1
Interior		
Bifurcating (B) link	½μ	<2μ
Tributary-bifurcating (TB) link	½μ	⩾2μ
Cis–trans (CT) link	≠½μ	<2μ
Tributary (T) link	≠½μ	⩾2μ

properties of exterior links. For six drainage systems in eastern Australia, Abrahams and Campbell (1976) showed that source (S) links are significantly shorter than tributary-source (TS) links, a distinction which becomes sharper further downstream in the network as the latter join progressively higher-magnitude links. A locational influence is thus implied.

Link length studies are very dependent on the accuracy with which the networks are initially delineated, particularly in the case of exterior links whose sources are rarely well-defined on topographic maps, but this problem should not detract from the main implication of these results: the geometric structure of channel networks is more complex than the random model would at first suggest. In terms of both major categories, exterior and interior links, link lengths are not independent random variables drawn from common populations. Significant differences in link length distributions occur between link types, related partly to ground slope and space-filling constraints (Abrahams, 1984a).

The development of the random model represented a profound change in methodology as regards the analysis of drainage network composition. Many of the empirical relationships previously obtained were explained and new relationships predicted. Topological studies of river systems inspired the application of network analysis to dendritic patterns in biology, notably brain cells (Berry *et al.*, 1975). In effect, the random topology and random link length models provide explicit null hypotheses or statistical standards against which natural networks can be compared. However, the equiprobability hypothesis may not be a true test of randomness because of geometric constraints that affect how basins must pack on to a three-dimensional surface (Goodchild and Klinkenberg, 1993). Also, despite the support provided by early tests, significant departures from expectation have been identified, particularly at the detailed link level, which indicate pronounced regularities in network structure associated with various controlling factors. The cost of a more realistic model which is less dependent on random elements and takes more account of systematic tendencies is, however, greater complexity. One possibility is to relax the condition that link lengths are independent random variables (Karlinger and Troutman, 1989), but such developments inevitably sacrifice some of the simplicity which is a hallmark of the random model.

Attention has recently focused on the fractal nature of channel networks. *Fractals* are geometrical structures which have irregular shapes and which retain the same degree of irregularity at all scales (i.e. they are *self-similar*). Fractal dimension (D) is a measure of that complexity, with a minimum value of 1 (smooth lines) and a maximum value of 2 as far as linear features are concerned. In the context of linear networks, the more thoroughly river channels fill the drainage area, the closer D is to 2. Values close to 2 (Tarboton *et al.*, 1988) and in the range 1.5–2 (La Barbera and Rosso, 1989) have been obtained from analyses of natural networks, but few networks are completely space-filling. Also, attempts to correlate fractal dimension with the exponent in Hack's length–area relationship and with Horton's ratios (bifurcation and stream length ratios) have not been entirely convincing (Phillips, 1993). Fractal dimension represents one means of quantifying form and as such has general applicability far beyond fluvial geomorphology, but it seems to add relatively little to what the probabilistic–topologic approach already offers. None the less, analysis in terms of fractal structures has been used to argue that natural networks develop towards an optimal form which minimizes the total rate of energy expenditure in the system (Rinaldo *et al.*, 1992).

An optimality principle has also been applied to the angular properties of river networks (Roy, 1985), serving to illustrate that there are still avenues to explore as regards network analysis either within or without the confines of the random model. For example, the random model does not deal directly with the hydraulic geometries of river channels or with the elevational characteristics of networks, and yet available relief and slope determine the potential energy which drives the fluvial system. As an initial step towards

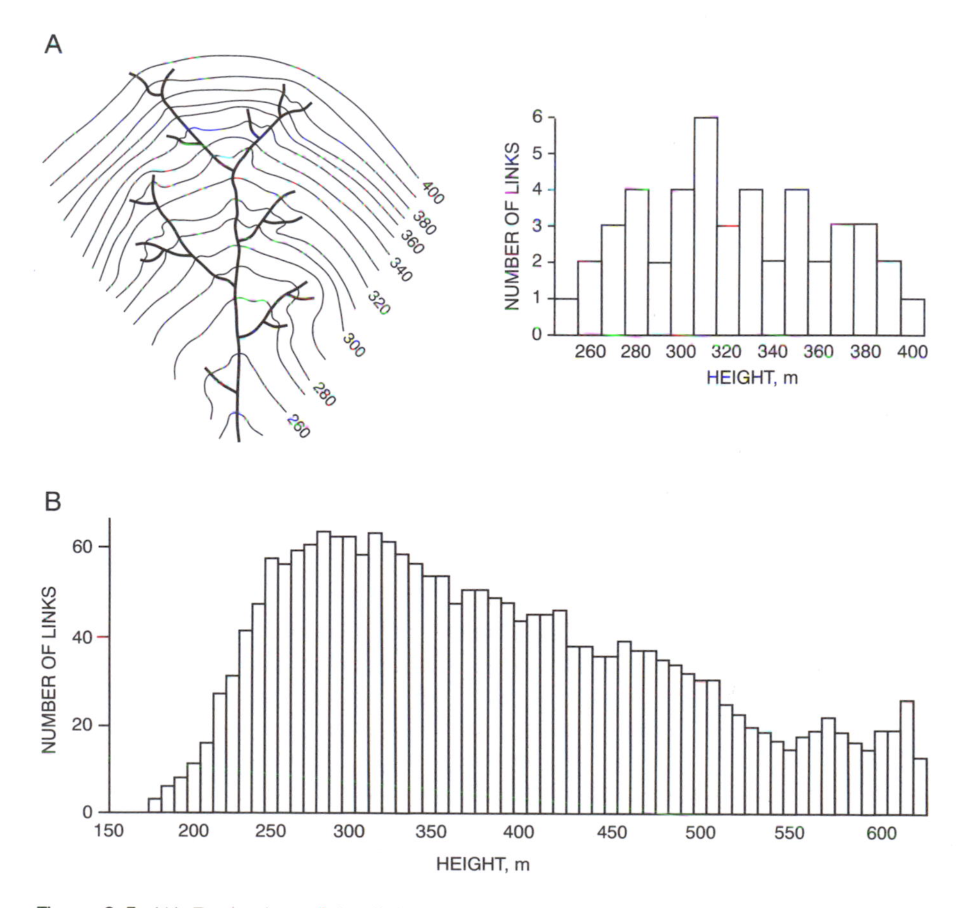

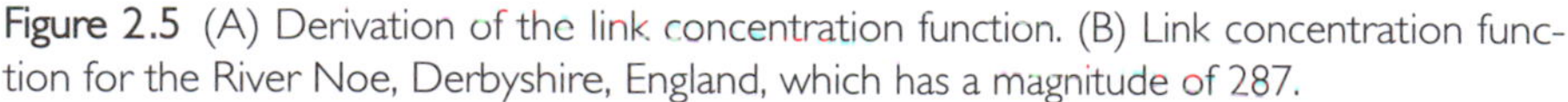

Figure 2.5 (A) Derivation of the link concentration function. (B) Link concentration function for the River Noe, Derbyshire, England, which has a magnitude of 287.

understanding the three-dimensional structure of drainage networks, Gupta and Mesa (1988) have devised a *link concentration function* which represents the number of links at various elevations above the basin outlet (Fig. 2.5). Since this function cannot adequately be predicted from the postulates of the random model, an appropriate theoretical basis for describing its variability is needed if empirical results are to be properly interpreted. There is also a need to understand how the elevational and planimetric structures of channel networks are interrelated, and how those structures are linked to the hydrologic processes which ultimately control them.

Drainage Density

Few attempts have been made to integrate the stream ordering and drainage density concepts introduced by Horton (1945). Drainage density (D_d), defined by

$$D_d = \frac{\Sigma L}{A_d} \tag{2.4}$$

where ΣL is the total channel length in a basin of area A_d, is regarded as the most important areal measure of network geometry in that it expresses the degree of basin dissection by surface streams. For a given valley network, the closer channel heads are to the drainage divides, the greater is the total length of channel and therefore the drainage density. It should be no surprise, therefore, that drainage density can be approximated by $1/l_s$, where l_s is the mean length from channel heads to divides (Montgomery and Dietrich, 1989).

Drainage density can have different connotations. In one sense it reflects the interaction between incisive channelization processes, ground surface resistance and diffusive infilling processes. In another it influences the efficiency with which water is discharged from an area during individual storms. Eight-fold variations in drainage density have been reported from a single catchment as the network expanded and contracted in response to short-term fluctuations in precipitation (Gregory and Walling, 1968). Dealing with the permanent rather than instantaneous wet-weather network, the main problem is to explain what controls drainage density. Unfortunately the theoretical background is somewhat limited and the approach has been largely empirical, with emphasis on how factors of assumed physical significance are spatially related to drainage density, much of the data coming from mid-latitude areas. The range of drainage density values is wider than Horton originally supposed, but of 42 areas listed by Gregory (1976), only 13 had densities greater than 15 km km^{-2} and only five densities greater than 20 km km^{-2}.

Two sets of factors determine drainage density: those which govern the amount and quality of water received at the surface, and those which control the subsequent distribution of that water, its availability for channel cutting, and erodibility. The first is climatic, while the second includes a complex mix of lithologic, vegetational, edaphic and topographic influences.

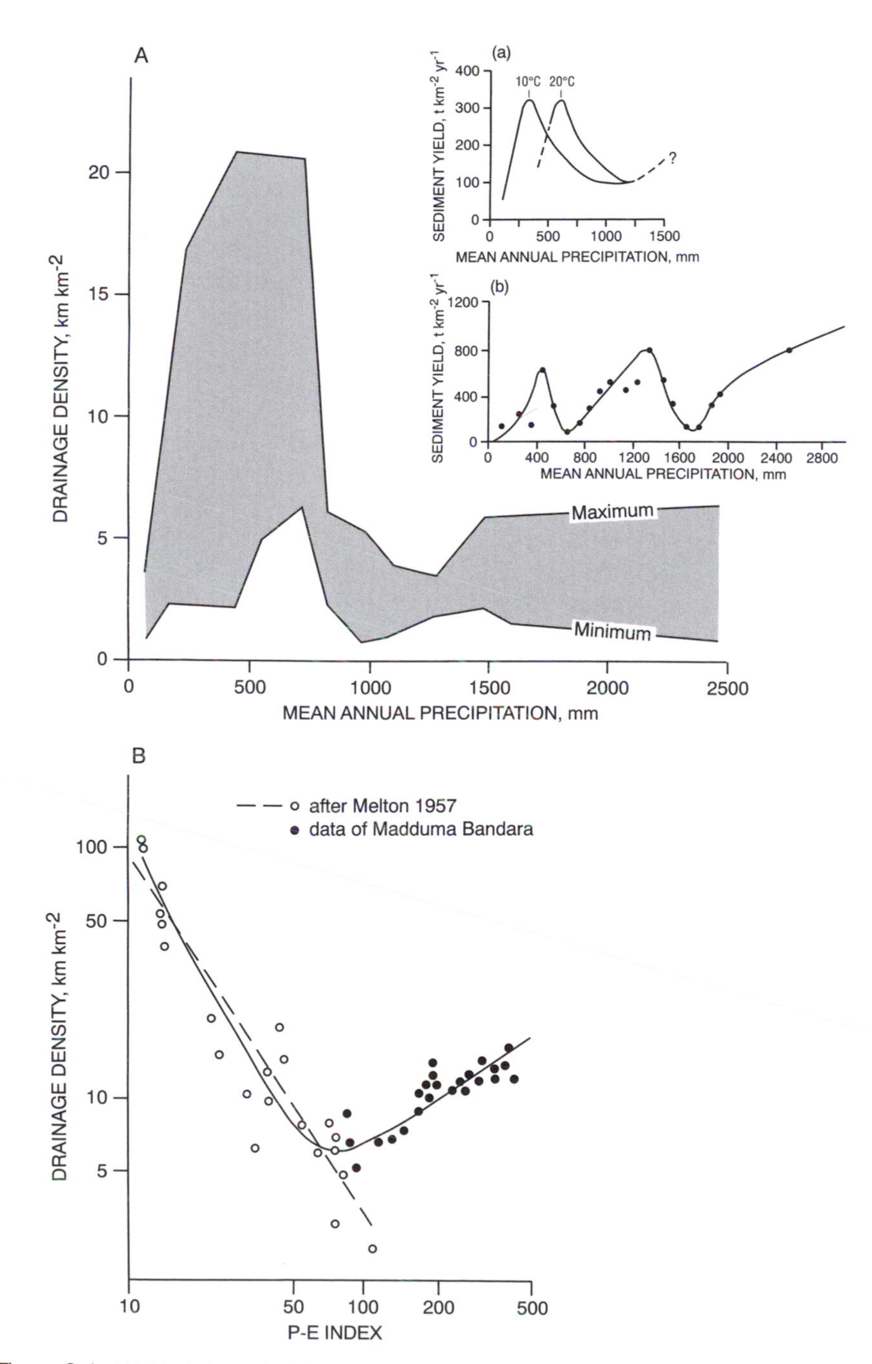

Figure 2.6 (A) Variation of drainage density in relation to mean annual precipitation, with corrections for basin size (after Gregory, 1976). The insets show relationships between sediment yield and mean annual precipitation: (a) for two mean annual temperatures (after Langbein and Schumm, 1958); (b) averaged global data (after Walling and Kleo, 1979). (B) Variation of drainage density as a function of precipitation-effectiveness (P-E) (after Madduma Bandara, 1974).

Drainage density is broadly correlated with mean annual precipitation (Fig. 2.6A). Maximum values occur in semi-arid areas where the range is also greatest. In those areas the proportion of precipitation receipt that flows as surface runoff, and which is therefore immediately available for erosion, is greater than in more humid environments. Drainage density decreases in more arid and more humid areas because of lower runoff and the impeding effects of vegetation respectively, although values probably increase again to a secondary peak in the humid tropics where mean annual precipitation exceeds 1500 mm (Abrahams, 1972; Walsh, 1985). It is no coincidence that this global pattern of drainage density variation is similar to that proposed for sediment yield (Langbein and Schumm, 1958; Fig. 2.6A), especially as regards their respective maxima. High sediment yields reflect increased channel development and a more efficient transport system.

The choice of mean annual precipitation reflects more the availability of data than causative significance, since there is no *a priori* reason why it is physically the most relevant rainfall variable. Rainfall intensity may be a more appropriate parameter in that it influences the short-term availability of water for channel cutting, especially during infrequent, high-magnitude storms. However, there is very little consensus as to the return period or duration of rainfall events which exert a critical control on drainage density. Indeed, the appropriate duration may vary with environment, tending to be short (hourly rainfalls) in impermeable areas dominated by Horton overland flow, and longer (daily rainfalls) in vegetated areas with permeable soils (Walsh, 1985). Seasonality is another possible factor, especially in combination with rainfall intensity, since markedly seasonal climatic regimes with high rainfall totals tend to have large drainage density values. Clearly the hydrologic input having the dominant influence on drainage density is difficult to isolate empirically when the several rainfall variables are so highly intercorrelated.

The effectiveness of precipitation in producing runoff can be expressed by Thornthwaite's P-E index, a measure of moisture availability for plant growth. In the most complete empirical study of drainage density variation, Melton (1957) identified the P-E index as the single most important variable. As the index increased over the range of 12–110, the drainage density decreased (Fig. 2.6B), which Melton interpreted as indicating the efficacy of an increasing vegetation cover to control erosion. In contrast, Madduma Bandara (1974) found a positive correlation between the two variables for Sri Lankan basins where the index varies from 86 to 373. The implication is that, beyond a critical value of effective precipitation (80–90), the relationship changes from negative to positive (Fig. 2.6B) because further increases in precipitation cannot be countered by similar increases in the vegetation cover, a result which has an obvious parallel in the tendency for drainage density to rise where mean annual precipitation exceeds 1500 mm.

The effects of vegetation are difficult to isolate because they are bound up with those of climate and soil cover. Certainly the presence or absence of vegetation has a major influence on the rate of erosion by surface water, and drainage density is highest in areas of sparse vegetation. However, once a more or less complete cover exists, differences in vegetation density appear to have relatively little effect on drainage density, provided the infiltration capacity is high enough to preclude Horton overland flow (Dunne, 1980).

Climatic characteristics, operating partly through their influence on vegetation, have a dominant control on drainage density variation at the global and regional scales. Inter- and intra-regional differences, which tend to be smaller than those between climatic provinces (Gregory and Gardiner, 1975), can be attributed to lithologic and topographic factors, especially in so far as they affect the infiltration–runoff relationship. Bedrock permeability influences the relative amounts of surface and subsurface flow, so that less permeable rock types are commonly associated with higher drainage densities (Gregory and Gardiner, 1975; Walsh, 1985). Increasing aridity may accentuate the effect of bedrock differences, since empirical studies have tended to show a strong dependence of drainage density on infiltration capacity in semi-arid areas but not in humid temperate ones (Kirkby, 1994). Positive correlations have been obtained between drainage density and both basin relief and slope (Abrahams, 1976; Roberts, 1978), suggesting that densities will be highest in steep headwater areas and when incision is at a maximum during landscape development. From relationships that show that the area draining to the channel head varies inversely with local valley slope (Montgomery and Dietrich, 1989, 1994; Fig. 2.9A), it follows that drainage density should decrease with slope.

The distinction between wet-weather and perennial networks underlines not only the dynamic quality of drainage density but also the need to establish the magnitude–frequency characteristics of rainfall and runoff events controlling channel initiation. While drainage density influences flow in the short term, flow determines drainage density in the longer term through environmental factors whose individual effects are difficult to isolate in empirical analyses because of intercorrelations. Correlative structures of the kind envisaged by Melton (1958) suggest that drainage networks are capable of achieving an equilibrium density adjusted to prevailing conditions of climate, vegetation, lithology and soils. Drainage densities attain maximum values in semi-arid areas (Fig. 2.6A), where surface runoff rates are high because of intense rainfall over surfaces with sparse vegetation cover and limited soil development. A change in environmental conditions may instigate a change in the network through expansion or contraction, the direction of which depends on the existing climatic regime. The peakedness of the drainage density and sediment yield curves (Fig. 2.6A) indicates the particular sensitivity of semi-arid environments where even small changes in precipitation or surface conditions can induce a marked response (Cooke and Reeves, 1976). However, despite the many studies of how environmental factors are related to drainage density, there have been few analyses of the processes by which those factors influence network form. Since half of the channel links in any network are exterior (i.e. first-order streams), approximately half of the total channel length lies in the headwaters.

Consequently the position and stability of channel heads are important determinants of drainage density in that they reflect the relative dominance of cut and fill processes. Given the spatially variable structure of existing networks, the logical step is to ask how that structure may have developed, beginning with the processes which operate on hillslopes.

HILLSLOPE PROCESSES

Hillslope Hydrology

The main processes acting on hillslopes, namely raindrop impact, surface water flow, subsurface water flow and mass wasting, involve water to a greater or lesser extent. Mass wasting is the downslope movement of soil or rock material by fall, slide, flow and creep processes. Although it takes place largely under the influence of gravity without the direct action of water, air or ice, water is frequently involved by reducing the strength and increasing the mobility of slope materials. Indeed, one process of mass wasting, shallow landsliding, has been identified as a possible mechanism of channel initiation (Dietrich *et al.*, 1986; Montgomery and Dietrich, 1988).

Water reaching the ground surface via rainfall or snowmelt can follow several paths on its way downslope (Fig. 2.7A). Where the maximum rate of absorption (infiltration capacity) exceeds the rate of receipt (normally expressed by rainfall intensity), water infiltrates the surface and either moves downward to replenish the *groundwater* reservoirs (route 1) or flows laterally as *throughflow* (route 2). Lateral subsurface flow occurs in soils where the hydraulic conductivity decreases with depth (Zaslavsky and Sinai, 1981), most notably in the presence of an impeding horizon of lower permeability which produces saturation in the more permeable soil above and causes flow to follow a path nearly parallel to the surface (Fig. 2.7B). Where permeability does not change significantly with depth, water movement is largely vertical and little or no lateral flow occurs within the soil layers. Subsurface flow may be diffuse when water flows through the soil matrix, or concentrated in soil pipes (Gilman and Newson, 1980; Jones, 1981) or macropores (Germann, 1990). Percolines, zones of deeper soil and better hydraulic conductivity (Bunting, 1961), represent a transitional type in that the flow takes place largely through diffuse seepage but is concentrated by dint of topography or the greater soil depth. The boundary between pipes and macropores is difficult to define precisely, but the former tend to be erosionally enlarged and show a higher degree of connectivity in the downslope direction (Anderson and Burt, 1990). This potential for subsurface flow convergence can speed up soil drainage rates quite considerably and, of greater importance geomorphologically, has implications for channel initiation. Percolines often occupy topographic depressions in the headwater area of existing streams and therefore represent preferential zones for channel extension, while there is now clear evidence of a direct link between piping and the development of surface channels.

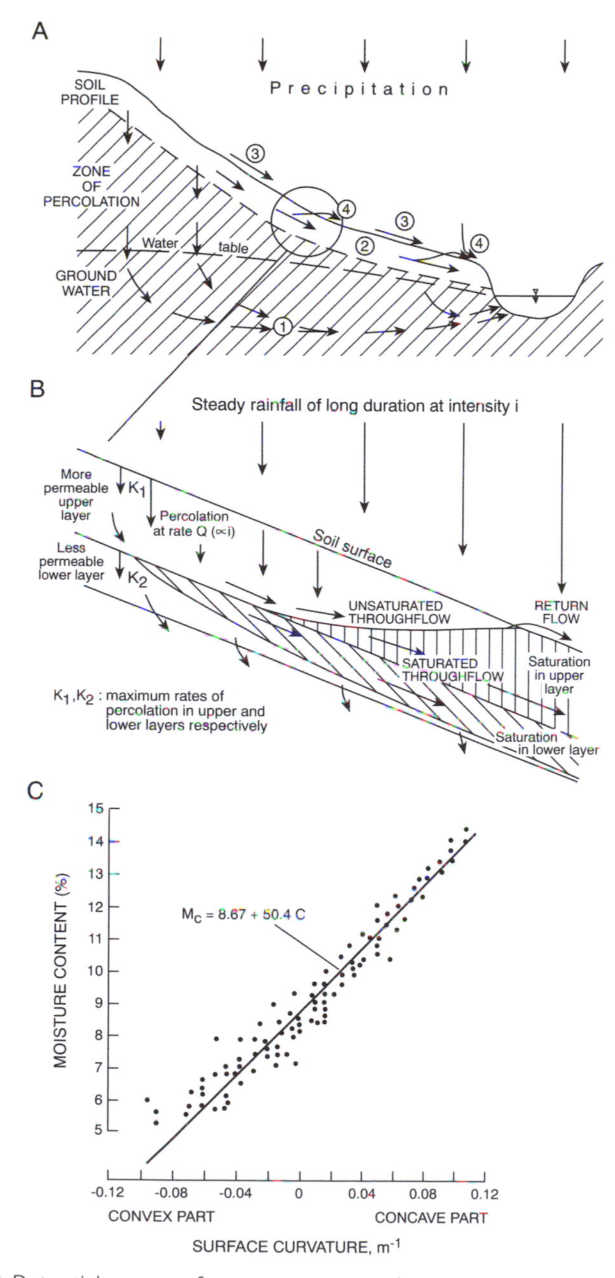

Figure 2.7 (A) Potential routes of water movement: 1, groundwater flow; 2, throughflow; 3, Horton overland flow; 4, saturation overland flow (comprising return flow and direct precipitation on to saturated areas). The unshaded zone represents more permeable topsoil, and the shaded zone less permeable subsoil or bedrock. (B) Schematic diagram of shallow subsurface flow and the generation of saturated conditions in a two-layer soil where a more permeable layer overlies a less permeable one. (C) Relationship of moisture content at depths of 0.2 m and 0.4 m (10 days after rain) with soil surface curvature (after Zaslavsky and Sinai, 1981).

The development of surface channels is more often associated with overland flow, which can be generated in two main ways. *Horton* or *infiltration-excess overland flow* (route 3) occurs when rainfall intensity exceeds infiltration capacity, the subsequent discharge increasing linearly downslope after sufficient time has elapsed for a steady state to be achieved (rainfall excess = outflow). Horton's (1933, 1945) infiltration theory of runoff predicts that this type of flow will be produced more or less instantaneously and simultaneously over a basin during heavy rain, implying a uniformity of rainfall and infiltration conditions which can at best be expected only in small basins. Infiltration characteristics in particular are highly variable, both spatially and temporally. Infiltration capacity tends to decrease asymptotically during the course of a storm as a result of surface compaction by rainbeat, inwashing of fine particles into pores, and swelling of clay particles, so that the requisite condition (rainfall intensity > infiltration capacity) may be attained only after some delay. In humid areas with a dense vegetation cover and well-developed soils, that condition may never or only rarely be attained, either because intensities are never high enough or infiltration capacities never low enough. Indeed, Zaslavsky and Sinai (1981) have gone so far as to suggest that all rain, be it of high or low intensity, must first enter the soil before part re-emerges on the slope. Clearly this type of overland flow is much more limited than Horton originally supposed (Table 2.3). It is rarely widespread over a drainage basin, a tendency which led Betson (1964) to propose the 'partial area' concept as a modification to the Horton runoff model. It is absent or uncommon in many humid areas, being generally associated with semi-arid ones where the vegetation is sparse and soils are thin. However, whether or not Horton overland flow is regarded as a zonal phenomenon, its occurrence is strongly related to soil properties, especially in so far as they affect the hydraulic conductivity of the surface layer.

The generation of *saturation overland flow* (route 4) depends on the moisture content of the soil before, during and after rainfall events. Provided rainfall continues long enough for deeper and less permeable soil layers to become saturated ($Q > K_2$ in Fig. 2.7B), throughflow will be deflected closer and closer to the surface as the level of saturation rises through the soil. If the soil eventually becomes saturated to the surface, saturation overland flow will occur. It consists of two components: *return flow*, which emerges at those points on the hillslope where surface saturation has been reached; and *direct precipitation onto saturated areas*, which is most often associated with the footslope zone. This type of flow can occur at rainfall intensities much lower than those required for Horton overland flow. It is also spatially more limited (Table 2.3), being preferentially generated in those localities susceptible to saturation (Kirkby and Chorley, 1967): at the base of slopes, especially those that are concave in profile; where soils are locally thin and less permeable; and in hillslope hollows. Hillslope hollows are particularly important for runoff generation, since the associated convergence of flow favours the accumulation of soil water and keeps the concavities wetter for

Table 2.3 Principal differences between types of overland flow

Characteristic	Horton overland flow	Saturation overland flow
Rainfall	Strongly related to rainfall intensity	More dependent on rainfall duration
Infiltration	Surface infiltration capacity of critical importance	Transmissibility of lower soil horizons more important
Distribution:		
Temporal	Begins soon after the storm when rainfall intensity and soil moisture content are high enough	Starts only when the underlying soil layers are saturated
Spatial		
(i) environmental	Especially semi-arid areas with sparse vegetation and thin soils	Humid areas with good soil and vegetation development
(ii) local	Can be extensive in small basins	Limited to zones of preferential saturation, often topographically influenced
Downslope variation	Linear increase of runoff volume	More complex pattern of change, related to slope, profile form, soil depth and hydraulic conductivity

longer periods, thereby hastening the onset of runoff when the next rain falls (Fig. 2.7C). Being strongly dependent on local topography and incident soil wetness, the size and extent of saturated areas vary seasonally and during individual storms.

Whether or not a spectrum of runoff and erosion models based on these four components of hillslope water movement can be recognized, with dominantly groundwater flow at one end and dominantly Horton overland flow at the other, all four do contribute to streamflow generation and channel development, their relative contributions varying with climate, vegetation and soils. Clearly the range of conditions is much broader than Horton originally envisaged.

Processes of Water Erosion

Various processes of erosion are associated with these different modes of flow, where erosion, which includes both the detachment and transport of material, is a function of: (a) erosivity, or the potential ability of a process to produce erosion; and (b) erodibility, or the susceptibility of a material to erosion. At the surface, erosion takes place largely by raindrop impact, overland flow or sheetwash, rill flow in small ephemeral channels, and gullying. Distinctions can sometimes be difficult to make, notably between sheet and rill erosion since the former often includes the effects of the latter, but the given order appears to represent a sequence of increasing rates of erosion. Over 900 days on an 11° slope in mid-Bedfordshire, sediment transport rates associated with raindrops, overland flow and rill flow were 20 g cm^{-1}, 400 g cm^{-1} and 19 000 g cm^{-1} respectively (Morgan, 1986).

Raindrop impact has several effects on the soil surface, including consolidation, breaking of physical and chemical bonds (detachment), and transport of soil particles by splash. The erosivity of rainfall is often expressed in terms of its kinetic energy, which is a function of drop size and velocity, but a remarkably small amount of that energy (< 0.3 per cent) is used in transport (Brandt and Thornes, 1987). Energy is used to compact the soil surface and to form the splash crater. The consolidation effect is best seen in the development of a surface seal, with finer particles being washed into pores to form a relatively impermeable layer up to 10 mm thick, which reduces infiltration capacity and thereby promotes greater surface runoff. Reductions in infiltration capacity of as much as 50 per cent have been reported during a single storm (Hoogmoed and Stroosnijder, 1984). Loamy sands are particularly susceptible to surface sealing so that, despite their considerable hydraulic conductivity, they can generate quite large overland-flow volumes even during low-intensity rainfall (Poesen, 1992).

Raindrops can exert locally intense vertical and lateral stresses (where stress is force per unit area), the latter of which are most important in soil detachment. In modelling the rate of detachment by raindrop impact (D_r), the effects of rainfall intensity (I) and slope angle (s) are often incorporated in the resultant equations:

$$D_r = a\, I^b s^c \tag{2.5}$$

where 2 is a typical value for b, indicating the sensitivity of detachment rate to variations in rainfall intensity. Although small positive values (0.2–0.3) are usually assigned to c, results on the influence of soil surface slope have not tended to be consistent. A theoretical model developed by Torri and Poesen (1992) confirms that slope has a positive effect on detachment, but shows that the effect becomes progressively less as the internal resistance of the soil increases. The use of equations is not straightforward because many factors influence rainfall erosivity and soil erodibility, the two elements to which detachment rate is ultimately related. As regards the latter, particles in the size range 0.04–0.25 mm seem to be most susceptible to detachment (Poesen, 1992). Also, the soil surface varies in erodibility through its degree of wetness. When dry, soil is detached by the compression of air ahead of a wetting front (*slaking*) in addition to the direct impact of raindrops; when wet, soil is most likely to be detached under saturated conditions since its shear strength is then at a minimum (Al-Durrah and Bradford, 1982); when inundated, detachment increases with water depth up to about 0.3 drop diameters because of increased turbulence, but declines thereafter, becoming negligible when depth exceeds 3 drop diameters. Nor is the effect of vegetation simple, since a high forest canopy can increase the kinetic energy of raindrops (Mosley, 1982). A low vegetation cover may afford the soil better protection, but the accumulation of a litter layer ultimately governs the amount of kinetic energy that raindrops impart directly to the soil surface.

Raindrop impact is principally a detachment mechanism preparatory to the removal of loosened particles by other surface processes, notably overland and rill flow. Its transportation role is often considered to be minor relative to that of surface runoff, although there is net transport downslope on inclined surfaces. Given that detachment is greatly influenced by the depth of surface water, raindrop impact is probably most effective in short, intense storms which generate little surface runoff or during the early stages of storms which do produce runoff, and on drainage divides where flow is typically shallowest.

Once water accumulates in sufficient quantity to exceed surface depression storage it starts to flow downhill as **overland flow**. Despite its alternative name, *sheetwash*, the flow is rarely a sheet of uniform depth; it commonly consists of a thin film of water with irregular threads of deeper, faster flow. On gentle slopes soil detachment is largely due to raindrop impact but, as the slope angle increases, so does detachment by overland flow. The relative contribution of rainfall and runoff to total detachment has been shown to depend on the power (= product of discharge and slope) of overland flow (Proffitt and Rose, 1991), indicating that not only land slope but also the depth and velocity of flow influence where and when overland flow becomes more significant than raindrop impact. In view of the generally shallow depths, detachment by overland flow has often been considered to be of minor importance, but this assumption is not always justified, particularly on steep slopes of highly erodible material.

In contrast to detachment, transport has long been regarded as a major role of overland flow. In attempts to predict the transporting capacity of overland flow, formulae developed for alluvial rivers have often been used, but this approach is questionable since the hydraulic conditions under the two regimes are very different (Govers, 1992). In particular, depths are much lower and slopes much higher in overland flow than they are in natural rivers, and impacting raindrops can render overland flow more erosive by increasing its turbulence. Nevertheless, material is moved in much the same way, as bed load and suspended load, although bed-load transport may be a more dominant component in overland flow. Also, sediment transport capacity is commonly expressed as some function of discharge and slope, indicating the importance of stream power considerations in the transport as well as the detachment phase of the erosion process. Vegetation cover is a critical factor governing the erosive ability of overland flow, but to be effective it needs to be close to the soil surface where its resistance effect is at a maximum. Grass stems can prevent significant sediment transport at all but the highest flows (Prosser and Dietrich, 1995).

Overland flow, particularly of the Hortonian type, acts in combination with the detachment ability of raindrops to erode soil particles and transfer them downslope where they frequently accumulate as fan-shaped deposits. The rate of erosion is controlled not only by the rate of detachment but also by the transport capacity of the flow, both of which are spatially varied. Close to the drainage divide, transport capacity is limited and detachment is at a maximum. With greater distance downslope, transport capacity increases with discharge while detachment remains constant or decreases.

Detachment and transportation ability increase substantially when flow is concentrated into **rills**, microchannels with typical dimensions of 50–300 mm wide and up to 300 mm deep. They are usually discontinuous and ephemeral features, often being obliterated between one storm and the next or even during a single storm when the supply of sediment from inter-rill erosion or rill-wall collapse exceeds the transporting capacity of the flow. The development of rills involves a change in flow state – from unconcentrated overland flow, via overland flow with concentrated flow paths, to microchannels without headcuts, and microchannels with headcuts when rills become better established. The greatest change probably occurs between the first and second stages, the disposition of stones, vegetation and other forms of macroscale roughness influencing the concentration process. A persistent problem has been to define the critical condition responsible for the onset of rilling, but there is disagreement as to the most suitable hydraulic parameter (Slattery and Bryan, 1992) – although, in line with Horton (1945), critical shear stress continues to be used (e.g. Gilley *et al.*, 1993). An alternative viewpoint regards rill initiation as a transitional rather than a threshold phenomenon, with no abrupt change in hydraulic conditions at the point of flow concentration (Slattery and Bryan, 1992).

Persistent rilling requires slopes steeper than 2–3° (Savat and De Ploey, 1982). As slope angle increases, so does the probability of rill formation, not only because of the increase in overland flow erosivity but also because of a decrease in topsoil shear strength, notably in those soils susceptible to surface sealing (Poesen, 1992). Erodibility is at a maximum in soils with a mean particle size in the range 0.016–0.063 mm, which is somewhat finer than that associated with minimum resistance to raindrop detachment (0.04–0.25 mm). However, mean particle size represents only one aspect of the soil texture effect. Rock fragments embedded in the topsoil increase sediment yield by concentrating erosive forces but, once exposed on the surface, they take on a more protective role and erodibility decreases (Poesen, 1992). Rilling is a potent form of erosion on hillslopes with little vegetation, generally contributing from 50 to 90 per cent of total sediment removal. It represents a form of concentrated action by surface flow which, under suitable conditions, can lead to the development of more permanent channels.

Gullies are relatively permanent water courses, but differ from stable river channels in having steep sides, low width : depth ratios and a stepped profile, characteristically with a headcut at the upslope end (Plates 2.1 and 2.2). Because they develop rapidly, gullies are usually regarded as indicative of accelerated erosion and landscape instability, brought about by an increase in the amount of flood runoff. The most common causes of that increase are climatic change, with accompanying variations in rainfall intensity and periodicity, and land-use change (Plate 2.2). Deforestation, excessive burning of the vegetation cover and overgrazing can all lead to greater runoff. Gullies have been associated with slopes in excess of 12–16° (Savat and De Ploey, 1982), but attempts to define the critical conditions for the entrenchment which characterizes gully development have not been entirely successful.

Three main processes have been identified as important in gully formation: surface flow, piping and mass movement. Under suitable circumstances gullies may develop from master rills, but their initiation by surface flow more commonly results from localized incision at a point or points where the vegetation cover is broken. Runoff concentrates in the depression and a small headcut forms, retreating upslope to leave a downvalley trench. Retreat is effected mainly by seepage erosion at the face of the headcut, or by overtopping and plunge-pool action which erode the lip and scour the base respectively (Plate 2.2). Maintenance of the headcut is important to the developmental process, and materials need to be sufficiently cohesive or be capped by a resistant band if the steep scarp is not to collapse. Where flow incises at several points downslope, discontinuous gullying develops, resulting in a more continuous form if conditions favour eventual coalescence. Many studies record gully initiation by subsurface pipe erosion and subsequent roof collapse, particularly in semi-arid badlands. The mechanism operates best where surface soils develop cracks down which stormwater can rapidly infiltrate, where a relatively impermeable horizon encourages

Plate 2.1 Gullying in marls, southeast Spain.

Plate 2.2 Gully headcut with erosion produced by overtopping flow and seepage. Gullying began after vegetation was cleared to facilitate tin mining, northeast Tasmania.

lateral flow, and where localized dispersion weakens the chemical bonds between particles. Once water breaks through the soil surface at some point downslope, headward retreat along the line of the pipe can be very rapid. Piping can also contribute to the third process of gully formation in which landslides leave deep, steep-sided scars that are preferentially exploited by surface water during subsequent storms. Indeed, this third process is more a reflection of combined than single action.

Gully erosion can present serious environmental problems, not only because vast quantities of soil may be removed but because the land becomes unfit for many forms of agricultural activity and the process is difficult to control once initiated. It is particularly severe in semi-arid areas where, by unhappy coincidence, sensitivity to change in environmental conditions and pressure on soil and vegetation resources are at a maximum.

Those factors which reduce surface runoff and encourage subsurface flow are generally the same ones that promote **solution** as an active process on hillslopes. Water moves much more slowly beneath the surface so that it is better able to come into chemical equilibrium with soil materials, removing plant nutrients in solution and weakening chemical bonds. The amount of solution depends on the volume and acidity of the moving water and on the solubility of the material, with effective transfer of solutes requiring a strong downward and lateral flow of water. The high dissolved load of many rivers (Table 3.2, p. 82) points to the effectiveness of solution as an agent, although part of the load may come from the deep solution of bedrock. The mechanical transport of soil material is generally much less important because of the very small flow velocities and the constraints imposed by the size of pore spaces through which particles must pass. Only where large void spaces develop, as in soil pipes, is subsurface sediment yield likely to be appreciable. In effect, piping can extend channelled processes beyond their obvious domain and provide one important mechanism for the formation and extension of surface channels (Jones, 1981, 1987).

Channel Initiation

Channels originate in a variety of ways, related in part to the history of the land surface on which they develop. They may form on a recently exposed surface or during a phase of network expansion. They may be influenced in their development by channel lines inherited from the past. Invariably they start out as exterior links or first-order streams. Understanding channel initiation involves two related problems: to understand the processes whereby water movement becomes sufficiently concentrated to cut a definable channel; and to identify the conditions under which that initial cut is maintained and enlarged to ensure initially a permanent channel, and eventually permanent flow within that channel (although numerous channels are maintained despite intermittent occupancy). The first is closely linked to the modes of hillslope water movement and their associated processes of erosion, while the second embraces the broader question of drainage network development, since permanence implies some level of integration between surface channels.

Channel Initiation by Overland Flow

The Horton (1945) model provides the obvious starting point, with channel initiation regarded as a threshold phenomenon. As it moves downslope, **Horton overland flow** exerts a shear stress (τ) on the surface given by

$$\tau = \gamma d \sin \theta \qquad (2.6)$$

where γ is the specific weight of water, d is the mean depth of flow and θ is the local slope angle. With increasing flow depth downslope and a constant or increasing gradient, shear stress and the potential for erosion will also increase with distance x from the divide. Horton argued that, at some critical point x_c, applied stress will equal surface resistance to give a 'belt of no erosion' upslope and a zone of potential sheetwash erosion downslope (Fig. 2.8). Once overland flow becomes erosive, Horton believed it to be inherently unstable and envisaged the immediate development of a system of subparallel rills which, through the processes of cross-grading and micropiracy, integrate to give a dominant rill capable of incision (Fig. 2.12, p. 47). Horton regarded the critical distance x_c as the most important factor controlling network evolution, since it sets a physically defined spatial limit on channel initiation. The question is, however, how valid are the assumptions on which the model is based?

For a start, Horton overland flow is much more limited than originally thought. Where it does occur, experiments have shown that it can be erosive without forming rills (Emmett, 1970; Pearce, 1976; Dunne and Dietrich, 1980) and that the diffusive action of rainsplash can stabilize a surface against incision by sheetwash (Dunne and Aubry, 1986). Thus, if the concept of a critical distance is accepted, the critical length for sheetwash erosion (x_c) is not necessarily the same as that for channel initiation (X_c). The onset of

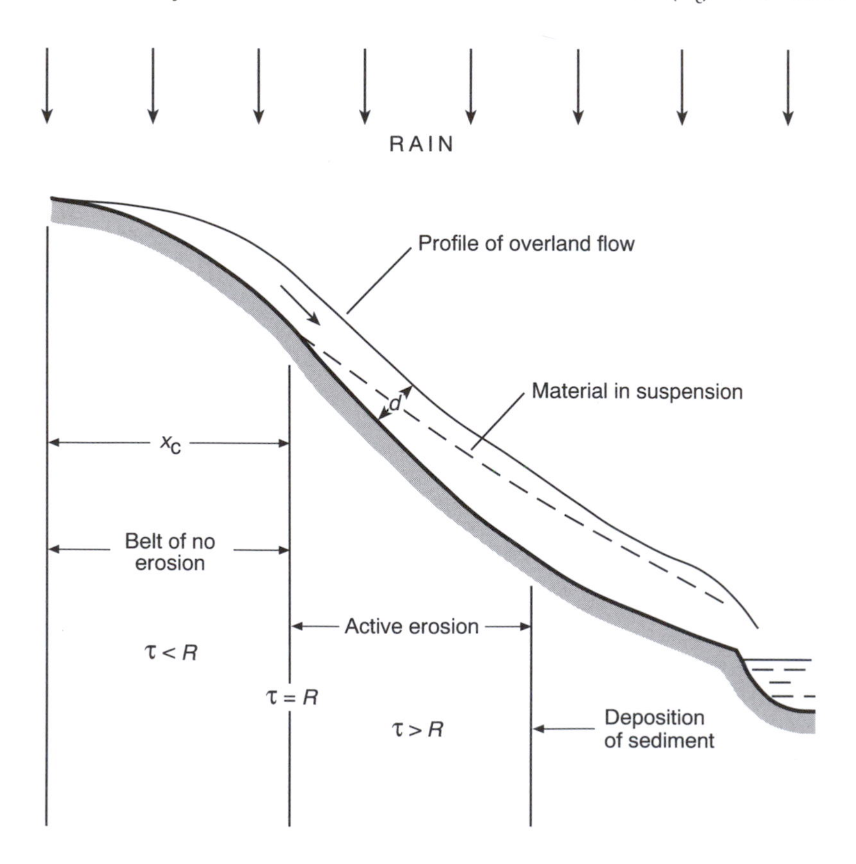

Figure 2.8 Hillslope profile showing the critical distance x_c (after Horton, 1945). τ, eroding stress; R, shear resistance of soil surface.

channelization requires another threshold to be attained, possibly related to critical conditions of slope geometry (angle, length, curvature) and flow dynamics. Whether the flow is laminar or turbulent could be important (Montgomery and Dietrich, 1994). In addition to critical length criteria, a critical area could be defined since channel initiation requires the accumulation of sufficient runoff, which is area-related (Schumm, 1956). Despite the limitations of the Horton model, the view persists that channel initiation reflects the exceedence of an erosional threshold, as the more elaborate model developed by Willgoose *et al.* (1991a, b, c) demonstrates. In their model, channelization occurs where the sediment transport rate, defined in terms of the product of discharge and slope, increases rapidly. In that way is the necessary incision produced.

An alternative and more formal approach based on linear stability analysis has shown that small-scale irregularities on a previously unchannelled surface will tend to grow only where the slope profile is concave (Smith and Bretherton, 1972). Surfaces that are straight or convex are stable against small perturbations, implying that any channel-like features will tend to disappear through the action of diffusive processes such as raindrop impact. Consequently if, during the long-term evolution of the landscape, profiles develop with convex upper and concave lower parts, channel heads would be located at inflection points where the profile form changes from convex to concave. Convergent zones (e.g. contour concavities or hollows) in the concave part of a slope are thus favoured loci for channel initiation, since there the local erosion rate is increased. Mosley's (1974) experimental results support this conclusion in showing that erosion tends to be greater on convergent as opposed to planar or divergent slopes.

Smith and Bretherton's (1972) model has been variously extended, by Kirkby (1987), whose analysis went beyond the infinitely small perturbations (cf. microrills) characteristic of the original, and by Loewenherz (1991), who identified for the first time what might control the lateral spacing of channels. Loewenherz's analysis suggests that, rather than the closely spaced instabilities which develop initially near the base of a hillslope, slower-growing, more widely-spaced channels ultimately determine the morphology of a drainage surface, and that they may extend upslope beyond the inflection point. Such deterministic modelling provides a theoretical explanation for the initiation of channels, regarding them as instability phenomena, but does not explicitly define the timescale involved or distinguish between an incision which expands into a valley or swale and one which develops into a channel with recognizable banks. Dietrich and Dunne (1993) provide insights into the likely conditions for channel incision and bank formation by overland flow, differentiating between cohesive and noncohesive materials, but as yet the approach is largely qualitative.

Like Horton overland flow, **saturation overland flow** exerts a shear stress on the underlying surface, but differs from it in a number of ways as regards channel initiation. The shear stress available for erosion is likely to be much less, since the resistance provided by a dense groundcover and root mat needs to be overcome before incision can take place. The relationship between discharge and distance from the divide is more complex since the spatial distribution of the overland flow is itself more variable, being strongly influenced by downslope variations in soil depth, hydraulic conductivity and topography. Consequently distance from divide becomes a less rigid criterion for determining both the occurrence of overland flow and the conditions conducive to channel formation. The short-term extensions of existing networks observed during intense storms (Gregory and Walling, 1968; Day, 1983) help to identify the most likely sites for channel initiation. They principally occur in downslope areas, where the vegetation is thinner, and in concavities, where moisture content usually remains higher for longer (Fig. 2.7C) and to which subsurface flow converges. Where return flow is a significant fraction of the saturation overland flow, an additional force associated with the exfiltrating water is applied, and this may help to trigger the initial incision through the vegetated surface. Morgan (1977) has observed rills developing from a sudden burst of water on to the surface of a slope in mid-Bedfordshire, resulting in the formation of a small cut which retreated headward as a channel. Channel initiation by saturation overland flow thus depends on both the shear applied by the surface flow and the seepage force of exfiltration, with the former probably having the major influence, although neither is well understood. The generation of that flow depends in turn on the movement of subsurface water, especially along preferred drainage lines, many of which are highly sensitive to local topography. Where saturation flow is the dominant mechanism, there is likely to be not only a greater distance from divide to channel head than in the Horton model but also a greater variability in that distance. Nevertheless, channel initiation still requires sufficient runoff, and the concept of a critical distance or contributing area remains relevant.

Channel Initiation Related to Subsurface Flow

That subsurface flow can follow well-defined and integrated paths suggests a genetic relationship between subsurface systems and surface channels. Those paths take two main forms: lines of diffuse seepage such as percolines, and lines of concentrated drainage such as pipes. The latter are particularly important. A factor common to both types is spatial heterogeneity in the permeability of the rock or soil through which the water is moving, especially when combined with across-slope irregularities in topography (Jones, 1987).

Seepage erosion involves the entrainment of material as individual particles or in bulk as a result of water flowing through and emerging from a permeable medium (Dietrich and Dunne, 1993). Because the hydraulic gradients required to produce failure are so steep (Dunne, 1990), it is probably a viable channel initiation mechanism only in cohesionless materials or in materials which have suffered considerable cohesion loss through weathering. A dense root mat can also provide sufficient cohesion to reinforce material against seepage erosion. When conditions allow, initial inhomogeneities in the porous medium or differential chemical weathering can perturb the subsurface flow field sufficiently to produce more permeable zones towards which water movement is concentrated. As the degree of flow convergence and the intensity of chemical weathering increase in a positive feedback, the critical conditions for seepage erosion are eventually reached, which may involve pipe development (Dunne, 1980). The excavation of a channel head and local lowering of the ground surface by the exfiltrating water intensifies the process, producing even greater convergence on the channel head which retreats upslope as seepage erosion undermines the headcut and triggers failure of the steepened margins. A limit to this self-perpetuating process is set by the minimum source-basin area required to supply sufficient water for continued seepage erosion (Dunne, 1980). Despite some regolith being too cohesive for this process to operate effectively (Dunne, 1990), it has been invoked as an initiating mechanism in a wide range of environments – from humid temperate Vermont (Dunne, 1980) to subhumid Colorado (Laity and Malin, 1985), and even extraterrestrial Mars (Kochel *et al.*, 1985).

Networks of voids exist at all scales within the soil, from interparticle gaps to macropores and pipes (Jones, 1987). Flow along the larger conduits (>5 mm in diameter) is now regarded as being more widespread than previously supposed and as having major hydrological and erosional significance (Jones, 1981). It is responsible for erosion that can eventually produce channels at the surface, for which Dunne (1990) proposed the general term **tunnel scour**. Some combination of fluid shear and collapse mobilizes large amounts of soil material and enlarges the conduits, but the detailed mechanics remain obscure because few measurements have yet been made. The growth of macropore/pipe networks is most likely where a horizon or surface of limited permeability has developed in the soil, where the substrate has a relatively high porosity, and where interparticle bonds have been considerably weakened (notably in the presence of soluble cements or dispersible clay minerals).

Macropore/pipe systems vary considerably in size, location and connectivity. They can exist at more than one level in the soil, with each level being activated by storms of different magnitude. Gilman and Newson (1980) identified two principal forms in the upper reaches of the River Wye: deep-seated pipes greater than 200 mm in diameter with seasonal or perennial flow, and pipes of about 40 mm in diameter located close to the surface with only ephemeral flow. Conduit networks can apparently attain a quasi-stable state without necessarily evolving into surface channels, but there is now sufficient evidence to establish such a link. Certainly conduits converge on the heads of many channels (Jones, 1981). Enlargement of the tunnels and especially thinning of their roofs can eventually lead to a line of pits which coalesce into surface channels with distinctively steep heads and banks (Plates 2.3 and 2.4). Whether channels formed in this way are maintained depends partly on the frequency with which runoff is generated thereafter, but the convergence which is symptomatic of conduit flow is often sufficient to ensure not only maintenance but also headward extension along the line of still-buried pipes. Indeed, channel networks formed by tunnel erosion and collapse tend to be widely spaced initially, but develop high drainage densities relatively quickly (Dietrich and Dunne, 1993).

In steep landscapes, subsurface flow can lead to the extension of channels through **shallow landsliding** (Dietrich *et al.*, 1986). The convergent topography associated with hollows causes colluvium eroded from adjacent hillslopes to accumulate there until it reaches a critical state. Of itself landsliding does not produce a channel, but failure exposes the underlying material, with subsequent erosion by saturation overland flow seemingly critical for channelization (Dietrich *et al.*, 1993). How widespread landsliding may be as a channel initiation mechanism is not known, but a model of the process developed by Montgomery and Dietrich (1989) is supported by data from the western United States (Fig. 2.9A). The model predicts an inverse relationship between drainage area to the channel head and local valley gradient, implying that the steeper the slope, the lower is the porewater pressure required to cause landsliding, and the smaller is the subsurface flow/drainage area needed to produce the subsequent channel incision. In effect the relationship defines a threshold between unchannelled and channelled regions as regards the landsliding mechanism, but the area–slope plot has wider implications in enabling the demarcation of various process regimes (Fig. 2.9B).

Plate 2.3 Roof collapse and headward extension along the line of a soil pipe, southeast Spain.

Plate 2.4 Development of a small-scale network of surface channels as a result of tunnel scour and roof collapse in a semi-arid area of southeast Spain.

Synthesis

The channel head is the upstream limit of concentrated water flow between definable banks (Dietrich and Dunne, 1993). It can be regarded either as a point of transition in the sediment transport process where incisive processes begin to dominate over diffusive ones, or as a point of change where processes not acting upslope become important. The first view comes from stability analyses of the appropriate process equations (Smith and Bretherton, 1972; Kirkby, 1987), which predict unstable growth in the form of a channel if the convergence of flow allows more sediment to be transported out of an indentation than is transported into it. The transition from a convex to a concave profile or from divergent to convergent topography has generally been equated with the transition from diffusive to incisive transport. The second, threshold-based view regards the channel head as a point where an erosional threshold is exceeded and a channel-forming process suddenly becomes effective. The resistance to channelization is overcome either by a high hydraulic gradient in the case of seepage (Dunne, 1990), by a high porewater pressure leading to landsliding (Dietrich *et al.*, 1986; Montgomery and Dietrich, 1989), or by an overland flow shear stress in excess of a critical value as in the original Horton (1945) model. At this point and downslope, concentrated surface runoff becomes a significant transport process. Although distinct, these views are not irreconcilable (Kirkby, 1994; Montgomery and Dietrich, 1994).

The transition from valley floor to channel may be gradual with no discernible topographic break, a common condition in headmost storm exten-

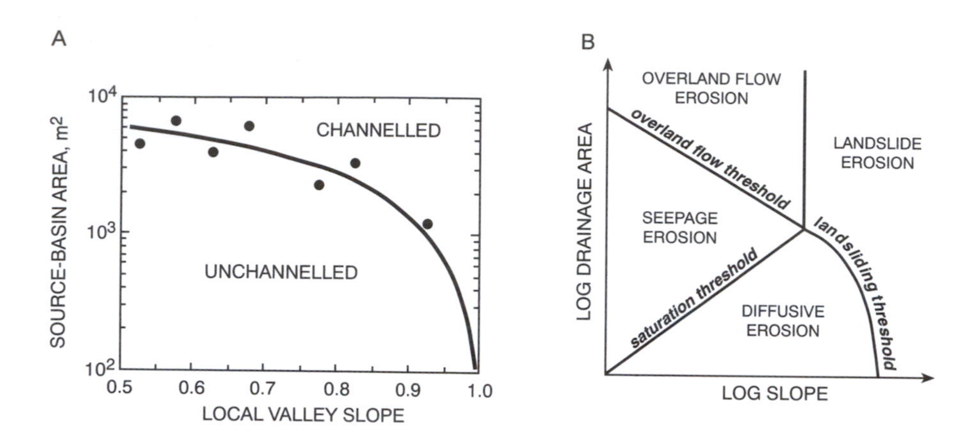

Figure 2.9 Source-basin area–slope relationships. (A) Averaged data for the Tennessee Valley, California, where channel head locations are maintained by recurrent landsliding (after Montgomery and Dietrich, 1989). The line represents a model prediction. (B) Demarcation of process regimes where different channel initiation mechanisms are dominant. The region of diffusive erosion should correspond to hillslopes stable to channelization tendencies (after Montgomery and Dietrich, 1994).

sions (Kirkby, 1994), or in the form of a step, the associated face being subject to erosion by plunge-pool activity and seepage forces. In providing a typology of channel heads (Fig. 2.10), Dietrich and Dunne (1993) emphasize the range of processes involved in channel initiation. In a relatively uniform landscape a single process may predominate, but the various processes responsible for channel initiation are not mutually exclusive, varying in their relative effectiveness with environmental conditions in so far as they determine hillslope hydrology and surface/subsurface resistivity. Even though Horton overland flow is usually associated with semi-arid areas, subsurface as well as surface processes operate there. Some of the most spectacular examples of piping come from those environments. In landscapes with variable slope and soil conditions, Fig. 2.9B suggests that the dominant mechanism is likely to be landsliding in steep areas and overland flow in lower-gradient ones with sufficient drainage area to support runoff, while between the thresholds for soil saturation and overland flow erosion is a zone subject to channelization by seepage erosion. Below the several thresholds is a zone where diffusive sediment transport is dominant, implying that hillslopes would be stable against channelization tendencies. Unfortunately, few data exist at present to provide an adequate test of the relationships suggested in Fig. 2.9B, which, nevertheless, provides a useful framework for investigating response to environmental change. If the drainage area necessary to initiate a channel is decreased, for example by the onset of wetter conditions, channel heads will migrate upslope, excavating valley fills and incising undissected hillslopes. In contrast, an increase in the drainage area needed to sustain a channel will result in local infilling of

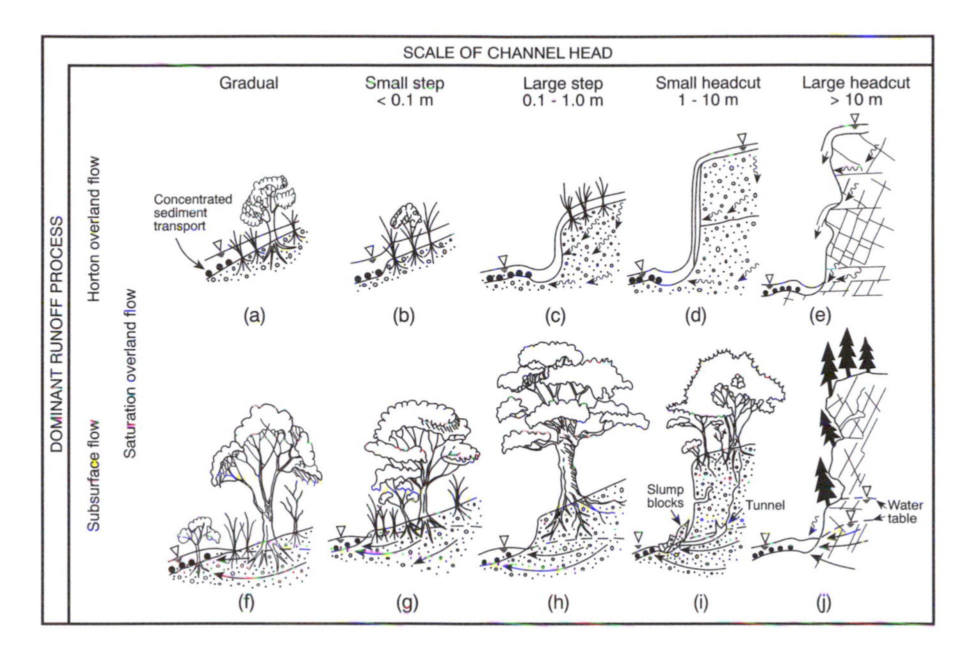

Figure 2.10 Typology of channel heads on the basis of incision depth and dominant runoff process. The smooth arrows indicate saturated flow, and the wiggly arrows indicate unsaturated percolation, including flow through macropores. Even in areas with significant Horton overland flow, the deep face of a large step or headcut can allow the emergence of erosive seepage. Plunge-pool erosion and slope-stability constraints become more important as the height of the feature increases (after Dietrich and Dunne, 1993).

channel heads and the development of dry valleys or hollows. Channel head location thus depends on the relative rates of incision and infilling, while at a larger scale the initiation and growth of surface channels are bound up with the long-term development of drainage networks.

Network Evolution

The drainage network is basically a transportation system for the movement of water and sediment. Given that a certain land area has to be drained and that surface runoff can be subdivided into unconcentrated and concentrated forms, various levels of drainage efficiency can be imagined (Chorley and Kennedy, 1971). From the point of view of the slower, unconcentrated form, maximum efficiency is achieved when basin slopes are as short as possible, which requires a complex bifurcating network of small channels. Experimental results (Parker, 1977) confirm that drainage efficiency (expressed by the ratio of discharge output to rainfall input) does improve as drainage density increases up to a maximum. From the point of view of concentrated runoff, water is removed more efficiently by larger channels since they offer less resistance, and maximum efficiency is achieved when one very large channel drains the area.

These ideal states are clearly incompatible, and in reality a compromise is achieved between the two extremes. Regarding the first point of view, basin slopes must be long enough to generate sufficient runoff for channel initiation. Also, beyond some maximum drainage density, efficiency may in fact decline as channel paths lengthen and become more circuitous (Parker, 1977). At the other end of the scale, a limit is set by the maximum length that a drained slope can attain before being subject to channel formation. The ideal scheme is complicated further: firstly, by the fact that not all water flows as surface runoff and in some environments subsurface flow is a major component; and secondly, by the presence of eroded debris which is also part of the transportation process. The nature of the compromise between the extremes of drainage density reflects the dominant hydrologic and sediment conditions existing within a basin, both of which are environmentally constrained and neither of which are independent.

In realizing the compromise, the question arises as to whether the network evolves to a form that achieves the most efficient removal of water and sediment for the given conditions. Optimality can be regarded as a fundamental principle governing system behaviour, and in this context has been applied to both particular characteristics such as junction angles (Roy, 1985) and the network as a whole (Howard, 1990; Rodriguez-Iturbe *et al.*, 1992; Rinaldo *et al.*, 1992). Rodriguez-Iturbe *et al.* (1992) argue that a sufficient explanation for the treelike structure of drainage networks is provided by a combination of three energy principles: (a) minimum energy expenditure in any link; (b) equal energy expenditure per unit area of bed anywhere in the network; and (c) minimum energy expenditure in the network as a whole, yielding the condition

$$\chi \Sigma Q_i^{0.5} L_i \rightarrow \text{minimum} \tag{2.7}$$

where Q_i and L_i are respectively the mean annual discharge and length of the *i*th link, and χ is assumed to be constant. Despite several debatable assumptions, the theory does reproduce well-known relationships such as Horton's laws of network composition (Rinaldo *et al.*, 1992), although Kirchner (1993) has shown that those laws provide a very weak test of theoretical models. Nevertheless, the idea of a minimum energy network has a certain appeal, especially as a basis for modelling, and is a logical development of earlier proposals (Leopold and Langbein, 1962). If optimality is a principle relevant to network evolution, then it must apply during the early stages of development before valley incision becomes so advanced that the potential for network form adjustment is severely constrained.

Evidence of Network Evolution

A complete specification of network evolution requires information on initial conditions, the relevant physical processes, the timescale involved, and the changing character of environmental controls. Such informatio rarely, if ever, available. Most natural networks have an unknown orig have had a long and complex history, and have been subject to many changes in environmental conditions. Evidence of evolution has been assembled from three main sources: direct observation, inference from existing networks, and theoretical modelling. Underlying all is the knowledge gained from the analysis of established networks, which has been used both as a basis for comparing networks of supposedly different age and as a basis for constructing and testing evolutionary models.

Direct observation of network development is necessarily restricted to a few suitable sites, such as surfaces that have been artificially produced (Schumm, 1956), recently exposed (Morisawa, 1964), or newly invaded (Plate 2.4). On the basis of his observations of a network (magnitude 214) that had developed between 1929 and 1953 in clay pits, Schumm (1956) proposed a schematic evolution of a third-order basin, which involved headward growth of a dominant channel with branching at the tip once a critical length was reached. Network expansion (elongation and addition of tributaries) and abstraction (elimination of tributaries) went on simultaneously, with lateral encroachment by larger streams being the main mode of abstraction.

Sequences can also be directly observed in laboratory experiments where the developmental process is accelerated (Flint, 1973; Parker, 1977). Parker's (1977) results from a test plot measuring 9 m by 15 m are by far the most comprehensive. Two sets of experiments were carried out:

(a) slope of 0.43° with base-level lowering before each run;
(b) slope of 1.83° with no base-level lowering except after the network had achieved maximum extension.

The networks grew primarily by headward extension and bifurcation of first-order streams (exterior links), but different modes of growth were identified in the two sets (Fig. 2.11A). In the first the network developed fully as it slowly extended headward with little subsequent addition of internal tributaries (Mode 1), while the second was characterized by the rapid growth of long first-order channels with tributaries being added later (Mode 2). Thus elongation and elaboration (infilling of the net) were more distinct processes in the second set. The network grew more rapidly in Mode 2, probably because of the steeper initial slope, but drainage density was lower at maximum extension (Fig. 2.15D, p. 51). As the network of Mode 2 continued to evolve beyond maximum extension, abstraction occurred, beginning near the outlet while the network was still expanding at its headwaters.

Network evolution observed directly in the field or laboratory has various shortcomings. It generally refers to small-scale development, and the problem is to apply results therefrom to the larger-scale, longer-term evolution of major networks. It may originate on a plane surface, but real networks are influenced in their development by variations in material type and initial topographic irregularities. Channels are invariably initiated by

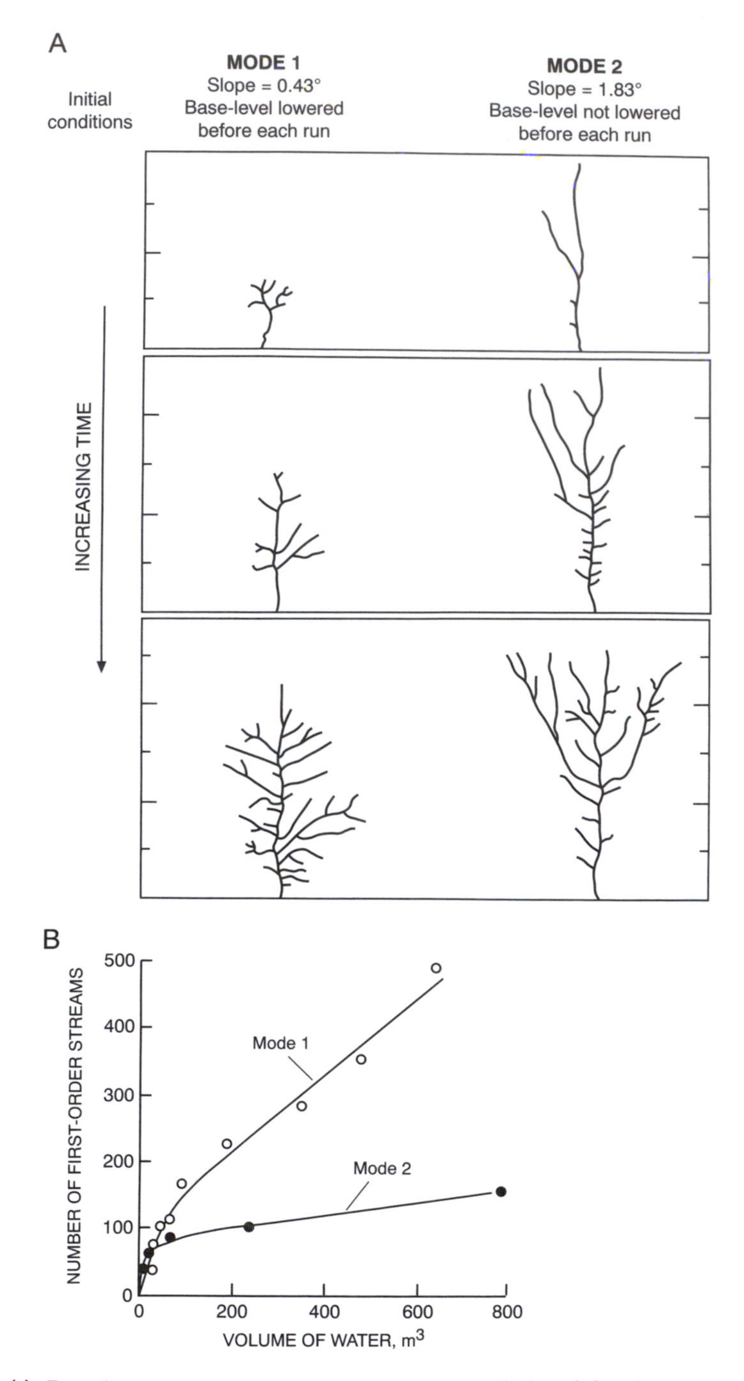

Figure 2.11 Experimental study of drainage network evolution (after Parker, 1977). (A) Examples of network growth by different modes. Networks are shown at equivalent times during the experiments. (B) Changes in the number of first-order streams during network development to maximum extension. Time is indexed by the volume of water over the system.

Horton overland flow but other processes may be of greater importance at the larger scale. Despite these drawbacks, small-scale studies are valuable particularly in suggesting detailed modes of network evolution.

Patterns of network evolution have been **inferred** either directly or indirectly on the assumption that time and space are interchangeable under certain circumstances (the ergodic hypothesis). Direct applications involve the arrangement of networks at supposedly different stages of development in a temporal sequence. By arranging selected topographic maps in such a way, Glock (1931) postulated several phases of development:

(a) *initiation* of a skeletal pattern;
(b) *elongation* by headward growth of the main streams;
(c) *elaboration* through the addition of small tributaries;
(d) *maximum extension*; and
(e) *abstraction* when tributaries are lost as relief is reduced through time.

Ruhe (1952) examined the networks developed on five glacial tills of variable age and, from the different degrees of dissection, inferred how drainage density changed through time (Fig. 2.15A). In a similar vein, Dohrenwend *et al.* (1987) have measured differences between networks developed on lava flows ranging in age from 0.015 to 1.95 Ma and argued that networks grow in a way that is consistent with the qualitative model proposed by Glock (Fig. 2.15B). Such studies tend to provide only broad indications of change and suffer from the problem that factors other than time may have been responsible for the differences.

Indirect applications involve relationships between selected network properties and a variable believed to change progressively through time. Relative relief has been a favourite candidate in this surrogate role. On the basis of a positive relationship between the proportion of tributary-source (TS) links and relative relief in 39 mature drainage basins, Abrahams (1977) proposed, firstly, that TS links rather than S (source) links develop as relief is increased by stream incision, and secondly, that TS links are preferentially abstracted with declining relief. The abstraction of exterior links tends to increase the mean lengths of both exterior and interior links as relief declines and channel infilling becomes more prevalent (Abrahams, 1976). These results imply that the planimetric properties of mature networks are controlled to a certain extent by evolutionary changes in relief. Although the problem of distinguishing evolutionary changes from spatial variations remains, controlled sampling can reduce the risk of incorporating environmental effects. Without using relative relief as a surrogate for time, other studies have also inferred modes of network growth at the detailed scale, on the basis of tests of the random topology model (James and Krumbein, 1969; Flint, 1980).

Theoretical models have been developed along both deterministic and probabilistic lines. Horton's (1945) deterministic model is an excellent example of how physical theory (the infiltration theory of runoff) and observed form (the laws of drainage network composition) can be rationalized in an evolutionary context. A system of subparallel rills initiated beyond the x_c

distance is modified by the processes of micropiracy and cross-grading to form a master rill (a–b in Fig. 2.12a), incision of which creates side-slopes on which new rill systems develop. This process continues until the available length of overland flow is less than x_c, a constraint on development which is best visualized within a diamond-shaped drainage area. Without such a shape it is difficult to determine which rill becomes dominant at each stage. The model is restricted in other, more fundamental respects. It suffers from the same environmental limitations already ascribed to Horton overland flow and the related concept of the x_c distance. Cross-grading and micropiracy are inconceivable processes of channel integration in large-scale networks. Governed by the principles of competition and the x_c distance, Horton's model is best applied to small-scale networks developed on unvegetated surfaces with low infiltration capacity and limited soil cover (Schumm, 1956).

Dunne (1980, 1990) envisaged network growth resulting from subsurface flow and seepage erosion (Fig. 2.13). Once the critical conditions for seepage erosion are reached, a channel is initiated and extends headward, which disrupts the groundwater flow pattern and encourages further flow convergence towards the channel head. Water emerging along valley sides may exploit some susceptible zone to form a tributary which also migrates headward and may eventually branch. The process of repeated failure, headward sapping and branching continues until the drainage area available at each spring head is too small to supply sufficient water for channel development. A balance is then struck between channelization and hillslope processes. This model, applicable at a larger scale than Horton's, explains why drainage patterns sometimes reflect geologic structure, and how stream capture can occur as neighbouring spring heads compete for water.

Willgoose *et al.* (1991a, b, c) have developed one of the most comprehensive deterministic models, which explicitly incorporates the interactive evolution of hillslopes and networks at the basin scale. A key element of the model is that erosion takes place at a faster rate in channels than on hillslopes. Channels continue to extend headward if a 'channel initiation function' (a), defined in terms of the product of discharge per unit width (q) and slope (s)

$$a = \beta_5 q^m s^n \qquad m, n > 0 \tag{2.8}$$

exceeds a critical threshold, where β_5 is a coefficient. Growth ceases when competition between advancing channel heads decreases their contributing area to such an extent that the channel initiation function falls below the channelization threshold (Fig. 2.14). The model has been used to explore how network properties vary during evolution, how different initial conditions influence development, and the circumstances under which a dynamic equilibrium might be attained. In effect the model represents a numerical laboratory in which issues such as the influence of spatial variability in erosivity (Moglen and Bras, 1995) can be addressed.

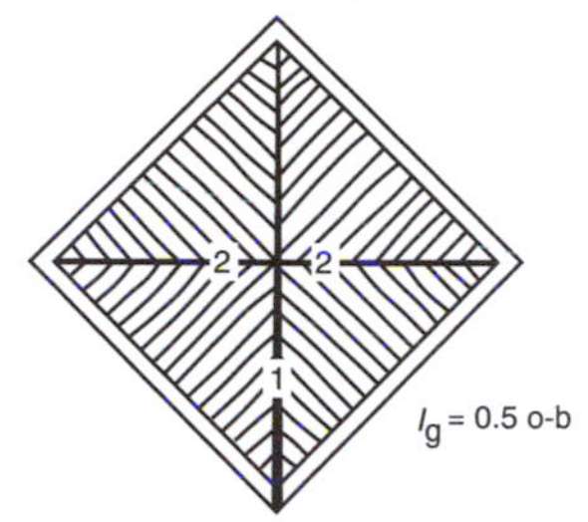

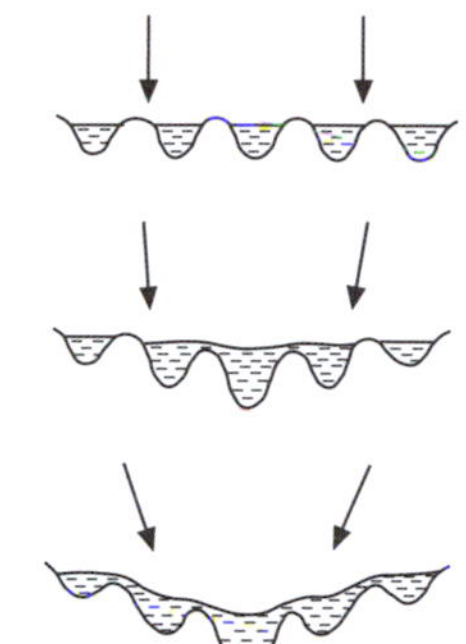

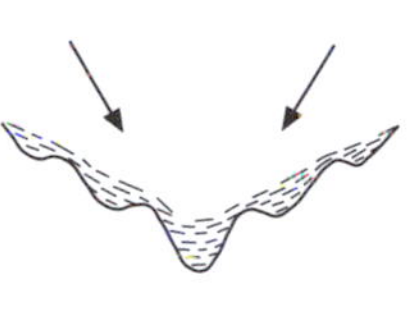

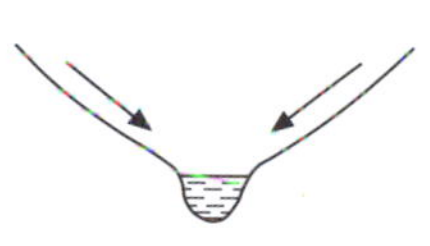

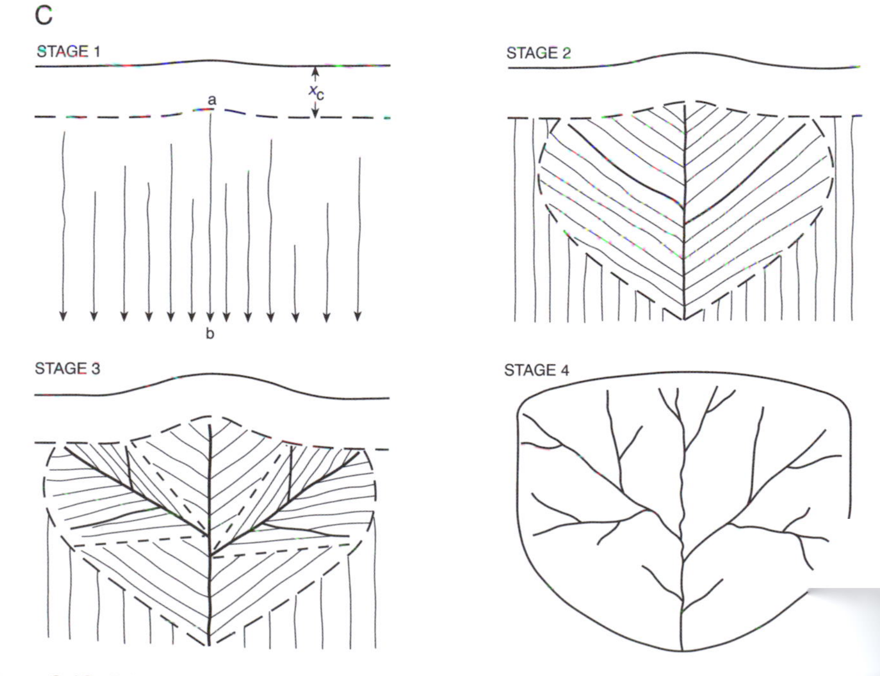

Figure 2.12 Schematic development of a drainage network (after Horton, 1945). (A) diamond-shaped basin. (B) Cross-sectional diagram of the process of cross-grading a the development of a master rill. Arrows indicate the resultant direction of overland flow (C) Development of a fourth-order network on a newly exposed surface.

STAGE 1: A single rapid tilt brings a smooth landsurface of permeable rocks above sea-level.

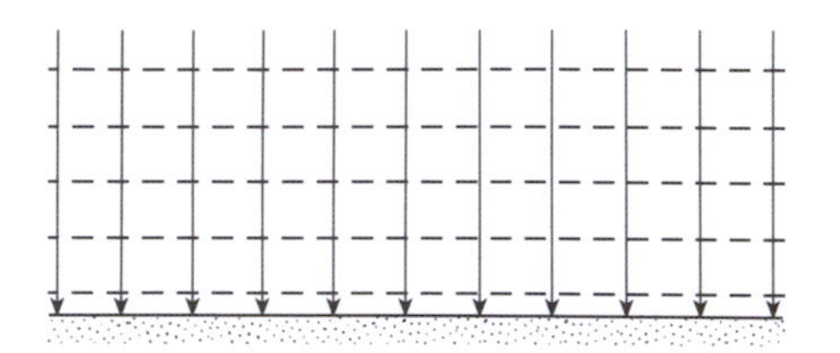

STAGE 2: Perturbation of the flow field leads to flow concentration towards a more permeable zone where piping erosion eventually occurs. A spring head is excavated, leading to further flow concentration, accelerated chemical weathering and repeated piping at the same site.

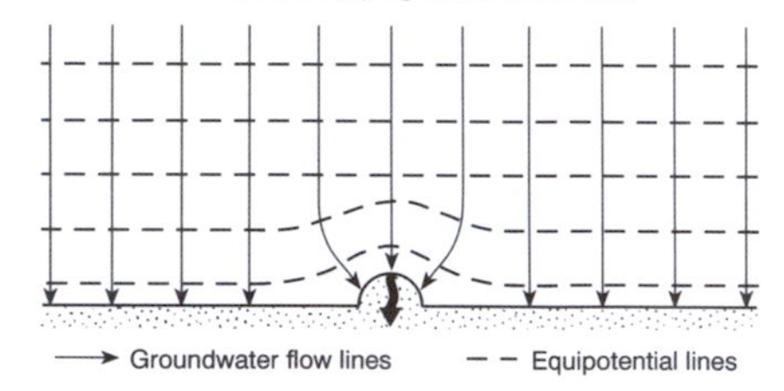

STAGE 3: Spring head retreat increases flow convergence and the potential for future seepage erosion. Water emerging along the valley sides exploits a susceptible zone to form a tributary which also undergoes headward retreat.

STAGE 4: The process of repeated failure, headward retreat and branching forms a network of valleys. The pattern stabilizes when the declining drainage area of each spring head is no longer large enough to supply enough groundwater to cause erosion of the weathered bedrock.

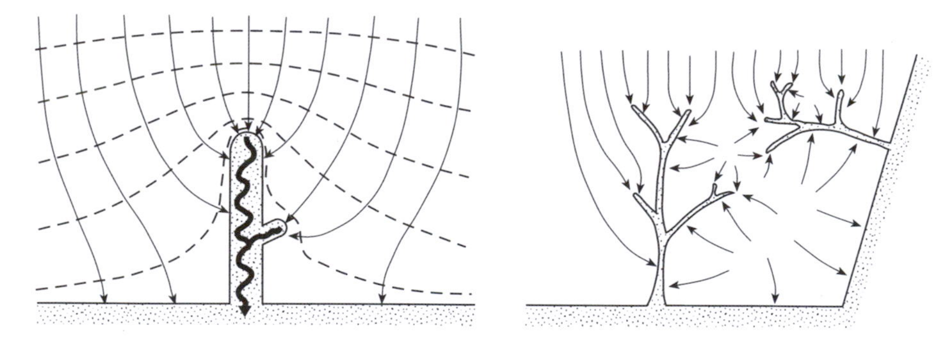

Figure 2.13 A model of drainage network development by seepage erosion (after Dunne, 1980).

Probabilistic tendencies can be incorporated in various ways: through the generation of random networks (Leopold and Langbein, 1962); the construction of random headward-growth models (Howard, 1971a; Stark, 1991); and the simulation of specific processes such as stream capture (Howard, 1971b). Stark's (1991) 'invasion percolation model' is a growth model in which stream heads branch and propagate in response to randomly assigned variations in substrate strength. In common with other statistical models, real and simulated networks are similar but Abrahams (1984a) regarded such success as largely illusory. He argued that simulation models have added relatively little to our understanding of network evolution, especially as regards the processes involved.

An important aspect of growth modelling is the probability of branching. Assuming that networks grow sequentially by the addition of an exterior link (first-order stream), three branching events are possible:

(i) bifurcation at the head of an exterior link;

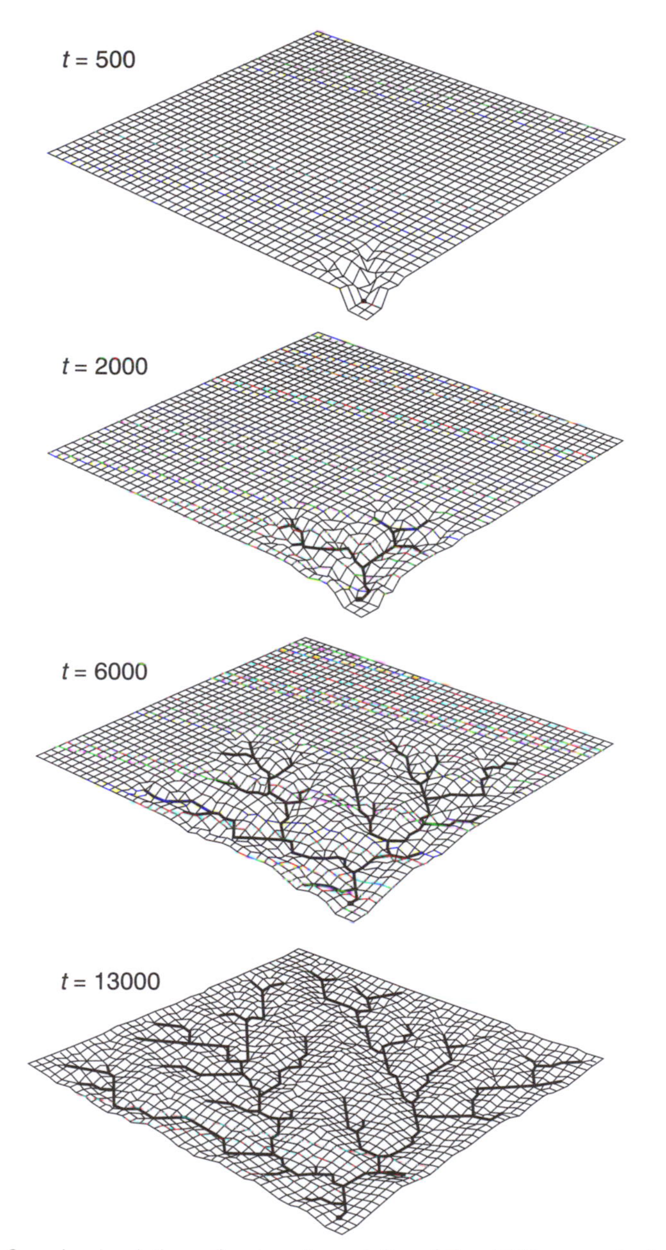

Figure 2.14 Sample simulation of network evolution (after Willgoose *et al.*, 1991a, b). Times (t) are non-dimensional.

(ii) right or left tributary development on an exterior link; or
(iii) right or left tributary development on an interior link.

Dacey and Krumbein (1976) considered three different combinations of these possibilities: model A = (i); model B = (ii) and (iii); model C = (i) and (iii). Empirical tests revealed that model A can be rejected in all cases and that models B and C fit the data best. These results have been confirmed elsewhere (van Pelt *et al.*, 1989; Knighton *et al.*, 1992), especially as regards the success of model B, suggesting firstly that simple bifurcation at the head of exterior links cannot account for the observed structure of natural networks, and secondly that branching occurs both at the edges and in the interior of networks. These implications conflict with contemporary opinion that growth is dominantly headward rather than sideways, at least in the initial phases, although model B does predict that exterior link branching is more likely during early stages of growth. Also, they do conform with both Dunne's (1980) spring-sapping model and Parker's (1977) first set of experiments (Mode 1, Fig. 2.11A). Again, however, there is a danger in arguing from simulated to natural processes, for the success of a model is no guarantee of its uniqueness as a means of explaining observed form.

This treatment of theoretical models is by no means exhaustive. However, it has illustrated how they vary in complexity and applicable scale, and has underlined the difficulty of providing definitive statements about network evolution. The testing of models is hampered by the dearth of historical observations and the need to work with a contemporary static base. As the scale at which network growth is considered decreases, the possibility improves of relating that growth to the physical processes involved. No single model can hope to be entirely satisfactory, but there is sufficient evidence to indicate general features of network evolution.

Modes of Evolution

The limited data available suggest that drainage density increases relatively rapidly at first as channels invade an unoccupied area, and then changes more slowly as a limit or equilibrium value is approached (Fig. 2.15). Initial increases in number of streams and total stream length may indeed be exponential in form (Flint, 1973; Knighton *et al.*, 1992). Parker's (1977) data indicate how different initial conditions can influence both the rate of change and maximum value of drainage density, with Mode 2 networks growing more rapidly but attaining a lower maximum (Fig. 2.15D). As channels extend and incise themselves, landscape mass is removed and the zone of maximum erosion tends to migrate headward. Sediment is derived from progressively further upstream, and downstream parts must continually adjust to this supply.

For convenience, network evolution can be subdivided into various phases similar to those proposed by Glock (1931): (a) *initiation*; (b) *extension* by headward growth (elongation) and tributary addition (elaboration); (c) *maximum extension*; and (d) *integration* by abstraction and capture. No historical inevitability should be assumed from this progression since one or more phases may be absent or vary in relative significance during evolution. Parker (1977), for example, could find no clear distinction between initiation and other phases of extension.

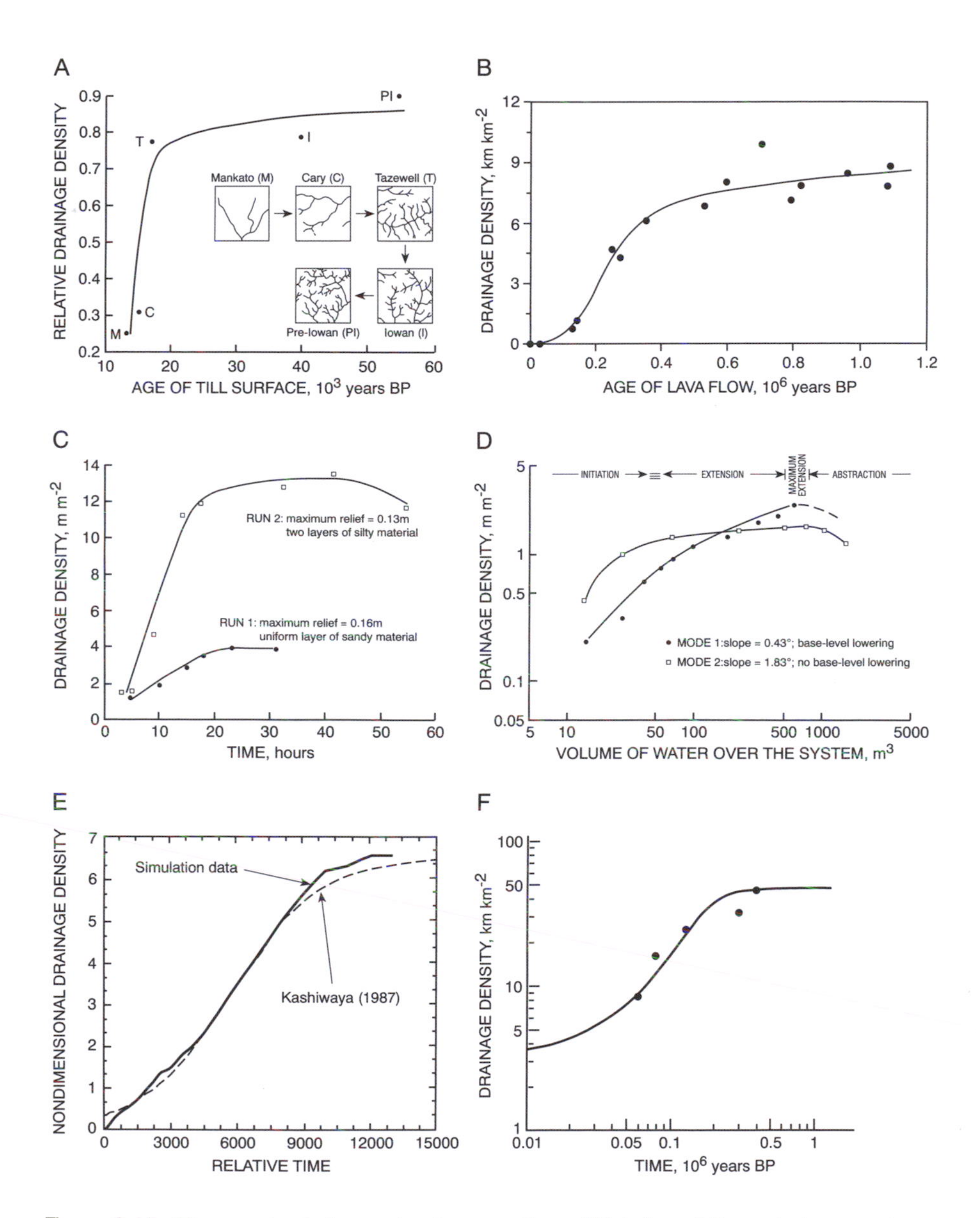

Figure 2.15 Changes in drainage density over time. (A) Inferred from drainage patterns developed on glacial tills of different ages in Iowa (after Ruhe, 1952, and Leopold *et al.*, 1964). (B) Inferred from drainage patterns developed on lava flows of different age in Nevada (after Dohrenwend *et al.*, 1987). (C) In small-scale laboratory experiments (data from Flint, 1973). (D) In laboratory experiments with different modes of growth (after Parker, 1977). Time is indexed by the volume of water over the system. (E) Based on a sample realization of a deterministic model (solid line) and Kashiwaya's (1987) model (dashed line) (after Willgoose *et al.*, 1991a). (F) Theoretical curve and data for South Kanto district, Japan (after Kashiwaya, 1987).

Three possible modes of *network growth* or *extension* have been proposed (Table 2.4). At one extreme is the Horton mode in which parallel systems of channels progressively develop and coalesce through capture. At the other is the 'expansion' mode in which lateral and headward growth occur together to give a more or less complete network which slowly extends headward. In between is the 'extension' mode, which is closest to that proposed by Glock (1931), with long low-order channels developing rapidly at first, before infilling by tributary addition later. Contemporary opinion favours a combination of the second and third modes, with the third probably being the most common (Schumm *et al.*, 1987). Assuming that the physical requirements for channel initiation are met, the essential problem is to determine where, how and at what rate exterior links (first-order streams) are added to the net. Such additions will clearly affect the topologic and link-length properties of evolving networks.

In Parker's (1977) experiments the number of exterior links increased rapidly at first (Fig. 2.11B), especially where the network developed fully during headward extension (Mode 1). Bifurcation ratios fluctuated considerably, tending to increase to a maximum of 7–11 before falling as higher-order streams were created. However, these oscillations were quickly damped out and bifurcation ratios became more stable as the network approached maximum extension, when values tallied more closely with Shreve's (1966) prediction of $R_B \sim 4$. Similar tendencies have been observed in experimental runs of the Willgoose *et al.* (1991c) numerical model.

As bifurcation ratios attain more improbable values or as the length of an exterior link increases beyond a certain limit, a branching event becomes more likely. The stability of a given length of stream presumably varies with environmental conditions. The length properties of exterior links differed for the two modes of growth in Parker's experiments. In Mode 1 the length distributions changed little during evolution, suggesting a degree of regularity where networks develop fully as they extend headward. In Mode 2, on the other hand, exterior link lengths tended to decrease with time. In particular, the S (source) links, which were primarily responsible for the rapid headward growth (Fig. 2.11A), became markedly shorter as tributaries

Table 2.4 Modes of network growth

Dominant mode of growth	Source
Parallel development and coalescence	Horton (1945); Leopold and Langbein (1962)
Headward growth and branching	Schumm (1956); Howard (1971a); Flint (1973); Parker (1977 – Mode 2); Dohrenwend *et al.* (1987)
Lateral and headward growth	Dacey and Krumbein (1976); Parker (1977 – Mode 1); Dunne (1980); Willgoose *et al.* (1991a, b)

developed after the initial phase. TS links, which appear to develop later (Abrahams, 1977) and be less subject to such changes, tended to have more stable length distributions.

The addition of exterior links is likely to be a discontinuous process conditioned partly by location in the network. The preferential development of TS rather than S links as relief increases (Abrahams, 1977) or as time elapses (Dohrenwend *et al.*, 1987) confirms that internal rather than headward growth dominates later phases (*elaboration*) when conditions for headward extension become more critical. Growth within networks cannot take place independently because of competition between adjacent streams for available drainage area (Fig. 2.16). Along major tributaries that join master streams at relatively small angles, the more limited area on the acute side may favour stream development on the opposite (obtuse) side (Flint, 1980). The first link at the outlet of such a tributary is therefore more likely to be trans than cis (Fig. 2.16B). The effect of this areal asymmetry may persist upstream to give further branching and the preferential development of larger streams on the obtuse side (Abrahams and Updegraph, 1987). Also,

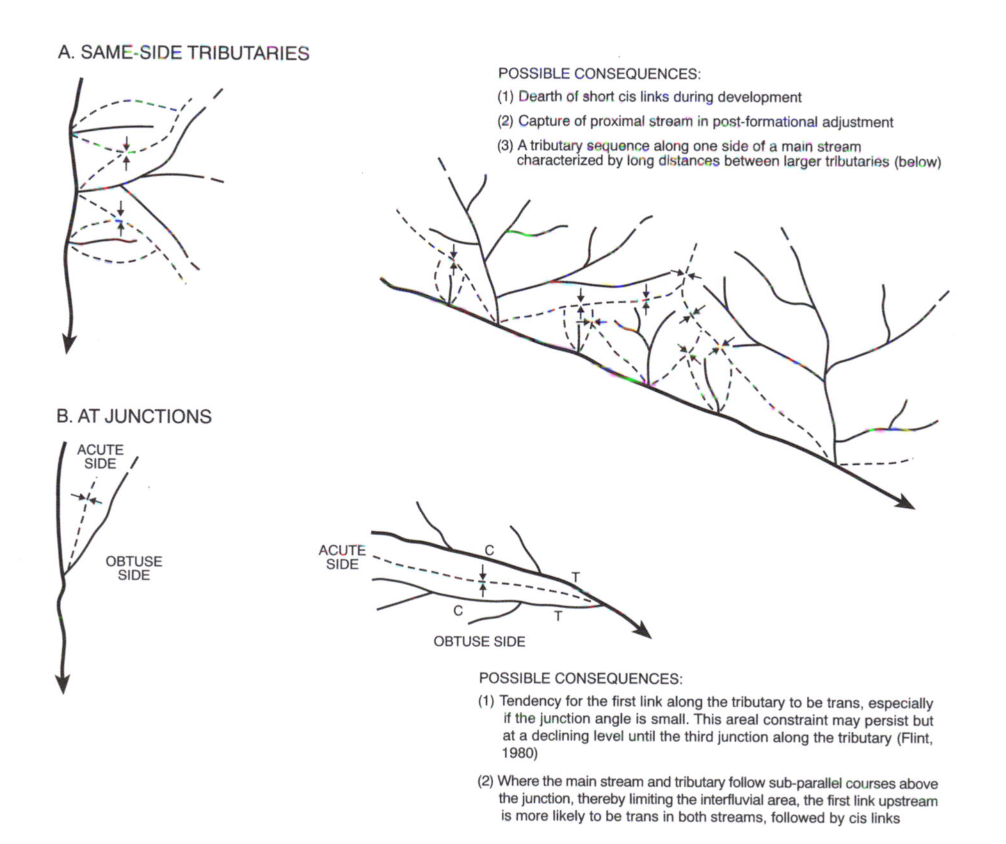

Figure 2.16 Competitive situations in network development. Double arrows indicate competition at inter-stream divides.

competition on one side of a stream (Fig. 2.16A) may induce successive tributaries to develop on opposite sides, leading to a preponderance of trans-links in networks. Although the frequencies of cis- and trans-links do not always differ (Dunkerley, 1977), the main point is that the constraint of available area can influence tributary arrangements during evolution. Indeed, any irregularity in basin shape or position of trunk streams may constrain tributary development (Flint, 1980).

Assuming that exterior links grow to a critical length before branching takes place, there should be an absence of short interior links. The fact that link lengths of all types and at all stages of growth appear to have gamma distributions in observed and simulated networks lends support to this argument (Dunkerley, 1977). However, the lengths of interior links are influenced by several factors in addition to the branching probability of exterior links. Where growth occurs other than by bifurcation at the headward end, the addition of tributaries in later phases would tend to shorten interior links, especially where links are bounded by tributaries on opposite sides. Abrahams (1980) found that smaller divide angles tend to result in longer interior links, presumably because of the more limited area available for tributary development above such junctions. In addition, the length of an interior link is not necessarily constant through time because of junction migration (Schumm, 1956), although such a process is less likely as channels become better established. The average spacing between bifurcations, which partly determines link-length properties, is controlled by those environmental factors which determine the density of drainage, but the problem remains of defining that average for given conditions.

Maximum extension signifies the imposition of limits to further expansion, which are determined by hillslope stability relative to the tendency for channel cutting. Channel-head position reflects an approximate balance between channelization processes on the one hand, and the counteracting effects of surface resistance and infilling on the other. Nevertheless, intermittent divideward extensions can occur at times of heavy rainfall (Morgan, 1972). Longer-term migration is dependent on the magnitude–frequency characteristics of channel-extending events and the morphology of the headward zone, with slope concavity in profile and plan being of critical importance. Calver (1978) demonstrated that in the long term, concavity can extend upslope into areas of convex slope profile without climatic or associated changes, leading to the preferential lowering of divides along valley axes. Although divideward limits may not be strictly constant through time, the fluvial system can attain an approximately stable drainage density, the magnitude of which is determined by the environmental conditions discussed previously (pp. 20–4).

Maximum extension signifies that some form of equilibrium has been attained. In terms of their numerical model, Willgoose *et al.* (1991a) showed that dynamic equilibrium is not possible without tectonic uplift, since only when the mean rate of erosion equals the mean rate of uplift will basin elevations and the average channel network be invariant with respect to time.

They also showed (1991c) that the topologic and geometric characteristics of evolving and maximum-extension networks are sensitive to initial conditions, an assertion which is supported by evidence from a wide variety of sources (Schumm *et al.*, 1987; Howard, 1994). The main streams of a mature network are essentially locked in place and have probably inherited conditions from a distant past. Minor tributaries, on the other hand, are not so restricted and, although they are strongly influenced by the morphology of the major valleys, are probably capable of relatively rapid adjustment. Given that about half of the channels in any network are exterior links, the potential for responding to change could be considerable. In effect, a distinction is drawn between longer-term valley dissection and shorter-term fluctuations in channel head position.

Competition is a major element in many evolutionary models, and further changes to network form (*integration*) may take place through the capture and abstraction of streams. In order to explain the relative dearth of short cis-links in their networks, James and Krumbein (1969) proposed that post-formational readjustments result in the elimination of proximal and less successful tributaries through lateral erosion (Fig. 2.16A). In Parker's (1977) experiments, abstraction occurred primarily by means of lateral migration or encroachment by larger streams. Although the reduction of inter-stream divides can, through time, lead to the diversion of surface waters (Miller, 1975), it is probably an effective process only in small-scale networks cut in erodible materials. Indeed, stream capture may be a relatively rare event in drainage network evolution (Bishop, 1995).

In larger networks, abstraction of fingertip streams is accomplished mainly by infilling, which reflects a shift in the balance between diffusive and incisive processes in favour of the former. It begins in the downstream part of a basin and proceeds headward as relative relief is reduced (Parker, 1977; Knighton *et al.*, 1992). Given an inverse relationship between stream-head area and slope (Fig. 2.9A), a decrease in drainage density is a consequence of declining relative relief. The upstream progression of abstraction explains the tendency for higher drainage densities in headwater areas. Indeed, downstream abstraction and headwater expansion can occur simultaneously, thereby prolonging the period of maximum drainage density. TS rather than S links may be preferentially abstracted as relief declines (Abrahams, 1977), but this need not always be so (Parker, 1977). The net effect of abstraction is a decrease in the number of exterior and, by definition, interior links. Such losses tend to increase the mean length of both types of link (Abrahams, 1976; Parker, 1977).

No explicit timescale has been given for this general picture of network evolution. While small-scale studies over a limited time can suggest the types of process and change which might be involved, extrapolation to large-scale networks is dangerous. Real networks have evolved over millennia and have been subject to complex variations in environmental and base-level conditions. Climate, in particular, has fluctuated considerably over the past 20 000 years (Fig. 6.2, p. 273). Even minor changes in precipitation characteristics can induce a marked network response in sensitive areas (Cooke and Reeves, 1976). Consequently, analyses of existing networks, and especially those that use relative relief as a surrogate for time, have the problem of differentiating between time-related and environment-related change. Any broad view of network evolution has to be considered against this background of changing external as well as internal conditions.

Hillslopes, Networks and Channels

Hillslopes and channels evolve together in response to prevailing climatic and physiographic conditions. The drainage network provides a fundamental link between hillslopes and channels in that it determines, largely through its density, the level of interaction between these two systems of water and sediment transfer. A network of magnitude M contains $(2M - 1)$ links and, since each link is situated between two hillslopes, $2(2M - 1)$ hillslopes. Each link can be visualized as a discrete subsystem with its own inputs, storage elements and outputs of water and sediment. Thus, even a moderately sized drainage basin will have a large number of hillslopes, with the potential for considerable variability not only in the soil, topographic and vegetation conditions that control the supply of water and sediment, but also in the level of interaction between hillslopes and their basal streams.

Hillslope–Basal Stream Interactions

On the one hand, hillslopes influence stream behaviour in that they represent source areas for the water, solutes and sediment which are supplied to streams and to which streams must adjust. Of particular hydrological significance is the way in which a hillslope responds to precipitation to produce the slope-base hydrograph, the characteristics of which vary with the relative contributions from the various hillslope flow routes (Fig. 2.7A) and with slope geometry, notably length and steepness. On the other hand, streams influence hillslopes in that they act as local base-levels and thereby affect slope stability via the processes of lateral undercutting and vertical incision or aggradation. Rapid stream downcutting is associated with the development of convex hillslope profiles, while aggradation tends to reduce slope-base gradients and leads to the development of concave profiles. Material for aggradation may come from the accelerated erosion of oversteepened slopes, so that a negative feedback can be rationalized which tends to maintain a dynamic equilibrium between hillslope and channel.

Maintenance of an approximate balance between the rates of debris transport in the two systems may be expected to result in a positive relationship between hillslope and channel gradients, the rationale being that a higher rate of sediment supply from steeper slopes is accommodated by gradient adjustment within the basal stream. Although good correlations have been obtained between valleyside and basal stream gradients (Strahler, 1950; Richards, 1977a), the relationship is more complicated than this simple model suggests. In particular, it begins to break down in higher-order valleys with extensive floodplains (Richards, 1977a, 1993).

The variable interaction between hillslopes and streams can be regarded as a continuum, the two ends of which are represented by 'strongly coupled links' and 'completely buffered links' (Rice, 1994). In the former, material is transferred from hillslopes to channel relatively rapidly and continuously, resulting in a sensitive and more or less synchronous adjustment between the two. In the latter, floodplain or valley-fill deposits are sufficiently extensive to protect hillslopes from basal erosion and to isolate the river from hillslope sediment supply. Consequently, the only sediment entering such links comes from upstream channels; as alluvial controls become dominant, colluvial controls are eclipsed. In between is a wide range of 'intermittently coupled links' where the degree of interaction is discontinuous in time and/or space. Non-alluvial material may be introduced to a stream where active hillslopes abut the channel, or where the basal stream undercuts the valley side in reaches with incomplete floodplains.

This continuum can be given spatial expression. Headwater areas are commonly regarded as the main production zones for water and sediment in a basin. There, hillslopes and channels are likely to be strongly coupled, with rapid stream response to precipitation on hillslopes and relatively frequent inputs of non-alluvial material. With increasing distance downstream the degree of coupling can be expected to decline. Larger discharges increase the significance of fluvial relative to hillslope activity, and wider floodplains progressively buffer the active channel from hillslope inputs. The rate at which the degree of coupling changes downstream will vary from basin to basin, depending on geomorphological history, basin morphometry and biophysical conditions (Rice, 1994). Caine and Swanson (1989) have evaluated the efficiency of coupling between hillslopes and channels in two small headwater basins, one in the Cascade Range and the other in the Rocky Mountains. The first is characterized by an approximate balance between the delivery and export of material, while the second discharges only 10 per cent of the sediment supplied to the channel, the remainder being stored on the valley floor. Thus, even headwater catchments can display inefficient coupling, with the implication that downstream change is not necessarily as simple as at first supposed.

The degree of coupling can also vary over time. The progressive evolution of gully systems in the Howgill Fells of northwest England illustrates how decoupling can develop over a period of about 200 years (Harvey, 1992, 1994). Gullies are initiated as streamside scars in response to lateral undercutting, feeding sediment directly to the basal stream. As the gullies retreat headward and develop linear channels, sediment accumulates at their downstream end in the form of debris cones which partially buffer the active slopes from basal stream activity. Basal removal becomes progressively less frequent, especially if the channel migrates away, until the eroding slopes become completely disconnected from the main stream, vegetation colonization eventually stabilizing the debris slopes. Systematic studies of coupling at this level of detail are rare, and yet coupling between hillslopes and channels is important as it constitutes one element in the throughput of water and sediment within drainage basins.

Networks and Channels

Networks and channels have traditionally been studied separately in fluvial geomorphology, with relatively few attempts to link the two. Leopold and Miller (1956) sought to establish a suitable methodology based on Horton's laws and a discharge–area relationship, arguing that stream order is related to discharge and thence to hydraulic variables such as width, depth and velocity. However, stream order provides little more than a crude index of discharge or position within the network hierarchy. Pizzuto (1992) combined hydraulic geometry equations (which relate fluvial morphology to discharge, bedload transport rate and grain size) with a watershed model which assumes that sediment is supplied only at channel heads, in order to predict how fluvial morphology varies as a function of position within the network. Although downstream changes in width and depth are reasonably well reproduced, results are of variable quality and the sediment supply assumption is unreasonable, particularly in view of hillslope–channel interactions.

Rivers adjust their morphology in the downstream direction partly in response to the water and sediment supplied by tributaries. However, the effect is not unidirectional. Tributaries interact with the main stream and such interactions provide one means of integrating networks and channels. Much like the channel relative to the hillslope, the main stream provides the base-level for the tributary, and changes in its vertical or horizontal position can initiate adjustment within the tributary. An explanation of cut-and-fill cycles in semi-arid basins invokes such an effect (Schumm and Hadley, 1957). Incision in the main valley lowers the base-level for tributary streams which degrade their beds, the degradation beginning at the junctions and extending upstream by headcut migration. This bed erosion along the tributaries delivers sediment to the main valley, where it accumulates and reverses the degradation. Aggradation continues until the gradient becomes steep enough for a new phase of main-valley incision to begin, and the whole cycle is repeated. A mechanism is required to trigger the initial incision, and under suitable circumstances river regulation could provide the means. Germanoski and Ritter (1988) described how bed degradation below a dam initiated very rapid downcutting along tributaries entering close to the dam, the incision also leading to bank instability and channel widening.

The introduction of water and sediment at tributary junctions can produce change both at the confluence itself and in the downstream channel.

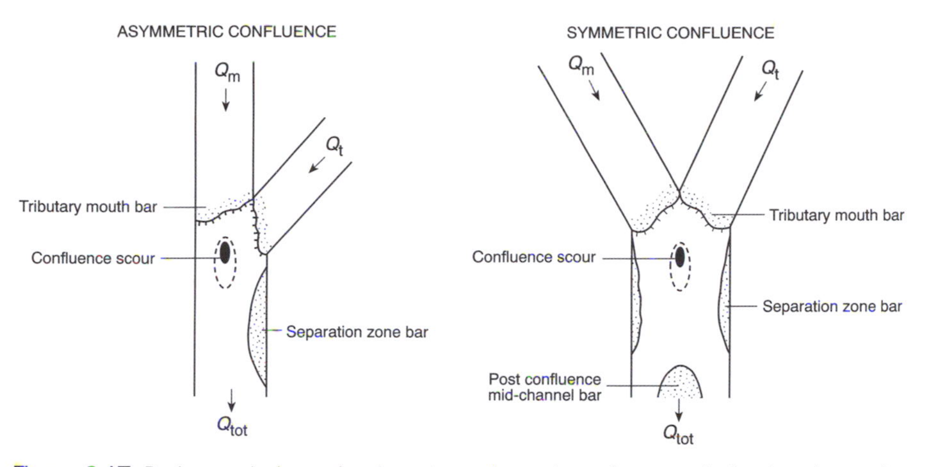

Figure 2.17 Bed morphology developed at channel confluences, indicating the major depositional and erosional elements within symmetric and asymmetric confluences (after Bristow *et al.*, 1993). Q_m, mainstream discharge; Q_t, tributary discharge; Q_r, discharge ratio $= Q_t/Q_m$.

Four morphological elements are commonly associated with confluences (Best, 1986; Bristow *et al.*, 1993; Fig. 2.17):

(a) bars across the mouth of each channel;
(b) a zone of scour where the confluent flows combine;
(c) bars formed in the separation zone downstream from the junction; and
(d) mid-channel bars in the downstream channel.

A basic distinction is drawn between asymmetric and symmetric confluences (Mosley, 1976). The specific location and morphology of these elements is controlled largely by the junction angle and discharge ratio $Q_r = Q_t/Q_m$, where Q_t is tributary discharge and Q_m mainstream discharge. As junction angle and discharge ratio increase, the tributary influence penetrates further into the main channel, producing deeper scour and larger bars within the separation zone (Best, 1988).

At a broader scale, downstream changes in channel properties have often been treated as continuous but, given the abrupt increase in discharge at tributary junctions, they could reasonably be expected to be discontinuous, especially in reaches where tributary contributions dominate the supply of water to the main stream. On the basis of a relationship between the channel width of link j (w_j) and link magnitude (M_j):

$$w_j = w_1 M_j^k \qquad (2.9)$$

Richards (1980) devised an expression for the width downstream from a junction (w_D) in terms of a ratio of the magnitudes of the downstream (M_D) and upstream (M_U) links:

$$\frac{w_D}{w_U} = \left(\frac{M_D}{M_U}\right)^k \Rightarrow w_D = w_U\left(\frac{M_D}{M_U}\right)^k \quad (2.10)$$

Where the incoming tributary is small, $M_D/M_U \sim 1$ and there should be little or no change in width. Roy and Woldenberg (1986) have argued that magnitude ratio is a poor predictor of width and criticized the model on the grounds that it explicitly ignores conditions in the incoming tributary (although $M_D = M_U + M_t$, where M_t is tributary magnitude). They developed an alternative formulation which rectifies this shortcoming, but although it describes average changes in channel form adequately, the changes at individual junctions are so variable and complex that they remain unpredictable. Rhoads (1987) contended that systematic changes in channel characteristics downstream from tributary junctions have yet to be convincingly established, and tentatively suggested that $Q_t/Q_m \sim 0.7$ represents a threshold separating variable behaviour below (< 0.7) from systematic behaviour above (> 0.7). It is not only the absolute magnitude of a tributary that determines its degree of influence on the main stream but also its relative magnitude, both of which vary considerably throughout a drainage network. Clearly tributary effects remain ill-defined, but they do represent an important avenue for integrating networks and channels.

Networks and Flows

Networks are from a long-term perspective a product of runoff processes, but their form, together with that of the hillslopes, influences runoff response in the short term. Catchment form and hydrologic response can be regarded as interdependent, becoming progressively more accordant over time provided environmental conditions remain reasonably stable for a long enough period (Kirkby, 1993). Rarely has the hydrologic evolution of a drainage basin been observed, but Ritter and Gardner (1993) exemplified the process in small basins disturbed by surface mining. In basins where Horton overland flow was dominant, stormflow hydrographs were characterized by increasing peak and total runoff and decreasing time to peak runoff as drainage density increased during network extension, and by opposite tendencies during network abstraction, reflecting changes in the relative efficiency of overland versus channelled flow as the network evolved. Whether networks are adjusted to provide the most efficient removal of water (and sediment) as optimality models suggest (Rodriguez-Iturbe *et al.*, 1992), their hydrologic relevance cannot be denied.

Attempts to establish a link between network form and hydrologic output from a basin have used a wide variety of indices, representing such network attributes as density (e.g. drainage density, stream frequency), size (e.g. network magnitude, total stream length) and internal structure (e.g. network width function). Given that drainage density is a measure of the efficiency with which water is discharged from a basin, its widespread application in empirical relationships is hardly surprising. For 15 basins in the United States Carlston (1963) related peak flow represented by the mean annual flood ($Q_{2.33}$) to D_d^2, treating drainage density as a static feature. Herein lies a problem. Since networks are dynamic with the potential for manifold variations in density during storms (Gregory and Walling, 1968; Day, 1983), the definition and measurement of a representative net is an important consideration. Digital elevation models (DEMs) can alleviate the problem by providing a means for the rapid and reliable delineation of both the perennial and extended networks (Wharton, 1994). However, as drainage area increases, ephemeral contractions and expansions assume less overall importance and the perennial net comes to dominate hydrologic response, at least in humid climates.

The basin characteristic that correlates most consistently with flood parameters is drainage area. In the most comprehensive study of British rivers where equations were successively obtained for estimating the mean annual flood (Natural Environment Research Council, 1975), drainage area proved to be the single most important independent variable at both national and regional levels. It basically operates as a scale variable, with magnitude as its network equivalent. Magnitude can be superior to drainage area as a predictor of mean annual flood, particularly in large basins where the precise delineation of the net is less critical (Knighton, 1987a; Fig. 2.18A). The use of magnitude has an additional advantage. The discharge of each link and the contributions made by individual tributaries can be readily estimated, enabling the downstream pattern of flow addition to be reproduced (Fig. 2.18B). Such patterns underline the discontinuity present in the fluvial system, which has implications for downstream channel adjustment.

The network width function (N(x)), defined by the number of links at successive distances x upstream from the basin outlet (Fig. 2.19A, B), provides a more direct representation of the spatial pattern of water courses within a catchment. Its use in hydrologic modelling stems from the idea that catchment response to rainfall input can be separated into a hillslope response function and a network response function, with the latter being based on N(x) (Naden, 1992; Fig. 3.1B, p. 66). The relative importance of these two elements will vary with soil and topographic conditions, and with basin size. In small basins (say < 50 km^2) catchment response will be dominated by the hillslope component, but in larger basins, where travel times through the network become increasingly important, it will come to be more affected by the network component and even resemble N(x) in basic form (Kirkby, 1976; Gupta and Mesa, 1988). Channel network geometry can significantly influence catchment response even in a basin as small as 1.24 km^2 (Mesa and Mifflin, 1986), but its main contribution will be at larger scales, as Naden (1993) has demonstrated for the rivers Severn and Thames, which have drainage areas of the order of 10 000 km^2 (Fig. 2.19C).

A major motivation for analysing network structure has been the prospect of being able to predict hydrograph characteristics from network variables. The network width function appears to provide better opportunities for hydrologic prediction than many of the network indices used in the

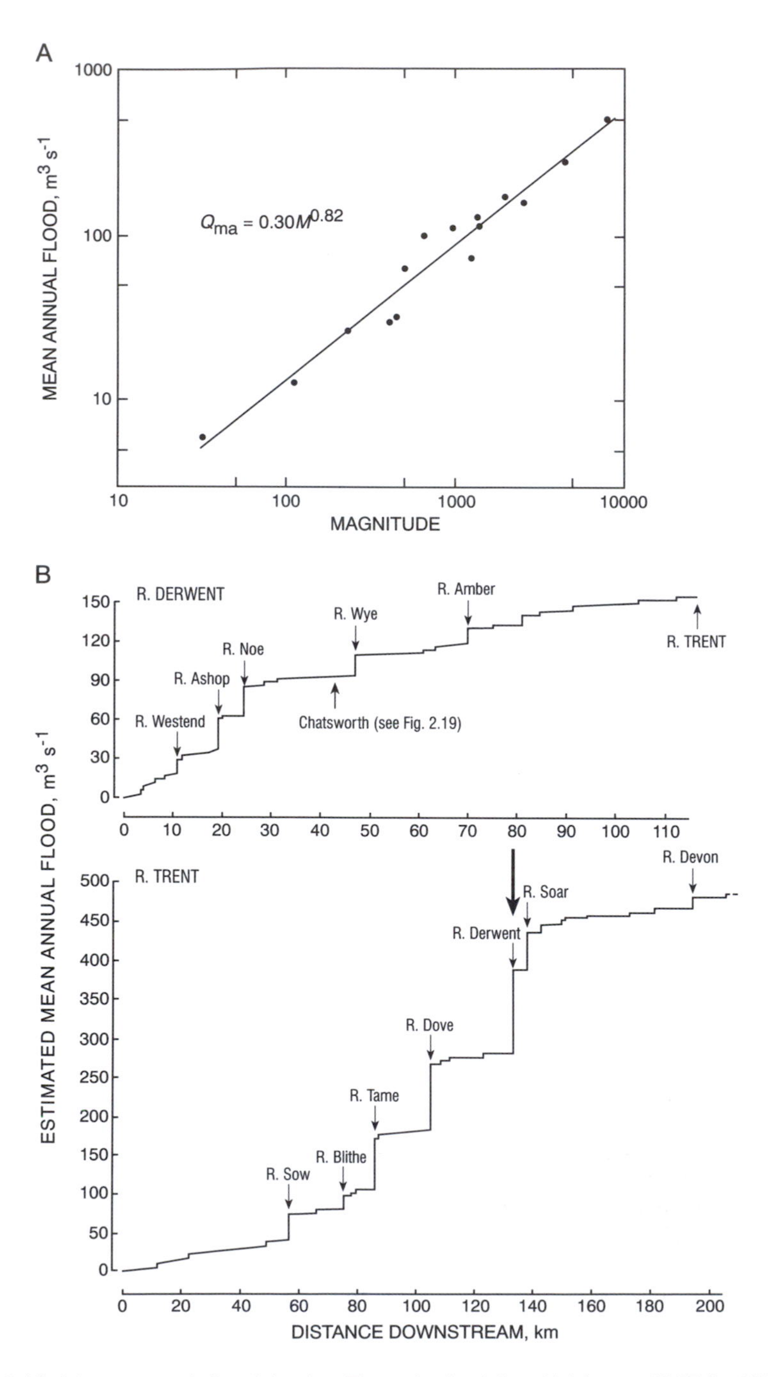

Figure 2.18 Mean annual flood in the Trent basin (after Knighton, 1987a). (A) Relationship between mean annual flood and link magnitude. (B) Downstream pattern of flow addition along the rivers Trent and Derwent, estimated from the equation in (A).

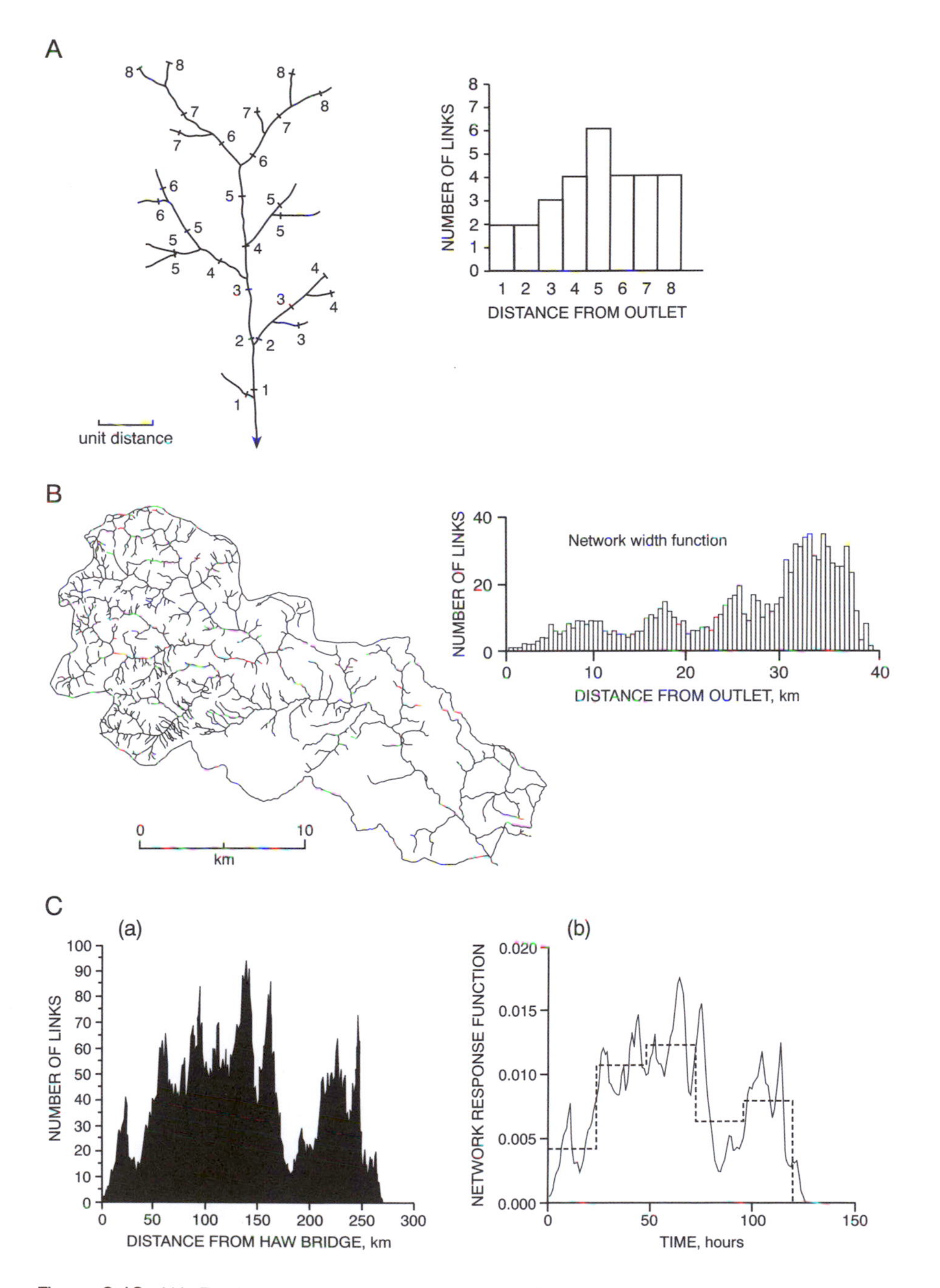

Figure 2.19 (A) Derivation of the network width function. (B) Network width function for the River Derwent at Chatsworth (see Fig. 2.18) (after Naden and Polarski, 1990). (C) River Severn at Haw Bridge: (a) network width function; (b) predicted hydrologic response for a 1-hour (solid) and 24-hour (dashed) effective rainfall (after Naden, 1993).

past. It explicitly recognizes the spatial distribution of channels, with application to the routing of flow through a network. It has direct links to the Geomorphological Unit Hydrograph (GUH), an approach pioneered by Rodriguez-Iturbe and Valdes (1979) in which the probabilities of water movement from a stream of one order to another are expressed in terms of Hortonian bifurcation and area ratios (R_B and R_A). However, the network width function does not incorporate the dynamic quality of drainage networks, although Beven and Wood (1993) have shown how it might vary with different levels of saturation in a catchment. Also, the altitudinal dimension is missing. Gupta and Mesa (1988) argue that the link concentration function (Fig. 2.5) coupled with the network width function could provide an important basis for describing the three-dimensional structure of channel networks and for analysing their association with river basin hydrology.

3

CATCHMENT PROCESSES

The river catchment provides a clearly defined physical unit for hydrologic and related studies. It contains the recognizable elements of an open system, with inputs, throughputs and outputs of energy and matter. A spatially diffuse input in the form of precipitation and weathered material is combined via a system of hillslopes and a network of channels into a single output at the mouth of the catchment. In effect, the transfer of water and sediment from land surface to ocean is characterized by a tendency towards increasing concentration and organization. As regards the sediment component, Schumm (1977) envisaged an idealized scheme consisting of three zones arranged in downstream sequence: an upper zone of sediment production, a middle zone of sediment transfer, and a lower zone of sediment deposition. These subdivisions are artificial since sediment is eroded, transported and deposited throughout a drainage basin, but they do underline how process dominance may vary longitudinally. This simple model also has hydrologic connotations, especially in view of the importance ascribed to headwaters as source areas and in view of the emphasis on the downstream continuity of the catchment, with the implication that downstream consequences can have upstream causes. This latter point is one reason why the catchment has come to be regarded as a basic unit not only for the physical study of process but also for the management of land and water resources (Newson, 1992).

CATCHMENT HYDROLOGY

Drainage basins, via their constituent hillslopes and channel networks, transform precipitation into outflow hydrographs. A large part of the incoming precipitation is lost through evapotranspiration, and additional losses may occur as a result of groundwater seepage. The remaining precipitation is added to water already stored in the basin and is released, usually with delay, as river discharge at the basin outlet. Figure 3.1A illustrates a typical storm hydrograph with its arbitrary separation into storm runoff and

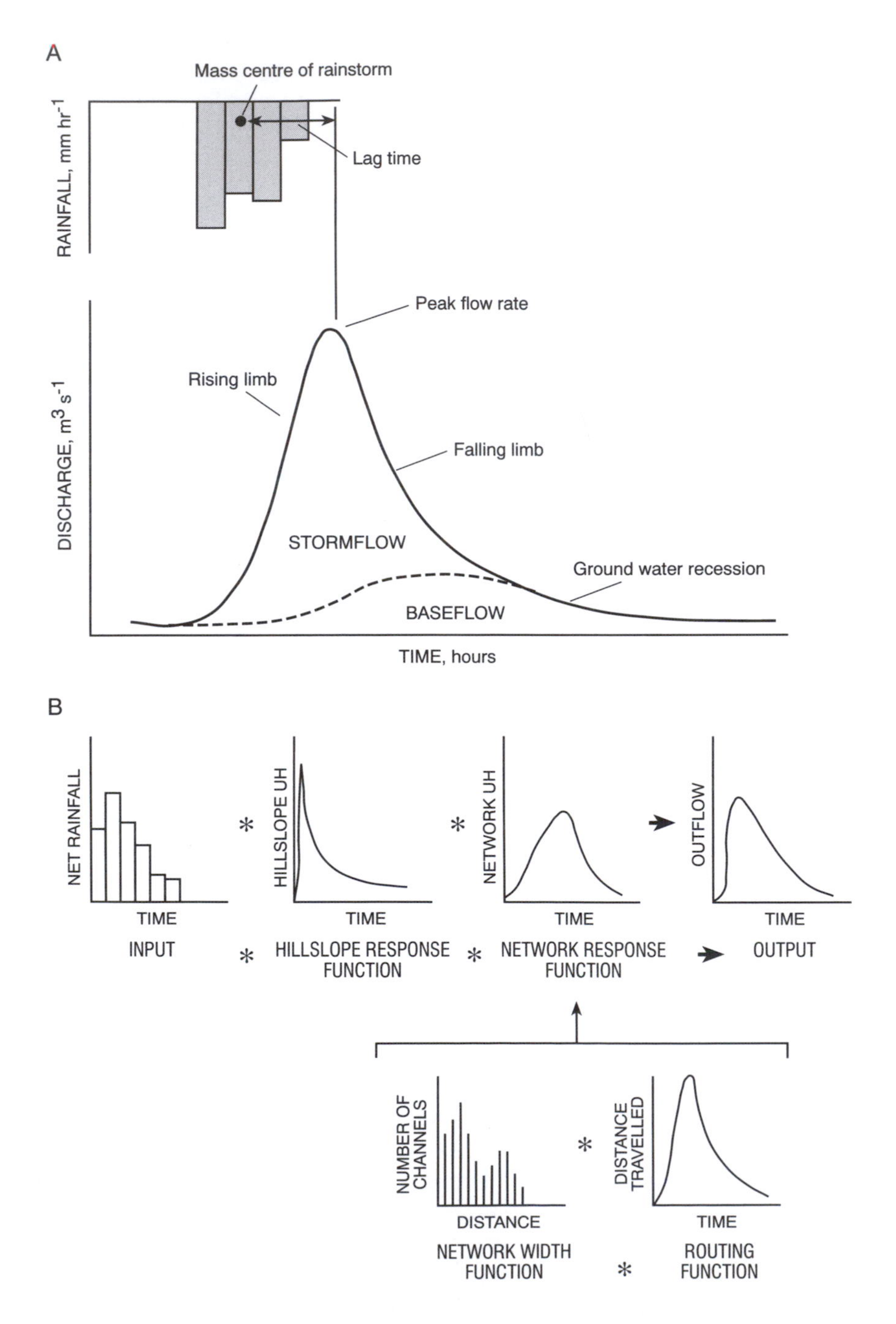

Figure 3.1 The outflow hydrograph. (A) Storm hydrograph characteristics. (B) The outflow hydrograph as a combination of a hillslope response function and a network response function, with the latter related to the network width function (after Naden, 1992).

baseflow components. Lag time and peak flow rate are two of the commonest parameters used to represent hydrographs.

The outflow hydrograph, $Q(t)$, can be expressed as a combination of a hillslope response function ($q_h(t)$) and a network response function ($q_n(t)$) (Gupta and Mesa, 1988; Fig. 3.1B), representing respectively the land and channel phases of water movement:

$$Q(t) = \int_0^t q_h(t - \tau)\, q_n(\tau)\, d\tau \qquad (3.1)$$

As outlined in the previous chapter, the network element can be based on the network width function (Fig. 3.1B), although the network area function constructed in a similar way provides an alternative (Troutman and Karlinger, 1985; Robinson *et al.*, 1995). The hillslope element can be subdivided into a fast component, most often associated with overland flow, and a slow component:

$$q_h(t) = p_f \delta(t) + p_s q(t) \qquad (3.2)$$

where $\delta(t)$ is a delta function, and p_f and p_s are respectively the proportions of the total flow in fast and slow response. Substituting equation (3.2) into (3.1) gives:

$$Q(t) = p_f q_n(t) + p_s \int_0^t q(t - s)\, q_n(s)\, ds \qquad (3.3)$$

the most important aspect of which is the separation of outflow into two parts. The first depends entirely on network structure and the dynamics of flow through the network, while the second depends in addition on the rate of runoff production from hillslopes. Equation (3.1) assumes that all hillslopes contribute flow to the channel network at the same rate ($q_h(t)$) but, given the number of hillslopes in a catchment and the variability of soil, vegetation and topographic conditions on those hillslopes, the rate of water supply is unlikely to be constant except in small, relatively homogeneous basins. The variety of hillslope hydrologic responses presents a particular problem for the development of physically distributed models which seek to describe each small sub-area of the catchment in terms of its detailed physics.

The relative importance of the two terms in equation (3.3) depends on the travel times of water moving through the hillslope and network phases. Typical hillslope flow velocities range from 30 to 500 m h^{-1} for Horton overland flow, and from 0.3 to 100 m h^{-1} for saturation overland flow, while those for subsurface flow are considerably less. Since the velocity of flow increases by an order of magnitude or more once water reaches the channel, in small catchments (say < 50 km^2) the delays within the network are very short compared with those on the hillslopes, so that the channels have a minimal impact on the outflow hydrograph. Troch *et al.* (1994) found that hillslope flow velocity had a dominant effect on the timing aspects of flood discharges in a small (7.2 km^2) Pennsylvanian basin. As catchment area increases, the influence of channel-routing times becomes increasingly important, as do storm movement patterns. Therefore, in headwater catchments which are primary source areas for runoff, a key element of hydrologic response is how far water has to travel in order to reach a channel (related to slope length and drainage density) and the route(s) by which it has been transferred (Troendle, 1985).

Streamflow Generation

More than 95 per cent of the water in streamflow has passed over or through a hillslope before reaching the channel network, following one or more of the routes identified previously (Fig. 2.7A, p. 25). Each route gives a different response to rainfall or snowmelt in terms of the volume of flow produced and the timing of contributions to the channel. Consequently the character of flow variation at the slope base is determined by the relative contribution from each source, which in turn is influenced by climatic regime, regolith and topographic form. For convenience, total flow is traditionally subdivided into two parts: direct runoff or stormflow, and baseflow (Fig. 3.2). The distinction is based on the time of arrival in the stream rather than the route followed, although the two are necessarily related.

Stormflow is produced by overland flow (of either the Horton or saturation variety) and rapid subsurface flow. *Horton overland flow* requires that rainfall intensity exceeds infiltration capacity, a condition which is most often achieved in semi-arid and arid areas with sparse vegetation and thin soils. Arid-zone hydrographs characteristically rise very steeply, have a sharp peak and are relatively short-lived (Fig. 3.6, p. 79). This type of overland flow is rarely as widespread over a basin as Horton (1933, 1945) originally supposed, being preferentially produced on relatively impermeable surfaces whose location is identified in Betson's (1964) partial area model, but can be a major contributor to storm runoff where surface hydraulic conductivity is limited. The velocity of Horton overland flow varies with gradient and surface roughness. With typical values in the range of 30–500 m h^{-1}, all parts of a 100 m slope should be contributing to the basal stream within 0.2–3 hours.

Saturation overland flow is produced when the soil becomes saturated by the addition of storm rainfall, not necessarily of high intensity. It is a mixture of precipitation that falls directly onto saturated areas, and infiltrated (or 'old') water which flows downslope and returns to the surface at a point of saturation (Fig. 2.7B). Its source areas are usually different from those for Horton overland flow, the preferred locations being at the base of slopes (especially those that are concave in profile), in areas of thin soil where soil moisture storage is limited, and in hillslope hollows towards which both surface and subsurface flows converge. Topography has a much more important role to play in governing the occurrence of saturation overland flow than it does in the case of Horton overland flow, a fact recognized in the development of topographically based hydrological models such as TOPMODEL (Beven and Kirkby, 1979). Depending on soil wetness, the size and extent of saturated areas vary both seasonally and during individual

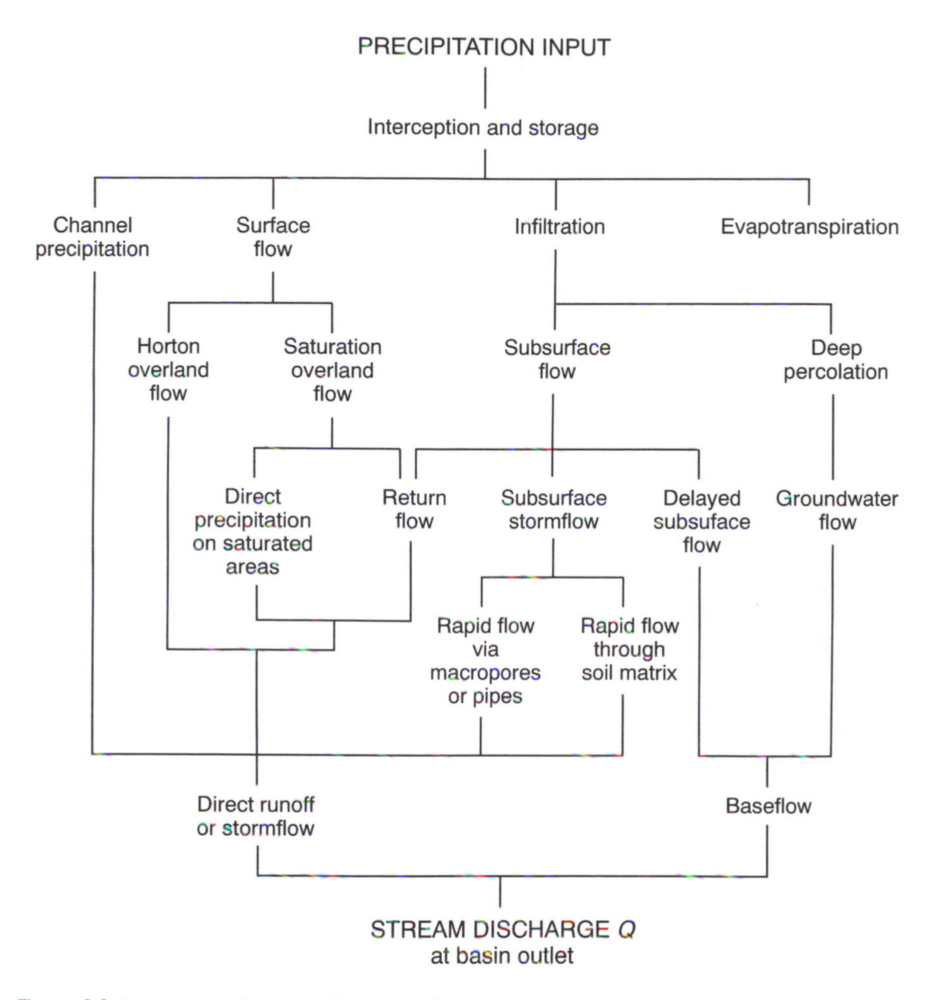

Figure 3.2 Schematic diagram of the runoff process.

storms. That variability provides the basis for the 'variable source area' concept of surface runoff (Hewlett, 1961), which is now more widely applied than any other, particularly in humid areas. Compared with Horton overland flow, saturation overland flow has smaller peak runoff rates and takes longer to generate those peaks (Fig. 3.3), largely because of its lower velocity and more limited spatial extent.

On hillslopes with permeable soils which become less permeable with depth or which overlie impermeable bedrock, *subsurface stormflow* can make a substantial contribution to the storm runoff component. It is generated by two main mechanisms: by non-Darcian flow through large voids such as macropores or pipes; and by Darcian flow through the micropores of the soil matrix. In well-structured soils macropores can provide readily accessible pathways which bypass the entire soil profile (Germann, 1990), but

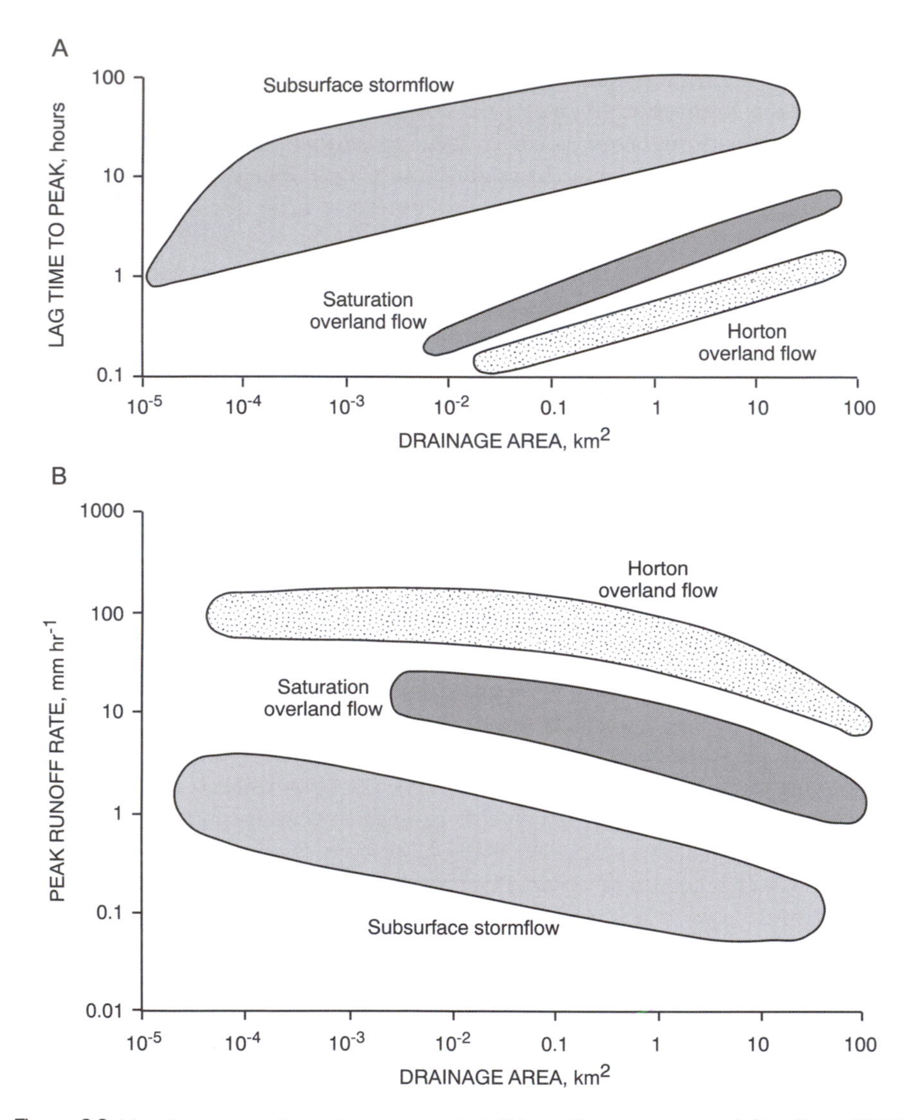

Figure 3.3 Headwater catchment response to hillslope flow processes (after Burt, 1992). (A) Lag times. (B) Peak runoff rates.

whether macropore flow contributes directly to subsurface outflow is not entirely clear. A more likely explanation is that it recharges the water table rather than providing the actual outflow, and causes a form of translatory flow to occur, in effect shunting water out of the soil matrix (Fawcett *et al.*, 1995). Soil pipes also accelerate soil drainage rates, but they tend to be larger and show a greater degree of connectivity than macropores. Consequently pipeflow can be a significant contributor to stormflow, especially when moderately heavy rain falls on a basin which is already quite wet (Anderson and Burt, 1990).

Contrary to previous assumptions, lateral subsurface flow through the soil matrix can generate subsurface stormflow, provided that the upper soil horizon has a high enough hydraulic conductivity, and there is a less permeable layer at depth capable of producing saturation at the base of that horizon and deflecting flow laterally downslope. Since its occurrence is closely connected with that of saturation overland flow, the source areas for these two types of stormflow are essentially the same, with hillslope hollows being of particular significance. At the base of slopes where the water table is close to the surface, the addition of a small amount of water can cause a disproportionate rise in the water table, leading to the lateral displacement and rapid drainage of 'old' water. Thus subsurface flow can contribute to the stream discharge peak occurring at the time of rainfall input. In addition to this immediate effect, it can produce a delayed hydrograph peak occurring several hours or even days after the storm, associated with the drainage of areas further upslope. Delayed subsurface stormflow may dominate the runoff response in basins where deep, permeable soils overlie impermeable bedrock and steep hillslopes border a narrow valley floor (Anderson and Burt, 1990). That subsurface stormflow can contribute to two peaks is reflected in the wide range of lag times depicted in Fig. 3.3A.

The variety of mechanisms responsible for producing storm runoff presents problems for the modeller seeking to predict catchment output, since they cannot all be represented satisfactorily in equation form. This applies particularly to subsurface stormflow. Climate, soil condition and bedrock lithology determine which runoff mechanism will dominate in a particular catchment, with vegetation cover and topography as important secondary controls at the hillslope scale (Burt, 1992). Horton overland flow will dominate where rainfall intensities are high relative to the hydraulic conductivity of the upper soil layer, while the occurrence of subsurface stormflow depends on the balance between the rate of infiltration and the hydraulic conductivity of the lower layer. Saturation overland flow occupies an intermediate position, being most likely in soils of medium to low hydraulic conductivity where drainage is slow, or in more permeable soils where flow convergence favours soil saturation. The particular mix of runoff mechanisms operating during a rainfall event determines the magnitude and timing of its storm runoff, the largest and fastest responses being produced by Horton overland flow (Fig. 3.3).

In permeable soils where there is no appreciable change in permeability with depth, percolating rainwater or snowmelt moves vertically downward to recharge the groundwater reservoirs and raise the level of the regional water table. **Baseflow** is mainly supplied from such sources, with additional contributions from delayed subsurface flow as water drains slowly from shallow soils (Fig. 3.2). Since water at depth moves at much lower velocities and by longer paths, its outflow into the stream lags behind the occurrence of rainfall, sustaining streamflow during rainless periods by spring flow and diffuse seepage. Darcy's Law describes how the velocity of groundwater flow (v_g) depends on hydraulic gradient ($-dh/dx$) and hydraulic conductivity (K):

$$v_g = -K\frac{dh}{dx} \tag{3.4}$$

Of particular significance is the hydraulic conductivity, which describes the ability of a porous medium to transmit water. K has high values for coarse sands and gravels (10^{-1}–10^{1} m h^{-1}) but low ones for compact clays and consolidated bedrock (10^{-6}–10^{-2} m h^{-1}). Thus the contribution of groundwater to stream discharge varies with the height of the water table (which influences the hydraulic gradient) and the geological composition of the catchment, including both its structure and its lithology.

Impermeable rocks are associated with sharply peaked hydrographs because they provide little or no baseflow to the river. On the other hand, permeable aquifers such as limestones and sandstones tend to produce flatter, broader and longer-delayed responses to precipitation input, as well as maintaining higher baseflow contributions overall. Indeed, such geologies may be largely incapable of producing flood hydrographs, so that high flows are basically groundwater-generated. Usually, however, high discharges are generated by surface and near-surface processes, with baseflow providing the recession flow on the falling limb of the hydrograph (Fig. 3.1A). During the course of a single event, the baseflow division curve will begin to rise only when the rainfall has had time to percolate down to the water table, but will usually finish at a higher level than it was at the start of the storm provided the rainfall has been large enough. The height of the water table and the baseflow contribution vary not only in the short term but also seasonally in tune with the slowly changing water balance of the catchment. Groundwater supplies the regular basal flow that maintains most humid-zone rivers throughout the year, so that any natural or artificial disruption of that supply can have serious consequences for the health of the river. A series of drier years coupled with increased abstraction has caused pronounced downstream migration of the heads of many streams flowing on the chalk in southern England.

Streamflow Throughput

The hydrograph depicted in Fig. 3.1A represents the outflow from a catchment. It combines the effects of multiple hillslope hydrographs and the transmission of flow through the drainage network. The hillslope hydrograph, shown as a sharply peaked entity in Fig. 3.1B, is conceptual, but in reality its form will be largely unknown and highly variable depending on the relative contributions from the various runoff sources considered above. Each link in the network can be regarded as a discrete flow subsystem, with inputs from direct precipitation (usually $< 5\%$), a pair of hillslopes and the upstream channel. As catchment area increases, channel travel times become longer relative to those on hillslopes, and hillslope supply becomes increasingly dislocated as progressively wider floodplains develop. In larger catchments, therefore, supply from the upstream channel assumes greater significance, and the form of the drainage network has an increasingly important influence on hydrograph shape until, at some unspecified drainage area, that influence becomes dominant. Robinson *et al.* (1995) attempted to define the transition between small (hillslope-dominated) and large (network-dominated) catchments using different perspectives, concluding that hillslope contributions become relatively small at catchment sizes greater than 50–100 km^2. Their results also suggested that the combined effect of hillslope and network contributions reaches a maximum at a drainage area of 2 km^2 where hillslope response is dominated by overland flow, and 20 km^2 where it is dominated by subsurface flow. Wherever the transition may lie, in large catchments much of the variety on hillslopes can be ignored or averaged out (Kirkby, 1988), as assumed in the Geomorphological Unit Hydrograph approach, which treats the hillslope as a single linear store with a constant rate of supply (Rodriguez-Iturbe and Valdes, 1979).

Runoff supplied to a channel moves downstream as a wave of increasing and then decreasing discharge. The transmission of a flood wave is subject to two main effects:

(a) *translation*, in which the wave moves downstream without any significant change in shape, although the time of peak flow rate will obviously occur later at downstream points; and
(b) *reservoir action*, whereby the time base of the wave is lengthened and the magnitude of the peak flow rate per unit area is decreased through temporary storage in the channel and valley bottom.

The simplest approach to flood routing assumes translation at a constant wave velocity, which gives reasonable approximations for small to medium catchments. However, reservoir action becomes more important with distance downstream as channel cross-sectional area increases and a wider floodplain causes additional attenuation of the flood wave due to the effects of floodplain storage.

Flood routing calculations are most easily carried out for a single reach of river without tributary contributions. With the river channel regarded as a linear store, the continuity equation expresses the rate of change in storage (dS/dt) as the difference between the input to (I) and the output from (O) the reach:

$$\frac{dS}{dt} = I - O \tag{3.5}$$

The routing problem consists of finding O as a function of time, given I and suitable assumptions about the form of S. One solution which is widely used to route a flood from one point to another is the Muskingum–Cunge method (Cunge, 1969), wherein the outflow hydrograph at time t (O_2) is expressed as a linear combination of the inputs at times t (I_2) and $t-1$ (I_1) and the outflow at time $t-1$ (O_1):

$$O_2 = c_1 I_1 + c_2 I_2 + c_3 O_1 \tag{3.6}$$

where $\Sigma c_i = 1$. The accuracy of the method depends on the choice of time interval (represented by the subscripts 1 and 2) and the length of reach. In most rivers the routing reaches have to be rather short, usually terminating at tributary junctions. O'Donnell (1985) has generalized the Muskingum model to deal with the case of significant lateral inflows, assuming that the total rate of lateral inflow is directly proportional to the upstream inflow rate. The method, which avoids the need for multiple routings over many sub-reaches, has been used successfully to estimate the outflow hydrographs of a 40 km stretch of an arid-zone river (Knighton and Nanson, 1994a).

The application of many flood routing methods depends very much on the availability of reliable discharge records and on the character of the river, in both its channel and floodplain forms. Cross-sectional area, slope and boundary roughness vary in the downstream direction, with major differences between within-bank and overbank flows, all of which affect floodwave velocity. Most methods route flow over single-channel reaches without direct reference to network structure or the effects of tributary inflows. A linear method which does recognize network structure is the Geomorphological Unit Hydrograph (GUH) approach (Rodriguez-Iturbe and Valdes, 1979), which expresses the hydrograph parameters, peak discharge (q_p) and time to peak (t_p) in terms of Hortonian bifurcation (R_B), stream length (R_L) and drainage area (R_A) ratios:

$$q_p = \frac{1.31}{L_\Omega} R_L^{0.43} V$$

$$t_p = \frac{0.44 L_\Omega}{V} \left(\frac{R_B}{R_A}\right)^{0.55} R_L^{-0.38} \qquad (3.7)$$

where L_Ω is the length of the highest-order stream and V is the peak velocity of the flood wave. However, in generalizing the characteristics of the network in terms of a few geomorphological parameters, which may be of dubious significance anyway (Kirchner, 1993), information about the shape of the network may be lost. The network width function retains more information about network structure. Its use in flood estimation requires that a suitable routing function be defined (Fig. 3.1B) but, unlike in the GIUH approach, it is not necessary to assume a constant wave velocity. Given that the effects of channel routing are to delay and reduce the hydrograph peak, it has been shown that the shape of the outflow hydrograph actually comes to resemble the network width function when the catchment is very large (Kirkby, 1976; Naden, 1992) or when the floodwave velocity is relatively low (Beven and Wood, 1993). However, the drainage network has yet to fulfil its potential as a basis for runoff prediction (Wharton, 1994).

Tributary inflows whose magnitude and timing may differ from those of the mainstream flow add to the complexity of the routing problem. Estimating the downstream pattern of flow addition (Fig. 2.18B, p. 62) may help to alleviate part of that problem. A further complication is the loss of water through the channel bed and banks, which occurs to some extent in all channels where the boundary is permeable, but is particularly prevalent in arid-zone rivers. There, transmission losses produced by seepage and evaporation can amount to more than 75 per cent of the total flow (Knighton and Nanson, 1994a), leading to a significant downstream decrease in flood discharge in the absence of appreciable tributary inflows. This is in complete contrast to perennial rivers and can have important implications for downstream changes in channel form and sediment transport.

Streamflow Output

Most measurements of stream discharge are made at gauging stations, where discharge (Q) is defined by the continuity equation as the product of cross-sectional area (A) and mean velocity (v):

$$Q = A.v = w.d.v \tag{3.8}$$

where w is width and d is mean depth. Provided a long enough record exists, daily mean discharges can be grouped into selected discharge classes to yield a flow–duration curve which shows the frequency with which discharges of different magnitude are equalled or exceeded. Thus a discharge of 7.6 $m^3 s^{-1}$ is equalled or exceeded about 10 per cent of the time at the Bollin gauging station (Fig. 3.4A). Mean annual discharge generally has a frequency of about 25 per cent, occupying approximately 40 per cent of the total capacity (bankfull cross-sectional area) of the channel (Dunne and Leopold, 1978). The shape of the flow–duration curve provides a good indication of a catchment's average response to precipitation. A steeply sloping curve reflects very variable flow and is usually associated with small catchments having much quickflow and little baseflow. In contrast, a very flat curve indicates little variation in flow, largely as a result of the damping effects of high infiltration and groundwater storage. The flow–duration curve for the River Bollin (Fig. 3.4A) is intermediate between these extremes, but is steeper than might have been expected given the dominantly lowland character of the catchment. If a sediment rating curve relating transport rate to discharge is also available, the flow–duration curve can be converted into a cumulative sediment transport curve to show the contributions made by various discharges to the total load transported.

Flood discharges have obvious significance for river channel adjustment and floodplain management. They are caused by various combinations of extreme conditions, such as heavy and prolonged rainfall or warm rain falling on a deep snow cover. Various procedures exist for computing flood frequencies. One of the simplest is based on the annual maximum series, which takes the single largest discharge in each year of record, from which the recurrence interval or return period (T in years) can be calculated:

$$T = \frac{(n+1)}{N} \tag{3.9}$$

where n is the number of years of record and N is the rank of a particular event. The resultant flood-frequency curve shows the average time interval within which a flood of given size will occur as an annual maximum (Fig.

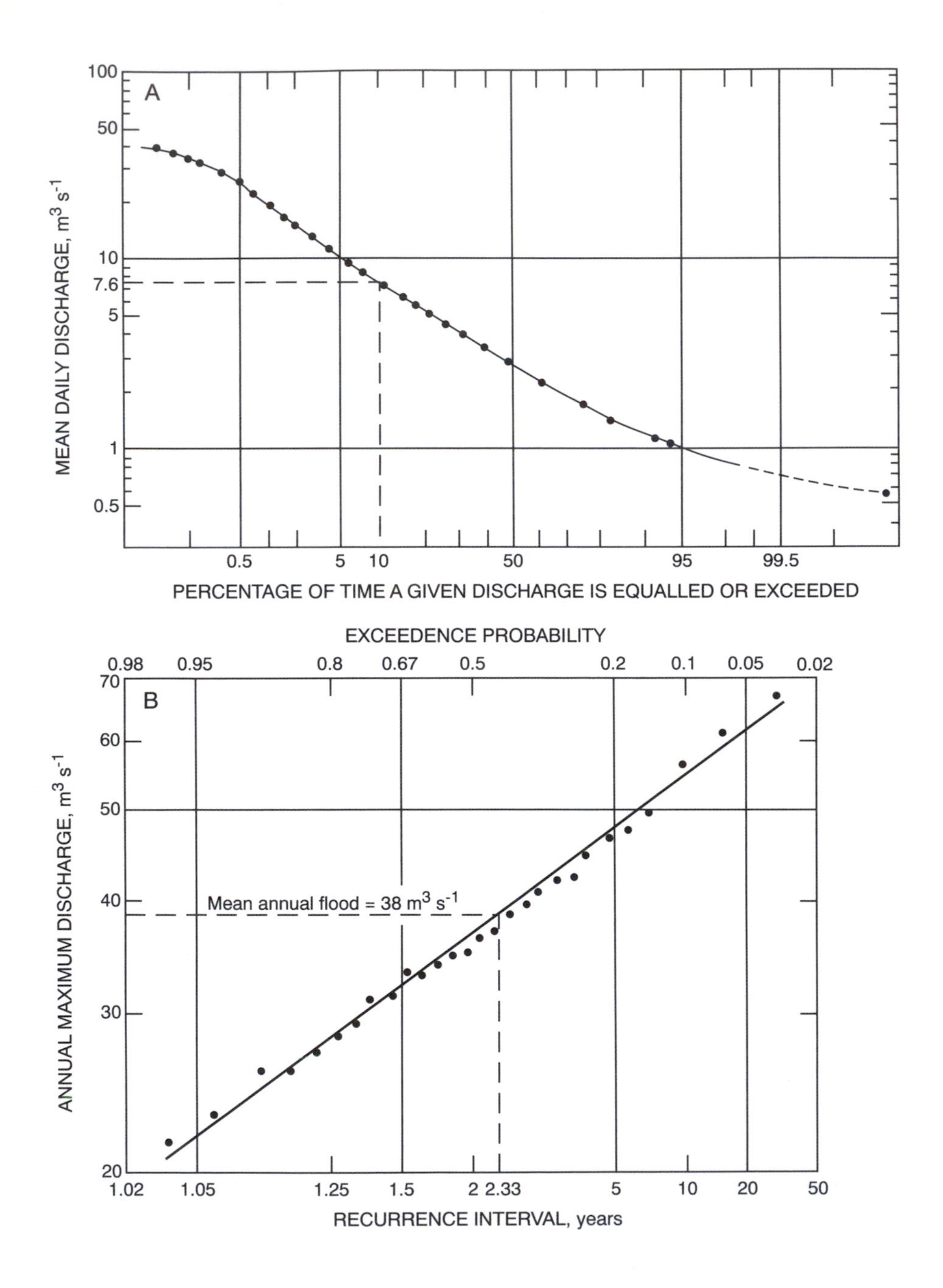

Figure 3.4 (A) Flow–duration and (B) flood-frequency curves for the River Bollin, Cheshire, England. The top scale of the flood-frequency curve gives the probability that the corresponding discharge is equalled or exceeded in any one year, while the bottom scale gives the average number of years in which the annual peak equals or exceeds the corresponding discharge (recurrence interval).

3.4B). Thus in any one year the probability of the annual maximum exceeding a flood with a recurrence interval of 10 years is 1/10. The mean annual flood with a recurrence interval of 2.33 years ($Q_{2.33}$) has often been used in relating channel form variables to discharge, a flow with a value of 38 m^3 s^{-1} at the Bollin gauge (Fig. 3.4B). Another flow with assumed morphologic significance is bankfull discharge (Q_b), defined as that discharge at which the channel is completely full, with a reported recurrence interval of 1.5 years in the United States (Leopold *et al.*, 1964). Reversing the dependency, Wharton (1995) showed how river channel dimensions can be used to estimate discharge characteristics at ungauged sites.

Not surprisingly, discharge is highly correlated with drainage area (Fig. 1.2A, p. 7). In many basins, discharge of a given frequency (f) increases less rapidly than drainage area (A_d) to give an exponent n in

$$Q_f = cA_d^n \tag{3.10}$$

which is less than 1 (Knighton, 1987b). The value of n is not independent of the frequency of flow considered or drainage area, tending to increase with more frequent flows and decrease with increasing basin size because of storage effects. Since drainage area is easier to measure, it has often been used as a surrogate for discharge in empirical studies of channel morphology.

In the *Flood Studies Report* (Natural Environment Research Council, 1975), where multiple regression equations were successively obtained for estimating the mean annual flood in British rivers, drainage area consistently proved to be the dominant independent variable at both national and regional levels. The size of the mean annual flood can be calculated from a multiplicative equation involving six catchment variables which can either be measured directly from topographic maps or be obtained from standard tables: drainage area (AREA), stream frequency (STMFRQ), effective rainfall (RSMD), soil type (SOIL), slope (S1085), and lake storage (LAKE):

$$Q_{2.33} = c\ \mathrm{AREA}^{0.94}\ \mathrm{STMFRQ}^{0.27}\ \mathrm{RSMD}^{1.03}\ \mathrm{SOIL}^{1.23}\ \mathrm{S1085}^{0.16}\ (\mathrm{LAKE} + 1)^{-0.85} \tag{3.11}$$

where c is a regional multiplier. Regional frequency curves for each of the 11 regions of the British Isles enable floods of other return periods to be estimated. The equations and techniques demonstrated in the report have been widely used for design purposes, both in Britain and elsewhere. However, given the empirical nature of the approach, the applicable geographical region needs to have a reasonable degree of physiographic homogeneity if the resultant equations are to give reliable estimates.

Equation (3.11) and its kin represent merely one example of a vast array of rainfall–runoff models which have been developed to predict streamflow output in one form or another, ranging from particular flow series parameters (such as $Q_{2.33}$) to the complete hydrograph for a given rainfall input. Paralleling the more intensive study in recent years of hillslope runoff processes, particularly subsurface ones, has been the development of physically based distributed models. They are termed 'distributed' because they explicitly represent the spatial distributions of catchment properties rather than treating the catchment as a lumped unit, and 'physically based' because they make use of equations describing the underlying physics of water movement (Fawcett *et al.*, 1995). TOPMODEL (Beven and Kirkby, 1979) subdivides the catchment into relatively homogeneous sub-areas and routes the separate outflows downstream to produce the final output. Using a topographic index, it can predict the fluctuating outflow caused by the expansion and contraction of variable source areas. The SHE (Système Hydrologique Européen) model incorporates the main elements of the land phase of water movement in both its surface and its subsurface forms (Abbott *et al.*, 1986). The spatial distribution of rainfall input, catchment variables and hydrologic output are represented in the horizontal direction by an orthogonal grid network of up to 2000 squares, and in the vertical direction by a column of horizontal layers at each grid square (Fig. 3.5). The SHE model has considerable potential for runoff prediction but, in common with other such models, a satisfactory representation of the physical processes involved is very difficult to achieve and the data requirements to drive the model are very demanding, particularly in larger catchments. Indeed, the practical difficul-

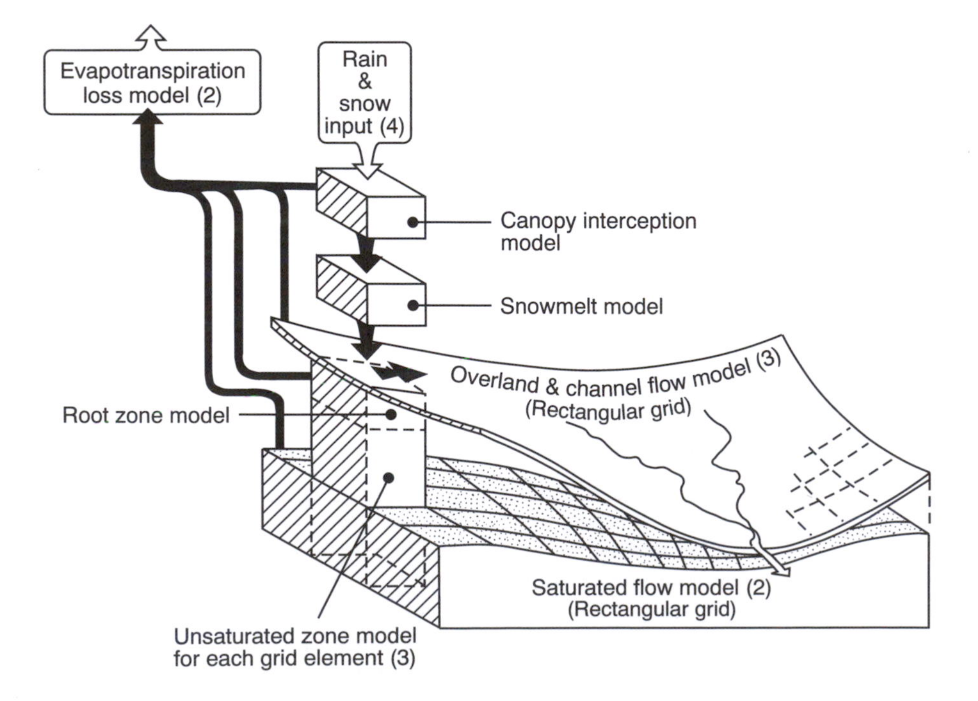

Figure 3.5 Schematic representation of the Système Hydrologique Européen (SHE) model, the bracketed numbers indicating the number of parameters involved in each component part (after Abbott *et al.*, 1986).

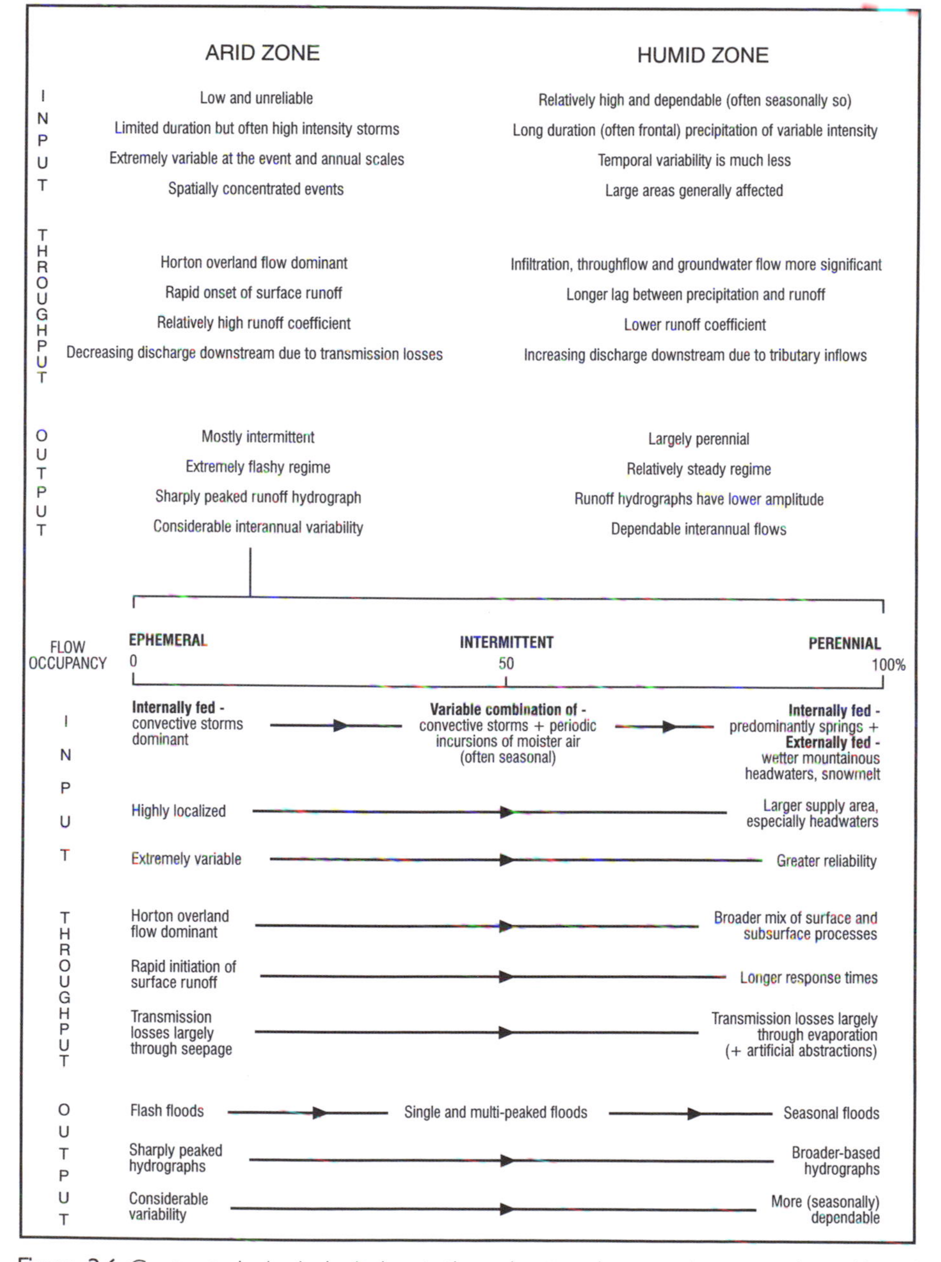

Figure 3.6 Contrasts in hydrologic input, throughput and output between the arid and humid zones (upper), and within the arid zone itself (lower).

ties of assembling an adequate database are so great that simpler approaches may provide a sufficient level of accuracy, especially in view of the rather limited range of forms that flood hydrographs appear to have.

Nevertheless, natural river flow is highly variable in both time and space. Thus, for example, a major distinction can be drawn between arid-zone and humid-zone hydrologic conditions, but even within the arid zone itself there is considerable diversity (Fig. 3.6). Partly because of the increasing availability of flow records, discharge has become a primary independent variable in geomorphological approaches to the description and analysis of river channel form (Figs 1.2B and 5.10, pp. 7 and 181). The importance of discharge can perhaps be best appreciated from the dominant-discharge concept and the hydraulic geometry approach pioneered by Leopold and Maddock (1953). However, discharge is a summary variable which does not express directly the forces involved in shaping channels. To that extent the role of discharge has possibly been overemphasized. A closer link is needed between channel form adjustment and the mechanical work performed by streams.

Catchment Denudation

The transport of material from land surface to ocean can be rationalized in terms of three process regimes (Statham, 1977):

(a) a *weathering regime* which involves the physical and chemical breakdown of solid bedrock and partially weathered material, prior to removal by other processes;
(b) a *slope regime* in which the products of weathering are moved down the gravity gradient in mass movements and by slope wash processes. These two sets of processes differ in the way that they entrain and move material – in mass movements the main driving force is the material's own weight and quite considerable depths can be moved at any one time, whereas slope wash processes operate largely through an agent which exerts the force and most movement is surficial;
(c) a set of *fluid-transfer regimes*, involving water, air and ice, of which the first is by far the most important; the driving force is primarily supplied in the form of a shear stress acting at the base of the flow.

These three process regimes have reasonably well-defined boundaries across which material is transmitted. In particular, all material entering a river system must cross the boundary between the slope and fluvial regimes provided by the channel banks and the channel head.

During the long-continued evolution of drainage basins, mechanical and chemical processes of denudation are involved in the removal of landscape mass. Estimates of the rate of denudation can be obtained from suspended sediment and dissolved load data available for many of the world's rivers (Tables 3.1 and 3.2). Care needs to be taken in interpreting these data, partly because they contain the impacts of human activities such as dam-building and forest clearance. Thus, Milliman *et al.* (1987) have estimated that the transport rate of the Huang Ho River in China has increased ten-fold, from 10^8 to 10^9 tonnes per year, since agricultural use of the interior loess plateau

Table 3.1 Material transport from the continents to the oceans

Continent	Land area, 10^6 km^2	Mean annual runoff, 10^3 km^3	Total annual suspended sediment load, 10^6 t y^{-1}	Total annual dissolved load, 10^6 t y^{-1}	Suspended sediment/ dissolved load ratio	Overall denudation rate*, mm (10^3 y)$^{-1}$
Africa	15.3	3.4	530	201	2.6	17
Asia	28.1	12.2	6433	1592	4.0	102
Europe	4.6	2.8	230	425	0.5	42
North and Central America	17.5	7.8	1462	758	1.9	43
Oceania/ Pacific Islands	5.2	2.4	3062	293	10.5	241
South America	17.9	11.0	1788	603	3.0	47

Sources: Degens *et al.* (1991), Meybeck (1979), Milliman and Meade (1983), and Walling (1987).
*Assumes a 65% denudation component for dissolved load and a rock density of 2.6 tonnes m^{-3}.

began over 2000 years ago. The contemporary sediment load of the world's rivers has been calculated at 15 × 10^9 tonnes y^{-1} (about 13.5 × 10^9 tonnes being carried in suspension and the remainder as bottom load) (Milliman and Meade, 1983), although the contributions from small rivers (A_d < 10 000 km^2) could be greatly underestimated (Milliman and Syvitski, 1992). With the total dissolved load amounting to about 4 × 10^9 tonnes y^{-1} and assuming that only 65 per cent of the dissolved load represents the products of chemical denudation (Meybeck, 1979), these data suggest an average rate of denudation of 65 mm per 1000 years. Mechanical processes account for approximately 55 mm/1000 years and chemical ones 9.5 mm/1000 years, giving a 6 : 1 ratio in the relative efficacy of mechanical and chemical processes of denudation (Walling, 1987). However, that only 65 mm of net removal is expected every 1000 years emphasizes the long periods of time required for significant landscape change.

Table 3.1 indicates considerable variability between continents as regards both the overall rate of denudation and the relative contributions of mechanical and chemical processes. In only one continent, Europe, does the dissolved load exceed the suspended sediment load, partly because of geological conditions and partly because of the intensive agricultural land use which promotes chemical weathering through soil disturbance (Walling, 1987). Elsewhere mechanical denudation is dominant, reaching a maximum relative to chemical denudation in Oceania and the Pacific Islands. However, those data themselves hide considerable variability. Australia, the largest single land mass in that area, contributes only a total load of about 80 × 10^6 tonnes y^{-1}. Although the data in Table 3.1 underline the dominant position of

Table 3.2 Suspended sediment and dissolved loads of major world rivers

River	Drainage area, 10^3 km^2	Mean discharge, 10^3 m^3 s^{-1}	Mean suspended sediment load, 10^6 t y^{-1}	Mean dissolved load, 10^6 t y^{-1}	Percentage of total load carried in solution
Africa					
Congo	3500	41.1	48	37	44
Niger	1200	4.9	25	14	36
Nile	3000	1.2	2	12	86
Orange	1000	0.4	0.7	1.6	70
Zambezi	540	2.4	20	25	56
Asia					
Ganges/Brahmaputra	1480	30.8	1670	151	8
Huang Ho	752	1.1	900	22	2
Indus	1170	7.5	100	79	44
Irrawaddy	430	13.6	265	91	26
Lena	2440	16.0	12	56	82
Mekong	795	21.1	160	59	27
Ob	2990	13.7	13	46	78
Yenisei	2500	17.6	15	60	80
Australia					
Murray–Darling	1060	0.7	30	8.2	21
Europe					
Danube	817	6.5	83	53	39
Dnieper	527	1.7	2.1	11	84
Rhone	99	1.9	40	56	58
Volga	1459	7.7	27	54	67
North America					
Columbia	670	5.8	14	21	60
Mackenzie	1810	7.9	100	44	31
Mississippi	3267	18.4	210	142	40
St Lawrence	1150	13.1	5.1	70	93
Yukon	840	6.7	60	34	36
South America					
Amazon	6150	200	900	290	24
Magdalena	260	6.8	220	20	8
Orinoco	990	36	150	31	17
Paraná	2800	15	80	38	32

Sources: Degens *et al.* (1991), Meybeck (1976), and Milliman and Meade (1983).

Asian rivers as contributors of dissolved and particulate matter, they provide only a very general picture of the solutes and sediments removed from the land to the oceans. There are large inter-catchment variations.

Solute Variability

Solutes in rivers are derived from a variety of sources – rock and soil weathering, the atmosphere, the biosphere, and the effects of human activity – the last of which is becoming increasingly important as the discharge of industrial effluents and the use of agricultural fertilizers increase. The composition and concentrations of solutes in rivers are determined by the mixing of water that follows different hydrologic and chemical pathways, but the dominant supply route is via some form of subsurface flow, the low velocity of which facilitates longer contact times between water and soluble materials.

Atmospheric contributions come in two main forms: dry fallout and wet precipitation. The former involves the deposition of relatively large particles (usually >20 μm in diameter) under the influence of gravity, whereas the latter consists of the chemical products of solution acting during droplet formation and condensation in clouds (washout) or during raindrop fall (rainout). Precipitation is naturally acidic, with pH values commonly in the range of 5–7, but can become even more so in the presence of atmospheric pollutants. Seawater spray is an important source of atmospheric aerosols, contributing high concentrations of Na^+, Cl^-, Mg^{2+} and K^+. With increasing distance from the coast, terrestrial aerosols assume greater significance, and since they tend to be rich in Ca^{2+}, NH_4^+, NO_3^-, NO_2^-, HCO_3^- and SO_4^{2-}, the chemical composition of precipitation input can show marked spatial variations. However, the concentrations of dissolved substances in precipitation (<10 mg L^{-1}) are generally much lower than they are in river water (>100 mg L^{-1}), so the atmospheric solute contribution tends to be relatively small except in specific instances such as Na^+ and Cl^-.

The *biosphere* occupies an intermediate position between the atmosphere and lithosphere. Biological processes contribute to the chemical characteristics of a catchment in a large variety of ways, ranging from the alteration of incoming precipitation by intercepting vegetation, to various microbial transformations which can strongly influence nutrient mobilization. The process of nutrient cycling provides inputs to the soil surface additional to those of the atmosphere, the first stage of which is litterfall where nitrogen, calcium and potassium are dominant elements. Solute uptake at or near the soil surface may be relatively rapid because of the ready availability of material derived from the decomposition of organic matter and the mineralization of humus. In forested catchments underlain by lithologies resistant to chemical weathering, the storage of solutes within the biomass, particularly nitrogen, may be more important in regulating streamwater chemistry than inputs from the atmosphere or from weathering.

However, chemical denudation is the result of *rock and soil weathering*, and much of the solute load in rivers comes from that source, estimated on a global scale at 60 per cent by Walling and Webb (1986). There are many types of chemical reaction involved in the weathering process, including solution, oxidation, hydration and hydrolysis. The release of solutes during weathering is influenced by the thermodynamic state of the reactions and by the residence time of water in the soil or rock body, where the latter affects whether weathering reactions can reach an equilibrium state. Elements vary considerably in their susceptibility to weathering and in their proportion in crustal rocks, which has consequences for the chemical composition of surface waters. On the basis of characteristic water analyses and calculations of the outcrop proportions of major rock types, Meybeck (1987) has estimated the relative contributions of those rock types to the dissolved load of rivers (Table 3.3). The data indicate the minor influence of crystalline (plutonic, metamorphic, volcanic) rocks on global weathering (11.5 per cent of solutes from 33.9 per cent of outcrop), whereas evaporites (about 1.3 per cent of outcrop) may contribute more than 18 per cent to the dissolved load

Table 3.3 Outcrop proportions and relative contributions to dissolved load of major rock types (after Meybeck, 1987)

Rock type	Outcrop (%)	Weathering products by rock type (% of total amount released)								Contribution to dissolved load (%)
		SiO_2	Ca^{2+}	Mg^{2+}	Na^+	K^+	Cl^-	SO_4^{2-}	HCO_3^-	
Plutonic rocks:										
granite	10.4	**10.9**	0.6	1.2	5.9	4.9		2.0	1.6	2.6
other	0.6	0.75	0.0	1.1	0.0	0.0		0.3	0.3	0.2
Metamorphic rocks:										
gneiss and schists	12.7	**11.7**	1.1	2.8	6.6	8.2		4.6	2.0	3.3
other	2.3	1.5	4.7	3.2	0.3	1.6		1.7	4.7	2.6
Volcanic rocks (basalt)	7.9 (4.2)	**11.1**	1.8	4.9	5.4	6.6		0.5	4.2	2.8
Sandstones	15.8	**16.6**	2.1	3.8	5.2	**19.6**		9.6	2.3	5.0
Shales	33.1	**35.1**	**19.9**	**30.7**	**22.6**	**41.0**	7.4	**30.3**	**22.5**	**20.0**
Carbonate sedimentary rocks (limestones, dolomites) (10)	15.9	**11.3**	**60.4**	**39.3**	3.5	**33.1**		8.6	**59.5**	**45.1**
Evaporites:										
gypsum	0.75	0.5	7.2	7.2	4.9	1.6	10	**32.7**	1.7	8.1
rock salt	0.55	0.4	2.2	5.8	**45.6**	3.3	**82.6**	9.6	1.2	**10.2**

Percentages ≥ 10% are highlighted in bold.

originating from chemical denudation. However, carbonate minerals found in sedimentary rocks are responsible for more than 50 per cent of the total load derived from denudation (67 per cent of calcium, 42 per cent of magnesium). Relative to granite, the rates of chemical weathering have been estimated as: granite 1, gneiss and schist 1, gabbro 1.3, sandstones 1.3, volcanic rocks 1.5, shales 2.5, selected metamorphic rocks 5, carbonate rocks 12, gypsum 40, rock salt 80. The lithology of a catchment can clearly exert a major influence on its river chemistry.

The average dissolved solids content of world river water has been calculated as 120 mg L^{-1}, more than 80 per cent of which is generally made up of just four components (HCO_3^-, SO_4^{2-}, Ca^{2+} and SiO_2) (Walling and Webb, 1986). Climatic and geologic factors are generally regarded as the most important environmental controls on the **spatial variation** of solute transport at the global scale. Climate affects the rates of weathering and associated solute release via its influence on water availability and, to a lesser extent, temperature. On the basis of data for 496 rivers, Walling and Webb (1986) obtained a positive relationship between mean annual dissolved load and mean annual runoff (Fig. 3.7A), which demonstrates that greater water availability increases the total quantity of dissolved material released or available for transport (although concentrations decrease with increasing runoff because of a dilution effect). However, there is a considerable amount of scatter about the regression line, reflecting the influence of other climatic and physiographic factors. Mean annual temperature has an important effect on the SiO_2 content of rivers. Differences in rock type often moderate the effects of climate (Fig. 3.7B), and the solute loads of basins underlain by sedimentary rocks are commonly about five times greater than those of basins developed on crystalline rocks. The relatively high dissolved loads of European rivers partly reflect the predominance of sedimentary rocks within their catchments, whereas the low values for Africa and Australia are associated with the existence of ancient basement rocks with a low susceptibility to chemical weathering (Tables 3.1 and 3.2) (Walling, 1987). However, Walling and Webb (1986) suggested that worldwide contrasts in solute content related to lithological differences are an order of magnitude less than those related to climatic ones, although geologic controls tend to become more important at smaller scales. Then also, local variations in topography and land use exert a greater influence on the composition and magnitude of dissolved loads. In the Upper Exe basin of Devon, total solute levels progressively decline as the percentage of moorland cover increases from less than 50 per cent to more than 90 per cent (Webb and Walling, 1983). The solute content of rivers can vary greatly in space depending on the sources, processes and pathways that characterize a particular drainage basin, a complexity which the inputs from anthropogenic activities will only serve to increase.

As the applicable scale decreases, the modelling of solute behaviour within catchments becomes a more realistic proposition. The development of a large array of **solute models** has characterized the last 15 years (Ball and

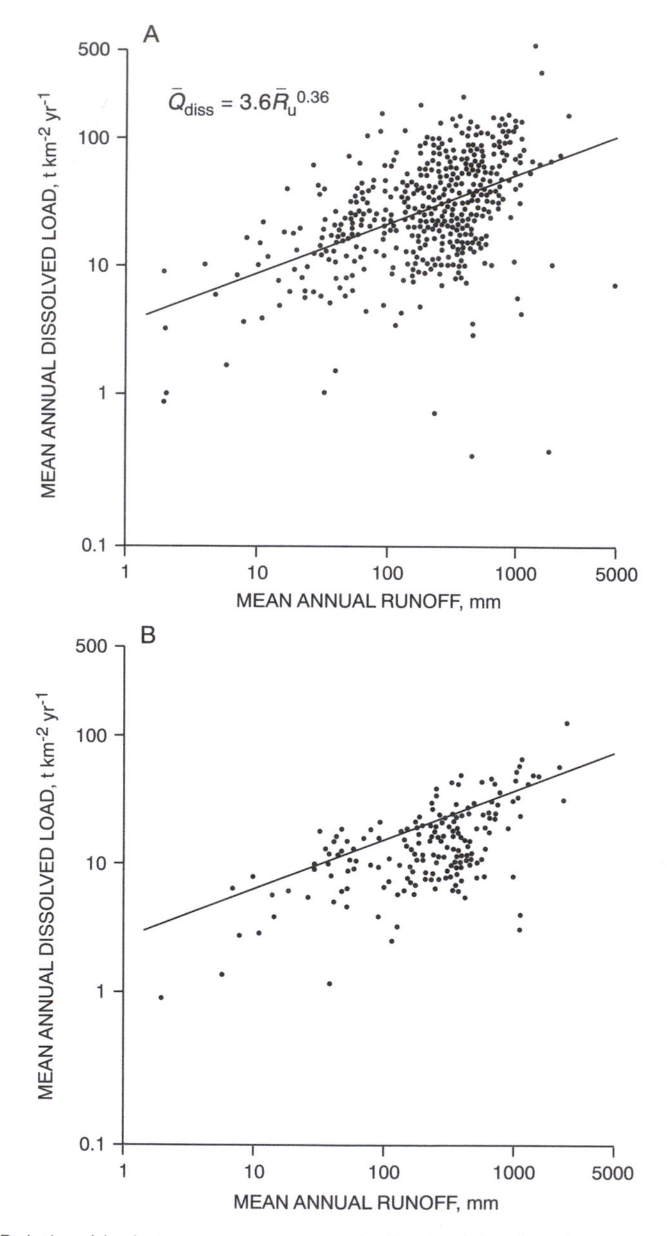

Figure 3.7 Relationship between mean annual dissolved load and mean annual runoff (after Walling and Webb, 1986). (A) A sample of 496 world rivers. (B) Basins underlain by igneous rocks, with the global regression line from (A) superimposed.

Trudgill, 1995), paralleling that of physically based hydrological models such as SHE (Fig. 3.5). The problems of acidification and pollution have provided the environmental context for model development, with the need to assess the sensitivity of solute behaviour to changes in atmospheric and surface inputs being an important stimulus. The consequences of 'acid rain', the loss of pesticides and fertilizers into groundwater, and the movement of industrial and radioactive waste have all been major causes of concern. However, the processes that control streamwater quality are complex and are far from being adequately understood. One aspect of that complexity is the interaction between dissolved and particulate matter, for substances in solution may become bonded to sediment and thereby experience a different transport history.

Sediment Yield and Sediment Budgets

The suspended sediment load is in most instances dominated by clay- and silt-sized particles (<0.062 mm in diameter). It not only provides an indication of the rate of mechanical denudation in a basin, but also has wider implications for the economic management of the fluvial system. Clark *et al.* (1985) estimated that the downstream impacts of soil erosion have cost the United States $6100 million at 1980 values, subdivided into off-stream effects of $1900 million (such as flood damage and water treatment) and in-stream ones of $4200 million (such as recreation, water storage and navigation). In addition, there is growing concern about the sediment-associated transport of contaminants, in which fine particles have a particularly important role given that the adsorptive capacity of sediment increases with decreasing grain size.

Suspended sediment comes from two main **sources**: surface and, to a much lesser extent, subsurface erosion of hillslopes and bottomlands by such processes as overland flow, rilling and gullying; and erosion of the channel boundary, particularly the channel banks, when fine material is sheared off by the flow or thrown into suspension after bank collapse. The products of natural erosion are supplemented by the inputs from a wide range of human activities which mobilize and supply sediment (agriculture, industry, mining, construction). In agricultural areas, soil loss from cultivated fields can account for a major part of the suspended sediment load. Indeed, Brown (1984) estimated that the world's croplands lose about 23 billion tonnes of soil in excess of new soil formation each year.

The relative contributions of non-channel and channel sources tend to vary with distance downstream. In general, the upper parts of catchments, with their steeper slopes and stronger coupling between hillslopes and channels, supply sediment largely from non-channel sources. Further downstream where slopes are gentler and floodplains wider, the potential for temporary storage of eroded material increases and the contribution from channel erosion becomes relatively more important. At least 65 per cent of the material carried by the lower Waimakariri River in New Zealand is supplied locally from the bed and banks of the channel (Griffiths, 1979). Various chemical and physical properties of sediment are now being used to 'fingerprint' suspended sediment sources. Obviously the property chosen must clearly differentiate between potential sources and exhibit conservative behaviour during transport. In an early application, Grimshaw and

Lewin (1980) distinguished channel from non-channel sources on the basis of sediment colour in the Ystwyth basin of central Wales, with just over half coming from the channel but the proportion of non-channel sediment increasing at very high discharges. However, it is becoming increasingly clear that, in order to improve the reliability of results and increase the number of potential sources which can be discriminated, a multi-property approach needs to be adopted. Walling *et al.* (1993) used a combination of two fallout radionuclides (caesium-137, lead-210), four mineral magnetic parameters, and the concentrations of organic carbon and nitrogen as fingerprint tracers to distinguish between suspended sediment derived from pastureland, cultivated fields and channel banks in Devon catchments. Radionuclides are particularly useful in that they are essentially independent of the underlying geology and soil type, and tend to be concentrated in the upper 10 cm of undisturbed soil profiles.

Sediment yield is defined as the total sediment outflow from a basin over a specified time period, with sediment in suspension as the dominant component. There are three main approaches used in **estimating sediment yield**, based respectively on direct measurements of suspended sediment concentration at the basin outlet, source-area measurements of erosion, and lake/reservoir surveys. The first requires the concurrent measurement of stream discharge and suspended sediment concentration, usually at a gauging station, so that a rating curve relating the two variables can be obtained (Fig. 3.8A(i)). In combination with a flow–duration curve (Fig. 3.4A), the rating curve enables a load–duration or sediment yield curve to be constructed (Fig. 3.8A(ii)), where load = concentration × discharge for a specified increment of discharge. Although arguably the most reliable method, it does have various problems. Because suspended sediment concentration is highly variable (Fig. 3.8A(i)), the sample period needs to be long enough to ensure a reliable concentration–discharge relationship. Bed load is not explicitly considered, although it is generally a minor component of the total (usually <15 per cent). In addition, the calculated value applies to the basin as a whole and is therefore not directly related to conditions within the drainage basin.

The second method gets round the last problem through on-site measurements of erosion, usually in experimental plots (Loughran, 1989). However, there need to be enough plots to provide a representative sample of intra-basin variations in erosion, which can present severe practical difficulties in basins of even moderate size. An alternative to plot measurements is to estimate erosion using a soil erosion model such as the widely applied Universal Soil-Loss Equation (Wischmeier and Smith, 1978) or the more recent WEPP (Water Erosion Prediction Project) model (Lane *et al.*, 1992). They have their own limitations and data requirements. Assuming that reasonable estimates of gross erosion (E) in the catchment can be made, the final but not inconsiderable step is to multiply those estimates by a *sediment-delivery ratio* (D_r) in order to obtain the sediment yield ($S_y = E.D_r$). It is well known that only a fraction, and perhaps even a small fraction, of the sediment eroded within a drainage basin will be transported to the outlet and be represented in the sediment yield, largely because of temporary or permanent storage. Although

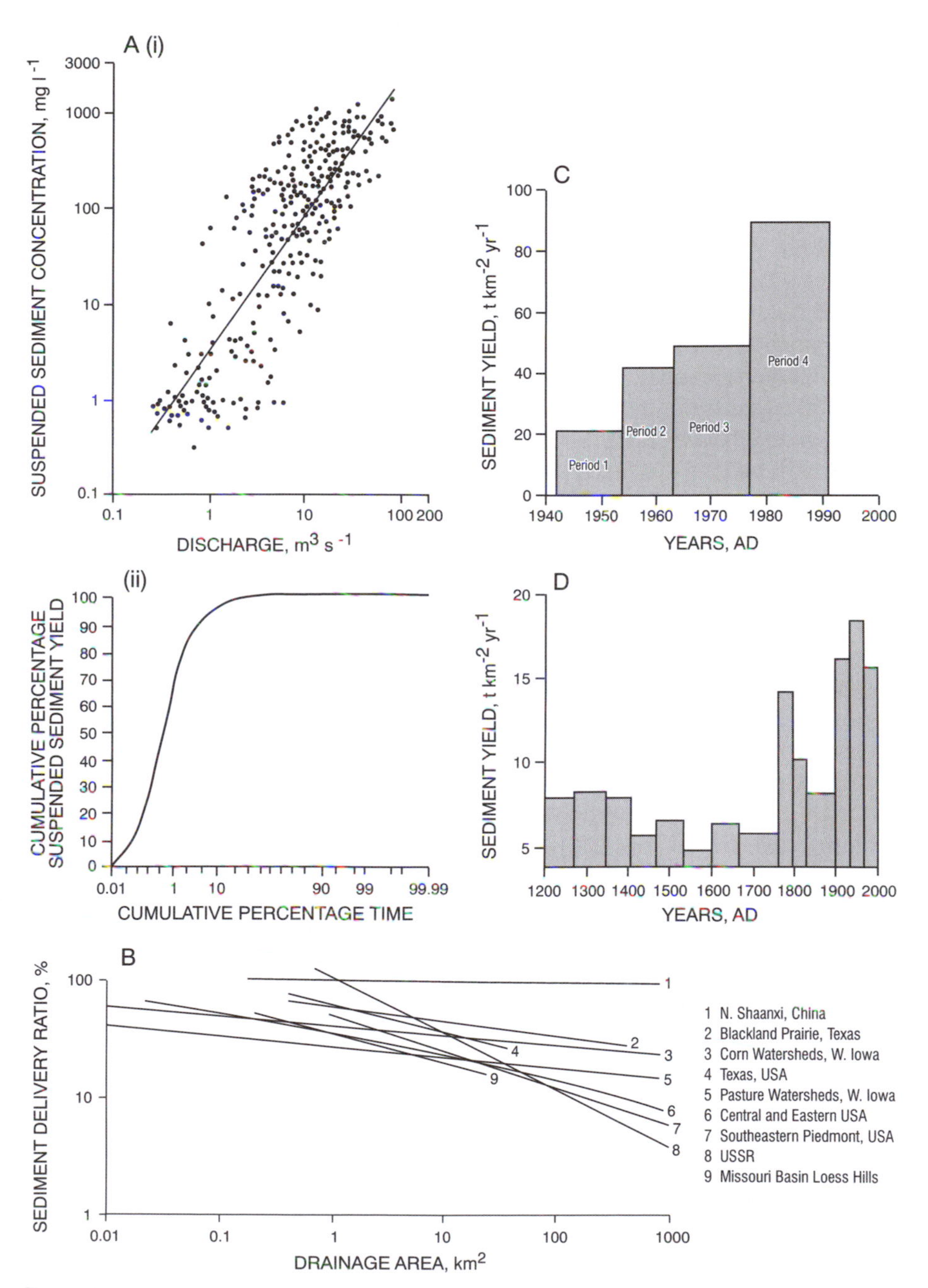

Figure 3.8 Sediment yield estimation. (A) A concentration–discharge relationship (i) and load–duration curve (ii), River Creedy, Devon, England (after Walling and Webb, 1992). (B) Delivery ratio in relation to drainage basin area for various parts of the world (after Walling, 1983). (C) Lake sediment records from a small Devon catchment indicate a four-fold increase in yield since 1942 (after Foster and Walling, 1994). (D) Lake sediment records from upland Wales show fluctuating sediment yields over 800 years (after Dearing, 1992).

likely to be less than 1, the sediment-delivery ratio is a highly variable quantity, being influenced by a wide range of geomorphological and environmental factors whose absolute and relative significance are difficult to establish (Walling, 1983; Richards, 1993). Basin area (A_d) has frequently been identified as a dominant control and the following relationship has been proposed for the United States (American Society of Civil Engineers, 1975):

$$D_r = kA_d^{-0.125} \tag{3.12}$$

the inverse form of which reflects the fact that larger basins have potentially more storage space, notably in the form of floodplains. However, considerable diversity exists in the way that even this one factor influences D_r (Fig. 3.8B). $S_y = E.D_r$ appears to offer a simple means of estimating sediment yield, but neither term on the right-hand side is determined easily. As regards D_r, the basic problem is to understand the linkages between on-site erosion and downstream yield, but the mechanisms of sediment transfer into storage and its subsequent remobilization are poorly understood.

One of the main advantages of lake surveys lies in their potential to provide longer-term records of variations in sediment yield, together with information on the changing character of sediment sources. Ideally, the lake should have a high trap efficiency and the sediment cores should contain sufficient datable material to enable the construction of an absolute chronology of sedimentation (Walling, 1988). The sediment record of a small Devon catchment indicated a four-fold increase in suspended sediment yield since the Second World War (Fig. 3.8C), attributable to increased livestock numbers and grazing intensity (Foster and Walling, 1994). In upland Wales a longer record covering the last 800 years has been assembled (Fig. 3.8D), high yields in the periods 1765–1830 and 1903–1985 corresponding to mining activity in the catchment (Dearing, 1992). Sediment yield chronologies analysed by Wilby *et al.* (1997) suggest that changes in winter storm patterns have been a significant factor in the UK since 1861, challenging the commonly held belief that anthropogenic activity has had a dominant influence over the last century or so. Given the lack of sediment-load data in the past, lake sediment studies provide a good opportunity for extending record lengths.

These three empirical approaches to sediment yield estimation are supplemented by various sediment routing models which, to a greater or lesser extent, describe in equation form the transport history of sediment from source to outlet. Physically-based modelling on a par with that in hydrology has been advocated as an ultimate objective (Morgan, 1986; Richards, 1993), involving submodels which represent the various entrainment and transport processes, and which route sediment through the drainage network. However, more so than in hydrology, the relevant delivery processes are not sufficiently well understood for adequate modelling of even moderately sized basins.

The data in Tables 3.1 and 3.2 suggest considerable **spatial variation** in sediment yield. Maximum values are associated with the highly erodible

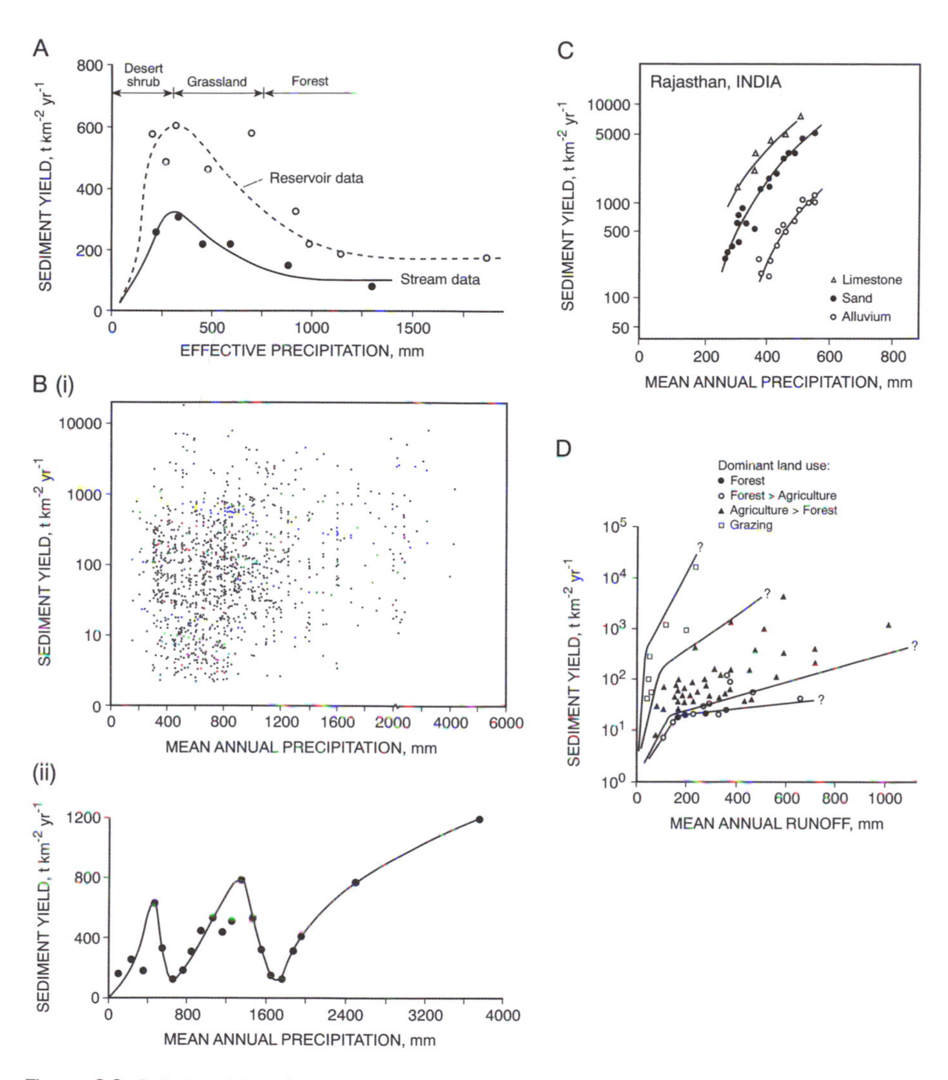

Figure 3.9 Relationship of sediment yield to: (A) Effective precipitation (after Langbein and Schumm, 1958). (B) Mean annual precipitation (after Walling and Kleo, 1979): (i) raw data, (ii) group-averaged data. (C) Mean annual precipitation for three different geologies (after Sharma and Chatterji, 1982). (D) Mean annual runoff for four land-use types (after Dunne, 1979).

loess soils of China and the Cenozoic mountain areas of the Pacific rim, while low values characterize much of northern Eurasia and North America where subdued relief and resistant basement geology are important controls (Walling and Webb, 1983). Four main groups of factors control that variation: precipitation and runoff characteristics; soil erodibility; basin topography; and the nature of the plant cover. The overall effect of climate is commonly defined by curves that relate sediment yield to mean annual precipitation or mean annual runoff (Fig. 3.9). For a mean annual temperature of 10°C, the Langbein and Schumm (1958) curve reaches a peak at an effective precipitation of about 300 mm (i.e. semi-arid conditions), trailing off at lower values because of lower runoff totals and at higher ones because an increasingly abundant vegetation cover affords better protection against erosion. However, this interpretation is limited in a number of respects, not least because the curve is based on group-averaged United States data in which tropical areas are not represented, and because sediment yield may again begin to increase for precipitation amounts over 1000 mm.

A much more comprehensive analysis by Walling and Kleo (1979) revealed not only considerable variability in annual sediment yields (Fig. 3.9Bi) but also a more complex average relationship with three peaks (Fig. 3.9Bii). Because of the scale of variability, Walling and Kleo questioned the justification for ascribing the three peaks to specific climatic conditions, but tentatively suggested that they represent respectively: continental semi-arid conditions (~ 450 mm); high-rainfall Mediterranean climates (1250–1350 mm); and tropical monsoon conditions (>2500 mm). The first corresponds approximately with the Langbein and Schumm (1958) peak (Fig. 3.9A), while the other two may indicate the effects of seasonal precipitation regimes. Indeed, sediment yield may be as much a function of the seasonal variability of precipitation as the gross amount (Wilson, 1973), since a seasonal regime partly restricts the growth of a vegetation cover. To some extent the Langbein and Schumm curve incorporates the interacting effects of precipitation/runoff magnitude and vegetation cover, but the former may be more significant than those authors supposed if the increase in potential erosivity associated with more humid areas is not offset by increased protection from a denser vegetation cover. However, the very wide scatter of points in Fig. 3.9B(i) suggests that no simple relationship exists between climate and sediment yield at the global scale. Data presented by Jansson (1988) indicate that variation in sediment yield may even be greater within than between climatic regimes. Small mountain rivers in Greece have annual sediment yields ranging from 150 to 4150 t km^{-2} (Poulos *et al.*, 1996). The lack of a simple climatic relationship can be attributed to the widespread impact of human activity which serves to distort any global pattern, especially one based on precipitation magnitude alone, and to the influence of other factors, notably relief.

Several recent studies have underlined the importance of relief. In assessing the relative influence of a wide range of factors on global soil loss, Phillips (1990a) found that slope gradient (the relief variable) accounted for roughly 70 per cent of the variation, with runoff production per unit of precipitation and rainfall erosivity (the climate variable) contributing approximately 15 and 14 per cent respectively. Milliman and Syvitski (1992) analysed the data from 280 rivers and concluded that basin elevation (expressed via maximum headwater elevation) is a major control of sediment yield, other factors such as climate and runoff being of only secondary significance. They emphasized that basin elevation is basically a surrogate variable for tectonism, large sediment yields being associated with rivers that drain active orogenic belts. Summerfield and Hulton (1994) correlated natural denudation rates with a range of morphometric, hydrologic and climatic variables, finding that both mechanical and chemical denudation rates, but particularly the former, were most strongly related to basin relief characteristics and mean annual runoff. Although the form and relative significance of the relief variable varies between these studies, they all identify the overall importance of relief in accounting for the variation of sediment yield at the global scale.

At smaller spatial scales where differences in climate and relief are more subdued, other factors assume greater significance, particularly in so far as they influence runoff production. Data from the arid zone of India suggest that differences in local lithology can produce considerable variation in sediment yield in an environment where the vegetation cover is relatively sparse (Fig. 3.9C). Interestingly, there is no downturn in sediment yield beyond 450 mm as global analyses suggest, possibly because human activity has reduced cover density to a level where increased precipitation does not lead to a decrease in sediment yield. Land use can significantly affect the magnitude of sediment yield through its influence on the degree of protection afforded by a vegetation cover, the physical characteristics of the soil and the potential for surface runoff generation. Dunne (1979) has analysed the sediment yields of 61 Kenyan catchments classified according to the land-use types:

(i) forest (F);
(ii) forest cover > 50 per cent of the basin with the remainder under cultivation (FA);
(iii) agricultural land > 50 per cent with the remainder forested (AF);
(iv) grazing land (G).

Within each type, sediment yield increases with mean annual runoff (related to mean annual precipitation by $Q = 0.000\,033 P_m^{2.27}$), but major differences exist in the rate of that increase (Fig. 3.9D). As the average cover density decreases from (i) to (iv), sediment loss becomes increasingly sensitive to runoff and there appears to be a progressive increase in the rate at which sediment yield varies. Dunne concluded that sediment yield, which varied considerably from 8 to 20 000 t km^{-2} y^{-1}, was largely controlled by land use, with climatic and topographic factors having subsidiary effects. These results emphasize the potential variability of sediment yield even within relatively small areas and the need for caution when using global syntheses. Relying on data of variable quality and geographical spread, analyses of sediment yield variation have overwhelmingly been empirical in character, with the attendant problem of sorting out relative influences from a multivariate situation.

Sediment yield at the outlet is commonly much less than gross erosion in the basin, largely because of storage. During 137 years of European settlement in three Minnesotan catchments, 38–73 per cent of all eroded sediment has travelled less than 4 km (Beach, 1994). In an attempt to clarify and quantify the linkages between upstream erosion and downstream yield, a **sediment budget** approach began to emerge in the late 1970s, in which the various sediment sources within a basin are defined, and the sediment supplied from those sources is routed to and through the channel system with due consideration given to the various opportunities for storage. Sediment may be temporarily stored and remobilized several times before reaching the basin outlet, adding to the complexity of the problem. Sediment budgeting can provide information for land management rapidly at low cost if reconnaissance techniques are used (Reid and Dunne, 1996).

The classic work of Trimble (1983) clearly demonstrates the importance of storage and remobilization in controlling sediment yield from the 360 km^2 Coon Creek basin in Wisconsin, US. Sediment budgets were prepared for two time periods, 1853–1938 and 1938–1975 (Fig. 3.10), the first being characterized by poor land management and severe soil erosion (especially in the upland parts of the basin) following the introduction of European farming practices, and the second by the application of soil conservation measures. Those measures reduced upland soil erosion by 26 per cent but the sediment yield did not change. The sediment budget approach showed that most of the material eroded during the first period had been stored on the lower slopes as colluvium and in the valley floors, only a small proportion reaching the outlet. Material stored in tributary valleys and in the upper part of the main valley was remobilized during the second period to give an unchanged value for sediment yield.

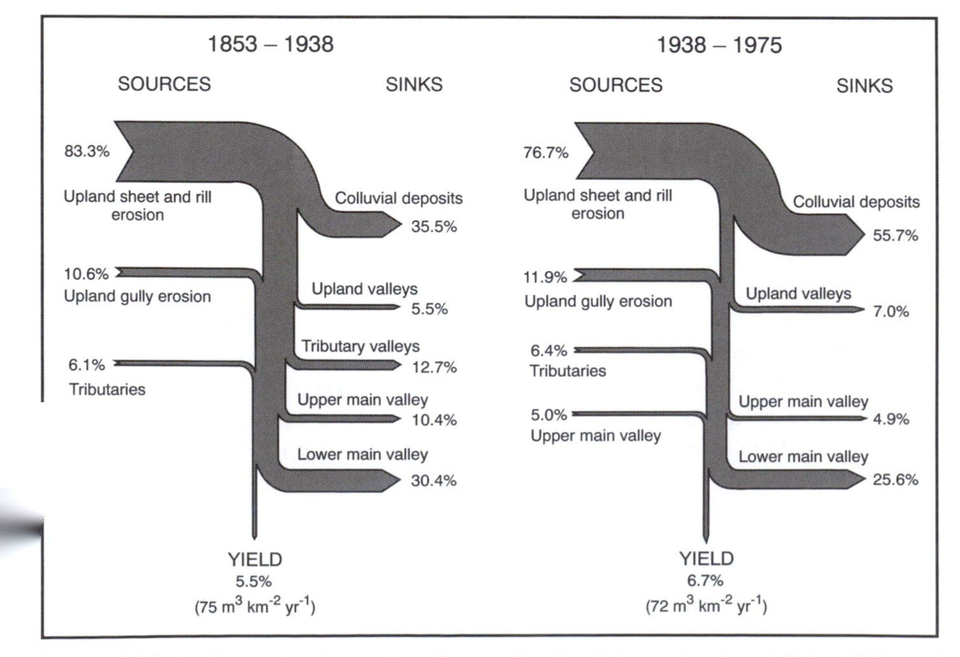

Figure 3.10 Sediment budgets for Coon Creek, Wisconsin, for 1853–1938 and 1938–1975 (after Trimble, 1983).

One of the most significant findings from budgetary studies such as Trimble's (1983) is the importance of storage in the history of sediment movement through a drainage basin, begging questions as to the size and location of the main sediment stores, the ease with which sediment can be stored and remobilized, and the residence times of sediment in storage. Trimble estimated that about 20 per cent of the sediment deposited in the upper tributary valleys during 1853–1938 has been remobilized since 1938, giving residence times of 10–100 years for those locations. Much longer residence times are evident in the glaciated valleys of British Columbia, where sediments deposited during the last glaciation are now being remobilized, reversing the common tendency for specific sediment yield (yield per unit area) to decrease downstream (Church and Slaymaker, 1989). Consideration of storage and residence times is particularly important in the case of the sediment-associated transport of contaminants, which may be stored for long periods in depositional sinks and released only slowly (Walling, 1988). Sediment budget analysis provides a useful framework for investigating the internal dynamics of sediment transfer, but accurate estimates of the rates and fluxes involved are difficult to obtain in anything but relatively small drainage basins. As drainage area increases, so do the opportunities for storage.

The potential for sediment storage over variable time periods not only underlines the discontinuous nature of sediment transfer but also raises doubts about the reliability of sediment yield as an indicator of catchment denudation. Sediment yield averaged only 10 per cent of mean annual gross erosion in four North Carolinan catchments (>1000 km^2), the remainder being stored as colluvium on hillslopes (76 per cent) or as alluvium in channels and floodplains (14 per cent) (Phillips, 1991a). Further complexity is added in catchments subject to human disturbance, which can have either a positive (e.g. forest clearance) or a negative (e.g. dam construction) effect on sediment yield. Sediment stores may build up over time until they reach a critical level when sediment is released 'catastrophically', producing sudden increases in sediment yield. Clearly there are many problems involved in attempting to derive denudation rates from measurements of sediment load, and the estimates in Table 3.1 must be regarded as very approximate. In comparing the relative magnitudes of the particulate and dissolved components of river load, and therefore the relative importance of mechanical and chemical denudation, there is also a scale effect to consider (Walling, 1987). Suspended sediment load tends to decrease with increasing basin area because of deposition *en route*, while dissolved load remains approximately constant because most of the dissolved material is transferred directly to the outlet. Thus the relative importance of the particulate component tends to decline as the size of the river basin increases.

4

FLUVIAL PROCESSES

The morphology of natural river channels depends on the interaction between the fluid flow and the erodible materials in the channel boundary. The underlying problem is to understand that interaction, given that it involves the distinct processes of entrainment, transport and deposition of sediment. The basic mechanical principles are well established but a complete analytical solution is still a long way off, largely because natural streams represent the movement of a fluid–solid mixture within boundaries that are themselves deformable. Fluid–boundary interactions occur over a wide variety of spatio-temporal scales, from the instantaneous motion of single particles to the long-term transport of bed material along a river. Given the highly variable character of natural river flow, the channel boundary has to withstand and adjust to a large range of forces if it is to be maintained as a coherent structure.

Mechanics of Flow

Water flowing in an open channel is subject to two principal forces: gravity, which acts in the downslope direction to move water at an acceleration of $g \sin \beta$ where g is gravitational acceleration (= 9.81 m s^{-2}) and β the angle of slope; and friction, which opposes downslope motion. The relationship between these two forces ultimately determines the ability of flowing water to erode and transport debris.

Open channel flow can be classified into various types based on four criteria (Table 4.1). If uniform flow exists, the gravitational and frictional components are in balance. Simple mathematical models can be constructed only if the flow is assumed to be uniform and steady, but flow in natural rivers is characteristically non-uniform and unsteady. The important distinction lies between laminar and turbulent flow.

Water is a viscous fluid that cannot resist stress, however small. In *laminar flow* each fluid element moves along a specific path with no significant mixing between adjacent layers. A very thin layer of fluid in contact with the

Table 4.1 Types of flow in open channels

Type of flow	Criterion
Uniform/non-uniform (varied)	Velocity is constant/variable with position
Steady/unsteady	Velocity is constant/variable with time
Laminar/turbulent	Reynolds number ($Re = vR\rho/\mu$) is <500/ >2500, with a transitional type when $500 < Re < 2500$
Tranquil/rapid	Froude number ($F = v/\sqrt{gd}$) is <1/>1, with critical flow when $F = 1$

Symbols: v, velocity; R ($=w.d/(2d + w) \sim d$ for wide channels), hydraulic radius; w, width; d, depth of flow; ρ, water density (~ 1000 kg m^{-3}); μ, molecular viscosity; g, gravity constant.

boundary is slowed so completely that it has no forward velocity, but resistance to motion along internal boundaries is less than at the bed and each successive layer of fluid away from the bed can slip past the one below to give a velocity profile which is parabolic in shape (Fig. 4.1A). In this way is shear stress,

$$\tau = \mu \frac{dv}{dy} \tag{4.1}$$

distributed throughout the flow, where dv/dy is the velocity gradient or rate of change of velocity (v) with depth (y).

When velocity or depth exceeds a critical value (Table 4.1), laminar flow becomes unstable and the parallel streamlines are destroyed. In *turbulent flow* the fluid elements follow irregular paths and mixing is no longer confined to molecular interactions between adjacent layers but involves the transfer of momentum by large-scale eddies. Accordingly the equation for shear stress must be modified to include an eddy viscosity term (η),

$$\tau = (\mu + \eta) \frac{dv}{dy} \tag{4.2}$$

Because $\eta \gg \mu$, turbulent flow exerts larger shear stresses than does laminar flow for the same velocity gradient (dv/dy). Also, mixing occurs much faster in turbulent flow, which accounts for the much steeper near-bed velocity gradient, and velocity tends to be more evenly distributed with depth because the larger-scale mixing slows faster bodies of water up the profile and speeds up slower ones below (Fig. 4.1Ai).

The mathematical analysis of laminar flow is well advanced but this type of flow rarely occurs in nature except as a very thin layer (the laminar sublayer) close to the channel boundary or as shallow overland flow (Emmett, 1970). In rough turbulent flows, even that layer is often disrupted and may be absent. At first sight equation (4.2) appears to offer a means of evaluating

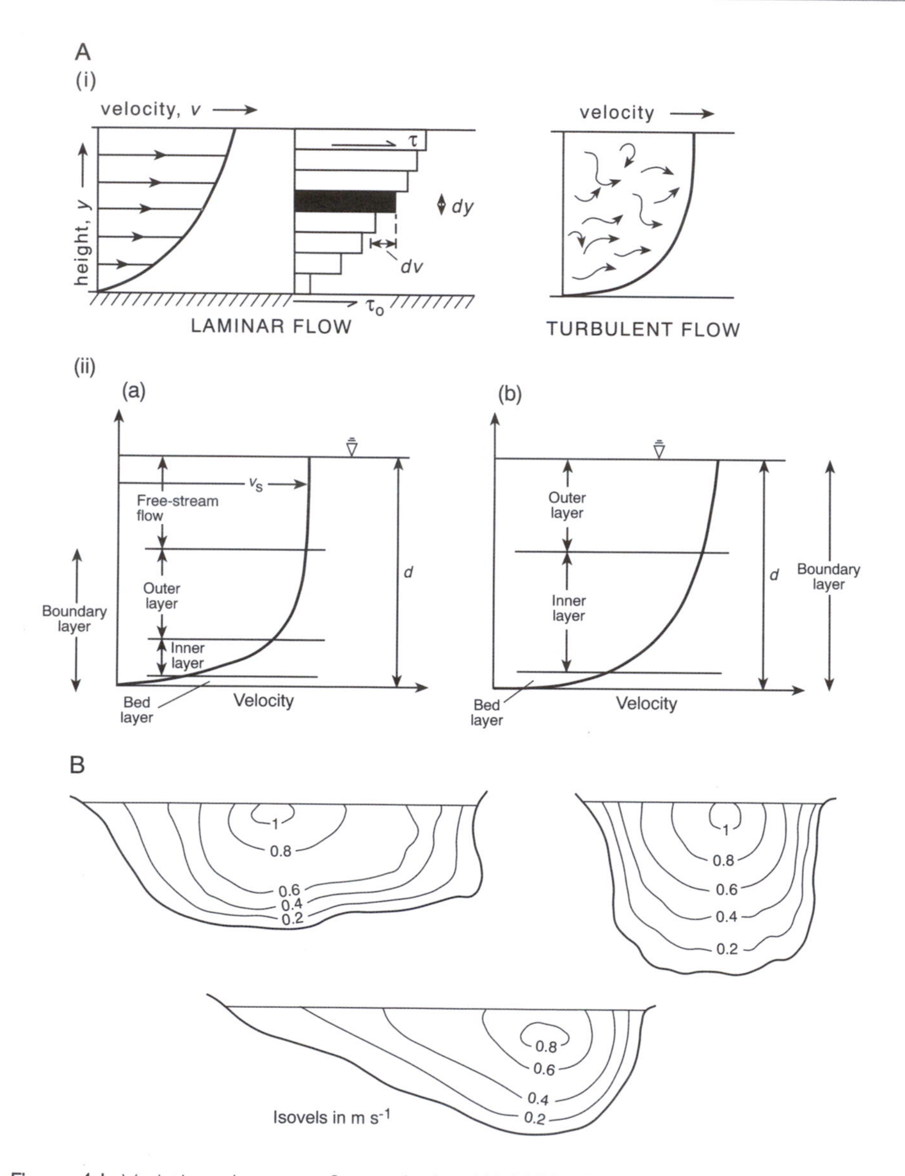

Figure 4.1 Variations in streamflow velocity. (A) With depth: (i) typical velocity profiles for laminar and turbulent flow; (ii) the structure of the boundary layer in deep (a) and shallow (b) flow. (B) At natural channel cross-sections. (C) (opposite) Downstream – relationship of velocity to discharge, Brandywine Creek (after Wolman, 1955). (D) With time: (i) velocity fluctuations at a point over a short time period; (ii) at-a-station changes in velocity with discharge measured over two years, River Bollin.

in theory the shear stress acting at the bed (τ_0), an important force in grain movement. However, representative values of eddy viscosity and the near-bed velocity gradient are very difficult to obtain because of the irregularity

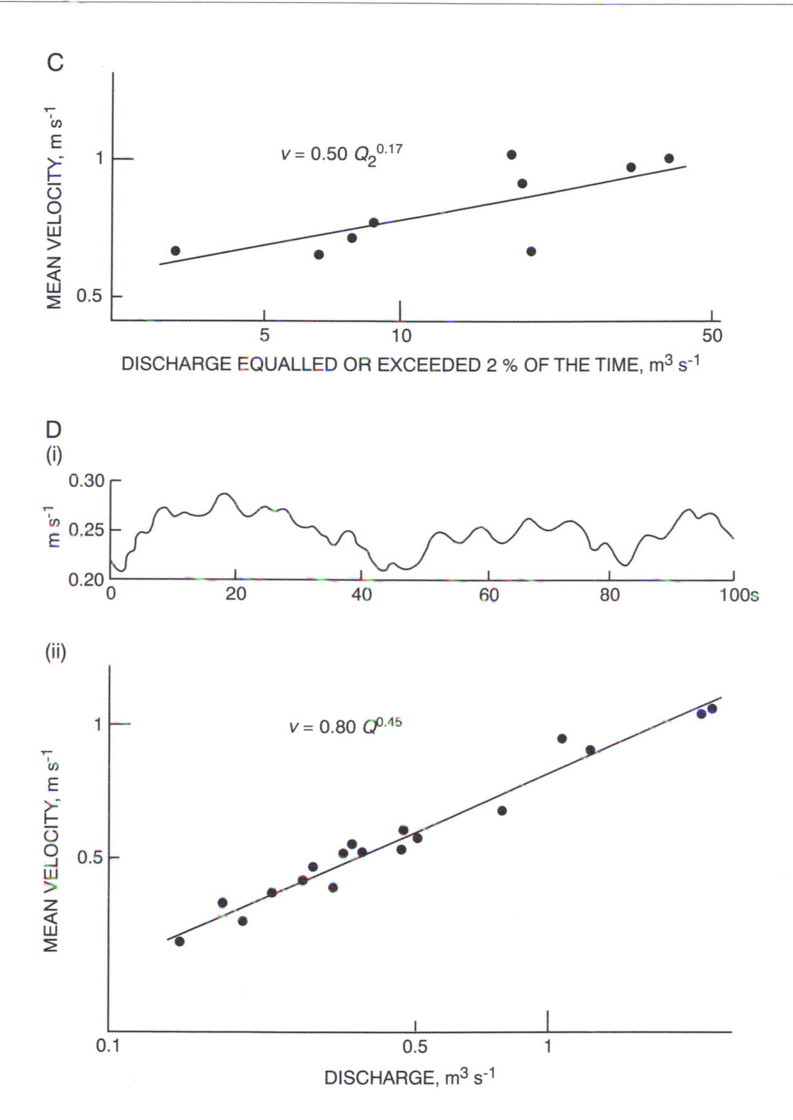

of turbulent flow. Eddies of varying size are generated and superimposed on the main downstream flow in both the vertical and the horizontal planes, constituting elements of a coherent structure of turbulence which is beginning to emerge. They give rise to unsteady flow events acting both towards ('sweeps') and away from ('ejections') the bed, which have important implications for the entrainment and suspension of non-cohesive sediment.

Velocity and Resistance

Velocity is a vector quantity having both magnitude and direction. It is one of the most sensitive and variable properties of open-channel flow because of its dependence on so many other factors.

Velocity varies in four dimensions (Fig. 4.1):

(i) *With distance from the stream bed.* The time-averaged velocity usually increases from zero at the bed to the free-stream velocity (v_s) at the edge of the boundary layer (Fig. 4.1Aii), the boundary layer being that part of the flow which is retarded by friction at the bed. In shallow flows the boundary layer may extend to the surface. It can in theory be divided into sublayers: a bed layer (or laminar sublayer), which may be disrupted or absent under natural conditions; an inner layer extending over 10–20 per cent of the depth, in which velocity varies semilogarithmically with depth; and an outer layer in the upper 80–90 per cent, where large-scale turbulence is dominant and the velocity profile diverges from a semilogarithmic form.

The shape of the velocity profile is strongly influenced by the depth of flow (d) and the size of roughness elements on the stream bed, where the latter includes both individual particles and forms moulded in the bed material. If the roughness is expressed in terms of bed material size (D), the two variables can be incorporated in a single index, the relative roughness ratio d/D. For a given depth of flow, the larger the roughness elements, the steeper is the velocity gradient towards the bed. Since stream beds generally contain a range of grain sizes, the problem is to select a grain diameter that best represents this component of resistance.

(ii) *Across the stream.* Velocity increases towards the centre of a stream as the frictional effect of the channel banks declines, an effect which becomes proportionately much less in channels with a high width/depth ratio (>15). The degree of symmetry in the cross-channel velocity distribution can vary considerably with the shape and alignment of the channel, becoming characteristically asymmetric in channel bends. The close relationship between velocity distribution, cross-sectional shape and erosive tendency is emphasized by the basic distinction between wide, shallow channels where the velocity gradient is steepest and the boundary shear stress greatest (equation (4.2)) against the bed, and narrow, deep sections where the velocity gradient is steepest against the banks, producing a greater tendency for bank erosion (Fig. 4.1B).

(iii) *Downstream.* Despite a declining slope along most rivers, velocity tends to remain constant or increase slightly as the channel becomes hydraulically more efficient and resistance decreases in the downstream direction (Fig. 4.1C). Variations in the rate of change of velocity do occur both along and between rivers, since velocity is merely one variable that can be adjusted to accommodate the downstream increase in discharge, but a value for m (the rate of change of velocity) in $v = kQ^m$ of 0.1 seems to be an appropriate average (Bathurst, 1993).

(iv) *With time.* Over time periods measured in seconds, point velocities may deviate by 60–70 per cent or more from the time-averaged velocity because of the inherent variability of turbulent flow. At the larger timescale of days, weeks or months, mean velocity at a section responds to fluctuations in discharge. The increase in depth with discharge tends to drown out roughness elements in the bed and thereby produce an increase in velocity (Fig. 4.1D(ii)), but the effect is not uniform and the velocity exponent *m* can vary considerably from section to section (Park, 1977). Values in the range 0.35–0.5 are not uncommon.

Velocity is thus a highly variable quantity in time and space. The character of that variation is important since velocity influences the processes of erosion, transportation and deposition (Fig. 4.5B, p. 110). Velocity is usually measured by current meter at selected points in the flow cross-section and expressed as an average value. Mean velocity at a cross-section is not the most relevant measure for defining the initiation of erosion, but it is still widely used, reflecting in part the difficulties of measurement close to the stream bed where the entrainment process has its origins.

Velocity is strongly related to **flow resistance**, one of the most important elements in the interaction between the fluid flow and the channel boundary. Several resistance equations have been developed (Table 4.2), of which the Darcy–Weisbach is recommended for its dimensional correctness and sounder theoretical basis (Task Force, 1963). All of the equations assume that resistance approximates that of a steady, uniform flow, but in natural channels with erodible boundaries the resistance problem is much more involved.

Total flow resistance consists of several components (Bathurst, 1993): *boundary resistance* resulting from the frictional effect of the bed material itself and the bed forms developed therein; *channel resistance* associated with bank irregularities and changes in channel alignment; and *free surface resistance* stemming from the distortion of the water surface by waves and hydraulic jumps. The effects of suspended sediment and aquatic vegetation may also be significant. Boundary resistance has been the main concern, with a basic subdivision into 'grain roughness' and 'form roughness' components.

Grain roughness is mainly a function of relative roughness (d/D or R/D), often expressed in the form

$$\frac{1}{\sqrt{ff}} = c \log\left(a\frac{R}{D_x}\right) \tag{4.3}$$

Table 4.2 Flow resistance equations

Chezy equation (1769)	$v = C\sqrt{Rs}$	
Manning equation (1889)	$v = k\frac{R^{2/3}s^{1/2}}{n}$	where $k = 1$ (SI units), $k = 1.49$ (imperial units)
Darcy–Weisbach equation	$ff = \frac{8gRs}{v^2}$	

Symbols: C, n, ff are the respective resistance coefficients; v, mean velocity; R, hydraulic radius; s, slope of the energy gradient; g, gravity constant.

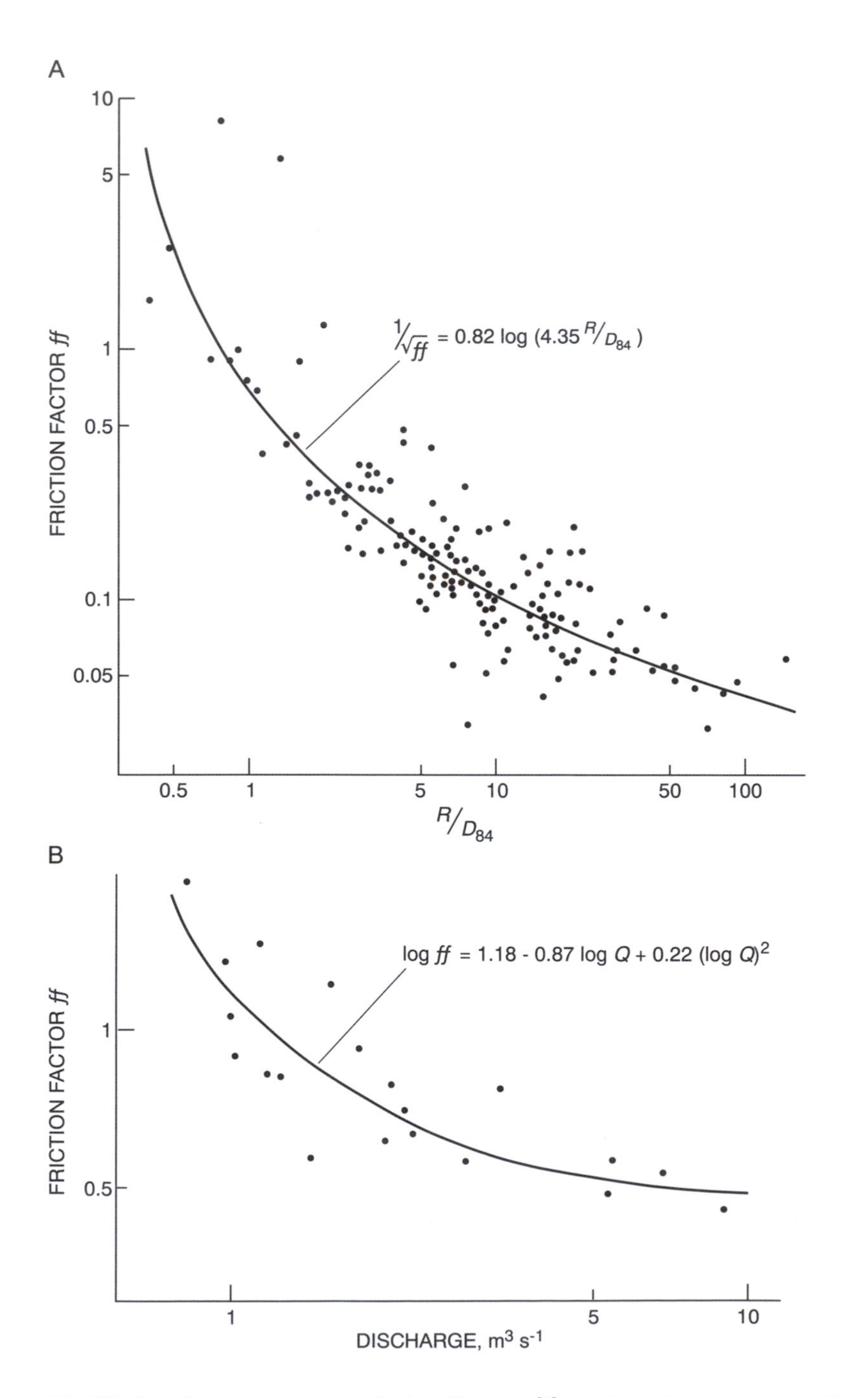

Figure 4.2 (A) Relationship between friction factor (ff) and relative roughness (data from Leopold and Wolman, 1957; Limerinos, 1970; Charlton *et al.*, 1978; Hey, 1979, 1988a; Prestegaard, 1983a). (B) Relationship between friction factor and discharge at a sand-bed cross-section, River Bollin.

where c and a are constants and D_x is a measure of the size of roughness elements (Fig. 4.2A). The roughness height of uniform material is simply taken as the grain diameter, but with non-uniform material the problem is to choose a representative grain diameter. D_{84} (the diameter at which 84 per cent of the material is finer) is commonly used since it takes account of the important influence on flow resistance of large particles, but the choice is somewhat arbitrary. Hey (1979) successfully applied a version of equation (4.3) to describe flow resistance over riffles in gravel-bed rivers, but the equation becomes less applicable where $R/D_{84} < 4$ and $w/d < 15$ (Bathurst, 1993).

Grain roughness can be the dominant component of resistance where stream beds consist of gravel (2–64 mm) or cobbles (64–256 mm). Equation (4.3) implies that, as depth increases with discharge at a cross-section, the effect of grain roughness is drowned out and flow resistance decreases, although possibly at a declining rate with higher discharge (Fig. 4.2B). Consequently velocity may also tend to change more slowly at higher flows, producing non-linearities in hydraulic geometry (Richards, 1973; Knighton, 1979; Fig. 5.10C, p. 181).

Form roughness stems from features developed in the bed material and, in sand-bed streams at least, often exceeds grain roughness in importance. It presents a particular problem in that, once grains are set in motion, the shape of the bed can be modified to give a variable form roughness dependent on flow conditions. In sand-bed streams where the bed is most readily moulded into different shapes, a sequence of bed forms correlated with increasing flow intensity has been defined (Fig. 4.3A), each form offering different levels of resistance (Simons and Richardson, 1966). Understanding of the relationship between bed-form geometry, sediment transport and hydraulic roughness remains incomplete, but changes in bed configuration represent an important self-regulating mechanism available to streams at the flow–bed interface. As discharge and sediment load increase with the passage of a flood wave, a transition from ripples to dunes may so increase the flow resistance as to offset the improved hydraulic efficiency associated with a greater depth, thereby slowing the rate of change of velocity (Richards, 1973, 1977b). In the upper regime the flow resistance remains relatively low until breaking waves develop, when considerable energy loss occurs (Fig. 4.3A).

The contribution of form roughness to total flow resistance over coarser beds has received much less attention until relatively recently. Bed forms of varying size have now been identified in this role, from microtopographic pebble clusters to the channel bars characteristic of riffle–pool sequences (Robert, 1990). Cluster bed forms, which generally consist of a single obstacle protruding slightly above neighbouring grains together with upstream and downstream accumulations of particles (Fig. 4.3B), are probably the predominant type of microtopography in gravel-bed rivers (Brayshaw, 1985). The analysis of detailed bed profiles can help to identify their effect on resistance, which may for practical purposes be scaled according to a characteristic grain size (Clifford *et al.*, 1992). The main resistance effect of a riffle–pool

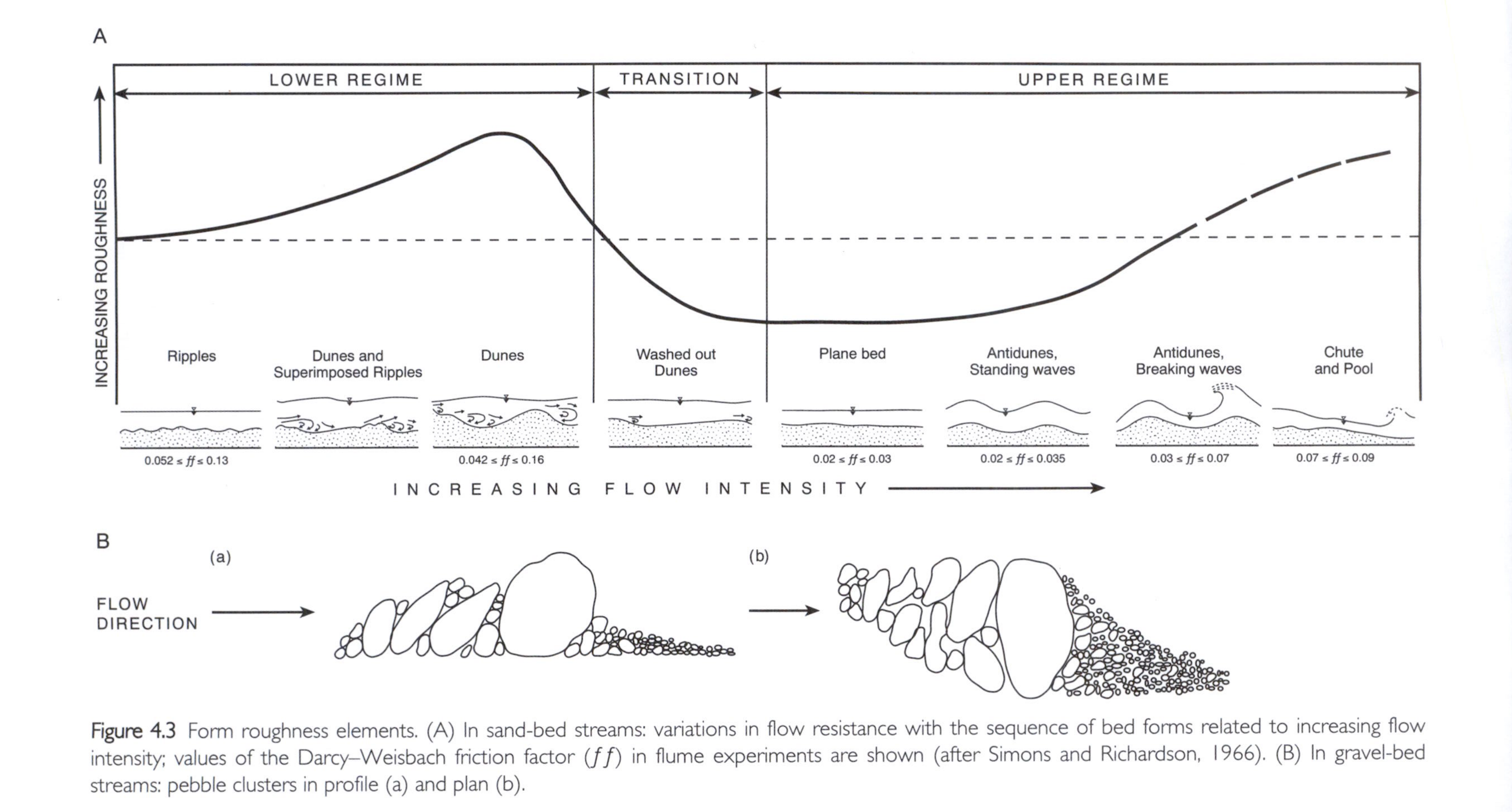

Figure 4.3 Form roughness elements. (A) In sand-bed streams: variations in flow resistance with the sequence of bed forms related to increasing flow intensity; values of the Darcy–Weisbach friction factor (ff) in flume experiments are shown (after Simons and Richardson, 1966). (B) In gravel-bed streams: pebble clusters in profile (a) and plan (b).

sequence is produced by the ponding upstream of the riffle or bar (Hey, 1988a), an effect of particular significance at low flows, although bar resistance can still account for 50–60 per cent of the total at higher bankfull stages (Prestegaard, 1983a). In boulder-bed streams, form roughness is associated not only with specific bed forms, such as step–pool systems, but also with individual particles which can exert a considerable form drag on the flow where $R/D_{84} \leq 4$. As yet the effects on flow resistance produced by the size, shape and spacing of these various bed forms have not been thoroughly investigated.

Even less is known about the remaining components of resistance. Local bank irregularities can produce large energy losses, while the additional resistance associated with channel curvature can exceed the contribution from grain roughness (Leopold *et al.*, 1960). Material suspended in the flow increases fluid viscosity and tends to damp down turbulence, thereby reducing resistance. Experiments by Vanoni and Nomicos (1960) showed that suspended load could decrease resistance by 5–28 per cent, but under natural conditions the effect is likely to be small except at times of very high suspended sediment concentration. Variations in the height, density and flexibility of aquatic vegetation can also influence resistance at the reach scale (Masterman and Thorne, 1994).

Neither the flow, form nor roughness within individual reaches is uniform. Values for resistance are usually obtained by calculation using one of the equations given in Table 4.2 or by visual comparison with representative reaches (e.g. Hicks and Mason, 1991) rather than by direct measurement, so that computed values tend to include the undifferentiated effects of all types of resistance. Attempts to separate total resistance into its component parts have achieved only qualified success, and knowledge of the resistance mechanism in natural streams remains far from complete. As regards boundary resistance, form roughness is the principal component in sand-bed streams, and in streams with coarser bed material where grain friction might be expected to dominate, it can still be the major contributor (Prestegaard, 1983a; Griffiths, 1989). Flow resistance is a primary element of stream behaviour through its link with bed material properties, sediment transport and the way in which a stream consumes its energy. Indeed, Davies and Sutherland (1980) have proposed that equilibrium channel development is governed by a principle of maximum resistance in which boundary deformation continues until the friction factor has attained a local maximum. Adjustment of the relative spacing of cluster bed forms may be one means whereby resistance maximization is brought about in gravel-bed streams (Hassan and Reid, 1990).

Stream Energy

The total energy input required to drive the global hydrological cycle is estimated at 1.3×10^{24} J y^{-1}, equivalent to the output of 40 million major power stations (Walling, 1987). In the fluvial system three types of energy are relevant, namely potential, kinetic and thermal (heat), only the first two of which can perform mechanical work. That work takes various forms:

(a) work done against viscous shear and turbulence (internal friction);
(b) work done against friction at the channel boundary;
(c) work done in eroding the channel boundary;
(d) work done in transporting the sediment load.

Since energy must first be used to maintain the flow against internal and boundary friction ((a) and (b)), a critical energy level must be reached before a stream can perform erosional and transportational work. The concept of an erosion threshold is therefore fundamental.

Water with a mass *m* entering a river at a height *h* above a given datum or base-level (the sea or the next tributary junction) has a potential energy

$$\mathrm{PE} = mgh \tag{4.4}$$

As water moves downslope that potential or position energy is gradually converted into kinetic energy:

$$\mathrm{KE} = \tfrac{1}{2} mv^2 \tag{4.5}$$

In conservative systems the principle of energy conservation

$$\mathrm{PE} + \mathrm{KE} = \text{constant} \tag{4.6}$$

applies and relatively simple models can be developed for describing system behaviour. Between any two points, a loss in potential energy is matched by an equivalent gain in kinetic energy. However, rivers are non-conservative systems. Friction causes much of the available mechanical energy to be dissipated in the form of heat which can perform no mechanical work. Consequently the well-established principles of mechanics cannot readily be used to solve flow problems without making assumptions.

Nevertheless, it is relevant to ask how potential energy could be distributed in the fluvial system since, for example, different distributions may be associated with different network topologies (Knighton, 1980a). Minimum energy principles have been used to reproduce well-known characteristics of drainage network structure (Rodriguez-Iturbe *et al.*, 1992), and to guide the search for a general principle governing stream behaviour and channel form adjustment. Yang and Song (1979) argued that, in moving towards a state of dynamic equilibrium, a natural stream chooses its course in order to minimize the rate of energy expenditure, one manifestation of which is a condition of minimum stream power. Stream power, defined by

$$\Omega = \gamma Q s \tag{4.7}$$

where γ ($=\rho g$) is the specific weight of water, Q is discharge and s is slope, is an expression for the rate of potential energy expenditure per unit length of channel (or rate of doing work). Chang (1979a, 1985) has used the concept of minimum stream power to derive relationships which describe the equilibrium geometry of alluvial rivers, and to explain discontinuities in the process of river channel adjustment. However, direct considerations of energy expenditure in natural streams are rare. There remains a need to relate more closely the activity of fluvial processes and the forms they develop to the physical concept of work.

Thresholds of Erosion

The movement of material depends on its physical properties, notably size, shape, density and structural arrangement. Grain size has a direct influence on mobility, and a typical classification is given in Table 4.3. A basic distinction exists between non-cohesive and cohesive (including solid rock) materials. In cohesive material, which generally consists of particles in the silt–clay range, resistance to erosion depends more on the strength of cohesive bonds between particles than on the physical properties of the particles themselves, adding considerably to the complexity of the erosion problem.

Entrainment and Bed Erosion

Most stream beds consist of **cohesionless** grains in the sand and gravel ranges. As the flow over a surface of loose grains gradually increases in intensity, a condition is reached where the forces tending to move a particle are in balance with those resisting motion. The problem is to define that threshold state.

Initial movement has commonly been specified in terms of either a critical shear stress (τ_{cr}) or a critical velocity (v_{cr}). An estimate of the mean boundary shear stress (τ_0) exerted by the fluid on the bed can be obtained from

$$\tau_0 = \gamma R s \tag{4.8}$$

where γ is the specific weight of water, R is hydraulic radius and s is slope, although this formulation assumes uniform flow and tends to overestimate the effective shear acting on a grain (Robert, 1990). Recognizing this limitation, one can then define the critical shear stress for spherical grains of

Table 4.3 Grain size classification

Class name	Size range	
	mm	Phi units (ϕ)
Boulders	≥ 256	≤ –9
Cobbles	64–256	–6 to –9
Gravel	2–64	–1 to –6
Sand	0.062–2	4 to –1
Silt	0.004–0.062	8 to 4
Clay	≤ 0.004	≥ 8

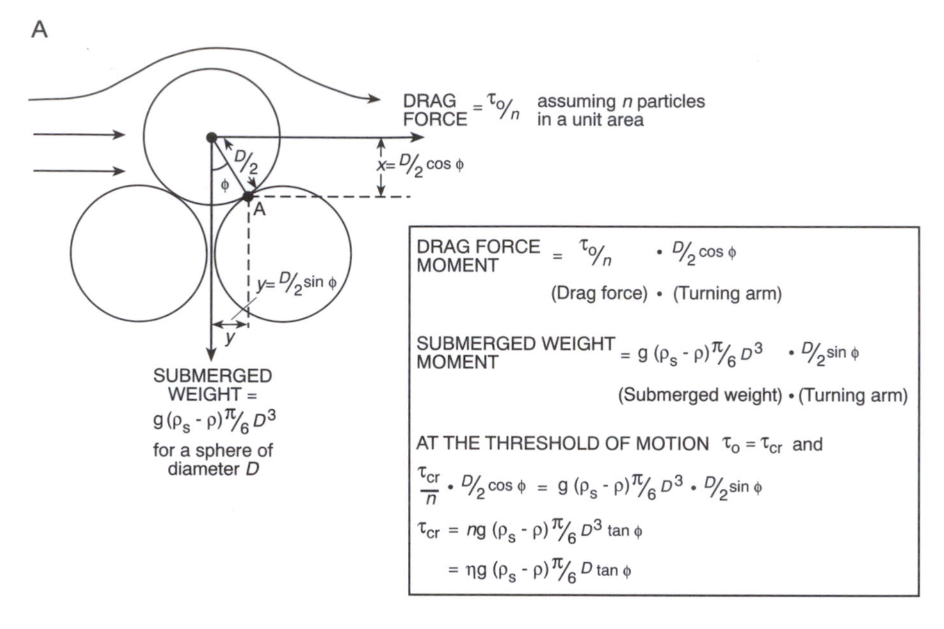

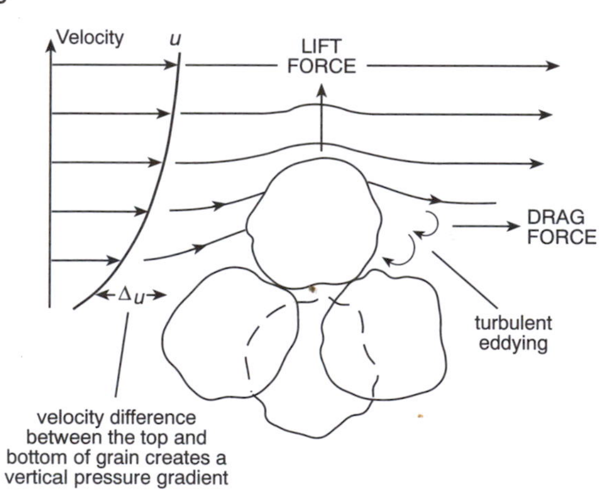

Figure 4.4 (A) The critical shear stress (τ_{cr}) defined for a grain resting on a horizontal bed, where ρ is fluid density, ρ_s sediment density, g the gravity constant, D the grain diameter and $\eta = nD^2$, a measure of grain packing. (B) Lift and drag forces acting on a submerged particle.

diameter D resting on a flat bed by equating the moments of forces acting about a downstream contact point (A in Fig. 4.4A), to give

$$\tau_{cr} = \eta g(\rho_s - \rho)\frac{\pi}{6} D \tan \phi \quad (4.9a)$$

This elementary deterministic model predicts that the shear stress needed to initiate movement depends principally on grain size (i.e. $\tau_{cr} \propto D$), with grain shape and the degree of packing (η) being additional influential factors. A major limitation of this model is that it ignores the additional force provided by lift. Acting normal to the bed (Fig. 4.4B), the *lift force* arises in at least two ways:

(a) the difference in flow velocity between the top and bottom of a grain sets up a pressure gradient which tends to move the particle vertically upwards;
(b) turbulent eddying may produce local velocity components which act directly upwards close to the bed.

The lift force decreases rapidly in magnitude away from the bed as the velocity and pressure gradients become less, but it may be crucial to the initial movement of grains, at least in sand-bed streams. Helley (1969) successfully predicted the initial motion of particles in the field by incorporating the additional effect of lift. The combination of lift and drag forces provides one explanation for the commonly observed process of *saltation* (Fig. 4.9B, p. 127), in which particles of sand size move along the bed in a series of hops.

Equation (4.9a) has been expressed in various ways, most notably by Shields (1936), who related a dimensionless critical shear stress (θ) to a particle Reynolds number ($\propto D/\delta_0$, where δ_0 is the thickness of the laminar sublayer). The plot (Fig. 4.5A) not only identifies the threshold state but reveals that it is not as straightforward as a simple resolution of forces would at first suggest. On hydraulically rough beds (the common condition in natural streams), θ rapidly attains a constant value k (reported values range from 0.03 to 0.06, with 0.045 now accepted as a good approximation (Komar, 1988), lower than the original Shields value of 0.06), to give thereafter

$$\theta = \frac{\tau_{cr}}{g(\rho_s - \rho)D} = k \Rightarrow \tau_{cr} = kg(\rho_s - \rho)D \qquad (4.9b)$$

which has a form akin to equation (4.9a). At lower values of D/δ_0 the function reaches a minimum of $\theta \sim 0.03$ in the smooth turbulent regime, corresponding approximately to grains in the size range 0.2–0.7 mm, before becoming inverse. Thus, for particle sizes less than about 0.2 mm, the threshold stress needed for entrainment must increase as particles get smaller because particles in that size range are submerged in the laminar sublayer and therefore not subject to the greater stresses associated with turbulent flow.

An alternative approach defines the critical condition in terms of velocity rather than shear stress, but the same basic trends are revealed (Fig. 4.5B), in particular that medium sand (0.25–0.5 mm) is the most easily eroded fraction. However, mean velocity is not the most relevant parameter in this context; the problem is to define and measure a representative near-bed velocity.

These approaches provide at best an average indication of the threshold state. Their limitations are reflected in the wide scatter of points on plots of field data (Fig. 4.5C). On the basis of these data for $D > 10$ mm, Williams

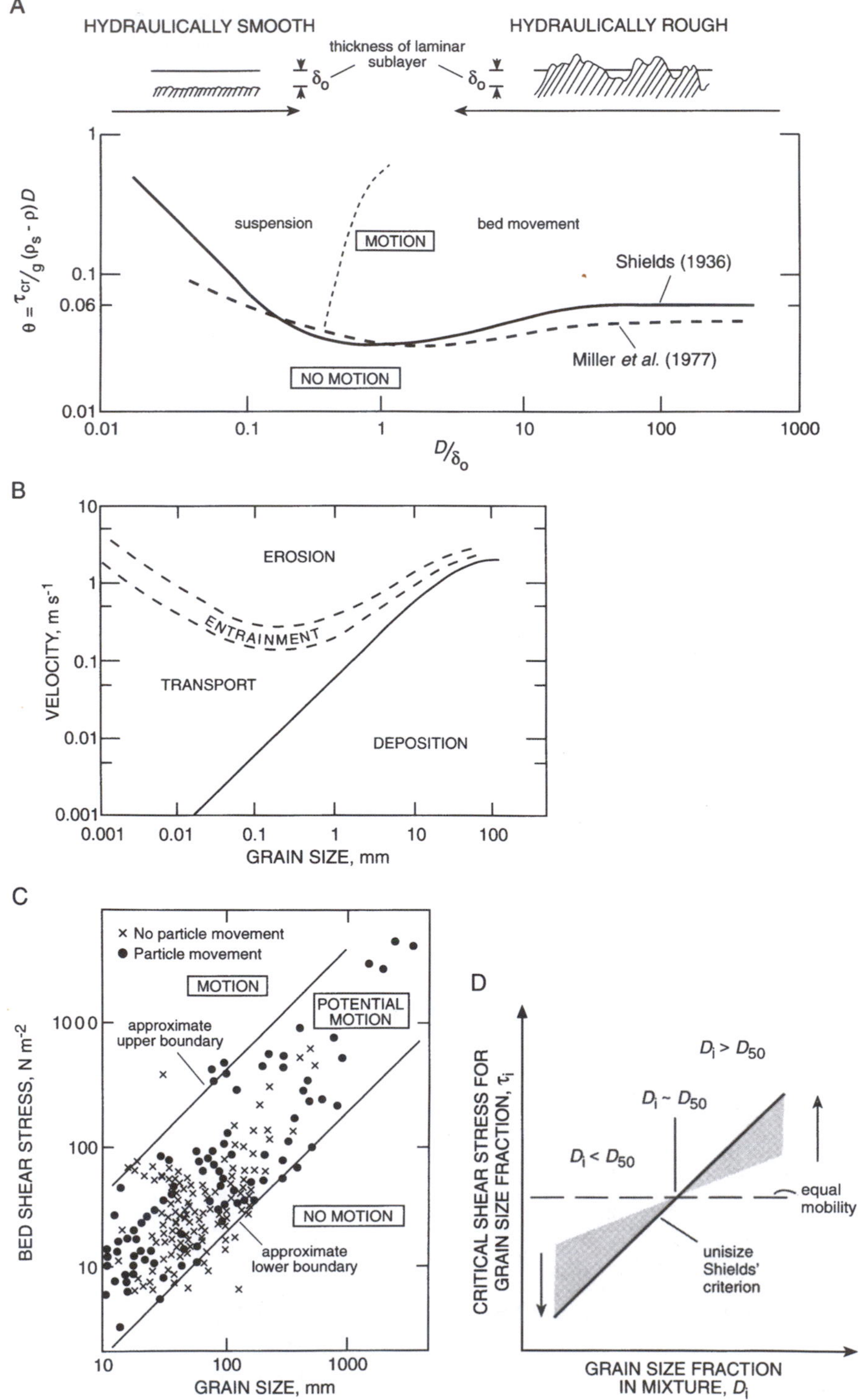
A
HYDRAULICALLY SMOOTH
HYDRAULICALLY ROUGH
thickness of laminar sublayer
δ_o
δ_o
$\theta = \tau_{cr}/g(\rho_s - \rho)D$
suspension
MOTION
bed movement
Shields (1936)
Miller et al. (1977)
NO MOTION
1
0.1
0.06
0.01
0.01
0.1
1
10
100
1000
D/δ_o
B
VELOCITY, m s^{-1}
EROSION
ENTRAINMENT
TRANSPORT
DEPOSITION
10
1
0.1
0.01
0.001
0.001
0.01
0.1
1
10
100
GRAIN SIZE, mm
C
× No particle movement
● Particle movement
MOTION
POTENTIAL MOTION
approximate upper boundary
NO MOTION
approximate lower boundary
BED SHEAR STRESS, N m^{-2}
1000
100
10
10
100
1000
GRAIN SIZE, mm
D
CRITICAL SHEAR STRESS FOR GRAIN SIZE FRACTION, τ_i
$D_i > D_{50}$
$D_i \sim D_{50}$
$D_i < D_{50}$
equal mobility
unisize Shields' criterion
GRAIN SIZE FRACTION IN MIXTURE, D_i

(1983) has suggested upper and lower limits for the Shields parameter θ of 0.25 and 0.01 respectively, a very wide range indeed. Two sets of factors are largely responsible for the scatter, related to the variability of flow conditions and bed material characteristics (Table 4.4). Problems exist in establishing a satisfactory definition of initial sediment movement and in obtaining appropriate measurements of bed shear stress or near-bed velocity. Short-term fluctuations in the flow can give rise to instantaneous stresses of many times the average, so that particles may be entrained at stresses much lower than predicted. Sediment entrainment is a function not only of the shear stress acting on the bed but also of the intensity of turbulence above it, which influences the magnitude and frequency of sweep and ejection events. Since eddy size and hence the energy available for moving grains are related to the dimensions of the flow, the size of the channel may affect the entrainment process. In two Pennine streams, critical shear stress in the narrower channel (w = 1.6 m) was 3–4 times higher than that in the wider one (w = 5.5 m) for a given size of material (Carling, 1983).

The bed material of natural streams is characteristically non-uniform, particularly at the coarser end of the scale. Properties other than size influence particle mobility, including pivoting angle (ϕ in equation (4.9a)) (Li and Komar, 1986), degree of grain exposure (Fenton and Abbott, 1977), and sediment fabric as seen in imbrication and cluster bed form structures. For a given grain size >8 mm, the Shields parameter may vary by nearly an order of magnitude depending on whether the bed is loosely or tightly packed

Table 4.4 Factors producing scatter about threshold curves (partly after Naden, 1988)

Flow conditions	Bed-material characteristics
Definition of the entrainment threshold	Degree of exposure or relative protrusion of grains
Use of average shear stress or velocity	Pivoting angles
Spatial variability of shear stress or velocity over the bed	Imbrication or clustering of particles
Irregularity of turbulent eddying	Degree of packing
Channel size	Grain shape Grain size distribution or relative size Microtopography

Figure 4.5 (opposite) (A) The Shields entrainment function. The dashed line represents the Shields curve modified by Miller *et al.* (1977). (B) Erosion and deposition criteria defined in terms of threshold velocities (after Hjulström, 1935). (C) Field-measured bed shear stresses as a function of grain size, from which zones of no motion, potential motion and motion can be defined (after Williams, 1983). (D) Summary diagram of empirical results on critical shear stress as a function of grain size. The solid line represents uniform material and the stippled areas represent the range of results for naturally sorted coarse-grained materials. The arrows signify increasing bimodality.

(Church, 1978). Many of these additional properties vary with the degree of heterogeneity in the bed material, implying that the critical shear stress will depend not only on the absolute size of a particle but also on its relative size. This led Parker *et al.* (1982) and Andrews (1983) to propose a hypothesis of 'equal mobility' for mixed beds ($\Rightarrow \tau_{cr} \propto D^0$): the hiding and protrusion effects of larger grains are regarded as sufficient to compensate for their greater submerged weight, so that all grain sizes become mobile at approximately the same shear stress. However, subsequent research (Ashworth and Ferguson, 1989; Komar and Shih, 1992) has indicated that size-selective entrainment does take place within coarse-grained alluvial channels, although not to the extent predicted by the Shields curve for unisize material (where $\tau_{cr} \propto D$). In grain mixtures the critical shear stress is reduced for coarser grains relative to the unisize Shields value (because of greater exposure and reduced resistance to motion) but increased for finer ones (because of reduced exposure and increased resistance to motion) (Fig. 4.5D). Entrainment of a specific grain does partly depend on its size relative to that of its neighbours. However, in strongly bimodal sediments this effect may disappear so that critical shear stress approaches its Shields equivalent (i.e. $b \rightarrow 1$ in $\tau_{cr} \propto D^b$), possibly because of size segregation on the bed surface (Wilcock, 1993).

The erosion of **cohesive** material presents a different and more complex problem. The forces resisting motion include not only those associated with particle geometry but also those electrochemical forces which bind the material together. The latter cannot be uniquely expressed as some function of particle size, so there is no critical shear stress for the erosion of cohesive material in the sense that one exists for cohesionless grains. Erosion takes place in aggregates rather than particle by particle.

Similar remarks apply to bedrock channels as regards the erosion threshold. Several processes have been identified in general terms:

(a) *corrosion*, or the chemical action of water;
(b) *corrasion*, or the mechanical (hydraulic and abrasive) action of water which can be effective when the flow is armed with particles, leading to surface abrasion and pothole development;
(c) *cavitation*, a process associated with the effects of shock waves generated through the collapse of vapour pockets in a flow with marked pressure changes.

Cavitation can cause severe erosion (Brown, 1963), but is of limited occurrence under natural conditions because of the high velocities required. These processes remain little more than descriptive terms, although Foley (1980) has attempted to model abrasion using engineering sandblast theory. Their effectiveness can be seen in the huge channels carved by Pleistocene floods in the Channeled Scablands of the northwestern United States (Baker, 1973).

The concept of a critical shear stress or velocity necessary for erosion is appealing, at least in theory. Theory and observation indicate that material in the size range 0.25–0.75 mm is the most susceptible to movement and, other conditions being equal, will be the first to be entrained. However, grain size is only one factor affecting the entrainment threshold, and the many other factors involved are difficult to incorporate in theoretical models, particularly since their relative importance remains unclear. Indeed, the heterogeneity of stream bed material and the inherent variability of natural flow conditions in the bed region are such that the applicability of deterministic models is limited, and the problem of defining the entrainment threshold takes on a statistical complexion in which the critical shear stress for a given grain size is characterized by a probability distribution rather than a single value. A further dimension to the problem is provided by the cohesivity of some stream beds where size-related threshold criteria appear to be less relevant.

Bank Erosion

The bank material of natural channels is also highly variable but does tend to become finer and more uniform downstream, with combinations of sands, silts and clays dominant, especially where a well-developed floodplain exists. Most channel banks possess some degree of cohesion because of finer material, so that the analysis of bank erosion is not a simple extension of the non-cohesive bed case with a downslope gravity component added. A further complication is provided by the effects of vegetation whose root systems can reinforce bank material and thereby increase resistance to erosion, particularly in the upper bank. Smith (1976) found that bank sediment with a root volume of 16–18 per cent and a 5 cm root mat was afforded 20 000 times more protection from erosion than comparable sediment without vegetation.

Two main groups of processes are involved: hydraulic action and mass failure. The removal of bank material by *hydraulic action* is closely related to near-bank velocity conditions (Odgaard, 1987; Hasegawa, 1989), and in particular to the velocity gradient close to the bank, which determines the magnitude of hydraulic shear. The high rates of bank retreat commonly associated with bend apices are explained by the steep velocity gradients and high shear stresses encountered against the outer bank of meander bends. The flow not only entrains material directly from a bank face but also scours the base of a bank, which leads to oversteepening and induces gravitational failure. The high shears generated within large-scale horizontal eddies can scour both bed and bank, enlarging existing embayments and increasing the amplitude of bank projections which become more susceptible to subsequent attack as the bank retreats discontinuously. Hydraulic action is probably the dominant process eroding non-cohesive banks where individual grains are dislodged or shallow slips occur along almost planar surfaces, while its effectiveness against cohesive banks depends upon the moisture content and degree of preconditioning of the material (Thorne, 1982). Hard, dry banks are very resistant, while wet ones are relatively easy to erode, especially if loosened by repeated wetting and drying or frost action. The distribution of velocity and shear stress and local turbulence characteristics have an important influence on the erosive potential of hydraulic action, which in many instances is the major process (Knighton, 1973; Hooke, 1979). In their study of the Connecticut River, Simons *et al.* (1979) estimated that flow forces were at least six times more effective than any other process. Those forces can be subdivided into:

(a) those that act close to the flow surface, including water waves induced by wind or passing boats (Nanson *et al.*, 1994);
(b) those that act near to the base of banks and lead to undercutting.

Significantly, velocity and boundary shear stress appear to be at a maximum in the lower bank region even when the flow is near bankfull (Bathurst *et al.*, 1979; Simons *et al.*, 1979).

The susceptibility of river banks to *mass failure* depends on their geometry, structure and material properties. Processes of weakening and weathering related in particular to soil moisture conditions reduce the strength of intact bank material and decrease bank stability (Thorne, 1982). Cycles of wetting and drying are especially important as they cause swelling and shrinkage of the soil, leading to the development of interpedal fissures and tension cracks which encourage failure. Seepage forces can reduce the cohesivity of bank material by removing clay particles and may promote the development of soil pipes in the lower bank. Shallow slips occur in cohesionless banks (Plate 4.1), while the dominant mechanisms in banks of high and low cohesivity seem to be deep-seated rotational slip and slab-type failure respectively (Thorne, 1982). The mass failure and retreat of near-vertical banks occurs primarily by slab failure in which blocks of soil topple for-

Plate 4.1 Shallow slips in sandy bank material, Wollombi Brook, New South Wales, Australia. Degradation has increased bank height and induced instability.

Plate 4.2 Slab failure of a cohesive river bank, River Bollin. Accumulation of failed blocks temporarily protects the lower bank from erosion.

wards into the channel (Plate 4.2), a process analysed theoretically by Osman and Thorne (1988).

The effectiveness of these several mechanisms is considerably enhanced by basal scour which seeks to increase bank angle. In composite river banks where cohesive material overlies non-cohesive sands or gravels, a relatively common condition along rivers with alluvial floodplains, Thorne and Tovey (1981) have proposed a mechanism of cantilever failure. Undercutting of the lower bank by hydraulic action generates an overhang or cantilever in the upper cohesive layer which then fails when a critical state is reached, the precise mode of failure depending on the geometry of the overhang. This process has been observed under permafrost conditions where river water thaws frozen sediments and cuts a thermo-erosional niche at the waterline (Walker *et al.*, 1987). The collapsed blocks produced by mass failure may break on impact and be removed, or they may remain intact to await removal by subsequent hydraulic action, meanwhile protecting the lower bank from further erosion (Plate 4.2). This pseudo-cyclic process (basal erosion, upper bank failure, lower bank accumulation, and removal of failed blocks) plays an important part in controlling the form, stability and retreat rate of all types of river bank.

Through the growth of ice wedges or ice crystals, *frost action* can be an important preconditioning process, widening pre-existing cracks and disaggregating surface material (Plate 4.3) to leave the bank more susceptible to subsequent attack. While accepting this indirect role, Lawler (1986, 1993) went one step further in suggesting that frost action in the form of needle ice is an effective process of erosion in its own right. In his study of a Welsh river

Plate 4.3 Disaggregation of surface material in a cohesive river bank. After a week of intense frost the surface became very friable to a depth of 3–5 cm.

it emerged as the strongest control on the amount and intensity of bank erosion, even in a temperate climate where freeze–thaw cycles are not particularly frequent. However, its direct effect is probably subsidiary to that of hydraulic action and mass failure, but its preparatory role emphasizes that the several processes are not mutually exclusive but frequently act in combination. *Trampling* by cattle can also contribute to bank retreat (Trimble, 1994).

The amount, periodicity and distribution of river bank erosion are highly variable because many factors are involved (Table 4.5), influencing these three aspects of erosion to differing extents. Average rates (Table 4.6) tend to mask that variability. Little erosion is likely to take place in the absence of high discharges, but similar flows need not be equally effective because bank wetting or frost action may not have reduced the strength of bank material to the same degree. Consequently correlations between flow volume and amount of erosion tend to be rather weak, although stream power (product of discharge and slope) accounted for 48 per cent of the variation in migration rate along meandering channels in western Canada (Nanson and Hickin, 1986). Wolman (1959) found that a large summer flood attacking dry banks produced little erosion, while lesser winter flows acting against thoroughly wetted banks caused considerable bank retreat. Multi-peaked flows which are more characteristic of winter months may be more effective than single flows of comparable or greater magnitude because of the increased incidence of bank wetting (Knighton, 1973). The degree of preparation of material can give a seasonal slant to the erosion process. Thus the amount of bank erosion is not solely a function of discharge magnitude or related shear stress conditions, so a threshold flow cannot reasonably be defined, although at some of her sites Hooke (1979) found little or no erosion at discharges less than the 5 per cent flow.

Table 4.5 Factors influencing bank erosion

Factor	Relevant characteristics
Flow properties	Magnitude-frequency and variability of stream discharge Magnitude and distribution of velocity and shear stress Degree of turbulence
Bank material composition	Size, gradation, cohesivity and stratification of bank sediments
Climate	Amount, intensity and duration of rainfall Frequency and duration of freezing
Subsurface conditions	Seepage forces, piping Soil moisture levels, porewater pressures
Channel geometry	Width, depth and slope of channel Height and angle of bank Bend curvature
Biology	Type, density and root system of vegetation Animal burrows, trampling
Man-induced factors	Urbanization, land drainage, reservoir development, boating, bank protection structures

Table 4.6 Measured rates of bank erosion

River and location	Drainage area, km^2	Average rate of bank retreat, m y^{-1}	Period of measurement, years	Source
Axe, Devon	288	0.15–0.46	2	Hooke (1980)
Bollin-Dean, Cheshire	12–120	0–0.9	2	Knighton (1973)
Colville, Alaska	53 000	0.1–4.0	3	Walker *et al.* (1987)
Des Monies, Iowa	32 320–36 360	2.4–3.7	37	Odgaard (1987)
East Nishnabotna, Iowa	1129–2314	2.1–3.2	6–8	Odgaard (1987)
Exe, Devon	620	0.62–1.18	2	Hooke (1980)
Ilston, Wales	7–13	0.04–0.31	2	Lawler (1986)
Mississippi, Louisiana	$>10^6$	4.5	17	Stanley *et al.* (1966)
Watts Branch, Maryland	10	0.5–0.6	2	Wolman (1959)
Western Canada	Various	0.57–7.26	21–33	Nanson and Hickin (1986)
Wisloka, Poland	–	8–11	2	Klimek (1974)

Average rates of bank erosion have been shown to increase with catchment size acting as a surrogate for discharge (Hooke, 1980; Brice, 1984), but the highest rates may occur in the middle reaches of rivers where stream power is at a maximum (Lawler, 1992). Certainly Nanson and Hickin (1986) found that migration rate correlated better with stream power than it did with discharge. Pursuing this downstream theme, Lawler (1992) suggested that the dominant process will change from hydraulic action to mass failure in the lower reaches of rivers because of decreasing stream power and increasing cohesivity and height of banks. Local site characteristics also influence the spatial distribution of erosion, the most notable being bank material composition, the degree of flow asymmetry and channel geometry. Materials in which sands and small gravels dominate are more liable to erosion than those with a high silt–clay content. In composite banks, stability is governed by the strength of the weakest component since its removal will eventually lead to failure in the rest of the bank (Thorne and Tovey, 1981; Pizzuto, 1984a). The degree of flow asymmetry has a profound effect on near-bank velocity conditions (Knighton, 1984), considered to be of crucial importance to the occurrence of erosion (Knighton, 1973; Odgaard, 1987; Hasegawa, 1989). In the medium-sized basin of Brandywine Creek, Pennsylvania, Pizzuto and Meckelnburg (1989) observed that near-bank velocity explained more than 90 per cent of the variation in bank erosion rates. Not only do the distribution and magnitude of velocity or shear vary with discharge but they also vary between reaches of different curvature. The migration rate in bends tends to reach a maximum where the ratio of radius of curvature to channel width falls in the range of 2–3 (Nanson and Hickin, 1986). Thus the volume of flow and the degree of bank strength reduction may provide the conditions necessary for erosion but by themselves they are not sufficient.

Streambank erosion poses a variety of riparian management problems, notably in the United States, where the annual bill for erosion protection amounts to more than a billion dollars (Newson, 1986). It is a significant source of the sediment load carried by streams, supplying over 50 per cent of the total input in many instances. The rate and pattern of erosion have an important role in controlling the form and migration of river channels, especially along meandering reaches, and in determining the style of floodplain development. Consequently there remains a need for detailed research into the dynamics of bank erosion in order to achieve a better understanding of the fluvial sediment transport system, and to improve the predictability of the amount and location of erosion, a not inconsiderable problem given the large number of factors involved (Table 4.5).

SEDIMENT TRANSPORT

Figure 4.6 illustrates the main elements in the movement of material through the fluvial system. The load carried by natural streams can be separated into three components:

(a) the *dissolved load*, consisting of material transported in solution;
(b) the *wash load*, comprising particles finer than those usually found in the bed and moving readily in suspension (< 0.062 mm); and
(c) the *bed-material load*, including all sizes of material found in appreciable quantities in the bed (generally > 0.062 mm).

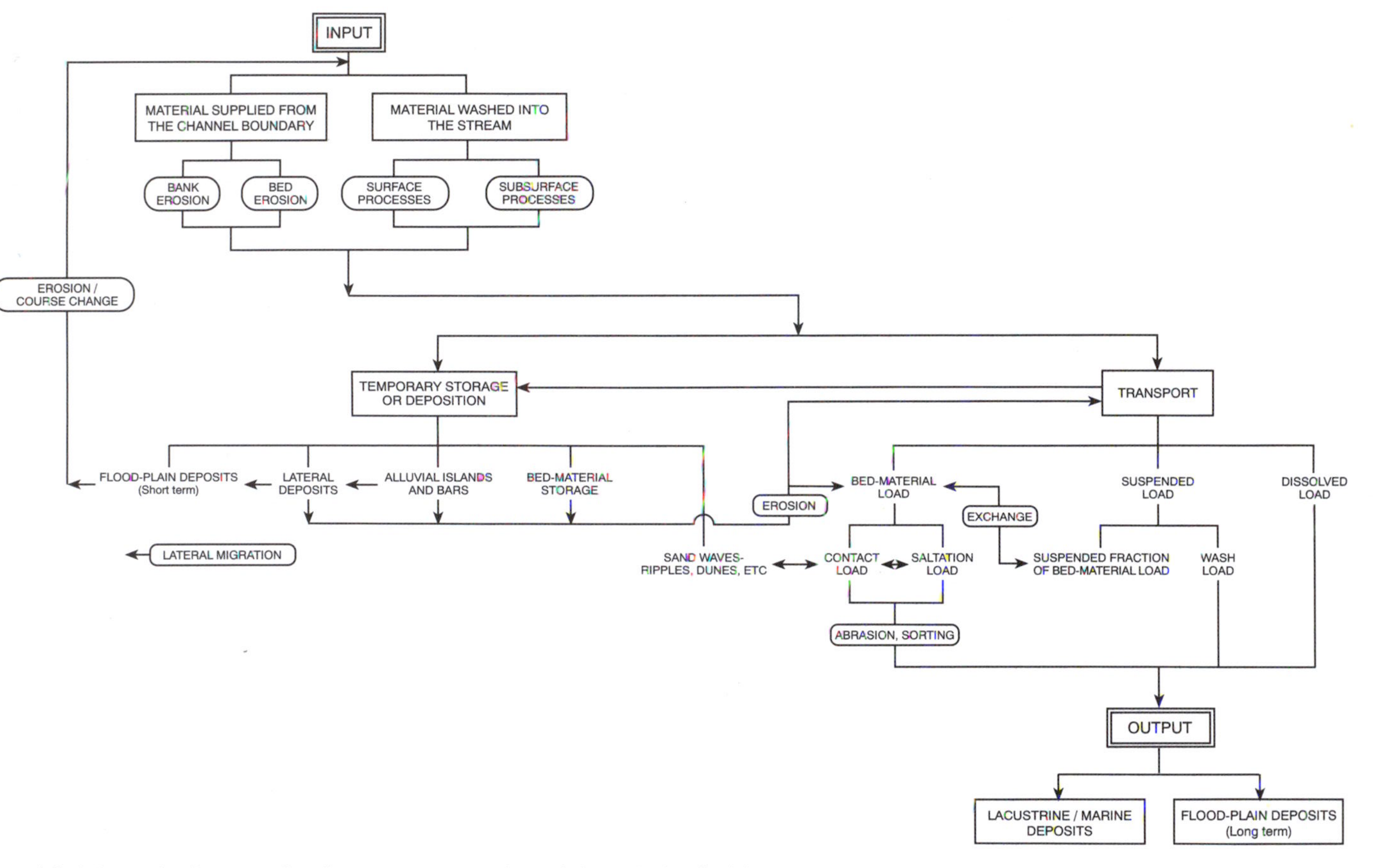

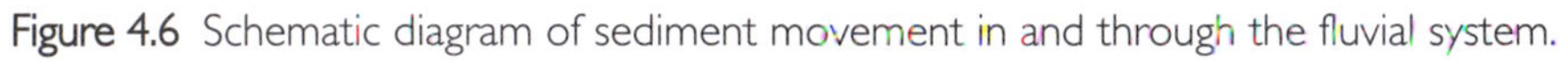

Figure 4.6 Schematic diagram of sediment movement in and through the fluvial system.

The bed-material load may be transported as *bed load*, when particles move by rolling, sliding or saltation at velocities less than those of the surrounding flow, or as *suspended load*, when particles are transported and temporarily maintained in the main body of the flow by turbulent mixing processes (Fig. 4.9B, p. 127). This distinction is somewhat arbitrary because there is an interchange of particles between the two modes of transport. As regards catchment denudation the dissolved and wash loads are the main components, but from a geomorphological standpoint the bed-material load is the principal concern because of its influence on the adjustment of river channel form.

A distinction is often made between *supply-limited* and *capacity-limited* transport. Much of the material supplied to streams is so fine that, provided it can be carried in suspension, almost any flow will transport it. Although an upper limit must exist in theory, the transport of this fine fraction is for the most part controlled by the rate of supply rather than the transport capacity of the flow. In contrast, the transport of coarser material (say >0.062 mm) is largely capacity-limited and therefore intermittent, becoming more so as grain size increases. The intermittency of bed-material transport and the possibility of prolonged deposition mean that the residence times of coarser sediments moving through even small drainage basins are likely to be very large.

The Dissolved Load

The dissolved load of a river varies in magnitude and composition according to the dominant sources, the rates of solute mobilization and the hydrological pathways operating within the catchment. Solute concentration (C) commonly declines with increasing discharge (Q) as a result of dilution to give a value for b in

$$C = aQ^b \tag{4.10}$$

which is less than 0. A global survey of 370 rivers revealed that 97 per cent had negative b values, the majority falling in the range 0 to –0.4 with an overall mean of –0.17 (Walling and Webb, 1986). At low flows solute concentrations are relatively high because runoff is supplied from the lower soil profile and groundwater reservoir where residence times are long, while at high flows more rapid pathways in the upper soil profile and at the surface are activated, providing less opportunity for solute uptake. However, since b is only slightly negative, the total dissolved load ($Q_{diss} = C \times Q$) continues to increase with discharge. Positive exponent values seem to arise only under special circumstances: for example, where atmospheric contributions are relatively large or where unusual geological or hydrological conditions cause near-surface runoff to be more highly mineralized than subsurface flow.

Equation (4.10) is a simple model of solute response to changing flow conditions, which can vary markedly in character for particular solutes and during individual storms. Water flowing by different routes through the soil at varying rates has differential access to soluble material. In a moorland catchment of northeast Scotland the products of chemical weathering (Ca^{2+}, Mg^{2+}, Na^{+}, SiO_2, HCO_3^-) are supplied in baseflow draining from the lower mineral soil horizons, but their concentrations are strongly diluted during storm events when runoff from the surface horizons is enriched with organic carbon and complexed Al, Fe and Mn (Reid *et al.*, 1981). Solute concentrations tend to be higher on the rising than the falling limb of a flood wave as a result of the mobilization of soluble material which has accumulated during the pre-storm period, producing a looped trend or clockwise hysteresis in the relationship between concentration and discharge (Walling and Webb, 1986). Consequently the dilution generally associated with greater water availability at the start of a storm may be delayed and even preceded by increased concentration. Such flushing effects are particularly pronounced in autumn storms after solute accumulation during the summer, giving a seasonal dimension to the variability. In contrast, solute stores may become so depleted during a period of multi-peaked flow, a characteristic of winter months, that an increasing concentration response in the first storm is replaced by a progressively enhanced dilution effect in later events. Additional complexity is provided during downstream transmission by the inputs from contrasting tributary sources and by the difference in velocity between the flood wave and the water that initiates it. The latter causes the clockwise hysteretic loop in the concentration–discharge relationship to open out gradually with distance downstream.

A large part of the dissolved load is carried by relatively frequent flows. In the Mississippi about 90 per cent is transported by flow events occurring monthly (Sedimentation Seminar, 1977), while in the smaller River Creedy in Devon, England, the most effective flows for solutional transport are at half-bankfull and below (Webb and Walling, 1982; Fig. 4.7). Dissolved load makes a significant contribution to the total load, implying that solution is an effective denudational agent, although measured values may include the effects of human activities. An annual amount of 3.7×10^9 tonnes represents about 20 per cent of the total load delivered by the world's rivers to the oceans (Walling and Webb, 1987a). However, the proportion of the total load carried in solution and the actual amounts vary considerably (Table 3.2, p. 82), depending on an array of climatic, geologic and other environmental factors which influence supply conditions at various spatial scales. Dissolved load is less dependent than sediment load on the quantity of flow and, although very important from an environmental standpoint, has little relevance to channel form adjustment.

The Wash Load

The wash load is the finest-grained fraction of the total sediment load, consisting of particles whose settling velocities are so low that they are transported in suspension at approximately the same speed as the flow and only

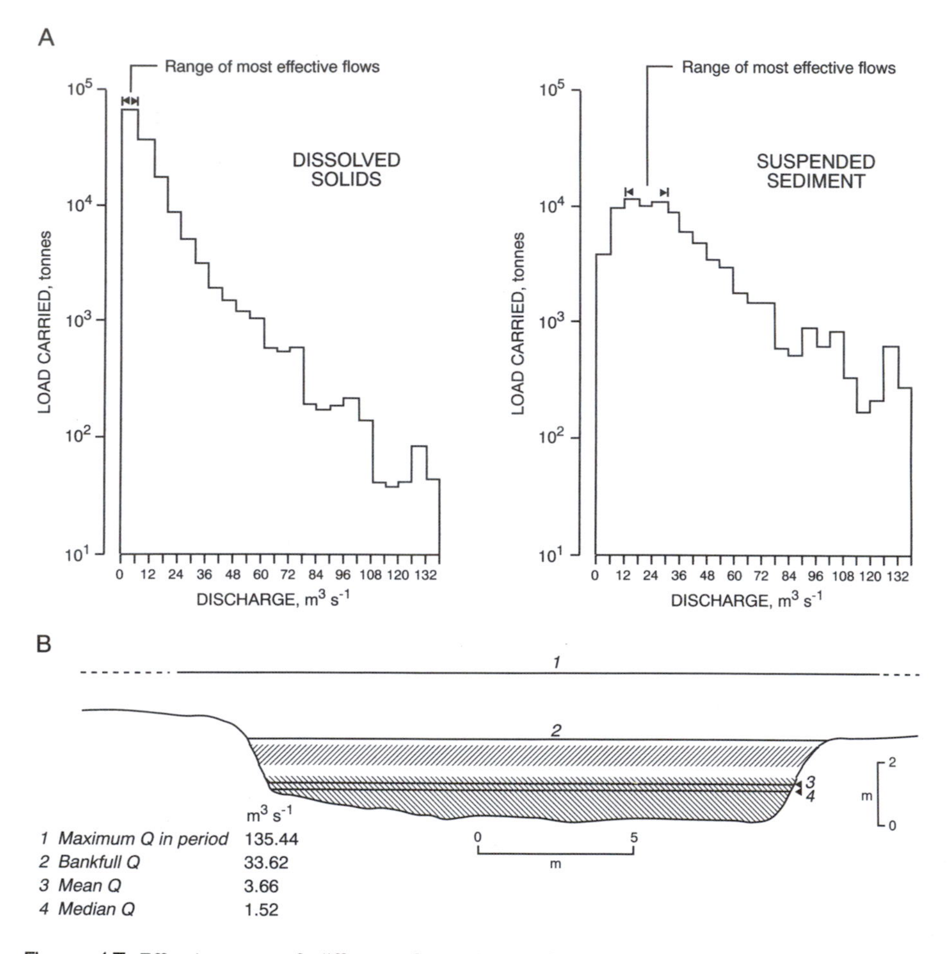

Figure 4.7 Effectiveness of different flow classes for transporting dissolved solids and suspended sediment in the River Creedy (A), and their relationship to stage levels at a natural channel cross-section (B) (after Webb and Walling, 1982).

settle out when flow velocities are much reduced. The fraction <0.062 mm accounts for more than 95 per cent of the suspended load carried by rivers in the Exe basin (Walling and Webb, 1987b), although the percentage may not always be this high (Walling, 1988). Vertical mixing is effected by turbulent eddies to give relatively little variation in concentration with depth, unlike the suspended fraction of the bed-material load. Concentrations can be very high if there are readily available sources in the drainage basin, as along the Huang Ho in China where much of the silty load is derived from highly erodible loess soils, and hyperconcentration levels in excess of 40 per cent by weight are reached during very large floods (Stoddart, 1978). However, only 24 per cent of the sediment that flows into the lower Huang Ho actually reaches the ocean, the remainder being deposited either on the alluvial plain (33 per cent) or in the delta region (43 per cent) (Milliman and

Meade, 1983; Fig. 4.8A). Ephemeral streams almost invariably carry high sediment loads in suspension.

The rate of wash load transport is determined principally by the rate of fine-grained sediment supply from the drainage basin rather than the transporting capacity of the flow. Most of the material is supplied from the erosion of cohesive river banks and from surface erosion in the catchment area

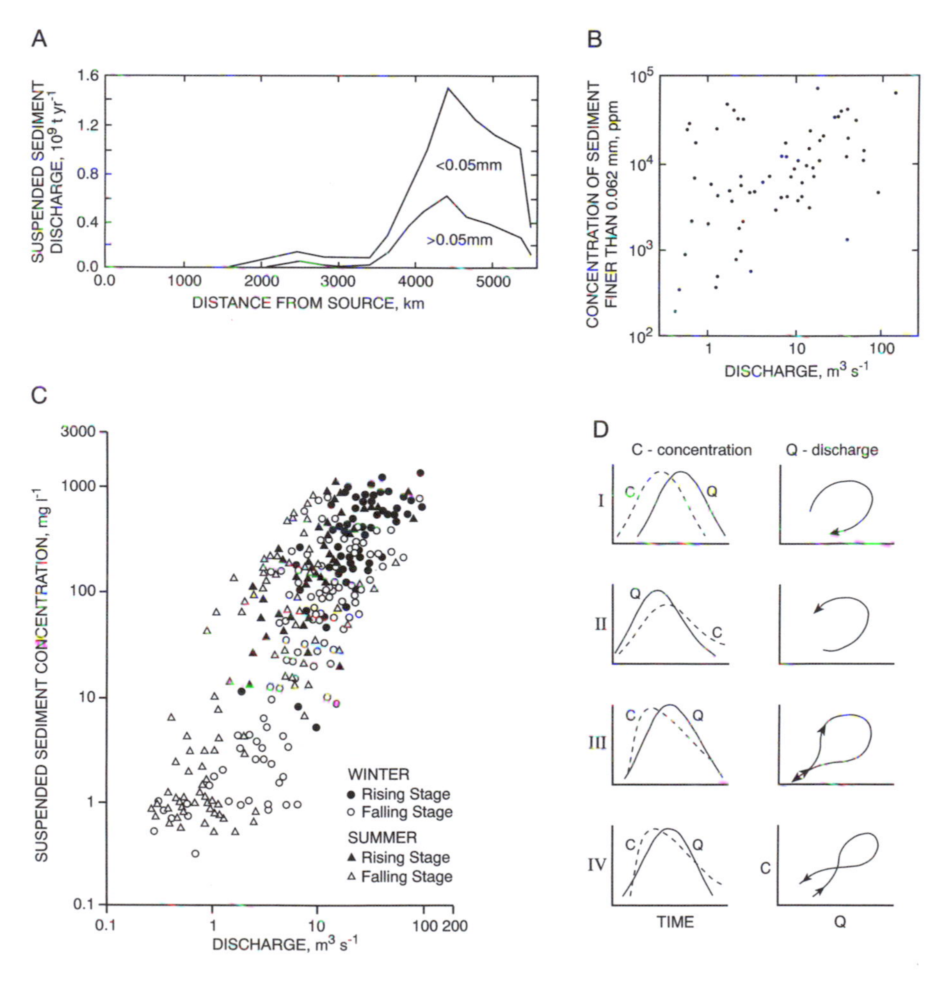

Figure 4.8 (A) Downstream variation of suspended sediment discharge along the Huang Ho River, China. Sediment discharge increases markedly as the river enters the loess region (about 3500 km from source), then decreases as the river crosses the alluvial plain (4500–5350 km) before entering the delta region (5350–5500 km) (after Milliman and Meade, 1983). (B) Lack of dependence of washload concentration on discharge, Powder River, Wyoming (after Colby, 1963). (C) Relationship of suspended-sediment concentration to discharge, differentiated with respect to season and stage, River Creedy, Devon (after Walling and Webb, 1987b). (D) Types of hysteretic loop in suspended-sediment concentration/discharge relationships when the two variables are not synchronous (after Williams, 1989).

by such processes as overland flow and gullying (see Chapter 3). While bank erosion is partly dependent on flow characteristics, the second source is independent of conditions within the stream. None the less, high suspended sediment concentrations are generally associated with periods of high discharge, and empirical concentration–discharge relationships having the same basic form as equation (4.10) have been established for many rivers, with the exponent *b* typically falling in the range 1–2 (Walling and Webb, 1992). However, such relationships should not be interpreted as simple transport functions because the amount of suspended sediment carried by a river is generally several orders of magnitude below its transport capacity. Rather, the dominant control on suspended sediment concentration is the rate of supply. Given the highly variable character of that supply, plots of concentration against discharge often show a very wide scatter of points (Fig. 4.8B and C).

Seasonal differences can contribute to the scatter, the higher concentrations during the summer period (April–September) in the River Creedy reflecting the greater availability of sediment at that time (Fig. 4.8C). Much of the scatter may be the result of hysteresis when the sediment wave is not synchronous with the water wave. The mechanisms behind hysteresis have not been adequately quantified, but Williams (1989) has attempted a qualitative classification of hysteretic relationships (Fig. 4.8D). The clockwise loop (Type I) where the sediment wave precedes the water wave to give higher concentrations on the rising than the falling limb at the same discharge is probably the most prevalent. It is symptomatic of the depletion of available sediment before the water discharge has peaked. Sediment yields from the catchment are often highest during the early part of a storm (or runoff season) when sediment is more available for transport, so that most of the sediment reaches the stream while the discharge is still rising. However, the form of the hysteresis may change with distance downstream. In small basins the sediment peak is much more likely to precede the discharge peak because sediment sources are nearby, but in large basins it may lag the discharge peak by a considerable time where upstream sources continue to supply the bulk of the load. This is because the flood wave moves 25–70 per cent faster than the water carrying most of the sediment. In such circumstances the clockwise loop (Type I) may give way to the anticlockwise loop (Type II). Williams (1989) could find no examples of Type III, while the commonness of the figure-of-eight loop (Type IV) is uncertain. Which streamflow-generating mechanisms are in operation, the nature and location of the sources, and the availability of sediment at those sources will influence in a largely unknown way how suspended sediment concentration will vary with discharge at a particular cross-section. Seasonal and hysteretic effects can explain part of the scatter on concentration–discharge plots, but improved estimation of suspended sediment load requires the development of supply-based models.

Improved estimation is needed as sediment contributions from such anthropogenic activities as deforestation, mining and construction work increase. Erosion within the catchment represents not only a loss of soil assets at source but also a potential liability further downstream. Reservoirs are very efficient sediment traps. Deposition therein can considerably reduce the storage capacity and life expectancy of a reservoir as well as adversely affecting the character of the downstream river. The closure of the Hoover Dam on the Colorado River reduced the downstream annual suspended sediment load from 125–150 million tonnes (pre-1930) to about 100 000 tonnes (Fig. 6.11A, p. 310), while the halving of sediment supply to the lower Mississippi through dam construction has contributed to rapid shoreline recession in the delta region (Meade and Parker, 1985). The silting-up of rivers can affect navigation and increase the risk of flooding in lower reaches. The silt concentration in the Huang Ho is so high that a reduction of 10–40 per cent could decrease peak discharge by about 20 per cent without any change in the actual volume of water (Stoddart, 1978). There also, levées originally built to contain the river have encouraged bed deposition, which has required the levées to be raised so that now the river is perched above the surrounding floodplain, presenting an even greater threat. The fate of toxic substances adsorbed onto sediment particles reinforces the need to understand the transport history of wash load.

Wash load transport has effects of more immediate relevance to stream behaviour and channel form adjustment. Very high concentrations (say >20%) damp down turbulence, increase the apparent viscosity of the flow and reduce settling velocities, enabling the transport of coarser grains and a larger bed-material load than would otherwise be the case (Simons *et al.*, 1963). The massive sediment input following the 1980 eruption of Mount St Helens has changed the character of the Toutle River from gravel/boulder to sand/gravel, and increased its sediment transport rate by two to three orders of magnitude for a given discharge (Bradley and McCutcheon, 1987). Where there is strong infiltration from the stream into the channel boundary, as in ephemeral streams, part of the wash load may be deposited as a caked layer which stabilizes the boundary and increases its resistance to erosion (Harrison and Clayton, 1970). Rivers with a large wash load may be morphologically different from those in which bed-material transport is dominant. Using the percentage of silt–clay in the channel boundary as an index (M) of the type of sediment load, Schumm (1960) argued that streams carrying predominantly wash load should have channels which are relatively narrow and deep. Large suspended loads also contribute to rapid changes in stream course and influence the style of floodplain development, particularly through overbank deposition.

Suspended sediment load is often the dominant component, contributing about 70 per cent of the total load delivered to the world's oceans each year. Together the Ganges/Brahmaputra and Huang Ho account for almost 20 per cent of that amount (Walling, 1987), emphasizing the high rates of mechanical denudation in Asian river basins. Suspended sediment transport is more erratic than dissolved load transport and is mostly associated with high-magnitude discharges, as evidenced by the fact that the value of b in the concentration–discharge relationship (equation (4.10)) is invariably positive for the former and negative for the latter. Consequently it occurs over a narrower range of higher flows. In the River Creedy, 90 per cent of the total suspended sediment load is transported in only 6 per cent of the time (22 days per year), while the equivalent figure for the dissolved load is 56 per cent of the time (204 days per year) (Webb and Walling, 1982; Fig. 4.7). Nevertheless, much of the sediment is carried by relatively frequent events of moderate magnitude as originally suggested by Wolman and Miller (1960), and the most effective discharge for suspended transport generally has a flow duration in the range of 0.3–5 per cent (1–20 days per year), although there can be quite wide variations (Ashmore and Day, 1988; Nash, 1994). In the lower Mississippi about 70 per cent of the sediment load is transported by flows with a duration in the range of 3–40 per cent (Biedenharn and Thorne, 1994), while the most effective flow in the Brahmaputra has a duration of 18 per cent (Thorne *et al.*, 1993). The effective discharge seems to become more frequent in larger basins (Ashmore and Day, 1988). Individual catastrophic events may carry larger amounts of sediment, but their contribution to suspended sediment transport over the entire range of discharges is less significant because they recur so infrequently.

The Bed-Material Load

Generally consisting of grains coarser than 0.062 mm, the channel bed is the principal source of material for bed-material transport. A distinction can be drawn between transport in sand-bed streams where grains tend to move in groups as migrating bed forms such as ripples, dunes and antidunes (Figs 4.3A and 4.9A), and transport in gravel-bed streams where grains move as individuals or in discontinuous sheets. Bimodal sediment may become separated into coarse and fine strips (Paola and Seal, 1995). As grain size increases into the boulder category (Table 4.3), movement as individual clasts becomes more distinct.

Unlike dissolved and wash load transport, the rate of bed-material transport is predominantly a function of the transporting capacity of the flow, although the development of an armour layer can restrict supply. Accurate field measurements are very difficult to make, errors being principally associated with the measuring devices themselves and with the extreme temporal variations in transport rate, which are a characteristic feature of bed material movement (Hubbell, 1987; Gomez, 1991). Measurement methods fall into two main categories: direct (pit traps, some form of sampling device placed on the bed) and indirect (tracers, repeated channel surveys). No one method is entirely satisfactory, but indirect channel surveys, provided they are detailed enough at the reach scale, can produce reliable results, and have the advantages of minimum disturbance to the flow and time-integrated sampling which averages out short-term fluctuations in the transport rate (Lane *et al.*, 1995).

Many variables influence the process of bed-material transport (Table 4.7). The basic problems are two-fold: to understand the dynamics of bed-

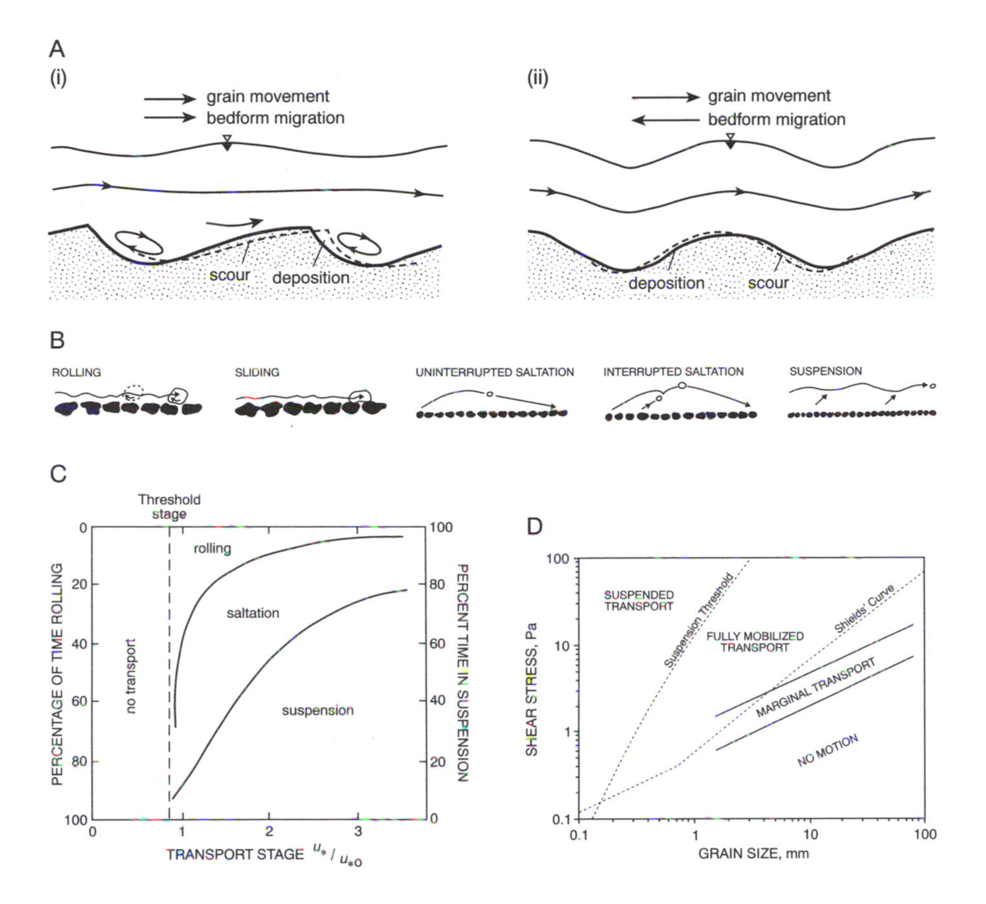

Figure 4.9 (A) Schematic diagram of the pattern of erosion and deposition over a dune (i) and antidune (ii) bed. (B) Modes of transport of bed-material load. (C) Percentage of time that single particles in a water stream experience rolling, saltation and suspension as a function of transport stage. Percentage of time in saltation is represented by the distance between the curves (after Abbott and Francis, 1977). (D) Domains of sediment transport in flume experiments, showing shear stress thresholds of incipient motion and fully mobilized transport (solid lines) as a function of grain size. The region of marginal transport lies between these thresholds (after Wilcock and McArdell, 1993).

material movement; and to establish a reliable relationship between sediment transport rate and relevant properties of the flow, fluid and sediment, thereby enabling the prediction of transporting capacity for given conditions.

As regards the **dynamics of movement**, particles roll, slide or saltate along the bed in a shallow zone only a few grain diameters thick once the entrainment threshold has been exceeded ($\tau_0 > \tau_{cr}$). Material transported in this way constitutes the *bed load*. Rolling is the primary mode of transport in gravel-bed streams, while saltation in which grains hop over the bed in a series of low trajectories is largely restricted to sands and small gravels. Less

Table 4.7 Variables pertinent to bed-material transport

Flow properties	Fluid properties	Sediment properties	Other properties
Discharge (Q)	Kinematic viscosity (ν)	Density (ρ_s)	Gravity (g)
Velocity (v)	Density (ρ)	Size (D)	Planform geometry
Flow depth (d)	Temperature (T)	Sorting (σ)	
Width (w)	Wash load concentration (C)	Fall velocity (v_s)	
Slope (s)			
Resistance (ff)			

massive particles may be carried upwards into the main body of the flow to be transported temporarily in suspension as *suspended load*, possibly once a second threshold has been reached. Movement in suspension is maintained against gravity by turbulent eddies of variable strength and direction so that particles do not follow predictable paths. Given these several modes of transport (Fig. 4.9B), the problem arises of identifying a criterion or set of criteria with which to distinguish them.

On the basis of flume experiments in which the movement of single grains (5–10 mm) was traced photographically, Abbott and Francis (1977) used transport stage (u_*/u_{*0}) in this role, where u_* ($= \sqrt{\tau_0/\rho}$) is shear velocity and u_{*0} is the threshold shear velocity at which motion begins. The development of suspension from saltation (regarded as an intermediate step) occurred much less rapidly than that of saltation from rolling (Fig. 4.9C). The time spent in the rolling mode decreased very rapidly from about 60 per cent close to the threshold stage ($u_*/u_{*0} \sim 1$) to 20 per cent at $u_*/u_{*0} \sim 1.4$. Saltation was a persistent mode of transport even above $u_*/u_{*0} = 1.9$ when suspension became dominant. Figure 4.9C suggests that all modes of bed-material transport can occur simultaneously over a wide range of flow conditions, even though one may predominate at any given stage.

Andrews and Smith (1992) defined different phases of transport in gravel-bed streams in terms of dimensionless shear stress ($\tau_* = \tau/g(\rho_s - \rho)D$). Phase I or marginal transport begins when the fluid forces are just sufficient to rotate gravel-sized particles out of the pockets in which they lie, the particles rolling or bouncing over their downstream neighbours until they settle into new resting places. As the dimensional shear stress increases, so does the number of particles in motion. Such partial movement occurs over the range $0.020 < \tau_* < 0.060$ (compare with the Shields curve, Fig. 4.5A). During Phase II transport at higher stresses ($\tau_* > 0.060$) most of the bed is mobile and significant saltation occurs. In many gravel-bed rivers, a substantial portion of the bed material will be carried under marginal transport conditions. From studies of 24 such rivers in Colorado, Andrews (1984) found that the dimensionless shear stress exceeded a value of 0.030 on an average of 12 days per year but rarely exceeded a value of 0.070 (about once in 100 years). The results of flume experiments confirm the existence of various domains of bed-material transport (Fig. 4.9D). Where a well-developed armour layer moderates the supply of sediment to the transport process, Phase I is characterized by the passage of the finer fraction over a basically stable coarser bed (Carling, 1987). Substantial bed mobility (Phase II transport) requires that the armour is breached, when significant changes in channel morphology can take place. According to the equal-mobility hypothesis, the particle size distributions of the bed load and the bed material should be approximately the same at all stages of flow. However, coarser fractions tend to be underrepresented during Phase I transport, notably over armoured beds, and the bed load becomes progressively coarser with increasing flow, approaching the bed material in size distribution characteristics only during Phase II transport when the bed is more fully mobilized (Komar and Shih, 1992; Lisle, 1995).

The onset of suspension transport, involving principally the sand fraction, has been linked to the turbulent properties of flow close to the bed of a stream, which are themselves related to the topography of the bed. Superimposed on the main downstream flow, turbulent eddies give rise to unsteady flow events which move fluid parcels towards ('sweeps') and away from ('ejections') the bed. From detailed field measurements 1 m above a dune field in the Fraser River, Lapointe (1992) showed that ejection events of a few seconds' duration had a dominant effect on suspended sediment concentrations, accounting for as much as 90 per cent of the vertical sediment mixing. Much remains uncertain about these burst-like motions and their overall significance to suspension transport. Once suspended, the fate of a particle depends on the balance between its fall velocity and the vertical component of flow associated with turbulent eddying. This interaction usually results in a vertical distribution of suspended sediment in which both the concentration and the average grain size decrease rapidly away from the bed.

Whether grains move in groups or as individuals, bed-load transport is characteristically intermittent. The transport velocities of entrained particles are so low relative to flow velocities (say 2–15 per cent) that travel distances are very short, particularly during an event where marginal transport conditions prevail. The motion of single particles is typically stochastic, consisting of a series of steps with intervening rest periods. In two gravel-bed Pennine streams, average particle velocities (including rest periods) varied from 2 cm h^{-1} at intermediate flows to 26 cm h^{-1} at bankfull discharge (Carling, 1987). On the basis of data from a range of fluvial environments, it has been shown that the mean distance of travel is weakly correlated with excess stream power, and that for relatively large particles in mixed beds it decreases rapidly with increasing grain size (Hassan and Church, 1992; Fig. 4.10). However, the travel distance of relatively small particles is largely insensitive to grain size because they tend to be trapped in the interstices and because bed surface structures inhibit their movement (Church and Hassan, 1992). Such effects become less prominent during Phase II transport when scour mobilizes more of the surface material. It is important to identify which factors have a controlling influence on the distance of particle transport, for the average movement characteristics of individual clasts ultimately determine the sediment transport rate.

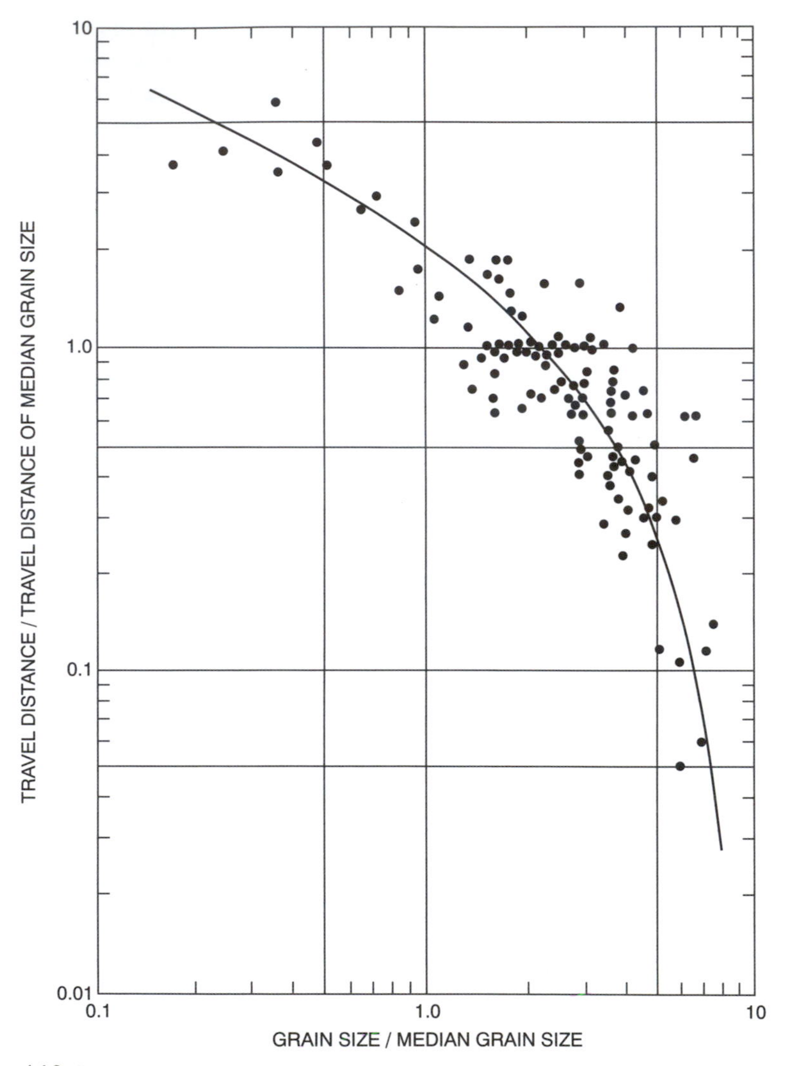

Figure 4.10 Relationship between scaled travel distance and scaled grain size (after Hassan and Church, 1992).

The **sediment transport rate** can be estimated from one of the many transport equations, which seek to express the maximum amount of sediment (capacity) that can be carried for given flow, fluid and sediment conditions. Many *bed-load formulae* can be classified according to whether they relate the sediment transport rate per unit width (q_{sb}) to either excess shear stress ($\tau_0 - \tau_{cr}$), excess discharge per unit width ($q - q_{cr}$) or excess stream power per unit width ($\omega - \omega_{cr}$):

$$q_{sb} = X' \tau_0 (\tau_0 - \tau_{cr}) \quad \text{Du Boys type} \quad (4.11)$$

$$q_{sb} = X'' s^k (q - q_{cr}) \quad \text{Schoklitsch type} \quad (4.12)$$

$$q_{sb} \cong (\omega - \omega_{cr})^{3/2} d^{-2/3} D^{-1/2} \quad \text{Bagnold (1980)} \quad (4.13)$$

where X' and X'' are sediment coefficients, d is flow depth, and D is grain size. All have the problem of determining the threshold flow (denoted by the subscript cr) at which incipient motion begins. The Einstein (1950) equation represents a somewhat different approach in that explicit recognition is given to the stochastic nature of bed particle movement and to the different rates at which individual size fractions in the load may be transported. Many of these equations rely on the empirical determination of coefficients and were developed for particular conditions, so that their overall applicability may be limited. When most material moves close to the bed, as at low transport rates, the total bed-material load can be estimated from a bed-load equation. At other times the total load can be obtained in one of two ways: indirectly, by addition of the bed and suspended components estimated separately; or directly, from a total load formula. Some procedures differentiate between wash load and the suspended fraction of the bed-material load, while others do not.

Gomez and Church (1989) evaluated the performance of ten bed-load formulae and concluded that none is capable of providing a generally satisfactory prediction of the bed-load transport rate. Stream power equations of the Bagnold type (equation (4.13)) are preferred but they can give poor results (Carson, 1987). Given the large number of factors involved (Table 4.7) and the lack of general agreement as to which should be included in predictive equations, it is perhaps not surprising that a widely accepted theory has not emerged. The underlying assumption that the maximum amount of bed-load is being transported can be challenged, particularly in gravel-bed streams where an armour layer has developed.

The *armour layer* is usually no more than one grain diameter thick and overlies a finer substrate to which it gives protection. Opinions differ as to the formative process, but downstream winnowing, which involves the selective removal of fine particles from the surface framework, has often been emphasized (Richards and Clifford, 1991). Armouring has also been regarded as a reflection of equal grain mobility (Parker *et al.*, 1982), or the result of 'traction clogging' in which coarse grains overpassing finer material are deposited and encourage the further deposition of rolling or sliding grains (Dunkerley, 1990). The significance of the armour layer is that it limits the supply of material so that the sediment transport rate, especially during marginal transport conditions, is lower than that predicted by capacity bed-load formulae. How ubiquitous or temporally stable is this coarse surface layer remains uncertain, but it does not seem to develop under ephemeral flood regimes (Laronne *et al.*, 1994).

The inadequacy of bed-load formulae is a consequence not only of fluctuations in sediment supply but also of the heterogeneity of most channel beds, especially gravel ones, and the inherent variability of bed-load transport rates. The prospect of transport rate being some unique function of hydraulic and related parameters may be unrealistic. *Spatial variations* in transport rate occur at a variety of scales. Bimodal sediment beds organized into coarse and fine strips may show differential transport, the finer strips transporting a larger amount and the coarser ones a smaller amount than a

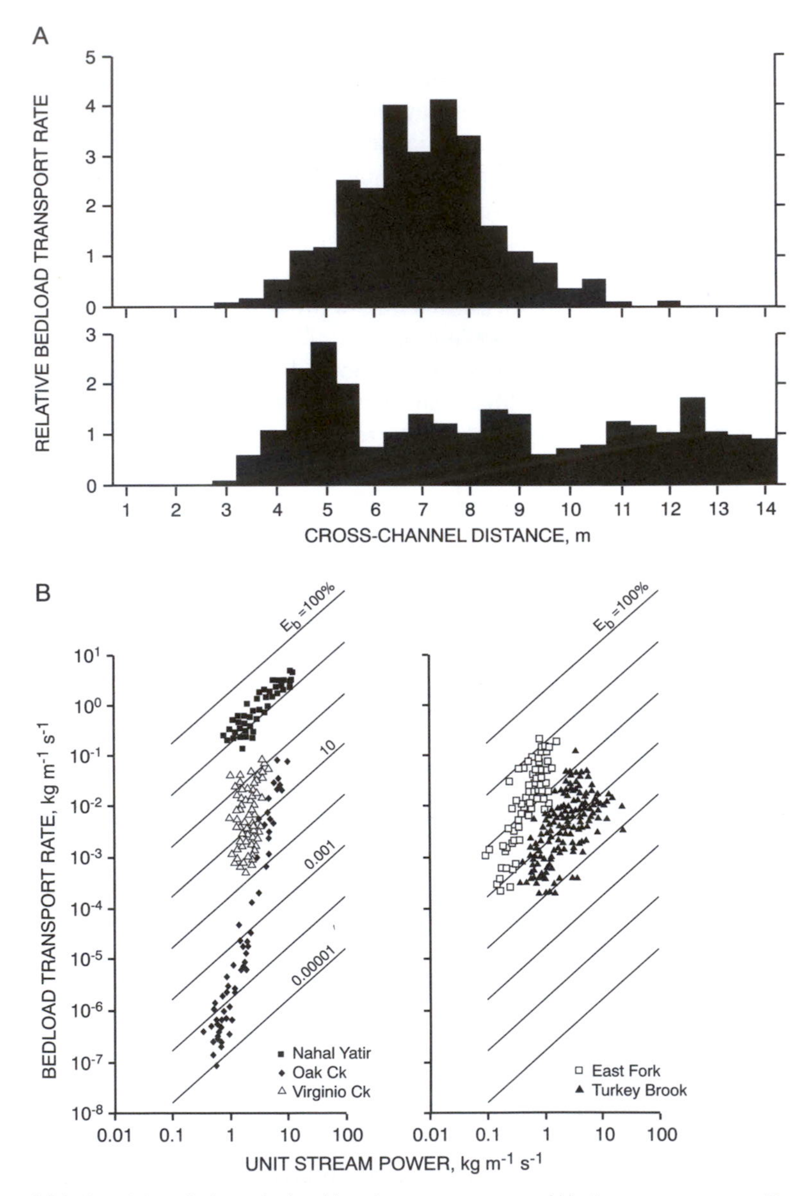

Figure 4.11 Spatial variations in bed-load transport rate. (A) Across stream – East Fork River, Wyoming (after Gomez, 1991). (B) Between streams – contrasts between the arid Nahal Yatir (Israel) and the perennial East Fork River (Wyoming), Oak Creek (Oregon), Turkey Brook (England) and Virginio Creek (Italy). E_b is Bagnold's percentage transport efficiency index (after Reid and Laronne, 1995).

bed without strips (Paola and Seal, 1995). The width of the active bed at a cross-section varies with river stage, but even at constant discharge the rate of transport is rarely uniform across a channel (Fig. 4.11A). This is particularly the case in meandering channels where the zone of maximum transport tends to follow the line of maximum bed shear stress through the bend, with a small but significant fraction of the load being transferred across rather than down the channel (Dietrich and Smith, 1984). Inter-stream comparisons reveal considerable disparities in both the efficiency of bed-load transport and the rate at which it responds to changes in flow conditions (Reid and Laronne, 1995; Fig. 4.11B). In Figure 4.11B the largest contrast exists between the very well-armoured Oak Creek and the ephemeral, unarmoured Nahal Yatir, where transport efficiency is respectively stage-dependent and stage-independent. In the latter, flash floods dominate the transport process and transport efficiency is much higher, which may be a general characteristic of arid-zone streams.

Temporal variations in the form of 'bed-load pulses' are now recognized as a common feature of bed-material transport (Gomez *et al.*, 1989; Hoey, 1992), under both quasi-steady (Fig. 4.12A) and variable (Fig. 4.12B) flow conditions. They have been linked with a variety of mechanisms (Table 4.8), but the migration of coherent bed forms or groups of particles is probably the most prevalent cause of variability. As dunes pass a given point, maximum amounts of transport are associated with the passage of dune peaks and smaller amounts with that of intervening troughs (Leopold and Emmett, 1976). Low-amplitude 'bed-load sheets' formed from the migration of heterogeneous bed load can also produce pulses (Whiting *et al.*, 1988). Reid *et al.* (1985) have speculated that such instability can occur in the absence of coherent bed forms as a result of the group movement of coarse particles in the form of a kinematic wave wherein the rate of transport is non-linearly related to particle concentration, reaching a maximum at intermediate concentrations. Over a longer timescale, large bars in gravel-bed streams may take a season or more to move through a cross-section. Indeed, Hoey (1992) identified a hierarchy of sedimentary features ranging from individual grains to megaforms that exceed several channel widths in length as being responsible for the generation of bed-load pulses in gravel-bed streams, each class having its characteristic wavelength and relaxation time. Such fluctuations have implications for the sampling strategies that need to be employed if reliable time-averaged estimates of bed-load discharge are to be obtained.

The process of *scour and fill*, which leads to short-term changes in stream-bed elevation, can accentuate instability in bed-load transport. During a flood it has commonly been observed, particularly in sand-bed streams, that the channel bed is scoured at high discharges on the rising stage and filled to approximately the pre-flood level on the falling stage (Leopold *et al.*, 1964). However, the spatial pattern of scour and fill is more variable than this statement suggests. Along a short reach (430 m) of the East Fork River, Andrews (1979) differentiated between two types of section:

(a) pool-like sections which scoured at discharges (Q) in excess of bankfull (Q_b) and filled at $Q < Q_b$;
(b) riffle-like sections with the opposite behaviour, filling at $Q > Q_b$ and scouring at $Q < Q_b$.

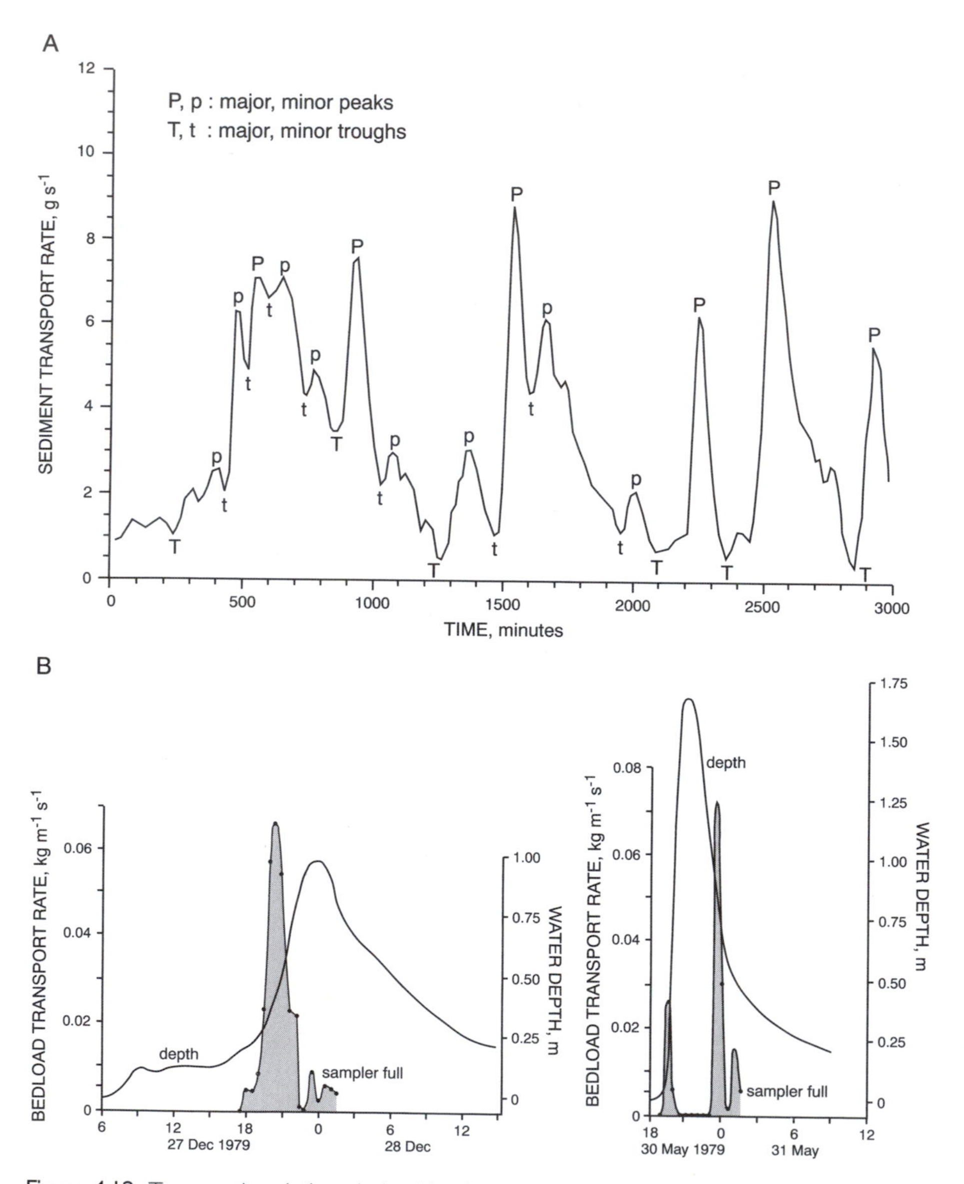

Figure 4.12 Temporal variations in bed-load transport rate. (A) During quasi-steady laboratory experiments (after Hoey, 1992). (B) During floods, Turkey Brook (after Reid *et al.*, 1985).

Thus the bed-load discharge alternated downreach between relatively high and relatively low as the sequence of sections respectively lost (scour) or accumulated (fill) material, the particular pattern depending on whether the flow was above or below bankfull. In effect, pool-like sections act as storage sites for finer material during low flows.

Table 4.8 Mechanisms responsible for generating temporal variations in bed-load transport rates that are independent of fluctuations in water discharge (after Gomez *et al.*, 1989)

Timescale	Probable cause of temporal variability
Long/intermediate-term	Wave-like translation of bed material through a reach Intraseasonal exhaustion
Short-term	Intra-event exhaustion Scour and fill Movement of bed forms (e.g. dunes, bars, sheets, particle clusters) Armouring
Instantaneous	Mechanics of particle entrainment and motion

Bed-load pulses of larger scale and longer duration have been associated with discrete sediment inputs which are translated downstream as waves (Nicholas *et al.*, 1995). Along rivers in British Columbia, Church and Jones (1982) identified alternating sequences of sedimentation and transportation zones. 'Sedimentation zones' characterized by wide, unstable, braided channels are regarded as large-scale sediment stores related in part to the downstream progression of large volumes of material introduced upstream by the late nineteenth-century erosion of moraines, most material moving quickly from one sedimentation zone to the next. Large inputs from tributary sources may produce a similar effect. The construction of a tributary fan constricts the main valley and induces upstream aggradation until such time that the river can break through and reactivate the stored sediment, which then migrates downstream in wave-like form. Sediment continuity can also be upset by human activities, notably mining. Gilbert (1917) likened the long-term movement of debris introduced from Californian gold mines to the passage of a wave, although the broadly symmetric wave envisaged by Gilbert (Fig. 6.14, p. 323) may be more complicated in basins with protracted storage and multiple input points (James, 1989; Knighton, 1989). Whatever the precise form of the wave, the introduction of large quantities of mining waste invariably results in channel bed adjustments, initial aggradation being followed by degradation once the sediment peak has passed, with accompanying modifications to other aspects of channel form. These and similar changes underline the fact that a close relationship exists between channel form and the input–output conditions of sediment load within a reach. The lack of a widely applicable bed-load transport theory is a drawback to modelling that relationship.

Of the three components of load, the bed-material load contributes least to the total but it is by far the most significant from the standpoint of channel form adjustment. An average contribution of about 10 per cent is often quoted, but there are too few data to be definite and contributions of over

80 per cent are not unknown, notably in steep mountain streams (Walling and Webb, 1987a). The order in which the three components of load have been considered – dissolved, wash, bed-material – can be thought of as a progression: of increasingly slower transport velocities, so that the load peak lags further and further behind the flow peak during any event; of increasingly discontinuous transport, involving more intermittent and short-lived movement; of increasing dependence on flow conditions and sources within the channel, so that capacity-limited transport assumes greater dominance over supply-limited transport. In addition, less frequent flows carry a higher proportion of the load, although the difference between suspended sediment and bed-material transport in this respect need not be large (Table 5.2, p. 166). For 17 gravel-bed streams in the western United States, Andrews and Nankervis (1995) estimated that the middle 80 per cent of the annual bed-material load is transported on average during 15.6 days per year (~ 4 per cent of the time), representing approximately 0.8–1.6 times bankfull discharge (Q_b). However, the frequency of the most effective flow is likely to decrease as the size of the bed material increases, an effect noted by Carling (1988) in neighbouring Pennine streams where median bed material size (D_{50}) differed by a factor of 2.5. The most effective flow was $0.94 \times Q_b$ where $D_{50} = 20$ mm, and $1.30 \times Q_b$ where $D_{50} = 50$ mm. Indeed, in boulder-bed streams only extreme floods may be able to move the larger clasts, which then act as constraints on channel form during subsequent flow conditions which are incompetent to move them.

The size and shape characteristics of bed material at any point along a river are determined by the initial supply conditions and the subsequent transport history of the constituent grains. **Downstream changes in bed material characteristics** represent an aspect of the transport process which has immediate geomorphological relevance because of the association with channel form, notably slope. It has generally been observed that particle size (D) decreases with distance downstream (L) and that the relationship has an exponential form (Fig. 4.13A):

$$D = D_0 \, e^{-\alpha L} \tag{4.14}$$

where D_0 is the initial grain size at $L = 0$. This functional form follows from the assumption that the decrease in particle size per unit distance is proportional to particle size:

$$\frac{dD}{dL} = -\alpha D \tag{4.15}$$

The rate of change (α) is highly variable but tends to become less where the bed material is finer (Table 4.9). Consequently particle size decreases rapidly in headwater areas where the initial material is usually coarse, and at a much slower rate further downstream, particularly if the bed there is composed of sand fractions. Average grain size decreased from 330 mm to 44 mm in the first 42 km of the Knik River in Alaska (Bradley *et al.*, 1972), while only a slight reduction was observed along 3300 km of the lower Amazon (Nordin *et al.*, 1980). The simple model represented by equation (4.14) does

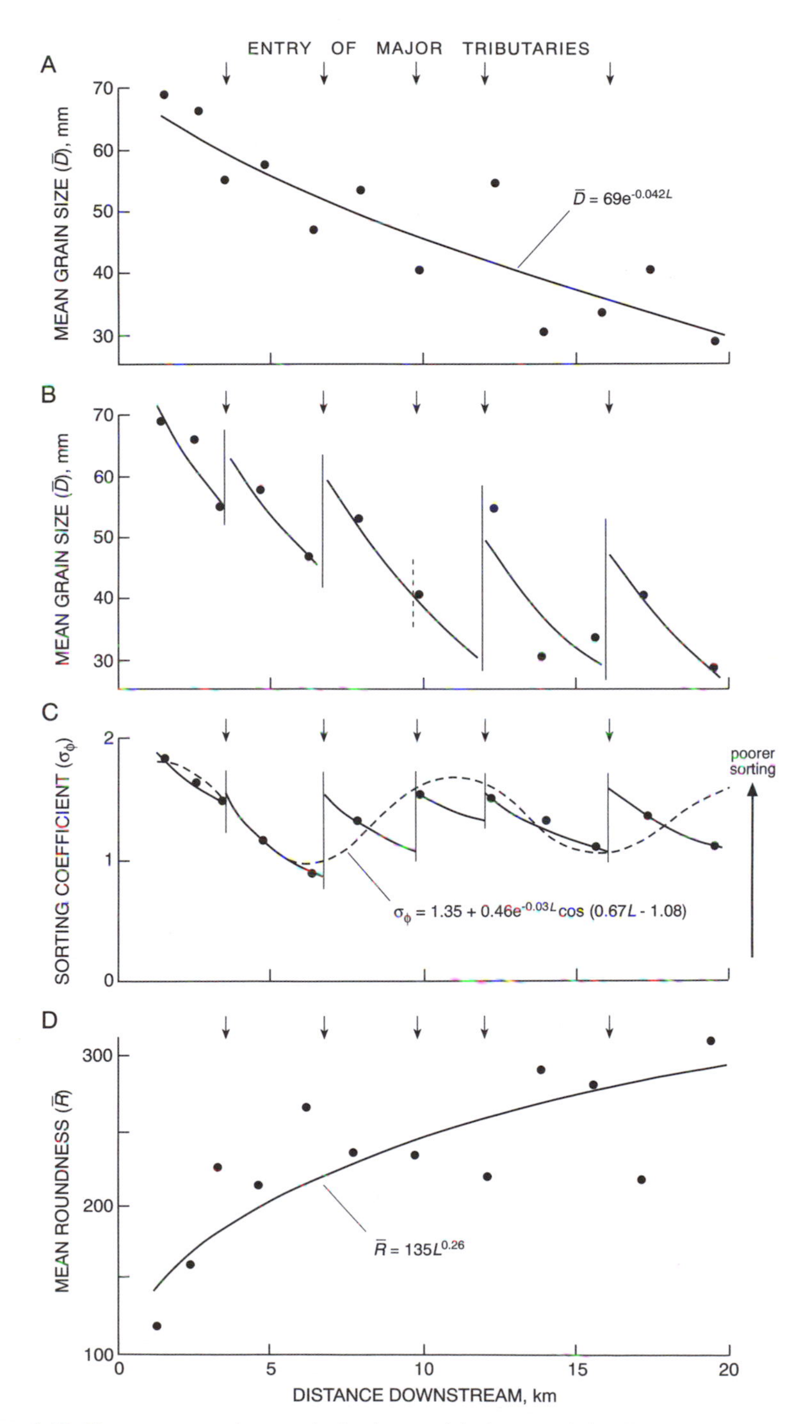

Figure 4.13 Downstream changes in bed material characteristics, River Noe, Derbyshire, England. (A) Exponential decrease of mean grain size. (B) Discontinuous exponential decrease of mean grain size to show the effect of tributary inflow. (C) The effect of tributary inflow on sediment sorting represented as discontinuous exponential change (solid lines) or pseudo-periodic variation (dashed line). (D) Increase in mean particle roundness.

Table 4.9 Downstream change in bed material size

Source	Location/ river	River distance, km	Parameter	Bed material range, mm	Type	α
Yatsu (1955)	Japan: Kinu	0–52	D_{50}	20–70	G	–0.0253
		60–100		0.4–0.9	S	–0.0238
	Kiso	0–15		35–70	G	–0.0348
		15–55		0.4–0.6	S	–0.0104
	Nagara	0–13		25–40	G	–0.0446
		13–49		0.7–1.2	S	–0.0173
Bradley *et al.* (1972)	Knik River, Alaska	0–26	$\bar{D}$	44–330	B/G	–0.081
Rana *et al.* (1973)	Mississippi, Vicksburg	440	D_{85}	0.3–0.9	S	–0.0010
			D_{50}	0.2–0.55	S	–0.00055
			D_{15}	0.18–0.33	S	–0.00045
Knighton (1980b)	Bollin-Dean, Cheshire	0–50	$\bar{D}$	0.33–67	G/S	–0.118
	Noe, Derbyshire	0–20		29–69	G	–0.042
Nordin *et al.* (1980)	Amazon, from Iquitos	3300	D_{50}	0.15–0.50	S	slightly negative
Brierley and Hickin (1985)	Squamish, British Columbia	40	D_{84}	53–610	B/G	–0.043
			D_{50}	37–300	B/G	–0.039
			D_{16}	18–120	B/G	–0.032
Kodama (1994)	Watarase, Japan	0–20	D_{50}	4–65	G	–0.089

Symbols: Bed material size defined as the mean ($\bar{D}$), median (D_{50}) or that size (D_x) at which *x* per cent is finer.
B, G, S refer to boulder-bed, gravel-bed and sand-bed streams respectively; α is the rate of change of bed material size in equation (4.14).

example, where there is a rapid transition from gravel to sand, or where sediment supply and movement are highly irregular as in low-order headwater streams (Rice and Church, 1996).

Three sets of processes contribute to downstream fining: abrasion, hydraulic sorting and weathering. Weathering by both chemical and physical means can cause substantial particle disintegration if material is stored for long periods in exposed sites, but its overall contribution, although uncertain, is probably small relative to that of the other two. Abrasion is a summary term covering such mechanical actions as grinding, breakage, impact and rubbing, which chip and fracture particles not only during transport but also in place when the combined effect of lift and drag causes particles to vibrate in pockets (Schumm and Stevens, 1973) or when static particles are sandblasted by an overpassing bed load (Brewer *et al.*, 1992). The multifaceted nature of the abrasion process makes it very difficult to model, either physically or mathematically, although Parker (1991) has developed a tentative theoretical model which deals with just one aspect of the process: binary collisions between saltating bed load and immobile bed particles. Particle lithology influences both the size of material supplied initially and the subsequent rate of wear, so that different lithologies may be expected to decrease at different rates and vary in their relative contributions to downstream samples (Werritty, 1992; Kodama, 1994).

Hydraulic sorting operates through selective entrainment (Fig. 4.5), differential transport (Fig. 4.10) and selective deposition. Although largely size-related, sorting is also affected by particle shape (Hattingh and Illenberger, 1995). The hypothesis that all sizes in a mixture have near-equal mobility (Parker *et al.*, 1982; Andrews, 1983) seems contrary to the argument that sorting is an effective mechanism of downstream fining. However, rapid downstream fining can be produced by selective transport alone in the presence of strong profile concavity (Hoey and Ferguson, 1994) and, even if equal mobility is assumed, by the preferential removal of finer material from patches formed locally through selective deposition (Paola and Seal, 1995). Local sorting is well known, both transversely in meander bends and streamwise along gravel bars (Bluck, 1987). Indeed, Bluck has suggested that sorting at the bar scale contributes substantially to more general downstream trends, an hypothesis for which there is some quantitative support (Clifford *et al.*, 1993). Certainly sorting at a large scale must reflect in some way the cumulative effect of many sortings at successive local scales.

Historically, abrasion was regarded as the dominant process responsible for downstream fining but, largely as a result of abrasion tank experiments (e.g. Kuenen, 1956), it was realized that observed rates were much larger than could be attributed to abrasion. Bradley *et al.* (1972) concluded that about 90 per cent of the downstream size reduction in Knik River gravels is caused by sorting, with the balance attributable to abrasion. However, the status of abrasion as an effective process of particle wear has been re-established via the 'abrasion in place' mechanism (Schumm and Stevens, 1973) and via recent tank experiments which replicate natural conditions more closely and produce diminution coefficients (α in equation (4.14)) one to two orders of magnitude higher than in previous experiments (Kodama, 1994). Kodama shows how abrasion depends strongly on the presence of different-size particles, hinting at the importance of mixed beds to the operation of the process. It may be that abrasion is most effective in upper reaches where slopes are steeper and particle size ranges larger, although Ferguson and Ashworth (1991) attribute the rapid downstream fining along a steep headwater stream almost entirely to sorting induced by an unusually concave longitudinal profile. The relative importance of abrasion and sorting is partly conditioned by the lithology and size of bed-material grains, and may be related to the type of vertical activity in the river, for abrasion has been associated with degradation and sorting with aggradation (Shaw and

Kellerhals, 1982). The exponent α in equation (4.14) contains the undifferentiated effects of both abrasion and sorting. If the factors which control the relative contributions of these processes to downstream fining are to be better understood, refined models of the kind proposed by Parker (1991) are needed.

Material supplied to the main stream from various sources – colluvial action, the channel boundary, tributary inputs – can disrupt downstream trends, particularly in upper reaches where there is strong slope–stream coupling and non-alluvial supplies are dominant (Rice, 1994; Rice and Church, 1996). Tributaries generally import coarser material so that both grain size and the sorting coefficient tend to increase below junctions (Fig. 4.13B and C). Where a sequence of tributaries enters a main stream, grain size and sorting may vary discontinuously in such a way that an exponential decrease below each junction is followed by a stepped increase at the next (Troutman, 1980), the magnitude of that increase being possibly related to the relative sizes of main stream and tributary at each confluence (Knighton, 1980b). Such behaviour is dependent on drainage network structure, which argues for a network-based approach to downstream fining (Rice, 1994; Pizzuto, 1995). In a small (A_d = 34.5 km^2) Pennsylvanian basin, Pizzuto attributed 80 per cent of the downstream decrease in mean grain size to spatial variations in supply, with reductions in size during transport only accounting for about 20 per cent.

In contrast to gradual downstream fining, abrupt gravel–sand transitions can occur under certain circumstances. Three causative mechanisms have been identified by Sambrook Smith and Ferguson (1995): local base-level control, lateral input of fine sediment, and abrasion/particle disintegration. Where a local base-level control such as a lake or debris fan induces a marked decrease in slope and selective deposition, only the sand fraction may remain mobile and any surviving gravel is swiftly inundated. Major inputs of sand are probably a less common cause, requiring more specialized conditions such as accelerated erosion somewhere in the basin or excess supply of mining waste (Fig. 6.15C, p. 327). The third mechanism involves the rapid breakdown by disintegration or crushing of grains in the 1–4 mm category, which may operate effectively only in high energy conditions. Abrupt transitions are often associated with a sharp break in channel slope and seemingly represent a threshold between different types of river (Howard, 1980, 1987), but their frequency of occurrence remains to be established.

Properties other than size will change during downstream transport. Particles tend to become rounder as a result of abrasion (Fig. 4.13D; Mills, 1979). Bed material is generally better sorted with distance downstream, but tributary inflow in particular can disrupt that tendency (Fig. 4.13C; Knighton, 1980b; Ichim and Radoane, 1990). The main point is that the bed material which has such an important influence on resistance and transport dynamics is not unchanging, either spatially or temporally. Observed at any one time, downstream changes in bed-material characteristics reflect a wide range of influences in the long-continued operation of the sediment transport system. Those changes are closely associated with channel slope, and may in fact be forced by profile concavity (Ferguson and Ashworth, 1991). Numerical modelling suggests that the timescale for sorting processes to produce downstream fining is much shorter than the timescale for bed slope adjustment (Deigaard, 1982), lending support to the notion that surface grain size should be regarded as a dependent variable in models of river channel adjustment (Hoey and Ferguson, 1994). Basically particles are reduced in size by abrasion and weathering, and assigned their position along a stream by sorting processes.

Sediment Deposition

The final element of the process triumvirate, deposition, has received comparatively little attention from geomorphologists and yet alluvial rivers build a wide range of depositional forms (Table 4.10). Deposition begins once the flow or shear velocity falls below the settling velocity of a particle, which for a given particle size is less than that required for entrainment, particularly within the finer fractions (Fig. 4.5B). Settling velocity is closely related to particle size, so that the coarsest fraction in motion should be deposited first, with progressively finer grains settling out as the flow velocity continues to fall. The net effect is a vertical and horizontal (both downstream and transverse) gradation in terms of sediment size.

Bed material can be moulded into coherent structures which may be broadly classified as 'hydraulic' (microscale and mesoscale), in that development is linked to local flow conditions over the bed, or 'sediment-storage' (macroscale and megascale) forms (Table 4.10; Church and Jones, 1982). All affect flow resistance and influence the dynamics of bed-load transport at a variety of scales. Thus, for example, particle clusters in gravel-bed streams (Fig. 4.3B) delay entrainment, inhibit bed-load movement, particularly during Phase I transport, and exert a form drag additional to that produced by the grains themselves. Naden and Brayshaw (1987) modelled the development of these hydraulic bed forms using a queuing model in which sediment movement is envisaged as a grain-by-grain process and the probability of deposition (i.e. joining the cluster queue) depends on grain size and the local shear velocity. Channel bars having wavelengths of the same order as the channel width or greater represent major storage places for the traction load that is moved only sporadically by high flows. Such features are more a reflection of sediment supply conditions and macroscale channel processes than local fluid dynamics. They both interact with and influence the mean flow pattern through a reach. Given their persistence, they can play an important role in determining the morphology of alluvial channels and have particular associations with channel patterns via such forms as alternate bars in straight channels, point bars on the inner banks of meanders, and braid bars in braided channels. Mid-channel bars formed initially from the lag deposition of coarse material may, in an actively meandering river, eventually become attached to and incorporated into the floodplain,

Table 4.10 Characteristic river deposits (after Task Committee, 1971; Church and Jones, 1982; Hoey, 1992)

Place of deposition	Name	Characteristics
Channel	Transitory channel deposits	Largely bed load temporarily at rest; coherent structures include micro-forms ($\lambda \sim 10^{-2}$–10^{0} m) such as ripples, and mesoforms ($\lambda \sim 10^{0}$–10^{2} m) such as dunes in sand-bed streams, pebble clusters and transverse ribs in gravel-bed streams, and boulder steps
	Alluvial bars	Formed initially from the lag deposition of coarser sediment; coherent structures include macro forms ($\lambda \sim 10^{1}$–10^{3} m) such as riffles and mid-channel bars, and megaforms ($\lambda > 10^{3}$ m) such as sedimentation zones
	Channel fills	Accumulations in abandoned or aggrading channel reaches
Channel margin	Lateral deposits	Point bars on the convex bank of meanders and marginal bars which may form an alternating sequence along straight reaches; may be preserved through channel shifting
Floodplain	Vertical accretion deposits	Usually fine-grained material deposited from the suspended load of overbank floodwaters; includes natural levée and backswamp deposits
	Splays	Local accumulations of predominantly sandy material, formed when water escapes from channels on to adjacent floodplains through breaks (crevasses) in natural levées
Piedmont	Alluvial fans	Formed by ephemeral or perennial streams emerging from steeply dissected terrain on to a lowland; sediments rapidly decrease in grain size with distance from the fan apex; several fans may coalesce to form an alluvial plain (bajada)
River mouth	Deltas	Formed where a stream deposits its load upon entering the sea or a lake; morphology depends on characteristics of sediment supply, and interplay between fluvial and marine processes

Symbol: λ is a characteristic wavelength.

so that the main sedimentary structures are preserved (Knighton, 1972; Hooke, 1985). In ancient deposits, bed forms are a key element in palaeohydraulic reconstruction, as demonstrated by Baker's (1973) interpretation of gravels deposited by catastrophic floods following the failure of Pleistocene lakes in the northwestern United States. The reliability of such reconstructions depends on an adequate understanding of the formative processes operating within present-day streams.

The most ubiquitous depositional feature is the **floodplain**, formed largely from a combination of within-channel and overbank deposits. Its evolution is conditioned by the rate of sediment supply, in terms of both volume and calibre, the availability of suitable sites for the accumulation of sedimentary units, and the energy environment of the river, which determines the amounts and sizes of material that can be transported. A change in one or more of these conditions may alter the dominant mode of floodplain construction. Thus, an increase in fine sediment supply for climatic and anthropogenic reasons caused a switch from within-channel sedimentation to overbank deposition as the dominant process in low-energy river systems of the English Midlands during the mid- to late Holocene (*c.* 4500–2500 BP) (Brown and Keough, 1992). The introduction of conservation measures to control soil erosion produced a change in the opposite direction from the 1930s onwards in Maryland floodplains, where lateral accretion of sand and gravel became the main process (Jacobson and Coleman, 1986).

Within-channel deposition principally takes place in the form of channel bars, although coarse lag deposits left behind on the channel bed can also make a significant contribution. As a river shifts its course laterally, erosion of one bank is approximately compensated by deposition against the other, often in the form of a point bar. With continuing migration, the bar is built streamward and increases in height as sediment is carried on to its surface by inundating flows (Fig. 4.14). Those flows become less frequent during construction so that progressively finer sediment is deposited on the bar as it grows surfaceward, to give a vertical gradation of sizes from coarsest to finest. Similar gradations are produced in other types of bar which, although originally in more central positions, may eventually become attached to one bank and incorporated into the floodplain (Fig. 4.14). Lateral migration of the braid-train is the main process of floodplain formation in the braided Waimakariri River (Plate 4.4), where the largest floodplains are predominantly composed of gravel bars capped with vertically accreted fines, the growth of vegetation aiding the accretion process by trapping and binding sediment (Reinfelds and Nanson, 1993). The more mature floodplain surfaces are relatively flat as a result of the infilling of abandoned channels, a common aspect of the later stages of floodplain construction in multichannel systems and one which involves overbank deposition. For floodplains to be built entirely of within-channel deposits, the channel must occupy all positions on the valley floor, the potential for which depends on the level of lateral and downvalley shifting. Within meandering channels the migration rate has been shown to depend largely

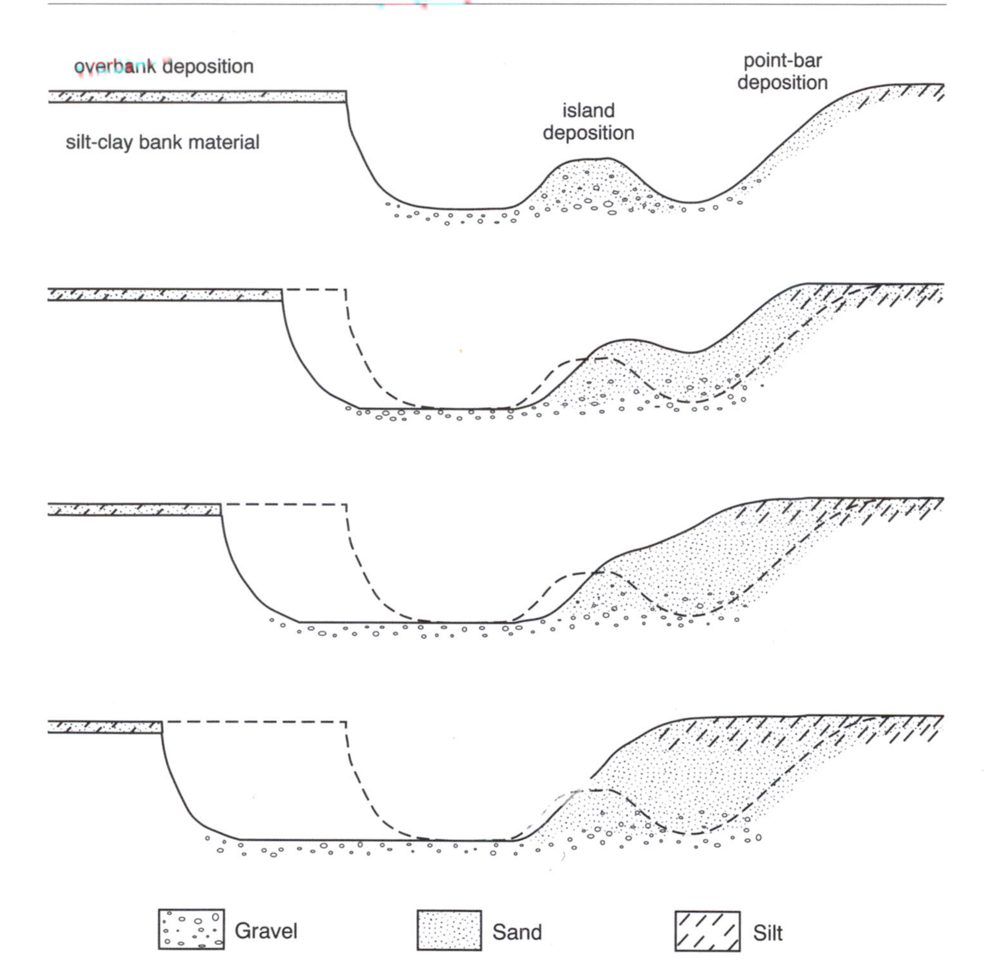

Figure 4.14 Diagrammatic representation of the progressive construction of a floodplain as a stream migrates laterally, based partly on observations along the River Bollin-Dean. As the stream erodes the left bank, a point bar is gradually constructed against the right bank and eventually incorporates alluvial island deposits. Deposits show both vertical and horizontal gradation of sediment size.

on stream power and bank material composition (Nanson and Hickin, 1986). The laterally active Waimakariri River is capable of reworking almost its entire valley floor in approximately 250 years (Reinfelds and Nanson, 1993), but a period of about 1300 years is required for the River Exe in Devon to traverse its 1.2 km wide floodplain (Hooke, 1980).

Flooding is a natural attribute of rivers. Although bankfull discharge has a variable recurrence interval, it is usually a relatively frequent event, about once or twice a year in Britain, suggesting that vertical accretion could be a major process of floodplain construction. During floods there is a strong interaction between the deep, fast flow of the channel and the shallow, slow

Plate 4.4 The braided Waimakariri River in the Southern Alps of New Zealand.

flow of the floodplain, with a transfer of momentum from one to the other. Consequently velocity and shear decrease in the former and increase in the latter. The momentum transfer is accompanied by a flux of suspended sediment from the channel to the floodplain, consisting of wash load (silts and clays) and the finer fractions of the bed-material load (particularly fine and medium sands). Since the transporting ability of the overbank flow tends to decrease away from the channel margins, both the amount and calibre of deposited sediment should similarly vary (Pizzuto, 1987), although the distribution and grain-size characteristics of overbank deposits depend not only on relative channel position and the composition of the suspended load but also on the topography of the floodplain (Walling *et al.*, 1992) and the nature of the vegetation cover. The range of depositional environments includes abandoned channel infills (a transitional type between within-channel and overbank forms); channel marginal deposits, notably levées and crevasse splays; and floodplain/backswamp deposits. In addition, the floodplain margins can be covered in colluvial deposits where hillslope erosion is active. The rate of overbank deposition depends on the frequency of inundating flows and their sediment load. Long-term rates reported by Bridge and Leeder (1979) range from 0.2 to 10 mm y^{-1} but, not unexpectedly, rates can be very variable even within short reaches of a river. Rates have varied between 0 and 10 mm y^{-1} over the last 35 years along the lower Severn (Walling *et al.*, 1992), and have reached almost 6000 mm y^{-1} along parts of the central Amazon (Mertes, 1994). When major floods have a low sediment content, extensive erosion rather than deposition may occur (Burkham, 1972).

Wolman and Leopold (1957) concluded that lateral accretion and within-channel deposition are the dominant processes of floodplain formation, accounting for up to 90 per cent of the deposits. If overbank deposition was the main process, then the channel would appear to become depressed within its own alluvium and vertical growth of the floodplain would eventually take it beyond the reach of all but the most infrequent events. However, the relative contributions of lateral and vertical accretion are not constant and overbank deposition becomes more significant where flooding occurs more frequently, fine-grained material is more readily available, and channels have greater lateral stability (Brackenridge, 1988). Not only may the timescales of the two processes be different, but the reworking of floodplain alluvium during channel migration tends to destroy the stratigraphic evidence for vertical accretion, leading to possible misinterpretation as to the original mode of formation. In the valley of the River Delaware, where the river shifts its course very slowly, floodplain sediments accumulated over 6000 years contain no point-bar deposits and are largely the result of overbank deposition (Ritter *et al.*, 1973). Along coastal rivers in New South Wales, vertical accretion gradually builds a floodplain over hundreds or thousands of years, following which extreme floods strip the floodplain to a basal lag deposit before it is re-formed by overbank deposition (Nanson, 1986).

Nanson and Croke (1992) identified three major types of floodplain, defined according to specific stream power ($\omega = \Omega/w$) at bankfull flow and sediment texture, in which the relative importance of lateral and vertical accretion varies (Fig. 4.15).

(a) High-energy ($\omega > 300$ W m^{-2}) non-cohesive floodplains where confinement by bedrock or boulders inhibits lateral migration, and the vertical accumulation of relatively coarse sands and gravels is the dominant process.
(b) Medium-energy ($\omega \in 10$–300 W m^{-2}) non-cohesive floodplains, typical of meandering and braided rivers, in which the main mechanism is lateral point-bar or braid-channel accretion.
(c) Low-energy ($\omega < 10$ W m^{-2}) cohesive floodplains, usually associated with laterally stable single-thread or anastomosing channels, whose development is governed largely by vertical accretion of fine-grained deposits and infrequent channel avulsions. Abrupt avulsions are an important mechanism in the evolution of larger floodplains whose cohesive bank material severely constrains lateral migration.

Explicit within this classification is the close association between the style of floodplain development and the type of channel pattern (notably, meandering, braided or anastomosing).

Floodplains provide one of the most important storage spaces for sediment as it moves discontinuously through the drainage basin. They can preserve an extended record of the changing climatic and anthropogenic conditions which have affected valley floor environments over long periods of time. Much of the soil eroded from the Piedmont uplands since European

HIGH ENERGY

Confined coarse-textured floodplain
ω >1000 W m^{-2}

Confined vertical-accretion sandy floodplain
ω = 300 – 1000 W m^{-2}

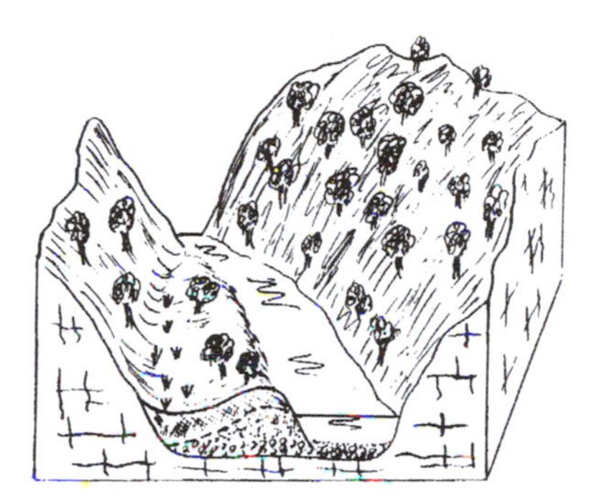

MEDIUM ENERGY

Braided river floodplain
ω = 50 – 300 W m^{-2}

Lateral migration, scrolled floodplain
ω = 10 – 60 W m^{-2}

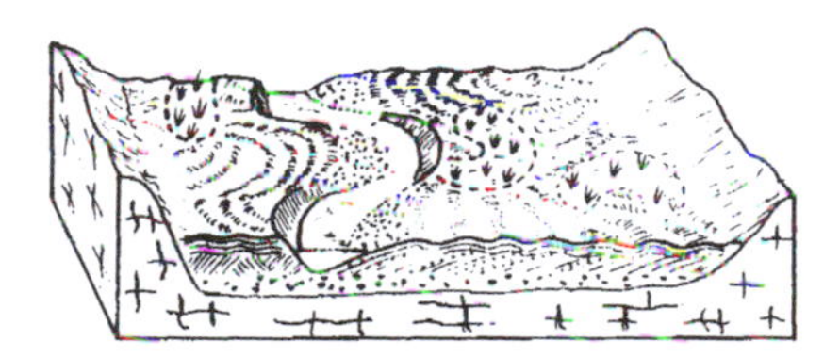

LOW ENERGY

Anastomosing river, inorganic floodplain
ω < 10 W m^{-2}

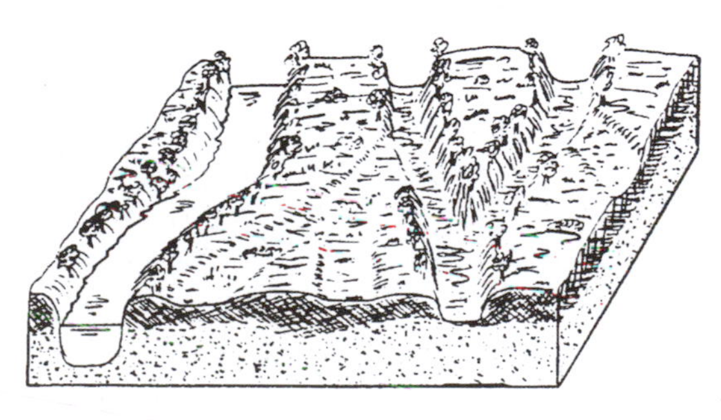

Figure 4.15 Examples of high-, medium- and low-energy floodplains (after Nanson and Croke, 1992).

farming began in the eighteenth century is still stored on hillslopes and in floodplains (Trimble, 1975). Sediment-associated pollutants that travel with the fine fraction of the suspended load can accumulate within floodplain deposits, leading not only to soil pollution on site but also to further river contamination if the deposits are remobilized by migrating rivers. Marron (1992) describes how arsenic-contaminated mine tailings introduced into a tributary stream have become incorporated into the floodplain of the main stream through overbank and point-bar deposition. Thus the floodplain can contain evidence about the environmental health of a catchment. It is also important to river-basin managers concerned with reducing the risk due to flooding, for many densely populated areas are concentrated on floodplains. In the United States 10 per cent of the population is exposed to flood hazards, and the annual bill for flood damage is well over a billion dollars. The floodplain is an integral part of the fluvial system, whose deposits influence channel form through the composition of the channel boundary and therefore the type of material supplied to a stream.

Alluvial fans are also major sediment stores, formed where a stream rapidly loses its transporting ability because of either an abrupt reduction in slope, which decreases stream power, or a sudden change from confined to unconfined status, which leads to flow divergence (Plate 4.5). Three main conditions are required for optimal fan development (Blair and McPherson, 1994):

(a) a topographic setting where a stream becomes unconfined as it emerges from an upland drainage basin onto a relatively flat lowland;
(b) the production of sufficient sediment for fan construction;
(c) a climatic environment capable of generating extreme stream discharges and mass wasting events, thereby giving high rates of sediment delivery from upland basin to fan site.

Plate 4.5 A series of alluvial fans along a mountain front in the tectonically active Southern Alps of New Zealand.

These conditions are best satisfied by dry-region mountains where block faulting maintains a steep mountain front, as in the Basin and Range province of the western United States or the East African rift valley. However, alluvial fans are not restricted to dry regions or mountain fronts, one location with an especially fluvial flavour being the high-angled junction of a steep tributary valley with a gentler main stream.

Alluvial fans generally have a conical surface form, with slopes radiating away from an apex at the point where the stream emerges from the upland basin (Plate 4.5). Not uncommonly a series of fans develops along a mountain front, which may eventually coalesce to form an alluvial apron or 'bajada'. As regards the evolution of these constructional landforms, a distinction can be made between primary and secondary processes (Blair and McPherson, 1994). The primary ones are responsible for supplying and transporting the bulk of the material involved in fan construction, and include both mass movement and fluvial processes. The main transport mechanisms range from debris flows to water-driven transport in the form of either unconfined sheetfloods or confined channels, although channels are usually wide, shallow and braided (Bull, 1977). Secondary processes, which are dominantly fluvial, remobilize and modify previously deposited sediment, so they are not involved in fan construction to any great extent but have a major influence on the development of surface features. These two sets of processes have different magnitude–frequency characteristics. The primary ones are infrequent, of short duration but of very large magnitude, while the secondary ones are associated with non-catastrophic conditions and operate between primary depositional episodes. Deposits tend to be coarse and poorly sorted because of the involvement of mass wasting processes and the short transport distance from source to fan, but facies characteristics do vary depending on which transport mechanism is dominant. Debris flows tend to dominate fans supplied from small, steep basins, while fluvial processes are relatively more active in large, less steep basins (Harvey, 1997).

The action of these processes and the related depositional environment are strongly influenced by the ambient lithologic, climatic and tectonic conditions. Bedrock lithology affects the size and quantity of material supplied initially. The different reactions of the various lithologies to physical and chemical weathering give rise to different styles of erosion and transport. Thus, cohesive debris flows make a more significant contribution where fans are supplied from clay-producing lithologies such as shales and mudstones (Harvey, 1990). Climatic factors have both direct and indirect effects, via the process regime and the vegetation cover respectively. The prevalence of alluvial fans in dry regions is a consequence of, firstly, the high rates of coarse sediment supply associated with the intense storm events which dominate the transport regime in dry regions, and secondly, the sparsity of the vegetation cover, which favours the rapid generation of surface runoff. The most favourable conditions for long-term fan growth exist in tectonically active areas. Active tectonism maintains relief, controls the overall context of fan development, and indirectly affects the rate of sediment supply (e.g. Plate 4.5 – high rates of uplift increase bedrock susceptibility to mechanical breakdown). These factors together with others, notably topography, control the rate and style of fan development, creating a range of fan morphologies.

The most thoroughly investigated aspect of fan morphology is the relationship between fan area (F) and drainage area (A_d), widely assumed to have a power functional form (Bull, 1977):

$$F = cA_d^n \quad (4.16)$$

Since the upland basin is the main source of sediment for fan growth, it is not unexpected that the relationship is positive, with the exponent n varying between 0.7 and 1.1 (Harvey, 1990). Larger fans are supplied from larger basins. The intercept c is more erratic, with values ranging from 0.1 to 2.1, and attempts have been made to relate that variability to the influence of environmental factors such as rock erodibility, precipitation regime and tectonic activity. It tends to be higher for fans supplied from areas of more easily eroded rock, and may therefore be used as an index of relative erodibility (Hooke and Rohrer, 1977), although results are not always consistent (Blair and McPherson, 1994).

Alluvial fans are clearly delimited systems, so the interaction between form and process should be more readily apparent than is the case with many other landforms. They can pose serious problems for settlement in arid environments because of the unpredictability of the flash floods which are the principal supply mechanisms. Their occurrence, evolution and morphology are controlled by the relationship between sediment production and sediment transport capacity, fans developing where sediment production is high relative to transport capacity. This is the case, above all, in arid mountain regions (Harvey, 1997).

ADJUSTMENT OF CHANNEL FORM

The interaction of the mobile fluid flow with the erodible solid boundary produces a wide variety of channel forms. In an attempt to impose some sort of rational order on that diversity, various classifications of river channels have been proposed. One of the earliest schemes classified rivers as youthful, mature or old according to their stage of development within the cycle of erosion (Davis, 1899), but time is not a readily applicable or useful basis for classification.

Both the magnitude of discharge and the character of the flow regime influence channel form (Harvey, 1969; Stevens *et al.*, 1975). Discharge has not proved to be an effective criterion for classification except where size is a primary consideration. Church's (1992) subdivision into small, intermediate and large channels is based only partly on discharge, the dimensions of the channel relative to the size of material and channel gradient being additional criteria. Characteristic river regimes have been identified (Beckinsale, 1969) but they have not been successfully related to channel form. Nevertheless, distinctions can be drawn between arid-zone and humid-zone rivers in terms of both their hydrology and geomorphology (see Fig. 3.6, p. 79), the former being associated with greater transport efficiency (Reid and Laronne, 1995; Fig. 4.11B, p. 132) and larger channel widths, at least in small to moderately sized basins (Wolman and Gerson, 1978). Gupta (1995) regards rivers in the seasonal tropics as falling between the arid and humid temperate cases, a channel-in-channel morphology being characteristic.

Schumm (1963a) subdivided channels into three types on the basis of the dominant mode of sediment transport, using the percentage silt–clay in the channel boundary (M) as his criterion:

(a) bed-load channels ($M \leq 5$);
(b) mixed-load channels ($5 < M < 20$);
(c) suspended-load (or wash-load) channels ($M \geq 20$).

The classification is based on data from a restricted range of sand-bed streams, which may limit its overall applicability, and rests on the assumption that M adequately reflects transport mode. A further dimension to Schumm's scheme is provided by a categorization in terms of channel stability (stable, eroding, depositing), reflecting the balance or otherwise between sediment supply and transportability, but natural channels cannot easily be assigned to one of these three categories.

From a practical standpoint, a more satisfactory material-related classification is based on boundary composition, a factor which influences sediment transport characteristics, resistance properties and therefore the adjustability of channel form (see Fig. 1.1, p. 2). An initial subdivision into *cohesive* and *non-cohesive* channels can be followed by further categorization (Table 5.1). The distinctions are not new, but Howard (1980, 1987) has argued that abrupt thresholds exist between bedrock, gravel-bed and sand-bed channels in particular, each type having a distinct hydraulic geometry. Bimodal-bed channels where significant quantities of both gravel and sand are present is a possible intermediate category (Sambrook Smith, 1996). Most natural channels are type B, with B1 and B2 dominant (see Table 5.1). Frequently the bed and banks of a channel are composed of different material, a common contrast being between cohesive banks and a non-cohesive bed.

Boundary composition is one of the criteria used by Rosgen (1994) in one of the most elaborate classification schemes yet devised, yielding 41 channel types (Fig. 5.1). The other criteria include an entrenchment ratio (flood-prone width/bankfull channel width), which expresses the degree of lateral confinement; sinuosity, which describes the degree of meandering; width : depth ratio, which summarizes channel shape; and channel slope.

Table 5.1 Channel classification based on boundary composition

Primary type	Secondary type	Characteristics
A. Cohesive	A1. Bedrock channels	No coherent cover of unconsolidated material; generally short segments, concentrated in steep headwater reaches
	A2. Silt–clay channels	Boundaries have a high silt–clay content, giving varying degrees of cohesion
B. Non-cohesive	B1. Sand-bed channels	'Live-bed' channels composed largely of sandy material which is transported over a wide range of discharges
	B2. Gravel-bed channels	Channels of coarse gravel or cobbles which are transported only at higher discharges
	B3. Boulder-bed channels	Composed of very large particles (>256 mm) which are moved infrequently; grades into A1)

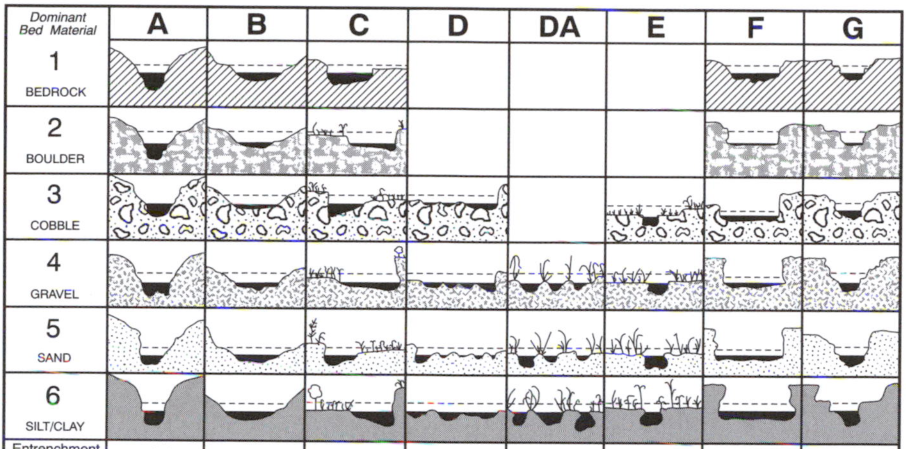

Dominant Bed Material	A	B	C	D	DA	E	F	G
1 BEDROCK								
2 BOULDER								
3 COBBLE								
4 GRAVEL								
5 SAND								
6 SILT/CLAY								
Entrenchment ratio	< 1.4	1.4 - 2.2	> 2.2	N/A	> 2.2	> 2.2	< 1.4	< 1.4
Sinuosity	< 1.2	> 1.2	> 1.4	< 1.1	1.1 - 1.6	> 1.5	> 1.4	> 1.2
Width : depth ratio	< 12	> 12	> 12	> 40	< 40	< 12	> 12	< 12
Water surface slope	0.04 - 0.099	0.02 - 0.039	< 0.02	< 0.02	< 0.005	< 0.02	< 0.02	0.02 - 0.039

Figure 5.1 Classification of channel types (after Rosgen, 1994).

The largest number of channels fall in types A to D (Leopold, 1994), representing relatively straight (A), low-sinuosity (B), meandering (C) and multiple (D, braided; DA, anastomosed) channels respectively.

Whatever classification is used, it is apparent that a wide range of channel types exists. Classifications generally apply to segments of channel rather than entire river systems because a stream of even moderate length will tend to have a range of types, indicating differential behaviour in the longitudinal direction. Transitions may be gradual or abrupt, as for example where a stream changes its bed material rapidly from gravel to sand. The application of a suitable classificatory scheme has implications for the management of the fluvial environment, including its biological and social as well as physical aspects, a point which Rosgen (1994) makes with respect to evaluating channel sensitivity to and potential recovery from disturbance.

Characteristics of Adjustment

A natural river can be thought of as an open system with inflows and outflows of energy and matter (Leopold and Langbein, 1962). Such a viewpoint emphasizes the important characteristics of stream behaviour: the external constraints or controls imposed on the system; the adjustments to the internal geometry of the system in response to those controls; and the nature of that adjustment, especially as regards any tendency towards system equilibrium.

The dominant **controls** of channel form adjustment are discharge and sediment load (notably bed-material load), independent variables which integrate the effects of climate, vegetation, soils, geology and basin physiography (see Fig. 1.1, p. 2). River channels not only exist to convey the variable amounts of water and sediment supplied to them, but are themselves the end product of that conveyance. Discharge data are relatively plentiful even though gauging stations are rarely close enough to ensure an adequate downstream sample, but information on bed-material load is rather meagre and there is the additional problem of determining which aspects of the load are important, be it the volume, size, composition or some other characteristic. Both controls vary considerably through space (along and between rivers) and with time. Along many rivers the addition of water and sediment from tributary sources of variable size produces discontinuous changes in the controls, with parallel discontinuities in channel morphology. Yet many empirical relationships are presented as if change is continuous.

The choice of a suitable timescale within which to study physical relationships has been a recurring theme among geomorphologists (Schumm and Lichty, 1965; Cullingford *et al.*, 1980). Time is a continuous variable, but representative time periods can be defined:

(a) instantaneous time ($< 10^{-1}$ years);
(b) short timescale (10^1–10^2 years);
(c) medium timescale (10^3–10^4 years);
(d) long timescale ($>10^5$ years).

In instantaneous time, stream properties possess single values. Channel form is not simply the product of instantaneous conditions and may itself influence hydraulic and related properties during the passage of floods, so that cause and effect are reversed. The short timescale is the most significant from an observational standpoint, and reasonably well-defined relationships can be expected between the independent variables and certain elements of channel form. At the medium timescale, the stream adjusts its internal geometry in such a way that the sediment supplied can be transported with the discharge available, so that material does not accumulate indefinitely. These two time periods are probably the most relevant as regards channel form adjustment, since mean water and sediment discharge are independent variables to which an average channel geometry is related. Over longer time periods in which landscape mass is removed in significant quantities and large climatic fluctuations occur (see Fig. 6.2, p. 273), discharge and load conditions are not constant in the mean and adjustment becomes both more complex and less definable.

Discharge and sediment load are not the only independent controls. Valley slope is an inherited characteristic which determines the overall rate of energy loss along a river, and can therefore modify the relationship between channel form and the primary variables, although at longer time periods it too becomes part of the adjustable system. Aside from its effect on the nature of sediment supply, geology acts as a constraint on channel adjustment through its influence on bed and bank material composition. Like geology, vegetation plays several roles: as a stabilizing influence on bed and bank materials, and as a constraint, particularly in forested environments where local adjustments are frequently required to accommodate the effects of downed timber (Heede, 1981). Within-channel accumulations of coarse woody debris can significantly affect channel morphology and sediment storage, particularly in lower-order streams (Nakamura and Swanson, 1993). Finally, the long history of anthropogenic activity makes it unlikely that rivers have remained unaffected by human-induced modifications, reservoir construction being the most conspicuous modern example. Thus, underlying the primary controls of discharge and sediment load are many factors influencing channel form at a variety of scales.

Adjustments to the internal geometry of the fluvial system involve a large number of variables whose interdependence is not always clear because the role of a single variable cannot easily be isolated. As before (see Table 4.7, p. 128), those variables can be grouped into flow properties, fluid properties and sediment-related characteristics (including sediment load), with the significant addition of channel form. A distinction can be drawn between:

(a) flow geometry, which includes the interactions among a set of dependent and semi-independent flow variables during temporal changes in discharge, with emphasis on the cross-sectional scale; and
(b) channel geometry, which refers to the three-dimensional form of the channel fashioned over a period of time to accommodate the mean condition of discharge and sediment load.

Clearly the two are related but not necessarily in a simple way. Although flow geometry is more concerned with short-term response, which the channel itself can influence, average patterns of behaviour are still sought that may have longer-term implications.

A stream must satisfy at least three physical relations in adjusting its *flow geometry*, namely continuity (equation (3.8, p. 75)), a flow resistance equation (see Table 4.2, p. 101), and a sediment transport equation. In a system with n variables, n independent equations are required for a unique solution to a given problem. With discharge and sediment load as the independent variables, resistance and debris size as semi-independent, and water-surface width, flow depth, velocity, slope and the pattern of flow as dependent, the requisite number of equations is not available and probably never will be (Maddock, 1969). There is therefore an element of indeterminacy in the behaviour of streams with mobile beds.

Certain aspects of adjustment can be approached through the technique known as 'hydraulic geometry' (Leopold and Maddock, 1953), which assumes that discharge (Q) is the dominant independent variable and that dependent variables are related to it in the form of simple power functions:

$$w = aQ^b \quad (5.1)$$

$$d = cQ^f \quad (5.2)$$

$$v = kQ^m \tag{5.3}$$

$$s = gQ^z \tag{5.4}$$

$$n = tQ^y \tag{5.5}$$

$$ff = hQ^p \tag{5.6}$$

$$Q_{susp} = rQ^j \tag{5.7}$$

where w, d, v, s, n, ff and Q_{susp} are respectively width, mean depth, mean velocity, slope, resistance (Manning's n or the Darcy–Weisbach ff) and suspended sediment load. From the continuity equation,

$$Q = w \times d \times v = aQ^b \times cQ^f \times kQ^m \tag{5.8}$$

it follows that

$$a \times c \times k = 1 \tag{5.9}$$

$$b + f + \mathrm{m} = 1 \tag{5.10}$$

The technique can be applied to both at-a-station and downstream adjustment. It is an essentially empirical procedure, experiencing many of the problems associated with statistical models. It does not give direct consideration to sediment load and assumes linearity in the relations (Richards, 1973). Nevertheless, hydraulic geometry has provided valuable insights into stream behaviour and represents one methodology for breaking into a system with more unknowns than independent equations. Extensions to the original formulation in the form of more sophisticated bivariate and multivariate models are being developed in an attempt to shed more light on the complex adjustment mechanisms which characterize stream behaviour (Bates, 1990; Rhoads, 1992; Ridenour and Giardino, 1991).

Channel geometry is the three-dimensional solid form, usually parameterized at some appropriate stage such as bankfull. 'Regime theory' assumes that it can be largely defined in terms of the adjustable variables, width, depth and slope, with meander form added as a fourth. Alternatively, the adjustment of channel geometry to external controls can be considered in terms of four degrees of freedom, representing different planes of adjustment (Fig. 5.2).

(a) Cross-sectional form – the size and shape of a channel in cross-profile either at a point or as a reach average.
(b) Bed configuration – the distinct forms moulded in the bed of particularly sand- and gravel-bed streams (the sequence of bed forms associated with the former (see Fig. 4.3A, p. 104) can also be regarded as part of the flow geometry).
(c) Planimetric geometry or channel pattern – the form of the channel when viewed from above, the commonest subdivision being into straight, meandering and braided (Leopold and Wolman, 1957).
(d) Channel bed slope – the gradient of a stream at the reach and longitudinal scales, where the latter also refers to the overall shape of the longitu-

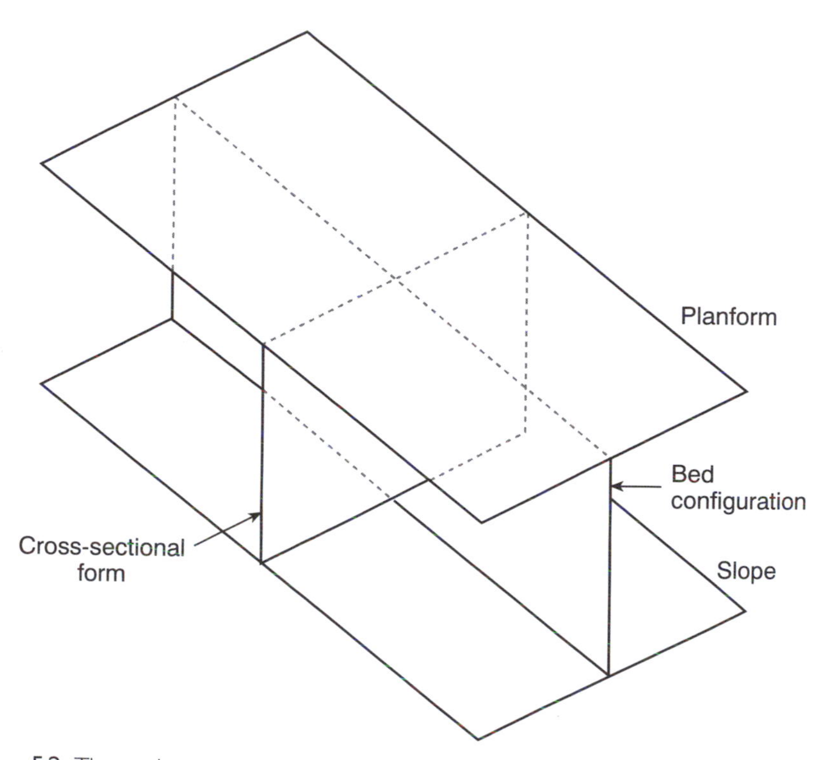

Figure 5.2 The major components of channel form represented as different planes of adjustment.

dinal profile. Slope and channel pattern occur in the same plane (Fig. 5.2), indicating that, in the short term at least, adjustment of the latter is one way of modifying slope.

These degrees of freedom provide the framework for this chapter but, although considered separately, they should not be thought of as independent of one another. Hey (1988b) identified nine degrees of freedom, specified in terms of particular parameters – bankfull width, mean depth, maximum depth, the height and wavelength of bed forms, slope, velocity, sinuosity and meander arc length – but at least one is a flow rather than a channel variable, and not all aspects of channel form are represented.

The four degrees of freedom outlined above are rather broad and each needs to be expressed in terms of a representative set of parameters suitable for model-building and obtaining relations, which is not in itself a trivial matter. Some are better defined in terms of measurable variables than others, cross-sectional form probably being the best represented. They are adjustable over a range of spatial and temporal scales (Fig. 5.3), implying a differential ability to absorb change or develop a characteristic form for given constraints of stream size and boundary resistance. The longitudinal profile has historically been of primary interest to geomorphologists, but it

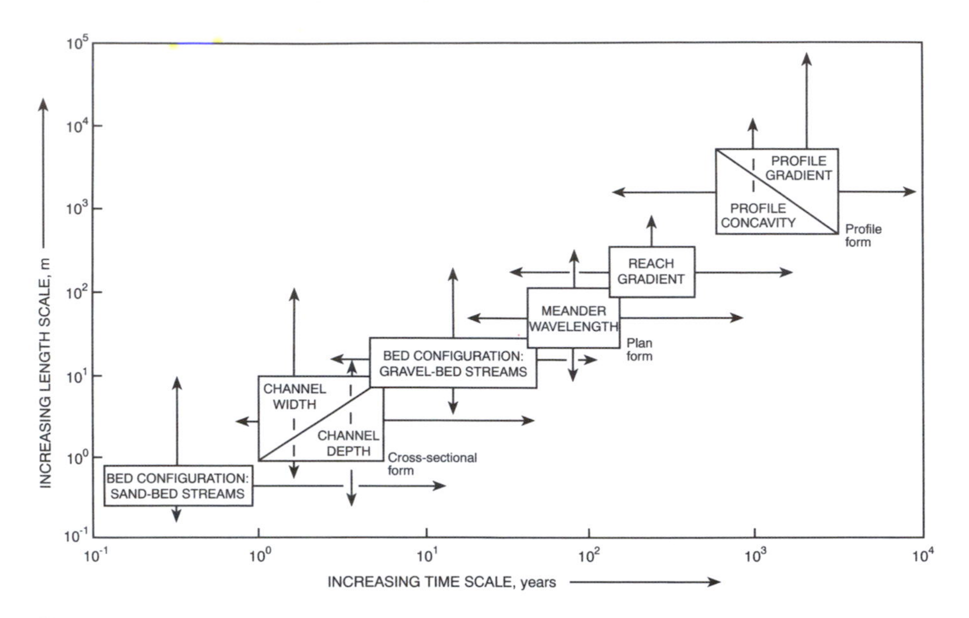

Figure 5.3 Schematic diagram of the timescales of adjustment of various channel form components with given length dimensions in a hypothetical basin of intermediate size.

is the least changeable component and can properly be regarded as an additional constraint on more adjustable ones. As the timescale of interest increases, so does the potential influence of past conditions on present form, with the attendant problem of assessing the extent to which present channel geometry is adjusted to prevailing levels of the control variables. Such a problem has implications not only for the reliability of empirical relationships (between form and control variables on the one hand and between different form variables on the other) but also for the question of whether a general physical principle is governing channel-form adjustment, the search for which has been a major concern over the past three decades (e.g. Langbein and Leopold, 1964; Yang, 1976; Chang, 1980; Jia, 1990).

The Concept of Equilibrium

An important characteristic of open systems is their ability for self-regulation. Negative feedback mechanisms moderate the effects of external factors in such a way that a system can maintain a state of equilibrium in which some degree of stability is established. True stability never exists in natural rivers, which frequently change their position and which must continue to pass a range of discharges and sediment loads. However, they can become relatively stable in the sense that, if disturbed, they will tend to return approximately to their previous state and the perturbation is damped down (stable equilibria in Fig. 5.4). Provided the controlling variables remain relatively constant in the mean, a natural river may develop characteristic or equilibrium forms, recognizable as statistical averages and associated with single-valued relationships to the control variables (Howard, 1988).

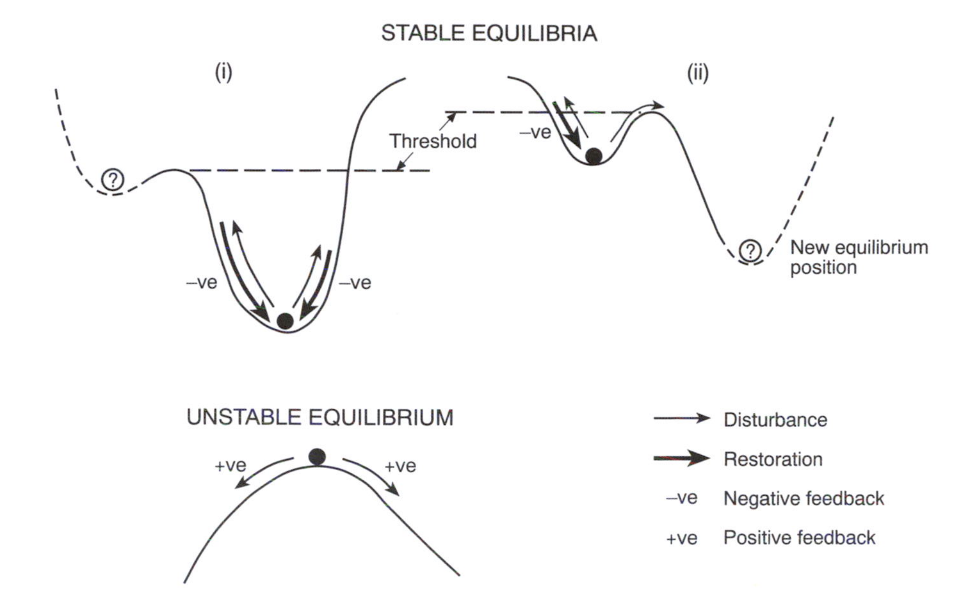

Figure 5.4 Diagrammatic representation of stable and unstable equilibria.

The diagrammatic representation of stable equilibria in Fig. 5.4 shows not only the restorative action of negative feedback mechanisms but also the consequence of threshold exceedence. If the disturbance reaches a critical or threshold level (large in (i) but relatively small in (ii)), the system will become unstable, with a possible return eventually to stable behaviour but in a new equilibrium position. Unstable equilibrium is associated with a situation where a small displacement leads to further displacement in the same direction, a characteristic of positive feedback. Such diagrams, which can be envisaged as potential energy surfaces, are useful representations of system status but they do not define the equilibrium states, the specific mechanisms involved in the maintenance of those states, or the threshold levels whose exceedence results in instability and possible approach to a new equilibrium position.

The concept of equilibrium has a long history in the engineering and geomorphological literature. Hydraulic engineers have been mainly concerned with stable channel design, as exemplified by 'regime theory' in which Anglo-Indian engineers developed an appropriate set of empirical equations for irrigation canals. Geomorphologists have been more interested in the behaviour of natural rivers as they approach and attain equilibrium, usually over longer time periods. Prior to 1950 the geomorphological approach was largely qualitative. Mackin's (1948, p. 471) definition of a graded stream as one

> in which, over a period of years, slope is delicately adjusted to provide, with available discharge and prevailing channel characteristics, just the velocity required for the transportation of the load supplied from the drainage basin

represents a watershed between traditional and modern ideas. It distinguishes between a long-term balance and short-term adjustment, but remains qualitative and retains the previous emphasis on channel slope as the principal means of adjustment.

The modern approach is more quantitative and stems directly from the empirical work on hydraulic geometry (Leopold and Maddock, 1953; Wolman, 1955). No exact equilibrium was implied but rather a tendency for many rivers to develop a recognizable average behaviour, although the empirical mean may not be as well defined in some contexts as originally supposed (Knighton, 1975a; Park, 1977). Adjustments to cross-sectional form (Wolman, 1955) and channel pattern (Leopold and Wolman, 1957) received greater emphasis as means of achieving equilibrium, and attention began to focus more on the dynamics of the adjustment process than on the stability of a particular channel shape or profile.

Extensions of these ideas into the realms of theory have involved the development of *extremal hypotheses*, postulates about river behaviour intended to solve the problem of obtaining equilibrium relationships which the physical equations of continuity, resistance and sediment transport cannot by themselves provide. Arguing by analogy between thermodynamic and fluvial systems, one approach proposes that rivers tend towards a state in which the energy distribution is most probable (the *most probable state*), attained through a compromise between two opposing tendencies: a uniform distribution of energy expenditure and minimum total work (Langbein and Leopold, 1964). The most probable state is characterized by a minimum variance of the system components, which, in the context of hydraulic geometry relations, is manifest in a minimization of the sum of squares of hydraulic exponents (i.e. b, f, m in equations (5.1), (5.2), (5.3)) (Langbein, 1964a; Maddock, 1969). The minimum variance hypothesis has been much criticized, not least because it has never been physically justified (Ferguson, 1986), but it does provide one rationale for river channel adjustment which combines deductive methodology with the probabilistic viewpoint that only average or most probable states can be defined.

Extremal hypotheses with a more physical basis have also been proposed, claiming that, in addition to the equations of continuity, resistance and sediment transport, a river must satisfy one of the following principles:

(1) minimum unit stream power (minimize vs) (Yang, 1976);
(2) minimum stream power (minimize γQs) (Chang, 1980);
(3) minimum energy dissipation rate (minimize $(\gamma Q + \gamma_s Q_s)Ls$) (Brebner and Wilson, 1967; Yang *et al.*, 1981);
(4) minimum Froude number (minimize $v/\sqrt{gd}$, which approximates to minimization of the kinetic energy/potential energy ratio (Yalin, 1992)) (Jia, 1990);
(5) maximum sediment transport rate (Kirkby, 1977; Ramette, 1979; White *et al.*, 1982);
(6) maximum friction factor (maximize $ff \Leftrightarrow$ minimize v^2/ds) (Davies and Sutherland, 1980).

where Q, Q_s, v, s, d, ff are hydraulic variables (discharge, sediment discharge, velocity, slope, depth, resistance), L is stream length, g is the acceleration due to gravity, and γ, γ_s are the specific weights of water and sediment, respectively. (2), (3) and (5) are generally regarded as equivalent (Bettess and White, 1987), implying that a river will minimize its slope for given water and sediment discharges (2) or develop a width which maximizes sediment load for given discharge, slope and bed material size (5). A broader level of equivalence has been surmised, with (6) as the most general of the extremal hypotheses (Phillips, 1991b), but that applies only for certain conditions. Despite their plausible predictions and apparent rational basis, extremal approaches have been strongly criticized because they do not consider adjustment mechanisms directly and because they can give rise to unrealistic implications (Griffiths, 1984; Ferguson, 1986). As White *et al.* (1982) freely admitted, they lack physical justification, which has also been one of the more persistent criticisms of the more probabilistic minimum variance hypothesis. Cao and Knight (1996) have proposed an extremal hypothesis which combines probabilistic and physically based viewpoints in that the probabilities of width and depth adjustment are assumed to be equal in the context of the downstream behaviour of stream power.

The concept of equilibrium is widely applied throughout science, but in geomorphology, where perspectives tend to be longer term than in most other disciplines, it has become something of a terminological minefield through which Thorn and Welford (1994) attempt to chart a course. They argue that it should be expressed in mathematical terms to avoid confusion, and be based on considerations of mass rather than energy. A definition in terms of a balance between the input and output of sediment load would satisfy the latter requirement, but it is usually beyond the wherewithal of direct testing. There is as yet no universally accepted set of criteria for determining whether all or part of a river system is in equilibrium. Renwick (1992) draws a distinction between (Fig. 5.5):

equilibrium – not a static state but form displays relatively stable characteristics to which it will return after disturbance;

disequilibrium – adjustment is towards equilibrium but, because response times are relatively long, there has not been sufficient time to reach such a state;

non-equilibrium – there is no net tendency toward equilibrium and therefore no possibility of identifying an average or characteristic condition.

Cyclic disequilibrium is well illustrated by New South Wales coastal rivers where alternating flood- and drought-dominated regimes lasting up to 50 years have caused considerable channel instability, with incomplete adjustment before the onset of the next regime (Warner, 1994). The non-equilibrium state is symptomatic of long-term transient behaviour which has been associated with river regimes where isolated events of large relative magnitude play the major role in channel form adjustment (Stevens *et al.*, 1975). Under such circumstances stable relationships between form and control variables are not expected.

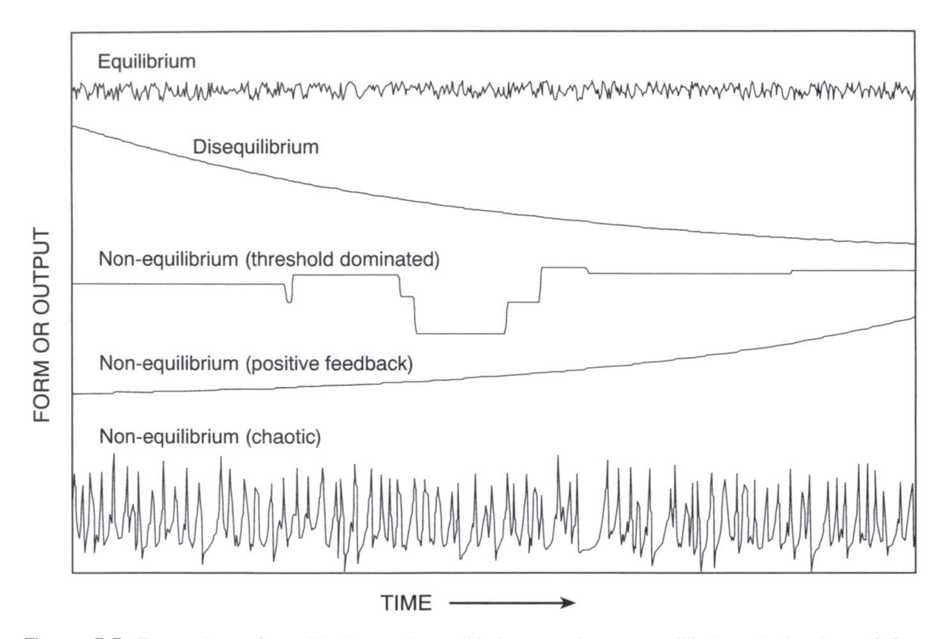

Figure 5.5 Examples of equilibrium, disequilibrium and non-equilibrium behaviour (after Renwick, 1992).

Rivers can at best attain an approximate equilibrium, manifest at some timescale intermediate between short-term fluctuations and long-term evolutionary tendencies in a regularity of channel geometry adjusted to external controls. In order to assess the ability of the fluvial system to make the necessary adjustments, there is a need to know the time period required for a stream to develop characteristic forms and the time period over which such forms are likely to persist. Different components of channel geometry adjust at different rates (Fig. 5.3) so that both time periods may be expected to vary from one component to another. The potential for adjustment also depends on the scale and resistance of the system, so that any tendency towards equilibrium may vary not only between river systems but also between different parts of the same system. Phillips (1992) rejected the traditional view of evolution towards a single end-state in favour of multiple equilibria which may be stable or unstable (Fig. 5.4), thereby denying the existence of a unique solution for given conditions of the control variables.

The Concept of Dominant Discharge

The frequent reference made above to a mean state which reflects the adjustment of channel geometry to imposed conditions suggests that a dominant or channel-forming discharge may be largely responsible for that geometry. Introduced originally to extend the application of regime theory from canals to natural rivers, which have a more variable flow regime, the concept of dominant discharge fulfils such a role. At the very least the concept is a useful analytical device in that the replacement of the frequency distribution of flows by a single discharge simplifies modelling strategies.

Dominant discharge has been defined in various ways: as the flow which determines particular channel parameters, such as cross-sectional capacity (Wolman and Leopold, 1957) or meander wavelength (Ackers and Charlton, 1970a); or as the flow which performs most work, where work is defined in terms of sediment transport (Wolman and Miller, 1960; Fig. 5.6A). Since it seems reasonable to suppose that river channels are adjusted on average to a flow which just fills the available cross-section, dominant discharge has been equated with bankfull flow, thereby giving it additional morphogenetic significance. This assertion was based on an apparent consistency in

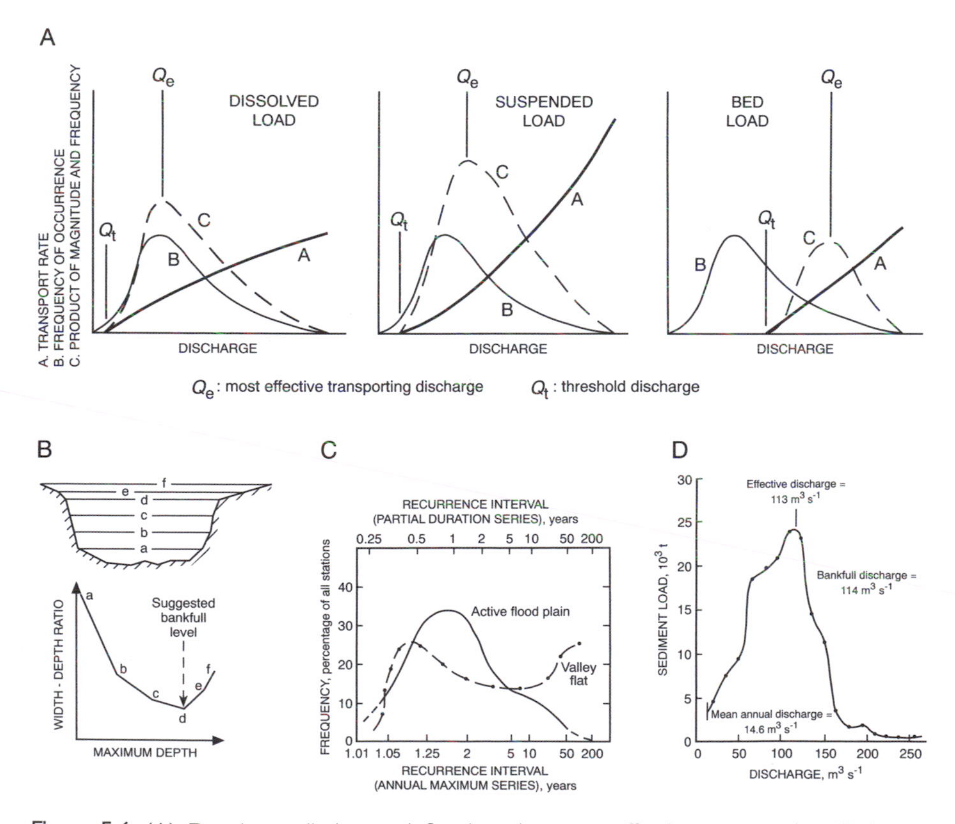

Figure 5.6 (A) Dominant discharge defined as the most effective transporting discharge. (B) Identification of the bankfull channel as the height at which the width : depth ratio becomes a minimum. (C) Frequency distributions of the recurrence interval of bankfull discharge where the bankfull channel was defined in terms of either an active floodplain (36 stations) or a valley flat (28 stations) (after Williams, 1978a). The modal recurrence interval for the former is about 1.5 years on the annual maximum series. (D) Mean annual sediment load transported by discharge increments, Snake River, Wyoming (after Andrews, 1980).

the frequency with which bankfull discharge occurs along streams (Wolman and Leopold, 1957), and on an approximate correspondence between the frequency of bankfull discharge and the frequency of that flow which cumulatively transports most sediment (Wolman and Miller, 1960). Flood flows may individually transport greater loads but recur too infrequently to have a greater cumulative effect. A link was thus established between dominant discharge, most effective discharge and bankfull discharge, with a supposed recurrence interval of about 1–2 years.

That link is limited in several respects:

(a) The bankfull channel cannot always be defined in the field, especially where the valley bottom is too narrow for an active floodplain or where several morphological levels exist. Nor is there a consistent method for specifying the bankfull channel, although many have been devised (Fig. 5.6B; Williams, 1978a). In channels with a readily erodible boundary, the 'bankfull' state (and hence inferred discharge) may reflect nothing more than the intensity of recent erosive flow events (Carson and Griffiths, 1987).
(b) Bankfull discharge is not necessarily of constant frequency even within a single basin (Pickup and Warner, 1976; Andrews, 1980). From data for 36 stations, Williams (1978a) found that the recurrence interval of bankfull discharge averaged 1.5 years as Leopold *et al.* (1964) had asserted, but the range was very wide (from 1.01 to 32 years) and only 62 per cent of the recurrence intervals lay between 1 and 2 years (Fig. 5.6C). Nevertheless, an unusually long (80 km) survey revealed a near-constant recurrence interval for bankfull discharge of about 1.6 years along the Murrumbidgee River (Page, 1988).
(c) Channel form parameters do not necessarily correlate best with bankfull discharge. On the basis of the argument that the best statistical correlation reflects the closest physical relationship, Carlston (1965) found that meander wavelength was better related to more frequent sub-bankfull flows than to bankfull discharge. Even assuming that a single discharge is largely responsible for shaping the channel, there is no guarantee that different form elements are adjusted to the same discharge, particularly in view of their variable rates of adjustment (Fig. 5.3).
(d) Bankfull discharge may not be the most effective flow as regards sediment transport. Indeed, most controversy has surrounded this issue. The overall shape of the effectiveness curve (C in Fig. 5.6A) and in particular the location of the peak depend firstly on the choice of transport rate function (A in Fig. 5.6A), whose threshold point and form vary with the calibre and type of load, and secondly on the discharge frequency distribution (B in Fig. 5.6A), the shape of which varies along a given stream as well as from one stream to another. Nash (1994) identified six basic types of effectiveness curve. In their original derivation Wolman and Miller (1960) considered suspended load, but it is bed load which is the most relevant from the standpoint of channel form adjustment, and

the implication of Fig. 5.6A is that its effectiveness peak is displaced towards less frequent discharges.

Much of the controversy concerns the frequency of the effective discharge, especially in relation to bankfull. On the one hand Baker (1977) has argued that rarer flows are more effective agents, notably in streams with a large proportion of high discharges and relatively resistant boundaries, while on the other Pickup and Warner (1976) found that the most effective discharge was more frequent than bankfull, recurring on average about 3–5 times each year. After examining a large number of, albeit, suspended load records, Ashmore and Day (1988) and Nash (1994) concluded that the recurrence interval of the effective flow is highly variable, in line with the considerable inter-basin diversity in discharge and sediment characteristics. Set against this are comprehensive results for bed-load transport which show very good agreement between the most effective flow and bankfull discharge (Andrews, 1980; Andrews and Nankervis, 1995; Batalla and Sala, 1995; Fig. 5.6D). In addition, recent laboratory work on floodplain–main channel interactions has indicated maximum transport at the bankfull stage, with reduced erosive tendencies at overspill discharges (Ackers, 1992). Clearly the issue is far from resolved. There is variability in the recurrence interval of the most effective flow which becomes less frequent in progressing from dissolved load through suspended load to bed load (Table 5.2), as the bed material coarsens and the threshold of sediment movement increases (Carling, 1988), and as drainage area decreases to give a higher relative proportion of large flows. Whether or not a flow defined as cumulatively transporting most material is the most relevant morphologically, it has been given formative significance in both the Brahmaputra (Thorne *et al.*, 1993) and the Lower Mississippi (Biedenharn and Thorne, 1994) where the relative effects of large floods have also been taken into account. There, dominant discharge corresponds to bar-full stage, the height to which mid-channel bars grow through accretion.

Bankfull discharge is not necessarily of constant frequency or the most effective flow. Channel form is the product not of a single formative discharge but of a range of discharges, which may include bankfull, and of the temporal sequence of flow events. The fluvial system has a memory for past events which, to a variable extent, influence current channel form and therefore the role of a subsequent discharge. In rivers or segments of river where the flow regime is very variable (i.e. the ratio of individual flood peaks to the mean annual flood is large) or the channel boundary is very resistant, high-magnitude, low-frequency floods may control channel form, so that the concept of dominant discharge and the related concept of channel equilibrium become less relevant (Stevens *et al.*, 1975; Baker, 1977). Such is the case in arid-zone streams (Graf, 1988a). Under such circumstances regression models which express a particular channel parameter (y) as a function of a particular discharge (Q) in the form $y = f(Q)$ are inappropriate. However, in many perennial, especially humid temperate, rivers, the

Table 5.2 Most effective discharge

Source	River/ location	Drainage area, km^2	Most effective discharge (Q_e)			
			Bankfull discharge (Q_b), $m^3 s^{-1}$	Duration of Q_e, %	Q_e/Q_b	Range of discharges relative to Q_b carrying (x%) of load
(1) *Dissolved load*						
Webb and Walling (1982)	Creedy, Devon	262	33.6	~ 20	~ 0.2	~ 0.05–0.4 (70%)
(2) *Suspended load*						
Webb and Walling (1982)	Creedy, Devon	262	33.6	~ 2	~ 0.65	~ 0.35–2 (70%)
Ashmore and Day (1988)	Saskatchewan River basin, 21 sites	10–>10^5		Mostly 1–10		
Thorne *et al.* (1993)	Brahmaputra, sand + silt		~ 62 500	18	~ 0.6	~ 0.4–1 (70%)
Biedenham and Thorne	Lower Mississippi, 464 km reach, sand		~ 36 000	13	~ 0.85	~ 0.5–1.1 (70%)
(3) *Bed load*						
Andrews (1980)	Yampa River, 15 sites	52–9660	0.3–255	0.4–3	0.71–1.11	
Carling (1988)	Carl Beck, D_{50}= 50 mm	2.2	2.35		1.3	0.6–1.4 (67%)
	Gt. Eggleshope Beck, D_{50}= 20 mm	11.7	5.6		0.94	0.8–1.4 (78%)
Andrews and Nankervis (1995)	17 gravel-bed rivers, western USA	4–3706	0.7–88.4		0.8–1.2	0.8–1.6 (80%)

bankfull channel is the one reference level which can reasonably be defined, and it remains intuitively appealing to attach morphologic significance to bankfull flow. Certainly it is advocated for design purposes (Hey, 1978, 1988b), and regression models of the form $y = f(Q)$ have provided a major basis for empirical studies of river channel form.

Cross-sectional Form

The cross-sectional form of natural channels is characteristically irregular in outline and locally variable. Table 5.3 lists the parameters commonly used to describe that form, measurements for which are usually obtained by stretching a tape or cable horizontally across the stream and measuring down to the channel boundary at set distances. Width and mean depth give the gross dimensions of the channel but do not uniquely define cross-sectional shape (Hey, 1978). Nevertheless, width : depth ratio remains the most commonly used index of channel shape even though it is not always the most appropriate (Pickup, 1976a). In particular it provides no indication of cross-sectional asymmetry, a recently quantified characteristic which is often linked with meanders (Knighton, 1982).

The parameters usually refer to the bankfull channel, although other reference levels such as the 'active channel' have been proposed (Osterkamp and Hedman, 1982). When flow geometry is the specific concern, they refer to the flow cross-section at a particular discharge. Both present problems of definition. In the case of the latter, a standard discharge index needs to be defined in order to compare upstream and downstream sections, but the choice is often arbitrary and may have limited channel-forming significance. Aside from bankfull discharge itself (Q_b), the choice has included the median (Q_2) and mean annual ($Q_{2.33}$) floods, and the flows which are equalled or exceeded 2 per cent, 15 per cent and 50 per cent of the time (Wolman, 1955). Reversing the depen-

Table 5.3 Cross-sectional parameters

Channel size	Channel shape
Width, w	Width : depth ratio, w/d
Mean depth, d	d_{max}/d (Fahnestock, 1963)
Cross-sectional area, $A = w \times d$	Asymmetry (Fig. 5.9A(i), p.178; Knighton, 1981a)
Wetted perimeter, P	$A^* = (A_r - A_l)/A$
Hydraulic radius, $R = A/P \sim d$	$A_2 = 2x(d_{max} - d)/A$
Maximum depth, d_{max}	
Bed width, w_b	

Definitions: A_r, A_l are respectively the cross-sectional areas to the right and left of the channel centreline; x is the horizontal distance from the channel centreline to the point or centroid of maximum depth.

dency, channel geometry parameters have been used as a basis for estimating peak discharge at ungauged sites, providing a methodology which is more reliable than traditional discharge–drainage area relationships (Wharton *et al.*, 1989).

The Equilibrium Cross-section

Rivers with erodible boundaries flow in self-formed channels which, when subject to relatively uniform controlling conditions, are expected to show a consistency of form, or average geometry, adjusted to transmit the imposed water and sediment discharges. The problem is to determine the nature of that geometry.

Various theoretical and empirical approaches have been used to define the equilibrium channel. **Theoretical** ones fall into two main categories:

(a) those which are deterministically based and rely on equations descriptive of the dominant processes, e.g. threshold theory, Parker's (1978) model of lateral diffusion;
(b) those which postulate an additional condition regarding the behaviour of stable rivers, e.g. minimum stream power (Chang, 1980), maximum sediment transport rate (White *et al.*, 1982).

Threshold theory is based on the fundamental assumption that all particles on the channel boundary are on the verge of movement at bankfull discharge, a resolution of the associated forces yielding a cross-section that is roughly parabolic in shape (Lane, 1955; Type B in Fig. 5.7A). With steeper and gentler slopes respectively, the Type A and Type C channels are needed for stability (Fig. 5.7A). Stevens (1989) extended threshold theory to take account not only of the shear stresses acting on the banks but also of the strength of the bank material, a vital consideration when attempting to define the stable width of alluvial channels. Threshold theory has limited applicability to natural channels which transport material at sub-bankfull flows and migrate laterally, but it does define a lower limit for cross-sectional form, with implications for stable channel design.

A more realistic approach is to consider the 'live-bed' case, but then, according to Parker (1978), the 'stable channel paradox' arises in which stable banks are incompatible with a mobile bed. He resolved the paradox by invoking a mechanism of lateral momentum transfer, which results in a bankwards decrease in shear stress away from the channel centreline where a moderate rate of transport is allowed. The associated stable channel resembles the Type A channel of threshold theory (Fig. 5.7A). Parker acknowledged his neglect of various real-world complications such as non-straight channel patterns and variable bank conditions, but the model has been tested with reasonable success (Pizzuto, 1984b) and subsequently used to show how bank material characteristics (Ikeda *et al.*, 1988) and bank vegetation (Ikeda and Izumi, 1990) can theoretically influence cross-sectional form.

For a given discharge and slope, Parker's (1978) model and Bagnold's (1980) equation (4.13) (p. 130) imply that a larger sediment load requires a

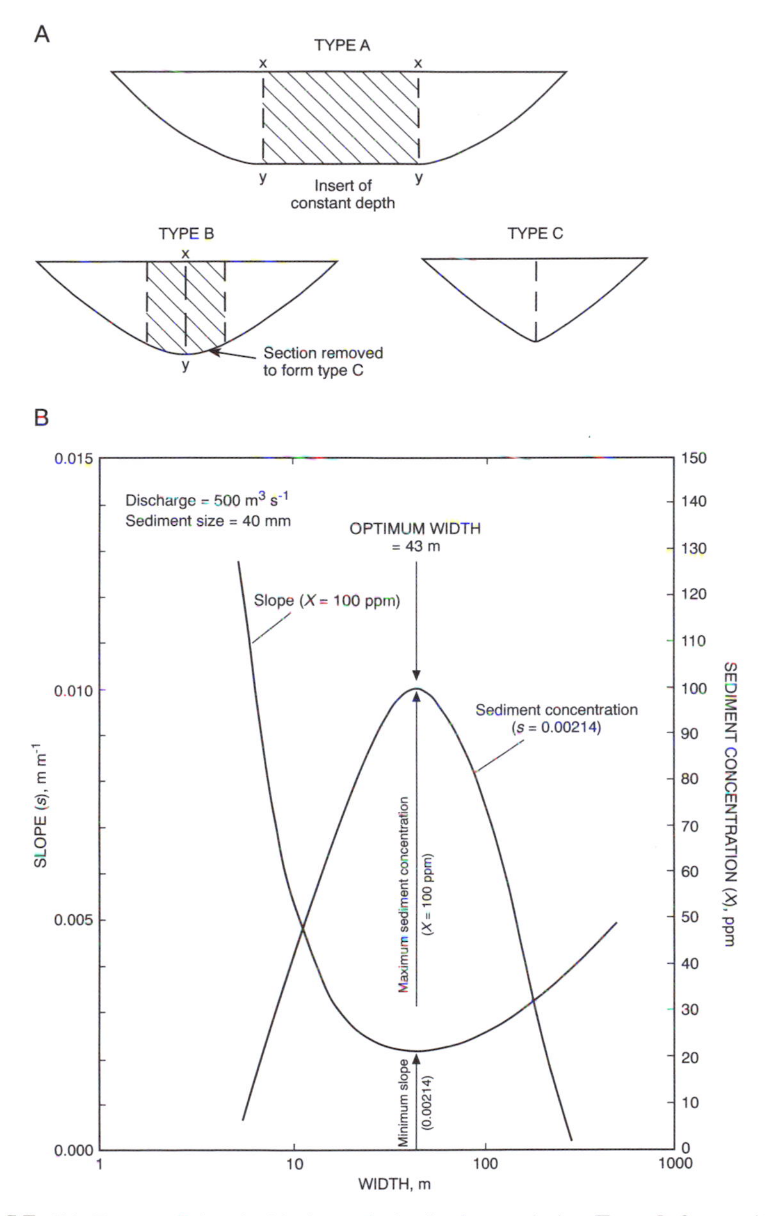

Figure 5.7 (A) Types of threshold channel, the basic one being Type B from which the others can be derived by insertion (Type A) or removal (Type C) of a central section. (B) Maximum transporting capacity (for a given slope) and minimum slope (for a given sediment concentration) achieved at an intermediate width of 43 m, assuming Q = 500 $m^3 s^{-1}$, D = 40 mm (after White *et al.*, 1982).

wider channel, whereas Henderson's (1966) threshold channel equations imply the contrary. A resolution of these contrasting arguments is achieved by the extremal hypothesis proposed by White *et al.* (1982) in which maximum transporting capacity is attained at some intermediate width

(Fig. 5.7B) and for which Carson and Griffiths (1987) provide guarded support. The optimum width : depth ratio can be shown to increase with discharge, slope and particle fineness. However, both this hypothesis and its equivalent, the minimum stream power hypothesis of Chang (1980), tend to underpredict the width of larger channels and neither considers bank resistance properties directly. Indeed, such is the natural variability in those properties that their incorporation in theoretical treatments presents considerable difficulties. Despite their proliferation and variety, theoretical approaches have not as yet satisfactorily defined the equilibrium cross-section, particularly for the live-bed case. Direct consideration of the distribution of shear stress over the channel boundary may be one way forward (Bettess *et al.*, 1988), an argument taken up by Huang and Warner (1995) in deriving empirical hydraulic geometry relations.

Empirical approaches revolve around the *regime theory* of the Anglo-Indian school of engineers and the hydraulic geometry methodology pioneered by Leopold and Maddock (1953). Derived originally for canals with a steady flow and fine sediment, regime 'theory' consists of a set of empirical equations which can be manipulated to give the width (w), depth (d) and slope (s) of an approximately stable live-bed channel whose cross-sectional form is maintained by a local balance between erosion and deposition. The more important equations are:

$$P = 4.84\, Q^{1/2} \tag{5.11}$$

$$R = 0.41\, Q^{1/3} D^{-1/6} \tag{5.12}$$

$$v = 0.50\, Q^{1/6} D^{1/6} \tag{5.13}$$

$$s = 0.00065\, Q^{-1/6}\, D^{5/6} \tag{5.14}$$

where v, Q and D are respectively mean velocity, discharge and bed material size (in mm), and $P \sim w$ and $R \sim d$ for large channels (Lacey, 1929). The equations were based on a narrow range of sediment sizes, which, coupled with the limited account taken of sediment discharge, restricts their application to natural streams. Indeed, large discrepancies have been noted between calculated and field data when the equations have been applied to conditions beyond those for which they were originally intended.

The hydraulic geometry approach encapsulated in equations (5.1)–(5.7) was intended to extend the regime concept to natural rivers. The choice of an appropriate discharge is a critical issue since exponent values are not independent of the selected discharge index (Wolman, 1955; Knighton, 1974). Bankfull discharge (Q_b) is an obvious candidate but it cannot always be defined and is not necessarily of constant frequency (Williams, 1978a; Fig. 5.6C). Most often relationships are obtained for a group of rivers within a given physiographic or geographic area rather than for single rivers, because there are rarely enough gauging stations along individual rivers to provide an adequate sample. Thus Bray (1982) obtained regime-type relationships for gravel-bed rivers in Alberta, while Charlton *et al.* (1978) and Hey and Thorne (1986) have undertaken a similar task for gravel-bed rivers in the UK. The work of Hey and Thorne (1986) epitomizes the empirical approach and extends the original Leopold and Maddock (1953) formulation by including independent variables additional to discharge. Based on data for 62 sites, the simple hydraulic geometry relations are:

$$w = 3.67\, Q_b^{0.45} \quad (5.15)$$

$$d = 0.33\, Q_b^{0.35} \quad (5.16)$$

$$(\Rightarrow v = 0.83\, Q_b^{0.20} \text{ by continuity; see equations (5.8)–(5.10))}$$

$$s = 0.008\, Q_b^{-0.20} \quad (5.17)$$

while the extended version includes provision for vegetation type (Type I = grassy banks, Type II = 1–5% tree/shrub cover, Type III = 5–50% tree/shrub cover, Type IV = >50% tree/shrub cover), bed-material size (D_{50}, D_{84}) and bed-load transport (Q_{sb}):

$$w = 4.33\, Q_b^{0.50} \quad \text{Type I} \quad (5.18a)$$

$$w = 3.33\, Q_b^{0.50} \quad \text{Type II} \quad (5.18b)$$

$$w = 2.73\, Q_b^{0.50} \quad \text{Type III} \quad (5.18c)$$

$$w = 2.34\, Q_b^{0.50} \quad \text{Type IV} \quad (5.18d)$$

$$d = 0.22\, Q_b^{0.37}\, D_{50}^{-0.11} \quad (5.19)$$

$$d_{max} = 0.20\, Q_b^{0.36}\, D_{50}^{-0.56}\, D_{84}^{0.35} \quad (5.20)$$

$$s = 0.087\, Q_b^{-0.43}\, D_{50}^{-0.09}\, D_{84}^{0.84}\, Q_{sb}^{0.10} \quad (5.21)$$

These equations were calculated so that the bankfull dimensions of stable, mobile gravel-bed rivers could be estimated for design purposes. They and their ilk provide a practical alternative to a theoretical problem, that of specifying the equilibrium geometry of channels with mobile beds, which has yet to be satisfactorily solved.

Spatial Variation – Downstream

Cross-sectional form adjusts to accommodate the discharge and sediment load supplied from the drainage basin, within the additional constraints imposed by boundary composition, bank vegetation and valley slope. Channel dimensions are not arbitrary but are adjusted, through the processes of erosion and deposition, to the quantity of water moving through the cross-section so that the channel contains all but the highest flows. Since discharge increases downstream with drainage area (Fig. 1.2A, p. 7), width and mean depth should similarly vary.

Channel width and depth are indeed systematically related to bankfull discharge as it varies over a very broad range of conditions (Fig. 5.8A; Table 5.4), suggesting that discharge is the dominant control of channel dimensions. The data support the oft-quoted opinion that width varies approximately as the square root of discharge (i.e. $b \sim 0.5$ in $w = aQ^b$), while the depth exponent (f in $d = cQ^f$) has a similarly narrow range of values with a mean of about 0.36 (Knighton, 1987b). Church (1980) suggested the possibil-

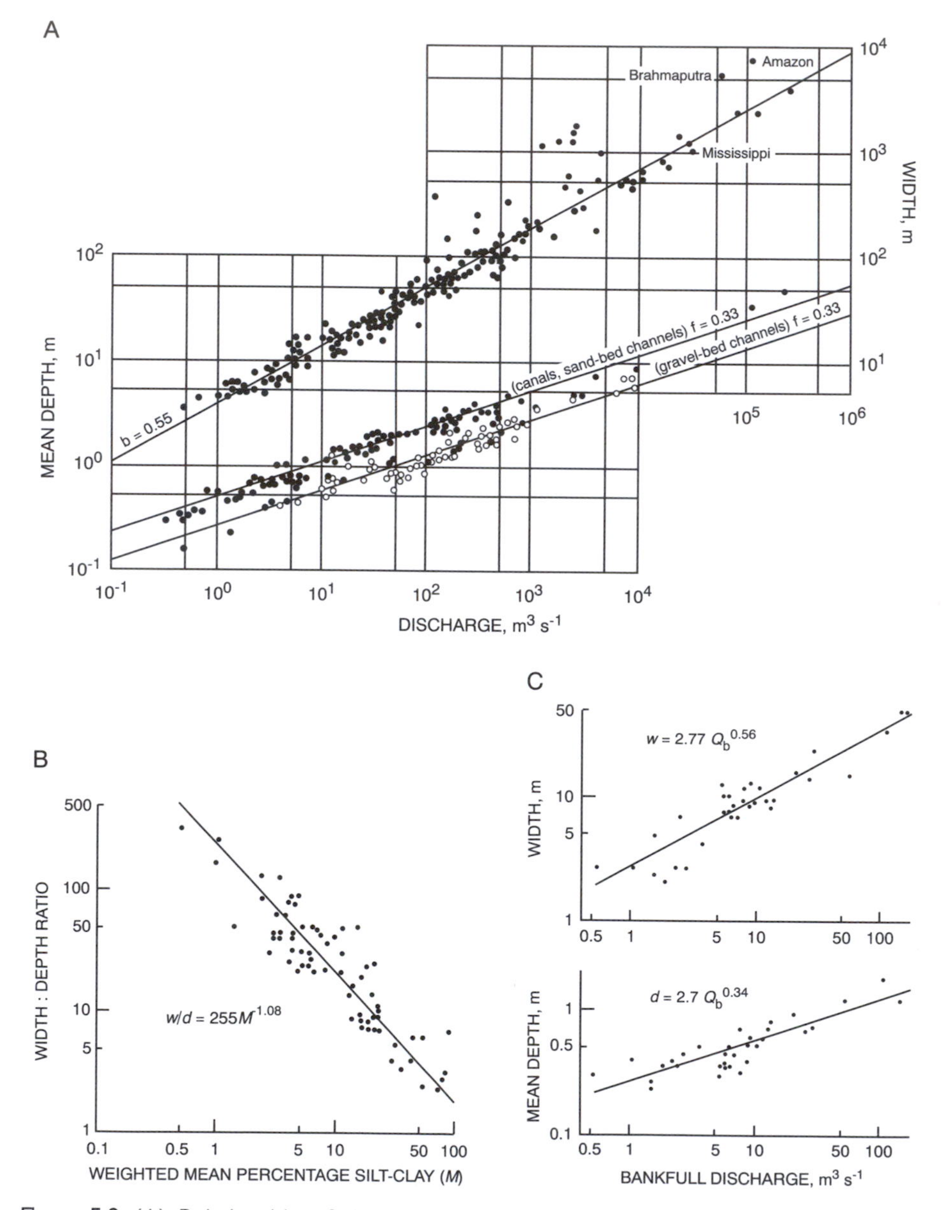

Figure 5.8 (A) Relationship of channel width and mean depth to discharge for a wide range of rivers (after Church, 1992). (B) Relationship of width : depth ratio to the weighted mean percentage of silt–clay in the channel boundary (after Schumm, 1960). (C) Downstream relationships of width and mean depth to bankfull discharge, Upper Salmon River, Idaho (data from Emmett, 1975).

ity of two regime types, $f = 0.33$ and $f = 0.40$, associated respectively with large and small sediment transport rates, but the data provide limited support for this distinction. Cao and Knight (1996) show theoretically that $b = (3/2)f$, so that if $f = 0.33$, b should equal 0.5. This apparent regularity must be tempered by the observations, firstly, that hydraulic geometry

Table 5.4 Downstream hydraulic geometry relations at bankfull or near-bankfull discharge

Source	Location/applicable condition	Discharge	Coefficients		Exponents			
			a	*c*	*b*	*f*	*m*	*z*
(1) *Threshold channels*								
Kellerhals (1967)	Western Canada	$\sim Q_3$	3.26	(0.42)	0.50	(0.40)	(0.10)	(–0.10)
Li et *al.* (1976)	Threshold theory	Q_b			0.46	0.46	0.08	–0.46
(2) *Gravel-bed rivers and canals*								
Nixon (1959)	Britain	Q_b	3.15		0.49			
Brush (1961)	Appalachians	$Q_{2.33}$	1.85	0.28	0.55	0.36	0.09	
Simons and Albertson (1963)	Indian and US canals, non-cohesive banks	$\sim Q_b$	2.61	0.31	0.50	0.36	0.14	–0.24
Emmett (1975)	Upper Salmon River, Idaho	Q_b	2.77	0.27	0.56	0.34	0.12	
Charlton et *al.* (1978)	Britain	Q_b	3.74	0.31	0.45	0.40	0.15	–0.24
Bray (1982)	Alberta	Q_2	4.79	0.26	0.53	0.33	0.14	–0.34
Andrews (1984)*	Colorado: thick bank vegetation	Q_b	3.91	0.49	0.48	0.37	0.14	–0.44
	thin bank vegetation	Q_b	4.94	0.48	0.48	0.38	0.14	–0.41
Hey and Thorne (1986)	Britain (see also equations (5.18)–(5.21))	Q_b	3.67	0.33	0.45	0.35	0.20	–0.20
Parker (1979)*	Theoretical – momentum diffusion	Q_b	4.40	0.25	0.50	0.42	0.08	–0.41
Chang (1980)	Theoretical – minimum stream powerr	Q_b			0.47	0.42	0.11	
(3) *Sand-bed rivers and canals*								
Lacey (1929)	Canals – Punjab	$\sim Q_b$	4.84	(0.39)	0.50	(0.33)	(0.17)	(–0.17)
Simons and Albertson (1963)	Indian and US canals: sandy banks	$\sim Q_b$	5.23	0.69	0.50	0.36	0.14	–0.30
	cohesive banks, small load	$\sim Q_b$	3.93	0.58	0.50	0.36	0.14	–0.30
	cohesive banks, large load	$\sim Q_b$	2.55	0.45	0.50	0.36	0.14	–0.24
Mahmood et *al.* (1979)	Canals – Pakistan	$\sim Q_b$	4.93	0.53	0.51	0.31	0.18	–0.09
(4) *Undifferentiated*								
Rundquist (1975)	Rivers and canals with gravel and sand beds	Q_b	4.39	0.38	0.52	0.32	0.16	–0.30
Langbein (1965)	Minimum variance theory	$\sim Q_b$			0.50	0.38	0.13	–0.55

Discharge defined as bankfull (Q_b) or the one having a recurrence interval of 2 (Q_2), 2.33 ($Q_{2.33}$) or 3 (Q_3) years.
Symbols: $w = aQ^b$, $d = cQ^f$, $v = kQ^m$, $s = gQ^z$; bracketed coefficient and exponent values come from multivariate relations.
* Analysis in terms of dimensionless variables.

exponents are conservative quantities constrained by the continuity requirement (equation (5.10)) and may therefore be rather insensitive in a similar way to the parameters in Horton's (1945) 'laws' of drainage network composition (Table 2.1); and, secondly, that there is quite a large amount of scatter about the trends in Fig. 5.8A, suggesting that variables other than discharge are relevant. Coefficient values (a, c) tend to vary more widely than exponents in Table 5.4, reflecting the multivariate character of channel form control.

Channel size is influenced not only by the magnitude of discharge but also by the hydrologic regime (Yu and Wolman, 1987). A river with a flashier regime and relatively high peak flows tends to develop wider channels (Osterkamp, 1980), which may help to explain why, relative to their perennial counterparts, arid-zone streams increase their width much faster in small to moderate-sized basins (Wolman and Gerson, 1978). Sand channels lacking sufficient fine or coarse material to form resistant banks are particularly susceptible to discharge variability (Osterkamp and Hedman, 1982). Although based on limited data, a comparison of two rivers in Venezuela reveals major differences in channel width and sinuosity which appear to be the result of different flood characteristics (Table 5.5). Ratios such as Q_p/Q_m provide one means of demonstrating the sensitivity of channel form to flow events and of assessing whether or not a river is likely to develop an equilibrium geometry. Any such tendency becomes progressively less probable as the ratio increases.

Discharge acts as a scale variable in determining the gross dimensions of the channel through its largely unknown relationship to the distribution of effective stresses at the channel boundary, a shortcoming which Huang and Warner (1995) attempt to rectify. Their study confirms that boundary composition has a significant influence on hydraulic geometry and in particular the shape of the channel. Data from the Missouri basin indicate that, as the bed material changes from a high silt–clay content to sand, there is a systematic increase in the values of the coefficient (a) and, to a lesser extent, the exponent (b) in the width–discharge relation (Osterkamp, 1980). Further increases in bed material size up to a boulder bed reverse the trend. These results suggest that rivers that transport relatively large amounts of sand require a wider channel. Certainly a larger bed load is often associated with wider, shallower channels, and Parker (1979) predicted that, for a given dis-

Table 5.5 Characteristics of Rios Guanipa and Tonoro immediately upstream of their confluence (after Stevens *et al.*, 1975)

River	Drainage area, km^2	Valley slope, $m\ m^{-1}$	Median bed material size, mm	Channel width, m	Sinuosity	Mean annual discharge (Q_m), $m^3\ s^{-1}$	Peak discharge (Q_p), $m^3\ s^{-1}$	Q_p/Q_m
Guanipa	2800	0.0013	0.35	15	2.3	17	105	6
Tonoro	1300	0.0015	0.35	183	1.1	11	535	47

charge, a 30 per cent increase in gravel load leads to a 40 per cent increase in width and a 25 per cent reduction in centre depth. Pickup (1976b) argued that a river adjusts its bed width to optimize bed-load transport, echoing the principle of maximum transport capacity put forward by White *et al.* (1982), who found that the condition was achieved at some intermediate width (Fig. 5.7B), although the optimal geometry is not always as clearly defined as depicted. How important are the effects of the size composition and amount of sediment load on cross-sectional form, both relatively and absolutely, remains uncertain.

The influence of bed-material size on depth can be deduced from multivariate relations, where the associated exponent typically lies in the range –0.10 to –0.17 (equations (5.12), (5.19)). Translated into a downstream context, a negative exponent implies that a decreasing bed-material size would tend to reinforce the effect of discharge on depth.

Bed-material size generally decreases downstream in conjunction with channel slope. Figure 5.7B shows that the optimum width for transport at a given slope coincides with the minimum slope required to convey a specified load, suggesting a close link between cross-sectional form and slope. To the extent that channel slope is constrained by valley slope and a long timescale of adjustment (Fig. 5.3), it can be regarded as an imposed variable, although the situation is complicated by the relatively rapid slope adjustment which can be achieved through a change in channel pattern. The relative effects of bed-material size and slope are difficult to disentangle, but steeper slopes tend to give rise to wider, shallower channels, an association predicted by both the minimum stream power and the maximum transport extremal hypotheses.

Bank material exerts an important sedimentological control on the strength and stability of channel banks, and therefore on the adjustment of channel width. Simons and Albertson's (1963) results for sand-bed canals illustrate the effect in that the width coefficient (a) for sandy banks is 33 per cent larger than for cohesive banks (Table 5.4). Although bank strength is not simply the function of any one material property, it does depend on the degree of cohesion, which can be expressed by the silt–clay content of the banks (B) or channel perimeter (M). For sand-bed rivers in the Great Plains, relations have been obtained by Ferguson (1973):

$$w = 33.1\, Q_{2.33}^{0.58}\, B^{-0.66} \tag{5.22}$$

and Schumm (1971):

$$w = 5.54\, Q_{2.33}^{0.58}\, M^{-0.37} \tag{5.23}$$

$$d = 0.12\, Q_{2.33}^{0.42}\, M^{0.35} \tag{5.24}$$

the implications for downstream hydraulic geometry being:

(a) well-defined width– and depth–discharge relations are expected if bank material remains uniform downstream;

(b) if banks become more cohesive downstream (B or M increases), which could be associated with an increasing dominance of suspended over bed-load transport, width will increase more slowly and depth more rapidly than expected from the effect of discharge alone, giving rise to a more box-like cross-section.

However, bank-material composition is highly variable and downstream trends are unlikely to be well-defined across a broad spectrum of rivers even though transported sediment tends to become finer downstream. Dury (1984) detected abrupt variations in channel width along part of the River Severn, which appeared to reflect local differences in bank strength. Channel banks are often vertically stratified with a basal gravel layer overlain by fine alluvium (Klimek, 1974; Andrews, 1982; Pizzuto, 1984a), a situation in which the strength of the weaker element controls bank stability and therefore the maintenance of channel width.

The resistance of channel banks to erosion is also influenced by the type and density of bank vegetation. Charlton *et al.* (1978) found that channels with grassed banks were on average 30 per cent wider, and those with tree-lined banks up to 30 per cent narrower, than the overall width–discharge relation would suggest, an effect confirmed both theoretically (Ikeda and Izumi, 1990; Millar and Quick, 1993) and empirically (Andrews, 1984; Table 5.4). Denser vegetation gives rise to a narrower and a deeper channel. Although real, the protective effect of vegetation is difficult to quantify. Hey and Thorne (1986) got round the problem by calculating separate hydraulic geometry equations for four major vegetation types (equation (5.18)), revealing the important role of vegetation in controlling the width and depth of alluvial channels. Channels with grassed banks were up to 1.8 times the width of those that were tree-lined.

The width and depth of a channel are not adjusted independently. Miller (1984) has addressed the question of their mutual adjustment and concluded that, whereas discharge has the dominant direct effect on cross-sectional form, sediment characteristics have the dominant indirect effect and therefore play the major role in the mutual adjustment mechanism. To some extent Schumm's (1960, 1971) data lend support to this idea. The original analysis produced the relation

$$w/d = 255\, M^{-1.08} \qquad (5.25)$$

which indicates a strong dependence of channel shape on the percentage of silt–clay in the channel boundary (Fig. 5.8B), while a re-analysis by Richards (1982) yielded

$$w/d = 800\, Q_{2.33}^{0.15}\, B^{-1.20} \qquad (5.26)$$

where the sediment variable has by far the larger influence on the form ratio. Links between width : depth ratio and boundary shear-stress distribution have been used as a basis for predicting downstream hydraulic geometry relations (Osterkamp *et al.*, 1983; Huang and Warner, 1995) but with only limited success.

The dominant controls on cross-sectional form are discharge, the absolute and relative amounts of bed-load transport, and the composition of the channel boundary particularly as it relates to bank stability. Given the multivariate character of those controls, it is perhaps too much to expect consistency in relations where discharge is the only independent variable. Nevertheless, discharge does vary systematically downstream along most rivers and is the major control of channel size. The effects of other factors are more difficult to assess, especially in a downstream context, and this applies particularly to sediment load. The extent to which downstream changes in cross-sectional geometry are consistent depends on the regularity with which flow and sediment conditions vary in that direction. However, downstream change along individual rivers is discontinuous, principally but not exclusively because of tributary inflow, which is the primary source of discharge addition (Fig. 2.18B, p. 62) and a major supplier of sediment (Knighton, 1987b). A network-channel model which accommodates the variable influence of tributaries, both as individuals and in sequence, could provide a basis for a more realistic assessment of the spatial variation of cross-sectional form in the downstream direction. Of the results considered in Table 5.4, only Emmett's (1975) apply to a single river system (Fig. 5.8C).

Spatial Variation – Local

Wolman (1955) recognized that local variations in cross-sectional form are a possible source of scatter in downstream hydraulic geometry relations. In particular, such variations can be related systematically to channel pattern and bed topography.

Divided or braided sections, which are often associated with coarse bedload transport, tend to have higher width : depth ratios than do comparable sections in meandering or straight reaches. Within the latter, cross-sectional asymmetry may oscillate in a pseudo-periodic way, dependent on the degree of riffle–pool or meander development (Fig. 5.9A), with maximum asymmetry close to bend apices (Knighton, 1982). Milne (1983) confirmed the greater asymmetry associated with sections in sharply curved as opposed to straight reaches. If channel pattern is locally variable, therefore, considerable fluctuations in cross-sectional shape can occur, with implications for the pattern of flow (Thompson, 1986). Markham and Thorne (1992) regard the width : depth ratio as the most important control over the flow pattern through river bends.

Similar in several respects to the contrast between braids and meanders is that between riffles and pools, with the former characterized by coarser bed material and higher width : depth ratios. Provided the contrast is consistent throughout a basin, downstream trends will be affected. On the River Fowey, riffle widths exceeded pool widths by about 12 per cent, leading to downstream width–discharge relationships with significantly different coefficient values (Richards, 1976a):

$$\text{riffles: } w = 4.54\, Q^{0.33} \tag{5.27}$$

$$\text{pools: } w = 3.85\, Q^{0.35} \tag{5.28}$$

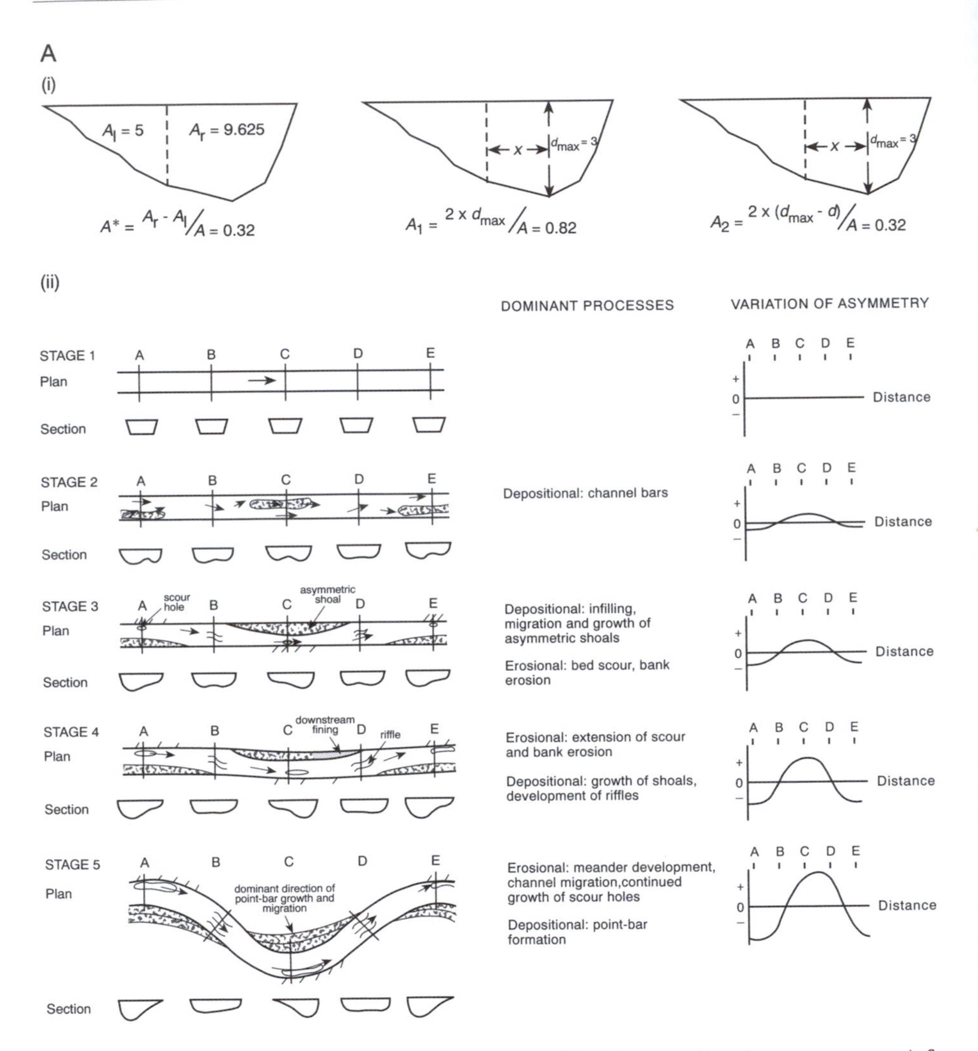

Figure 5.9 Local variations in cross-sectional form. (A) Cross-sectional asymmetry – definition of asymmetry indices (i) and suggested mode of development and variation (ii). (B) (opposite) Mean depth and width series of the bankfull channel along two lengths of a gravel-bed stream. The auto-regressive models for the detrended series are given, where b_s is a random term.

Compared with the downstream change in riffle depth, depth differences between adjacent riffles and pools were equivalent to a 350 per cent increase in drainage area.

The riffle–pool sequence is an oscillatory bed form (Keller and Melhorn, 1978), which raises the question of lag correlation in channel geometry series at the local scale where a given form variable (y) at some point s may be related to previous (upstream) values at $s-1$, $s-2$, . . . of either another form variable (x):

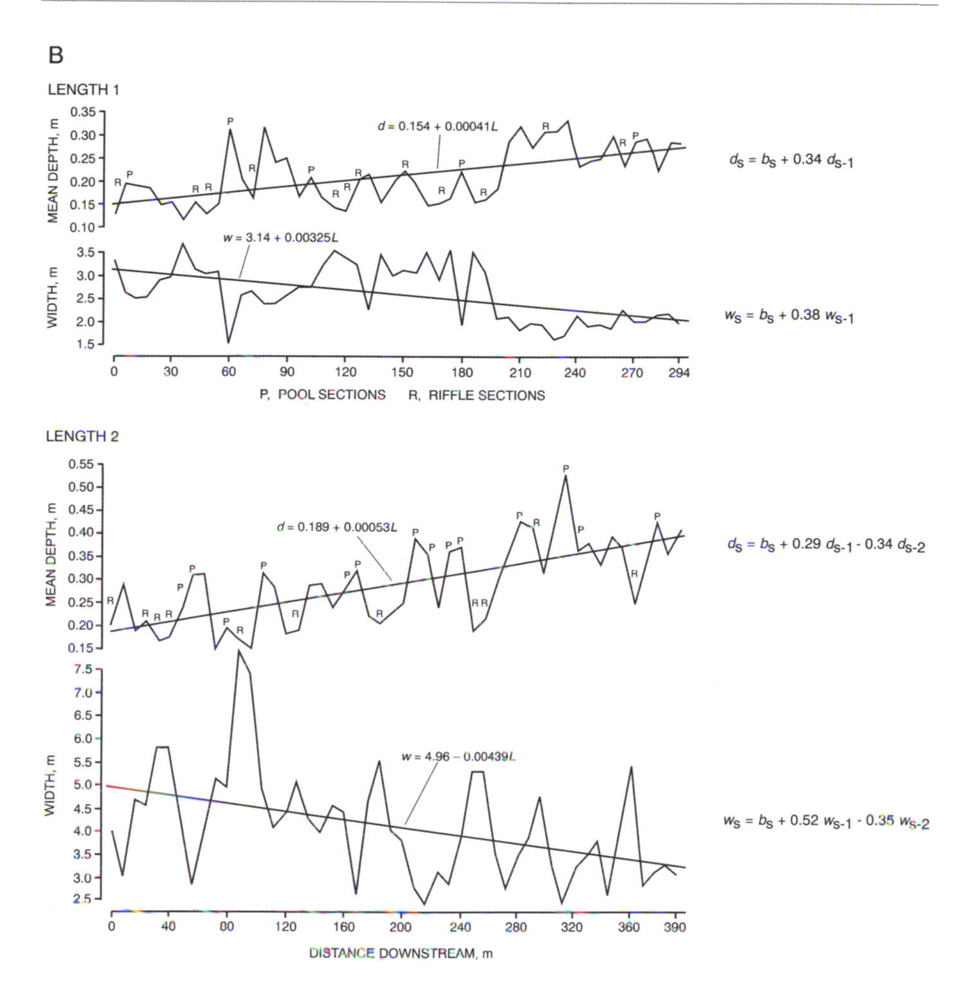

Figure 5.9 (B)

$$y_s = a_0 + a_1x_{s-1} + a_2x_{s-2} + \ldots + a_px_{s-p} + \text{(other terms)} \qquad (5.29)$$

or itself (autocorrelation):

$$y_s = a_0 + a_1y_{s-1} + a_2y_{s-2} + \ldots + a_py_{s-p} + \text{(other terms)} \qquad (5.30)$$

As regards the first equation (5.29), Richards (1976a) found that width was related to bed height (H) lagged by one increment:

$$w_s = 2.21\, H_{s-1} \qquad (5.31)$$

suggesting that width fluctuations are a product of flow characteristics induced by upstream changes in bed topography. It provides one explanation for the tendency of riffle widths to exceed pool widths since bed height increases over a riffle, deflects the flow towards one or both banks and thus promotes undercutting. Models of the second type (equation (5.30)) have been fitted to width and depth series measured along separate lengths of a small gravel-bed stream (Knighton, 1981b; Fig. 5.9B). Figure 5.9B shows not only the extent to which cross-sectional form might fluctuate over quite short river distances, but also the close association between that form and an oscillating bed topography. That association becomes better structured further downstream, as evidenced by the second-order models fitted to the Length 2 series. It is also interesting to note that along both lengths there are small but significant trends in width and depth, and that they are of opposite sign, width decreasing and mean depth increasing over lengths where there are no tributary inputs.

Underlying the main downstream trends are local fluctuations in cross-sectional form of both a random and a systematic nature. As regards the latter, those fluctuations can be linked to other channel form elements and to the action of physical processes. Thus, bank stability and the local adjustment of channel width have been associated with the process of scour and fill, sections which scour at high discharges, tending to become wider (Andrews, 1982). Lane (1995) advocates a shift in emphasis from the catchment-scale generalization of relationships between surrogate process variables such as discharge and channel form, which is largely descriptive, to a smaller-scale, more intensive study of physical processes and their link to channel adjustment, which provides a stronger explanatory focus.

At-a-Station Hydraulic Geometry

Unlike downstream hydraulic geometry, which deals with spatial variations in channel properties at some reference discharge, at-a-station hydraulic geometry deals with temporal variations in flow variables as discharge fluctuates at a cross-section, usually for a range of discharges up to bankfull (Fig. 5.10A). The relations retain the same basic form (equations (5.1)–(5.7)) but greater emphasis is placed on the adjustment of flow geometry. That adjustment can be subdivided into three main phases (Fig. 5.10B): a residual phase ($Q < Q_t$) below the threshold discharge for entrainment (defined by Ferguson, 1994), when flow characteristics are determined largely by the cross-sectional form left over from a previous high flow; an active phase ($Q_t < Q < Q_b$) when the bed becomes mobile (further subdivision could be undertaken here to reflect the different phases of bed-load transport identified in Chapter 4 and the possibility of bank erosion); and an overbank phase ($Q > Q_b$), when overspill on to the floodplain occurs and there is usually a marked discontinuity in the response of the hydraulic variables as stream width expands rapidly. The difference ($Q_b - Q_t$) will vary along and between rivers, equalling zero in a threshold channel, and being smaller in a gravel- than a sand-bed reach. An extensive period of observation is usually required to define the at-a-station hydraulic geometry adequately, covering a number of above-threshold discharges and their variable effect on the channel, so the relations do not generally describe adjustment

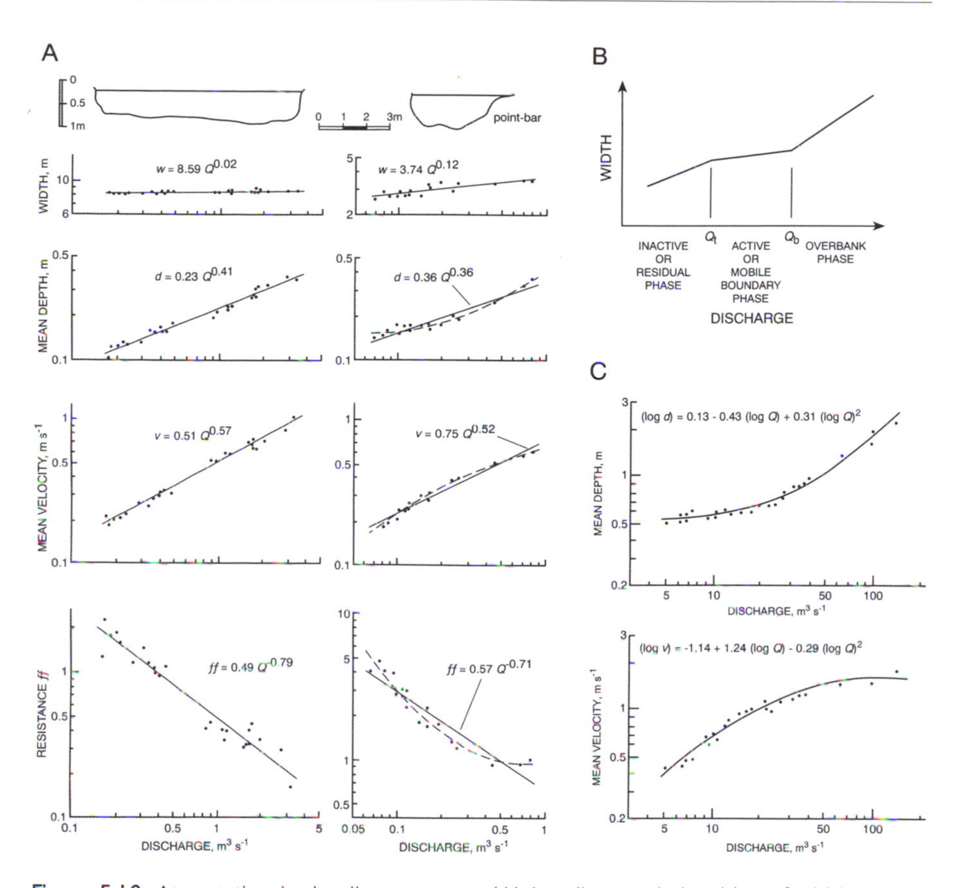

Figure 5.10 At-a-station hydraulic geometry. (A) Log-linear relationships of width, depth, velocity and resistance to discharge at two cross-sections, River Bollin-Dean. The cross-sections are shown at bankfull stage. The dashed lines in the second set of graphs indicate log-quadratic tendencies. (B) Hypothetical width adjustment to illustrate phases of adjustment. (C) Log-quadratic relationships of depth and velocity to discharge (after Richards, 1977b).

during a single large flow event, although they have occasionally been used for that purpose (Leopold *et al.*, 1964).

Only two theoretical developments deal explicitly with the at-a-station case: Li *et al.*'s (1976) threshold theory, which is restricted to small gravel streams, and Langbein's (1964a, 1965) minimum variance hypothesis. The latter is more flexible in that hydraulic exponents can be predicted for a wide range of channel types. It adopts a probabilistic standpoint in maintaining that only average or most probable relationships can be defined. The basic postulate is that, as well as satisfying the physical requirements of continuity (equation (5.8)), sediment transport and flow resistance, a stream adjusts to increasing discharge by minimizing the total variance of its dependent variables, where the relevant variances are the squares of the hydraulic exponents. Thus if width, depth and velocity are the only adjustable variables, the variance sum to be minimized is

$$b^2 + f^2 + m^2 \tag{5.32}$$

Table 5.6 illustrates a minimum variance calculation where width adjustment is constrained by $b = 0.19f$, an example applicable to a section with cohesive but non-vertical banks (Williams, 1978b).

The hypothesis has been criticized for its lack of direct consideration of sediment transport and for mathematical liberties in the original derivation, although alternative derivations are possible (Knighton, 1977a; Williams, 1978b). It depends on the log-linear form of hydraulic geometry, which may

Table 5.6 An example of a minimum variance calculation

(1) Define the variables, constraints and variances involved:

	Variable	Relation	Variance	Constraint
Independent	Discharge	$Q \propto Q^1$	1	
Dependent	Width	$w \propto Q^b$	b^2	$b = 0.19f$
	Depth	$d \propto Q^f$	f^2	
	Velocity	$v \propto Q^m$	m^2	
	Slope	$s \propto Q^z$	z^2	$z = 0$ (slope constant)
	Shear	$\tau \propto Q^{(f+z)}$	$(f + z)^2 = f^2$ since $z = 0$	
	Darcy-Weisbach friction	$ff \propto Q^{(f+z-2m)}$	$(f + z - 2m)^2 = (f - 2m)^2$ since $z = 0$	

(2) Define the variance sum to be minimized:

$b^2 + f^2 + m^2 + z^2 + (f + z)^2 + (f + z - 2m)^2$

$= (0.19f)^2 + f^2 + m^2 + f^2 + (f - 2m)^2$ since $b = 0.19f$, $z = 0$

$= (0.19f)^2 + f^2 + (1 - 1.19f)^2 + f^2 + (3.38f - 2)^2$ since $b + f + m = 1$ $\Leftrightarrow m = 1 - 1.19f$

$= 14.88f^2 - 15.82f + 5$

(3) Set the derivative equal to zero and solve to find the minimum:

$\frac{d}{df} = 29.76f - 15.82 = 0 \quad \therefore f = 0.53$

whence $m = 1 - 1.19f = 0.37$

$b = 0.19f = 0.10$

$(f - 2m) = -0.21$

The minimum variance adjustment for the given condition is:

$w \propto Q^{0.10}$ $d \propto Q^{0.53}$ $v \propto Q^{0.37}$

$\tau \propto Q^{0.53}$ $ff \propto Q^{-0.21}$

not be widely applicable (Richards, 1973; Ferguson, 1986). A further problem is the selection of variables to be included in a minimum variance adjustment, although Williams (1978b) has shown that Langbein's original choice of width, depth, velocity, shear and Darcy–Weisbach friction is the most reliable. Miller (1991a) considered an alternative interpretation of Langbein's hypothesis, that of equable change among the hydraulic variables which respond to changing discharge, but it is the underlying tenet of the hypothesis that has been dismissed by Ferguson (1986) as unhelpful, unjustified and unnecessary. Furthermore, Phillips (1990b) regards at-a-station hydraulic geometry as basically unstable, an argument which does not preclude equilibrium but suggests that a change or perturbation in one or more of the system components will result in a new equilibrium rather than restoration of the pre-disturbance condition. The equilibrium is therefore of the unstable kind (Fig. 5.4), implying that no single approach is likely to be universally applicable for predicting at-a-station relationships.

Numerous at-a-station analyses have now been carried out but no overall pattern has emerged regarding the variation in exponent values (Park, 1977), although Rhodes (1977) has identified the most common types of channel:

(i) $m > f > b$ and $m > f + b$
(ii) $f > m > \frac{2}{3}f$ and $f > b$

Such is the scale of variation that the relevance of an average or most probable flow geometry can be questioned, lending support to Phillips' (1990b) argument. However, local factors may explain at least part of the considerable variety. Differences may be related to channel pattern (Knighton, 1975a) or to bed topography (Richards, 1977b; Clifford and Richards, 1992), with adjacent riffle and pool sections being distinguishable by

$$f_{\text{riffle}} > f_{\text{pool}}$$

$$m_{\text{riffle}} < m_{\text{pool}}$$

Also, bank stability and the composition of the channel boundary seem to have a marked effect on at-a-station adjustment (Knighton, 1974; Williams, 1978b; Table 5.7).

Despite the variation in b, f and m values, Ferguson (1986) regards at-a-station hydraulic geometry as determinate. The cross-sectional shape of the channel inherited from the last flood determines the rate of change of stream width with depth ($\Rightarrow w=f(d)$), while standard flow resistance equations determine the rate of change of velocity with depth ($\Rightarrow v=f(d)$). Since $w=f(d)$ and $v=f(d)$ are not necessarily power functions, neither are the relationships between width, depth, velocity and discharge. Thus Ferguson finds the considerable variation in at-a-station exponent values unsurprising given the wide variety of channel shapes and frictional characteristics. The analysis assumes that the cross-section is inherited with a rigid boundary

Table 5.7 At-a-station hydraulic geometry relations

Source	Location/ applicable conditions	Average exponent values						Number of stations
		b	*f*	*m*	*y*	*p*	*j*	
(i) Empirical								
Leopold and Maddock (1953)	Mid-west USA	0.26	0.40	0.34			2	20
Wolman (1955)	Brandywine Creek, Pennsylvania	0.04	0.41	0.55	–0.20		1.88	7
Leopold and Miller (1956)	Ephemeral streams, semi-arid USA	0.25	0.41	0.33			1.3	10
Lewis (1969)	Rio Manati, Puerto Rico	0.17	0.33	0.49				10
Wilcock (1971)	R. Hodder, coarse bed, cohesive banks	0.09	0.36	0.53				9
Knighton (1975a)	R. Bollin-Dean, coarse bed, cohesive banks	0.12	0.40	0.48	–0.24	–0.61	2.04	12
Harvey (1975)	R. Ter, cohesive banks	0.14	0.42	0.43				8
(ii) Theoretical								
Li et *al.* (1976)	Threshold theory	0.24	0.46	0.30				
Langbein (1965)/ Williams (1978b)	Minimum variance theory:							
	(i) Vertical banks ($b = 0$)	0	0.57	0.43	–0.05	–0.29		
	(ii) Cohesive but non-vertical banks	0.10	0.53	0.37	–0.01	–0.21		
	(iii) Non-cohesive boundary	0.48	0.30	0.22	–0.02	–0.14		

Symbols: $w = aQ^b$, $d = cQ^f$, $v = kQ^m$, $n = tQ^y$, $ff = hQ^p$, $Q = rQ^j$.
Note: it is generally assumed that $s \propto Q^0$ at a cross-section.

which governs how the flow geometry will behave, but that rarely applies over the entire range of contained flow because the channel bed at least usually becomes mobile at sub-bankfull discharges (i.e. $Q_t < Q_b$ in Fig. 5.10B), giving rise to flow–form interactions. In large sand-bed rivers the most effective discharge, let alone the threshold discharge, is well below bankfull (Table 5.2; Eschner, 1983). At-a-station hydraulic geometry is rarely concerned only with a single large flow and its aftermath, or with sub-threshold discharges. Hydraulic geometry relationships can change over time as a result of flow–form interactions (Knighton, 1975a, 1977b; Eschner, 1983), providing another source of variability as well as adding grist to Phillips' (1990b) instability mill.

At-a-station width exponents are usually much less than their downstream counterparts, as well as being more variable (Tables 5.4 and 5.7). The composition of the channel banks can significantly influence the rate of width adjustment so that, where banks are particularly cohesive and almost vertical, water-surface width remains approximately constant with changing discharge (Fig. 5.10A; Knighton, 1974). Given the continuity requirement (equation (5.10)), any constraint on b will affect the rates of change of depth (f) and velocity (m). The velocity exponent (m) is generally much higher than in the downstream case.

The rates of change of depth and velocity depend on sediment load and flow resistance. Bed-load channels are characteristically wide and shallow with relatively small f and high m values, an apparent requirement for competence to increase with discharge (Wilcock, 1971). Certainly sediment load is sensitive to velocity changes (Colby, 1964). Leopold and Maddock (1953) obtained a direct correlation between the m/f ratio and the rate of change of suspended load (j in Table 5.7), suggesting that velocity must adjust more rapidly to accommodate higher rates of increase of suspended load. However, the links between at-a-station hydraulic geometry and sediment transport are poorly established.

Flow resistance is a major element in the interaction between the fluid flow and the channel boundary. It consists of several components, of which boundary resistance resulting from the frictional effects of the bed material itself and the bed forms developed therein is the most important. In general, resistance decreases with increasing discharge as the effects of grain roughness in particular are drowned out, giving a negative value for y in $n \propto Q^y$ (Fig. 5.10A; Table 5.7). However, changes in resistance with increasing discharge may be more complex than the simple log-linear model suggests, one alternative being log-quadratic relations for describing depth and velocity changes (Richards, 1973):

$$\log d = f_1 + f_2(\log Q) + f_3(\log Q)^2 \qquad (5.33)$$

$$\log v = m_1 + m_2(\log Q) + m_3(\log Q)^2 \qquad (5.34)$$

Subject to the requirement that $f_3 = -m_3$ (Knighton, 1979), depth curves tend to be concave-upward ($f_3 > 0$) and velocity curves concave-downward ($m_3 < 0$) (Fig. 5.10A and C). The form of the latter can be explained by the declining rate of resistance decrease at higher discharges (Figs 4.2B (p. 102) and 5.10A), which is caused by (a) the drowning-out effect being more successful at lower flows, and (b) the increase in form roughness as dunes develop and enlarge at higher flows in sand-bed streams. Both effects lead to a slowing in the rate of change of velocity.

The variability of at-a-station hydraulic geometry inhibits the drawing of simple conclusions. In broad outline, the width exponent (b) appears to be largely a function of channel geometry and therefore boundary composition, while the rates of change of depth and velocity are dependent partly on cross-sectional form and partly on transport- and resistance-related factors which tend to be more variable. It is perhaps significant that the most consistent set of at-a-station relations, that of Wolman (1955), was obtained for straight-reach sections along a single stream with relatively uniform bank material. In a channel with a rigid boundary or one with Q_t close to Q_b (Fig. 5.10B), Ferguson's (1986) determinate argument seems reasonable, but the situation becomes more complicated as (Q_b − Q_t) increases. The original log-linear formulation may not be capable of handling more complex situations, giving way to log-quadratic (Richards, 1973), segmented (Bates, 1990) and other forms of analysis (Rhoads, 1992). Discontinuities in relationships have been noted at the threshold discharge (Q_t in Fig. 5.10B), particularly in the velocity–discharge plot (Hickin, 1995). Nevertheless, at-a-station hydraulic geometry provides a valuable means of describing and analysing flow behaviour at the cross-sectional scale, with implications for in-stream ecology and river management (Mosley, 1983). Geomorphologically, it can help to explain how local variations in flow behaviour are involved in the development and maintenance of reach-scale channel forms such as the riffle–pool sequence (Ferguson, 1986).

Adjustability

Studies of rivers subject to high flood discharges or altered hydrologic conditions indicate that cross-sectional form is one of the most adjustable components of channel geometry, at least in the width dimension. Along the Gila River in Arizona between 1905 and 1917, a series of large winter floods carrying low sediment loads destroyed the floodplain and widened the channel from about 90 m to 610 m (Burkham, 1972; Fig. 6.9B, p. 299). From 1918 onwards the floodplain was reconstructed by smaller floods carrying large loads so that by the 1960s the channel had almost regained its former width. The ability of a river to recover from such extreme events depends on the supply of sufficient fine material for channel reconstruction and the rate of vegetative regeneration. Channels in more arid areas where vegetation growth is inhibited tend to have not only longer recovery times but also greater susceptibility to such events.

Rivers adjust their cross-sections in response not only to isolated events but also to more sustained changes of either a natural or anthropogenic origin. Downstream from reservoirs where flood magnitudes and sediment load are much reduced, decreases in bankfull cross-sectional area of over 50 per cent are not uncommon (Petts, 1979). In the Platte River system, where peak and mean annual discharges have decreased to 10–30 per cent of their pre-dam values, channel widths have been reduced by equivalent amounts over 40–60 years (Williams, 1978c), the narrowing process tending to lag behind the flow reduction by up to 15 years (Fig. 6.11D, p. 310). There also, large but inconsistent changes in channel depth have accompanied the complex regulation of water and sediment delivery to the rivers. Change in the opposite direction is illustrated by the Cheslatta River, which widened its channel from 5 m to 75–100 m and entrenched itself 10–15 m below the former floodplain over a period of 20 years after flow diversion had increased mean annual discharge fifty-fold (Kellerhals *et al.*, 1979).

Width and depth can clearly adjust rapidly to changing conditions, although the scale and rate of adjustment are likely to vary considerably. Thus, Williams and Wolman (1984) found instances of increasing, decreasing and constant width below dams in the western United States. There will be a time lag between cause and effect, especially in the case of channel narrowing, which requires the import and systematic redistribution of large amounts of material to form lateral berms and bars. The sensitivity of cross-sectional form raises the question of a river's ability to attain and maintain a stable width and depth, particularly where a single large event can produce substantial change. Whether or not transient behaviour is prevalent, it is interesting that the Gila River returned approximately to its pre-flood condition even though the process took 50 years (Fig. 6.9B). In more humid environments where vegetation and material properties give greater stability to channel banks, extreme floods appear to have less effect (Costa, 1974; Gupta and Fox, 1974), so that a mean channel geometry can more readily be maintained.

BED CONFIGURATION

Natural streams rarely have flat beds. Shear stresses above the critical for transport will mould cohesionless beds into discernible forms whose geometry depends on flow characteristics which in turn are influenced by those forms, resulting in complex feedback relationships. Bed forms are symptomatic of local variations in the sediment transport rate, which can give rise to quasi-regular sequences of erosions and depositions. In effect bed forms represent an important means of adjustment in the vertical dimension (Fig. 5.2), related nevertheless to the transverse and horizontal adjustments that natural streams can also make.

Numerous classifications of bed forms have been proposed, the main elements of which are summarized in Table 5.8. Bars are larger-scale features having lengths of the order of the channel width or greater and are generally classified according to their shape and position. Composed of a wide range of grain sizes, these several types of bar are usually exposed at certain stages of flow and have important links with channel pattern. At a smaller scale are various concealed bed forms most commonly associated with sand-bed streams, the most widely recognized sequence being ripples, dunes, plane bed and antidunes (Fig. 4.3A, p. 104). Because of its relationship to sediment transport and flow resistance, considerable effort has been invested in the study of this sequence, particularly in laboratory flumes. In contrast, the range of small-scale bed forms developed in gravel-bed streams has only recently been recognized, a basic classification distinguishing between those with long axes parallel to the flow (pebble clusters: Fig. 4.3B) and those with long axes transverse to the flow (transverse ribs, transverse clast dams, step–pool systems) (Robert, 1990). The latter appear to

Table 5.8 A classification of bed forms

Bed form	Dimensions	Shape	Behaviour and occurrence	
Small-scale forms	10^{-2}–10^{2} m			
(1) Sand-bed streams				
Ripples	Wavelength less than 0.6 m; height less than 0.04 m	Triangular profile; gentle upstream slope, sharp crest and steep downstream face	Generally restricted to sediment finer than 0.6 mm; discontinuous movement at velocities much less than those of the flow	Lower regime of roughness; form roughness dominant
Dunes	Wavelength of 4 to 8 times flow depth; height up to a third of flow depth; much larger than ripples	Similar to ripples but more variable	Upstream slope may be rippled; discontinuous movement; out of phase with water surface	
Plane bed			Bed surface devoid of bed forms; may not occur for some ranges of depth and bed material size	Upper regime of roughness; grain roughness dominant
Antidunes	Relatively low height, dependent on flow depth and velocity	Sinusoidal profile; more symmetrical than dunes	Less common than dunes, occurring in steep streams; in phase with surface water waves; bed form may move upstream, downstream or remain stationary	
(2) Gravel-bed streams				
Pebble clusters	10^{-1}–10^{0} m	Linear in flow direction	Streamlined; consist of a coarse obstacle clast (~ D_{90}) with collections of stoss- and lee-side particles	
Ribs	10^{0}–10^{1} m	Transverse to flow direction	Repeated ridges of coarse clasts, whose spacing is roughly proportional to the size of the largest particle in the ridge crest	
Steps	10^{0}–10^{1} m	Transverse to flow direction	Stair-like sequence formed by arrangement of boulders and cobbles across steep-gradient channels	
Large-scale forms	10^{1}–10^{3} m			
Bars	Lengths comparable to channel width	Variable	Five main types: (1) Point bars: form particularly on the inner bank of meanders (2) Alternate bars: distributed periodically along one and then the other bank of a channel	

Table 5.8 continued

Bed form	Dimensions	Shape	Behaviour and occurrence
			(3) Channel junction bars: develop where tributaries enter a main channel (4) Transverse bars (include riffles): may be diagonal to the flow (5) Mid-channel bars: typical of braided reaches

form a continuum related to clast size and are characteristic of steep channels with rapid, shallow flow (Richards and Clifford, 1991). The formation of these various bed features indicates the presence of systematic tendencies in the ability of natural streams to sort and transport material over a wide range of flow and bed material conditions.

Ripples, Dunes and Antidunes

Bed forms are classified into lower and upper flow regimes according to their shape, resistance to flow, and mode of sediment transport (Simons and Richardson, 1966): *lower flow regime* – plane bed, ripples, dunes; and *upper flow regime* – plane bed, antidunes, with a transition zone between the two where bed configuration ranges from washed-out dunes to plane bed (Fig. 4.3A). Phase diagrams have been prepared in an attempt to predict the occurrence of the various bed forms, defined most often in terms of particle size and some function of the flow such as unit stream power (= γqs) or boundary shear stress (Fig. 5.11). The diagrams indicate that, for a given bed material size in the sand range, the sequence of ripples, dunes, plane bed and antidunes is correlated with increasing flow intensity. Relying largely on the results of controlled flume experiments, such diagrams need to be applied with caution to natural streams where flow conditions can vary markedly over the stream bed and during the passage of a single flood.

Starting with a flat sandy bed (lower-stage plane bed), some sediment transport may take place over the surface at shear stresses just above the entrainment threshold. However, the bed is deformed at relatively low competent stresses into small wavelets instigated by the random accumulation of sediment (Coleman and Melville, 1996), and then into ripples which are roughly triangular in profile, with gentle upstream and steep downstream slopes. Rarely occurring in sediments coarser than 0.6 mm, ripples are usually less than 0.02 m in height and 0.6 m in wavelength, dimensions which are seemingly independent of the flow depth. Wavelength and, to a much lesser extent, height, increase with grain size (Raudkivi, 1997). Initiated by the turbulent bursting process, the small bed features translate at speeds inversely proportional to their height, which leads to coalescence and the development of a more orderly pattern. Ripple wavelength grows

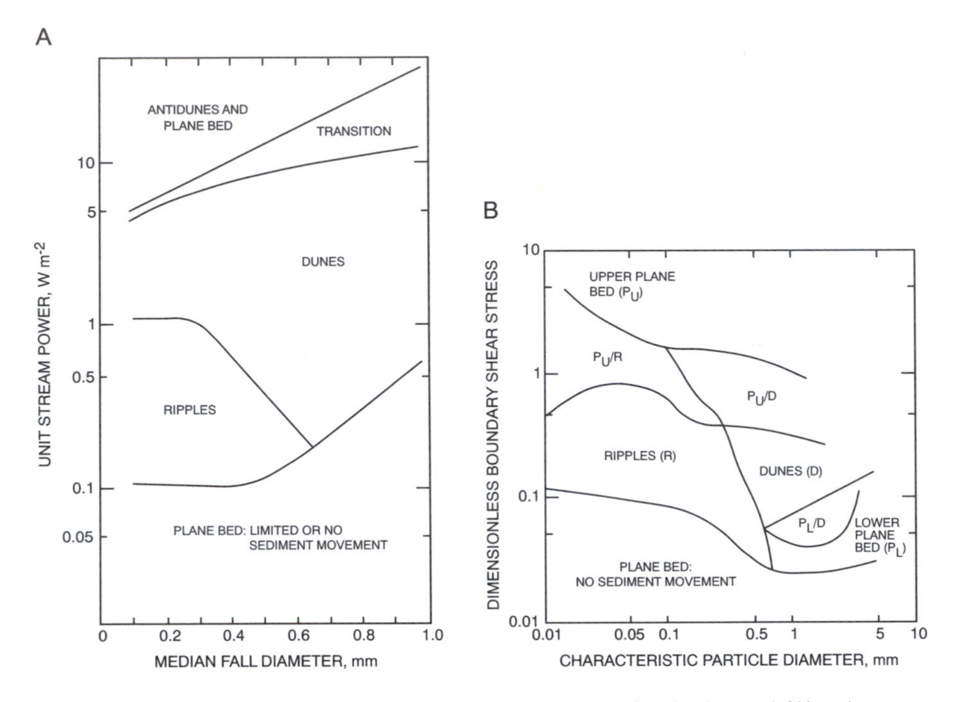

Figure 5.11 Existence fields for bed forms defined in terms of grain size and (A) unit stream power (after Simons and Richardson, 1966) or (B) boundary shear stress (after Allen, 1983).

with time until it reaches an equilibrium value, more than doubling in the process (Raudkivi, 1997).

As shear stress increases further, ripples are overtaken and eventually replaced by dunes, the commonest type of bed form. Although superficially similar, they can be distinguished from ripples by their larger height and wavelength, attaining values of over 10^1 m and 10^3 m respectively in large rivers. They differ from ripples also in that their height and wavelength are directly related to water depth. On the basis of his argument that the formation of dunes is related to the structure of macroturbulent flow and in particular to the burst length (L, the longitudinal distance between the birth and decay of a macroturbulent eddy; Fig. 5.12), Yalin (1992) obtained a relationship linking dune wavelength (Λ_d) and flow depth (d):

$$\Lambda_d \equiv L \approx 6d \tag{5.35}$$

which seems to fit measurements made during large-river floods quite well (Julien and Klaassen, 1995). The passage of dunes is a major cause of variability in bed-load transport rates in sand-bed streams (Gomez, 1991), as well as having implications for suspension transport.

Dunes are eventually washed out to leave an upper-stage plane bed characterized by intense bed-load transport, which prevents the patterns of erosion and deposition required for the formation of three-dimensional bed

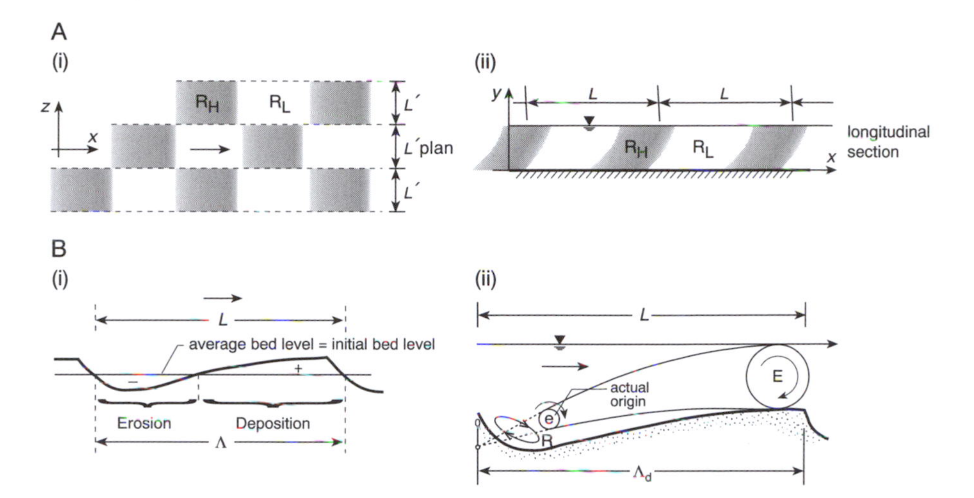

Figure 5.12 Burst cycles (after Yalin, 1992). (A) Alternating regions of high-speed (R_H) and low-speed (R_L) flow (i) in plan and (ii) in longitudinal section. (B) (i) Bed form represented as an erosion–deposition sequence of wavelength Λ; (ii) burst-forming eddies (e) over a dune bed – they originate at the interface between the main flow and the 'roller' (R) at the downstream dune face, and terminate (E) at the dune crest.

forms. As flow intensity again increases, standing waves develop at the water surface and the bed is remoulded into a train of sediment waves (antidunes) which mirror the surface forms. Antidunes are more transitory and much less common than dunes, forming in broad, shallow channels of relatively steep slope when the sediment transport rate and flow velocity are particularly high. The degree of regularity in these bed forms is variable and often a wide range of wavelengths and heights is present, even under the controlled conditions of a laboratory flume. However, there is usually enough regularity for a distinct pattern to be recognized.

Although the mechanisms responsible for their formation are not fully understood, the bed forms do possess certain common characteristics. All are the result of an orderly pattern of scour and deposition. In particular, ripples and dunes move downstream through erosion of their upstream slope and deposition on their downstream face at velocities which are small compared with those of the flow (Fig. 4.9A, p. 127). In the North Fork Toutle River, dunes composed of coarse sand and fine gravel migrated downstream at an average rate of 0.03 m s^{-1} with flow velocities in the range of 1.6–3.4 m s^{-1} (Dinehart, 1989). The migration of ripples and dunes generates structures in the form of cross-bedding which can be preserved and used in the palaeohydraulic interpretation of alluvial deposits. Antidunes can migrate upstream through scour on the downstream face and deposition on the upstream face (Fig. 4.9A), move downstream, or remain stationary, but develop under conditions of such rapid flow that the probability of definite structures being constructed and preserved is much lower.

The development of bed forms implies that a plane, cohesionless, granular surface is unstable where there is sediment transport. Leeder (1983) envisaged a strong interplay between turbulence, sediment transport and bed forms. Burst cycles are believed to be an inherent component of macroturbulent flow responsible for the initiation of sediment transport, with sweeps exerting high instantaneous stresses against the bed and ejections carrying sediment away from the bed. Because bed disturbance is not uniform, such cycles have an important role in the initiation and maintenance of bed forms. Each bed element represents one 'erosion–deposition' sequence with a wavelength dependent on the length of the burst process (see equation (5.35); Fig. 5.12B). Best's (1992) experiments on the initiation of ripples in relation to turbulence indicated the formation of multiple hairpin vortices which in turn generate multiple sweeps. The latter then define patches of entrainment from which local erosions are generated and ripple sequences created. Sediment transport rates vary across individual bed forms as a result of form-induced accelerations and decelerations, promoting scour in the troughs and deposition towards the crests. The amplitude of the initial bed undulations and the length of the burst process (or size of the macroturbulent eddy) increase through positive feedback until a limit is reached, which according to Yalin (1992) is defined by equation (5.35). Bagnold (1956) showed theoretically that the formation of ripples and dunes is necessary if some degree of stability is to be achieved during sediment transport. Without the additional resistance provided by these bed forms, the channel could be destroyed as a coherent structure for the conveyance of water and sediment. However, there are few data to suggest how that resistance affects the turbulent structure.

Bed forms exert a drag on the flow additional to that associated with the grains themselves. On the upstream side of an obstacle, the rising bed elevation causes an acceleration of the flow and a commensurate decrease in pressure. Beyond the crest of the obstacle, depth and pressure increase, and the flow decelerates. This effect, which produces pressure gradients, gives rise to form drag. Flume data obtained by Simons and Richardson (1966) show how resistance, expressed by the Darcy–Weisbach friction factor ff ($= 8gRs/v^2$), varies with the type of bed form:

Lower flow regime
Ripples: $0.052 \leq ff \leq 0.13$
Dunes: $0.042 \leq ff \leq 0.16$
Upper flow regime
Plane bed: $0.02 \leq ff \leq 0.03$
Antidunes: $0.02 \leq ff \leq 0.07$

Regarding plane-bed friction as indicative of the grain roughness component, the data show that form roughness can make a large contribution to total flow resistance, especially in the lower flow regime (Fig. 4.3A, p. 104). Flow separation occurs on the downstream side of dunes in particular, generating large-scale turbulence and causing considerable energy dissipation.

Form roughness depends on the size, shape and spacing of the bed waves but cannot easily be predicted (Allen, 1983). Van Rijn (1984) expressed it as a function of two parameters, bedform height (Δ) and bedform steepness ($\Psi = \Delta/\Lambda$) in:

$$k_{form} = 1.1\ \Delta(1 - e^{-25\Psi}) \tag{5.36}$$

which indicates that, relative to height, ripples produce greater form roughness than dunes because of their greater steepness. However, the model achieves only moderate success, underlining the severity of the prediction problem. These several results show, none the less, that a stream can alter the local resistance to flow and consequently the magnitude of any resistance-dependent variable by modifying its bed configuration. Via their extremal hypothesis, Davies and Sutherland (1980) believed that such deformation would continue until the friction factor was maximized for ambient flow conditions.

On the one hand, bedform adjustment represents a response to changing discharge and load conditions during, for example, the passage of a flood wave, although there is always a lag between a change of flow and a corresponding change of bed form because of the redistribution of sediment involved. That lag, which has a spatial as well as a temporal dimension, will increase with the size of both the bed form and the river (Allen, 1983). On the other hand, bedform adjustment represents a means of modifying, principally through its effect on resistance, hydraulic variables such as velocity and depth which influence the local transport rate. The introduction of log-quadratic relations (equations (5.33) and (5.34)) into at-a-station hydraulic geometry was largely in response to non-linear bedform effects (Richards, 1973). Under certain conditions a given discharge can be transmitted at two or more different depths and velocities, depending on the type and size of bed form produced. Bed configuration in sand-bed streams is one of the most adjustable components of channel morphology, with the potential for regulating short-term interactions between hydraulic variables.

The Riffle–Pool Sequence

The development of alternating deeps (pools) and shallows (riffles) is characteristic of both straight and meandering channels with heterogeneous bed material in the size range of 2–256 mm (Fig. 5.13A). Pools are especially associated with meander bends and often have a lateral bar even in straight reaches, giving the cross-section a distinctly asymmetric shape. The gravel accumulations which form the intervening riffles are generally lobate in shape and frequently slope alternately first towards one bank and then towards the other, so that the flow tends to follow a sinuous path even in a straight channel. The definition of individual pools and riffles in a sequence can be problematic. One simple method deals directly with the bed topography itself and defines pools and riffles as, respectively, negative and positive residuals from a trend line fitted to bed height data (Richards, 1976b).

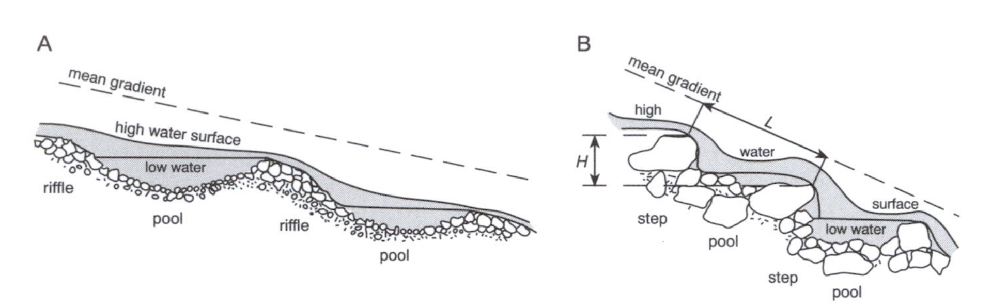

Figure 5.13 Definition sketches for riffle–pool (A) and step–pool (B) channels.

O'Neill and Abrahams (1984) devised a more objective technique based on the differences in height between successive points in a bed elevation survey.

A significant feature of riffle–pool geometry is the more or less regular spacing of successive pools or riffles at a distance of 5 to 7 times the channel width (Fig. 5.14A). The spacing distance is thus scale-related. This oft-quoted proportionality needs to be viewed against the background that both channel width and spacing distance are inherently variable even in short lengths of stream, and that both present problems of consistent definition and measurement. It describes at best an average condition. The most extensive data-set has values of pool-to-pool spacing ranging from 1.5 to 23.3 channel widths, with an overall mean of 5.9 (Keller and Melhorn, 1978). Despite these problems, the notion of rhythmic change in the bed topography of gravel-bed streams is now firmly entrenched. Even in a channel disturbed by channelization schemes and the introduction of woody debris, the inter-riffle distance generally falls within the range of 5 to 7 channel widths (Gregory *et al.*, 1994). Second-order autoregressive models (i.e. $p = 2$ in equation (5.30)) have been fitted to bed height data through riffle–pool sequences (Richards, 1976a; Clifford, 1993a), underlining the pseudo-cyclic character of this reach-scale bed form.

Riffle–pool development is traditionally associated with gravel-bed streams, although Keller and Melhorn (1978) regarded it as a fundamental characteristic independent of boundary composition. Riffles and pools tend to be absent from or poorly developed in boulder-bed streams (Miller, 1958; Leopold *et al.*, 1964), where they may be replaced by steps and pools (Whittaker and Jaeggi, 1982; Chin, 1989), and show little tendency to form in channels that carry uniform sand or silt. However, well-defined sequences have been found in bedrock and boulder-bed gorge courses (O'Connor *et al.*, 1986; Baker and Pickup, 1987; Wohl, 1992; Wohl *et al.*, 1993). Also, concentrations of coarser particles analogous to riffles and spaced at 5 to 7 times channel width have been observed in otherwise sandy ephemeral streams, even though the concentrations had no topographic expression (Leopold *et al.*, 1966). The degree of riffle–pool development probably varies with bed material size relative to the transport potential in a reach, since the ability of a stream to modify its bed depends upon the mobility of the available

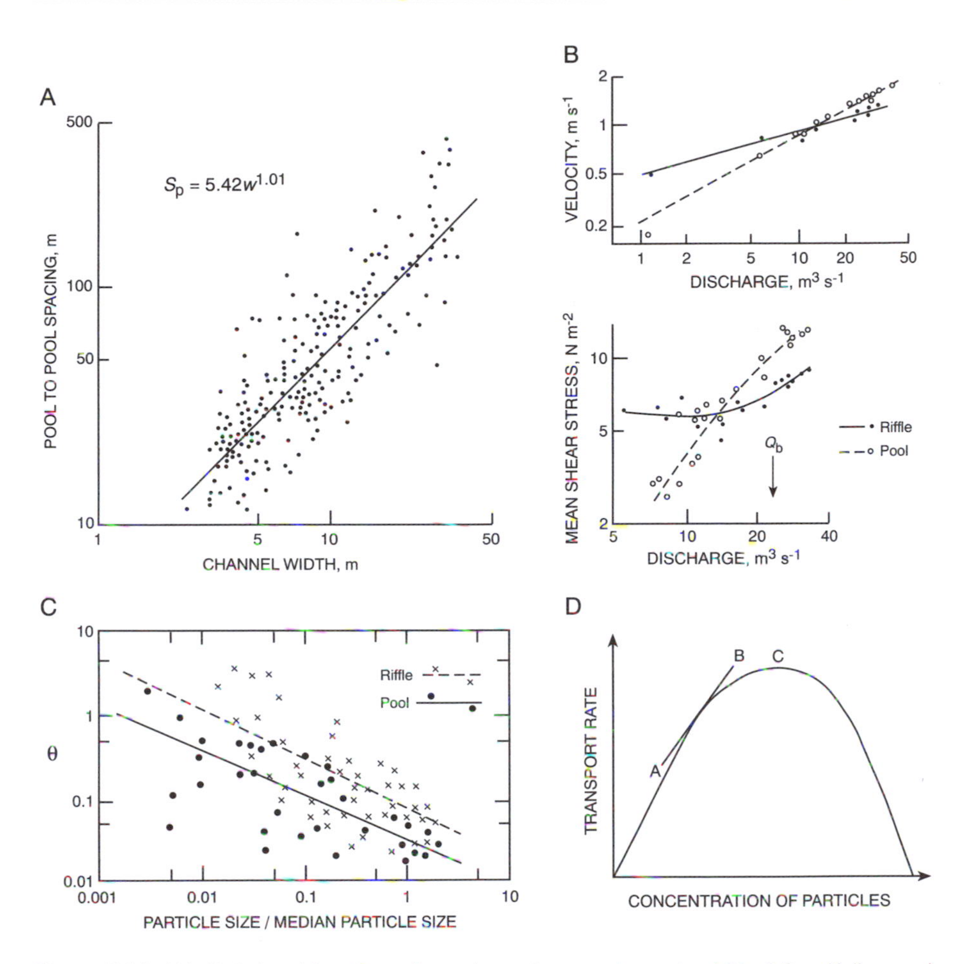

Figure 5.14 (A) Relationship of pool–pool spacing to channel width (after Keller and Melhorn, 1978). (B) Velocity and shear stress reversal at riffle and pool sections (after Andrews, 1979, and Lisle, 1979). (C) Shields entrainment function for riffle and pool data showing the higher entrainment threshold required to initiate particle movement on riffles (after Sear, 1996). (D) Relationship between transport rate and concentration in a kinematic wave, maximum transport being attained at an intermediate concentration (C) when wave velocity (given by the tangent AB) is zero.

material and the frequency of competent flows, which have a longitudinal dimension (Knighton, 1981b). Relative pool depth increases as gradient decreases along three Californian rivers because the channel boundaries become less resistant (Wohl *et al.*, 1993).

Riffles tend to have coarser bed material than do adjacent pools, suggesting the action of local sorting mechanisms (Keller, 1971; Lisle, 1979; O'Connor *et al.*, 1986). Thus bed topography and particle size characteristics are interrelated. Clifford (1993a) found that both bed height and particle size series could be described by second-order autoregressive models with very similar coefficients and an underlying spatial scale of 5–7 times channel width. However, distinctions in particle size may not always be consistent, or be small relative to lateral differences (Milne, 1982). On the basis of results from English rivers, riffles and pools have been differentiated in terms of sediment structure rather than sediment size, with tight/imbricate structures dominant at riffles and open structures dominant in pools (Clifford, 1993a; Sear, 1996). How consistent is this distinction remains to be seen but, given that structuring imparts greater mechanical stability to sediments, Clifford and Sear use it as a basis for explaining the maintenance of elevation contrasts between juxtaposed riffles and pools.

Evidence suggests that riffles and pools have distinctive channel and flow geometries. Riffles tend to be wider and shallower at all stages of flow (Richards, 1976a; Fig. 5.9B). As regards response to at-a-station increases in discharge, riffle and pool sections appear to be distinguishable by (Richards, 1976a; Lisle, 1979; Carling, 1991; Clifford and Richards, 1992):

$$f_{\text{riffle}} > f_{\text{pool}}$$

$$m_{\text{riffle}} < m_{\text{pool}}$$

$$z_{\text{riffle}} < 0, z_{\text{pool}} > 0$$

where f, m and z are respectively the rates of change of depth, velocity and water-surface slope. Since velocity and slope are greater and depth is less over a riffle than in a pool at low flows, the net effect of these differences in flow geometry is to produce a convergence or more even distribution of these flow variables along a reach at high flows. Competence, expressed by bed velocity or boundary shear stress ($= \gamma Rs$), will also tend to become more evenly distributed or may even be reversed so that, contrary to the low-flow condition, it is higher in the pools at those discharges which transport most material in gravel-bed streams (Fig. 5.14B; Keller, 1971; Lisle, 1979; Keller and Florsheim, 1993). In combining high-flow transport through pools and low-flow storage on riffles, such reversal provides a mechanism for the concentration of coarser material in riffles and for the maintenance of the riffle–pool sequence. However, the general validity of the reversal hypothesis has recently been questioned on a number of grounds, although a tendency towards convergence is not denied (Carling, 1991; Clifford and Richards, 1992). Given the lower water depths over riffles, continuity (equation (5.8)) requires that riffles should be significantly wider than pools at high flows for velocities to be lower and reversal to occur, but bank strength may prevent substantial widening at the riffles relative to the pools (Carling, 1991). Nevertheless, even without reversal, competence could still be higher at high flows in the pools if their bed sediment has a more open structure (Clifford, 1993a). Indeed, Sear (1996) found that particles in pools had a lower entrainment threshold than had particles on riffles (Fig. 5.14C).

A complete explanation of riffle–pool formation needs to consider not only how they develop but also why they develop within the broader context of stream behaviour. Basically, given an initially flat bed, a riffle–pool sequence forms through a combination of scour and deposition, organized spatially to give a more or less regular spacing between consecutive elements. In the initial stages at least, a key question is sediment mobility since that influences the extent to which coarser fractions can be concentrated into incipient riffles. Such concentrations probably occur at high discharges (bankfull and above), while lower flows (up to bankfull) may be sufficiently competent to amplify and maintain the initial undulations once they have reached a critical height. A distinction is thus drawn between those flows which form and those which maintain a riffle–pool sequence, the ranges of which will tend to vary with the quantity and grain size of sediment supplied. At what point there is a shift from formation (when spatial contrasts in flow and transporting ability continue to amplify the difference in height between the two bed elements) to maintenance (when the height difference assumes some degree of stability) remains uncertain.

Various mechanisms have been invoked to explain the development of riffle–pool sequences. In so far as riffles represent concentrations of units (coarser particles), Langbein and Leopold (1968) likened them to kinematic waves in traffic flow, the important properties of which stem from interactions between individuals in the flow. Those properties can be appreciated from a flux–concentration curve which shows that the transport rate attains a maximum at some intermediate concentration (C in Fig. 5.14D). Then also wave velocity (given by the tangent AB) is zero, implying a static waveform. Interestingly, maximization of sediment transport is the extremal condition which a stream in equilibrium must satisfy according to White *et al.* (1982). Thus, as a result of grain–grain interactions, any random influx of particles will not remain random during downstream transport but will tend to accumulate in groups having wave-like forms and a more or less regular spacing. Bed material continues to move downstream during competent flows but the position of these groups (riffles) remains fixed. Once the perturbations have been formed, they themselves can generate the flow conditions necessary for their continued development. On the basis of an extension of the kinematic wave model, Naden and Brayshaw (1987) have successfully simulated the development of two small-scale bed forms, similar in wavelength to pebble clusters and step–pool systems on the one hand and antidunes on the other. That the first of these can be so simulated ties in with observations that riffles have more tightly packed sediments than do pools (Clifford, 1993a; Sear, 1996). Clifford argues that riffles are maintained by the aggregate effect of microscale form–flow interactions. Kinematic wave theory thus provides an explanation of how material moves through riffle–pool sequences, why they are relatively stable and how they could be maintained.

Yang (1971) explained riffle formation as a combined process of dispersion and sorting, arguing that greater dispersive stresses at potential riffle sites force larger particles to the surface and thereby raise the bed. The finer grains are washed out to leave concentrations of coarse particles at the riffles which are further accentuated by this combined process until a stable state is reached. The hypothesis contains several unverified assumptions, casting doubt on the efficacy of the process.

Keller and Melhorn (1973) suggested that the regular pattern of scour and deposition required for the formation of a riffle–pool sequence may be effected by an alternation of convergent and divergent flow along the channel, combined with secondary circulation currents. Surface flow convergence at the pool induces a descending secondary current which increases the bed shear stress and encourages scour, while surface flow divergence at the riffle produces convergence at the bed and thereby favours deposition. Thompson's (1986) diagrammatic representation of the process (Fig. 5.15B) envisaged a repeated decay and regeneration of the secondary cells in association with, and as a consequence of, the developing bed forms. With pool development alternating from one side of the channel to the other, a link to meander initiation is provided. Although the approach is largely qualitative, flow patterns of alternating convergence and divergence have been observed under field conditions (Thompson, 1986; Petit, 1987).

Yalin's (1971, 1992) theoretical treatment of macroturbulent flow and the bursting process envisages alternating regions of high-speed and low-speed flow with wavelength L (Fig. 5.12A). In the case of vertical bursts, a discontinuity at some point $x = 0$ induces similar behaviour at $x = L, 2L, \ldots$, the break-up of one burst (of length L) triggering the 'birth' of the next. Such streamwise fluctuations in the velocity field could give rise to the necessary alternation of scour and deposition, the regions of high-speed flow being associated with pools. Although the original derivation applies to the formation of dunes with a characteristic wavelength of $L \approx 6d$ (equation (5.35)), it has been suggested that a fundamental velocity perturbation of $L \approx 6w$ is more appropriate to the development of riffles and pools (Richards, 1976b; Hey, 1976). On the basis of three-dimensional velocity measurements, Clifford (1993b) has argued that riffle–pool units are initiated by roller eddies following a disturbance to the turbulent flow field caused by a major obstacle, along similar lines to Yalin. Spatial contrasts in near-bed turbulence induced by incipient riffle–pool topography then create differences in entrainment which enhance and maintain the sequence through a form–process feedback. Thus the development of an alternating bed topography is related to longitudinal oscillations in the velocity field of turbulent flow.

No one explanation of riffle–pool formation is entirely satisfactory and many of the ideas remain rather speculative. Indeed, more than one process may be involved. An alliance of the longitudinal flow variations emphasized in the macroturbulent theory with the secondary circulation patterns explicit in alternating convergent–divergent flow provides one plausible possibility, for there is a correspondence between the horizontal version of the first with $w = 2L'$ (Fig. 5.12A) and the diagrammatic representation of the latter (Fig. 5.15B). Basically, the problem is to identify an initiating mechanism which will ultimately produce a quasi-regular alternation of bed topography. Once initiated, the bed perturbations interact with the flow to generate conditions necessary for their maintenance. The hypothesis of

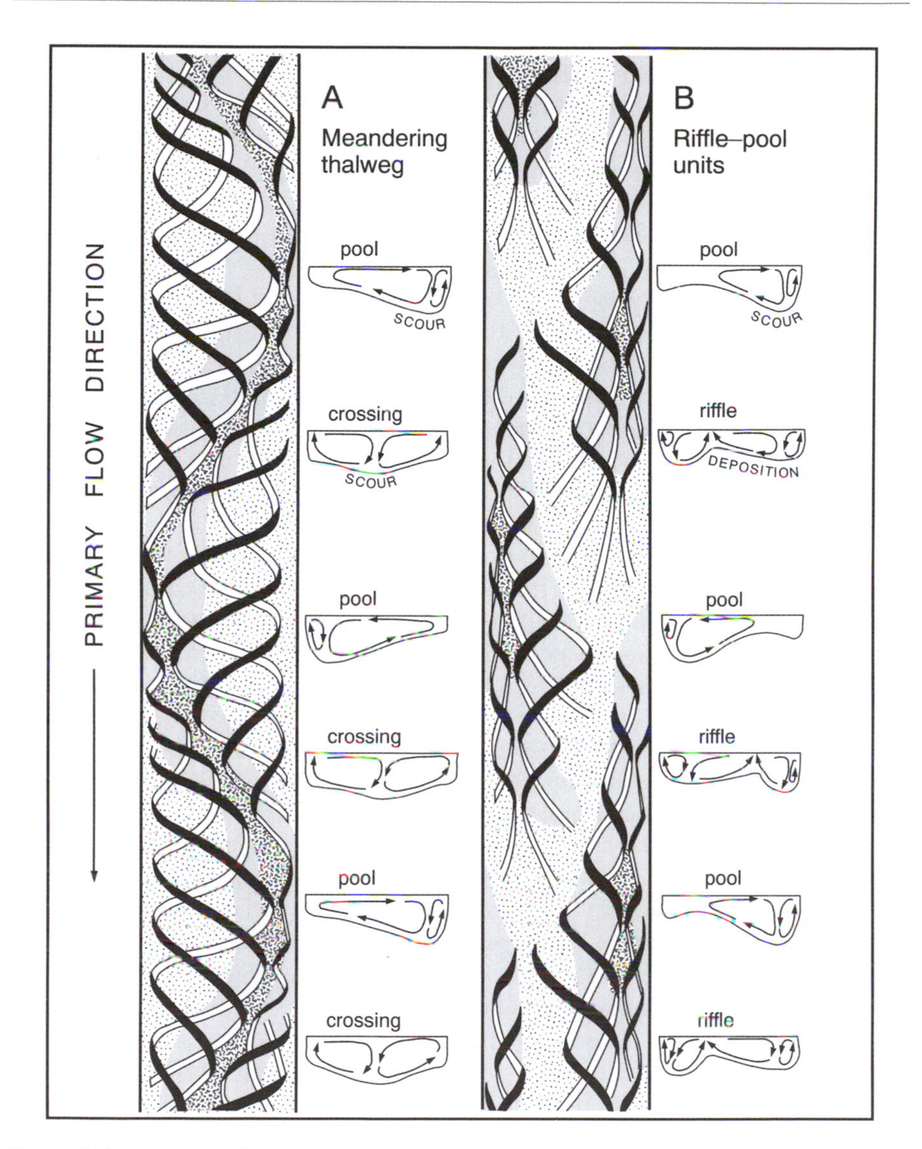

Figure 5.15 Models of flow structure and associated bed forms in straight alluvial channels. (A) Einstein and Shen's (1964) model of twin periodically reversing, surface-convergent helical cells. (B) Thompson's (1986) model of surface-convergent flow produced by interactions between the flow and a mobile bed, creating riffle–pool units of alternate asymmetry. Black lines indicate surface currents, and white lines near-bed currents.

velocity (Keller, 1971; Keller and Florsheim, 1993) or shear stress (Lisle, 1979) reversal, which gives greater flow competence in pools than over riffles above a certain range of flows (Fig. 5.14B), offers one mechanism for a pattern of scour and deposition capable of producing areal sorting of stream bed material and of maintaining the pre-existing bed topography. However, doubts have been expressed about the universality or necessity of reversal (Carling, 1991; Clifford and Richards, 1992; Sear, 1996). Flow simulations over a riffle–pool bed suggest that reversal is more likely with wide, well-spaced riffles and pools floored by coarse lag gravels (Carling and Wood, 1994), the bed material element of which is rather contrary to experience but not unknown. An alternative explanation is that greater competence can be established in pools because more tightly packed sediments delay entrainment and restrict transport over riffles (Clifford, 1993b: Sear, 1996).

As regards its broader implications, the riffle–pool sequence is seen as a means of self-adjustment in gravel-bed streams, with significance firstly for the attainment and maintenance of quasi-equilibrium, and secondly for the development of a meandering planform. Yang (1971) maintained that riffle–pool formation is one way in which a natural stream minimizes unit stream power, while the correspondence between kinematic wave stability and maximum transport (Fig. 5.14D) has interesting connotations in terms of another extremal hypothesis, that of White *et al.* (1982). Observation and theory testify to the relative stability of a riffle–pool bed (Leopold *et al.*, 1966; Langbein and Leopold, 1968; Keller and Melhorn, 1973, 1978), with riffle position remaining fixed during transport through a process of particle replacement in which the larger particles tend to move from riffle to riffle. Indeed, the largest particles in the riffles may be essentially static except during extreme flows (Lisle, 1979; Wohl, 1992). Dury (1970) found that little change had occurred in the riffle–pool morphology of a bedrock river over 100 years. Large floods can destroy a riffle–pool bed (Stewart and La Marche, 1967), but this need not necessarily be the case and the amplitude of the bed forms may indeed be increased (Gupta and Fox, 1974; Baker, 1977). In short, the riffle–pool sequence appears to be a valid equilibrium form.

Its link with meandering is based partly on the fact that the spacing of 5 to 7 times channel width is approximately half the straight-line meander wavelength and that riffles and pools in straight reaches have analogous points in meanders, namely points of inflection and pools at bend apices. Consequently several models of the transformation from a straight to a meandering pattern incorporate riffle–pool development as a significant element (Tinkler, 1970; Keller, 1972; Figs 5.9A and 5.20B, p. 222). The pattern of flow in meandering channels (Fig. 5.19C, p. 218) is regarded by Thompson (1986) as a natural development of that associated with riffle–pool units in straight channels (Fig. 5.15B). However, meanders do develop in channels without a riffle–pool bed. Also, straight reaches with a riffle–pool sequence can be stable against the tendency for meanders to form (Keller and Melhorn, 1973; Richards, 1976a).

Many of the various claims regarding the riffle–pool sequence have yet to be fully tested. Hydraulic data from gravel rivers at high discharge, a requisite for the effective testing of the reversal hypothesis, are especially meagre. Nevertheless, this bed form does illustrate the strong relationship between flow and channel geometry, and the interdependence of the several modes of channel form adjustment. In particular, the riffle–pool sequence and the meandering planform represent two sources of flow resistance capable of modifying the rate and distribution of energy loss at the reach scale, the difference between them being the plane in which the roughness principally operates. In addition to their geomorphological significance, riffle–pool sequences provide the habitats needed for the maintenance of viable populations of several fish species and a range of invertebrate fauna (Clifford and Richards, 1992).

Steps and Pools

Steep mountain streams differ from their lowland counterparts in a number of important respects. They are more strongly coupled to adjacent hillslopes which therefore become major sources of sediment supply. They are greatly influenced by large particles with diameters on the same scale as the channel depth or even width, which produce considerable energy losses. They have very episodic sediment transport which is to a large extent controlled by the availability of sediment from a limited number of sites (Whittaker, 1987). In consequence they develop a distinctive stepped-bed morphology (Grant *et al.*, 1990).

Alternating steps and pools having a stair-like appearance are a characteristic feature of mountain streams flowing over slopes greater than 3–5% (Chin, 1989; Abrahams *et al.*, 1995; Fig. 5.13B; Plates 5.1 and 5.2). Steps are typically formed from accumulations of boulders and cobbles, which span the channel in a more or less continuous line and separate a backwater pool upstream from a plunge pool downstream. The risers of individual steps are generally composed of several large boulders that act as a framework against which smaller boulders and cobbles are imbricated, creating a tightly interlocking structure with considerable stability. Given the need for one or more keystones, the development of steps is strongly influenced by local supply and transport conditions. The small pools between steps provide storage sites for finer bed material, creating a contrast in sediment size which is much sharper than that between riffles and pools. Although steps composed of boulders are the most common type, they can also form in bedrock (Hayward, 1980; Wohl and Grodek, 1994) and through accumulations of large woody debris in heavily forested catchments (Keller and Swanson, 1979). Step–pool sequences have been reported from a wide range of humid and arid environments (Chin, 1989), and analogous forms have even been observed in supraglacial streams (Knighton, 1981c). They thus appear to be a fundamental element of steep fluvial systems.

Step–pool morphology can be characterized by two variables – step wavelength L measured parallel to the average bed slope (s), and step height H (Fig. 5.13B) – so that H/L is an index of step steepness and bears a close relationship to the loss of head per unit length of channel (Abrahams *et al.*, 1995). As in the case of the riffle–pool sequence, the spacing (L) of consecutive elements is related to channel size, with average values of about 2–3 channel widths (Whittaker, 1987; Chin, 1989). Pool–pool spacing has a

Plate 5.1 Step–pool sequence, French Alps.

somewhat higher average of 2–4 widths in two Oregon streams and is also rather variable because of an uneven distribution of bedrock outcrops and boulder deposits along the channels (Grant *et al.*, 1990). Step structure appears to be better defined and more regular on steeper slopes (Judd and Peterson, 1969; Wohl and Grodek, 1994). Perhaps of more significance is the widely reported negative correlation between step length and slope (Judd and Peterson, 1969; Whittaker, 1987; Chin, 1989; Grant *et al.*, 1990; Wohl and Grodek, 1994), which implies that, if step height is largely controlled by the largest particles and their size is approximately constant along a reach, an increase in bed slope must be accommodated for the most part by more closely spaced steps. In terms of Whittaker's (1987) relationship

$$L = 0.31\ s^{-1.19} \tag{5.37}$$

the decrease in step length is particularly rapid as the slope increases up to about 0.15 m m^{-1}. The influence of bed slope on the adjustment of step–pool morphology is further illustrated by a relationship between average step steepness ($\overline{H/L}$) and slope obtained by Abrahams *et al.* (1995) from field and laboratory data,

Plate 5.2 Step–pool sequence, Corsica.

$$\overline{H/L} \sim 1.5\, s \qquad (5.38)$$

which indicates that the total elevation loss due to steps averages about 1.5 times the total elevation loss along a reach, so that about a third of the average step height is the result of pool scour. This approximate equality has implications for the resistance to flow produced by the step–pool bed form.

A pseudo-cyclic pattern of acceleration and deceleration characterizes the flow regime as water flows over or through the boulders forming each step before plunging into the pool below. Such 'tumbling flow' is supercritical (i.e. $F > 1$ in Table 4.1, p. 97) over the step and changes to subcritical ($F < 1$) in the pool, in the event causing a considerable amount of energy dissipation through turbulent mixing (Hayward, 1980; Whittaker and Jaeggi, 1982). In addition, energy is expended as a result of the form drag exerted by the large particles that make up the steps. Thus, step–pool sequences have an important resistance role, which is of particular significance in mountain streams where alternative forms of energy dissipation such as lateral adjustment are inhibited by the narrowness of the valleys, and where the large amounts of potential energy generated by the steep slopes could otherwise lead to extreme erosion. Similar structures are designed for steep ornamental streams with the express purpose of dissipating energy.

The results of flume experiments have extended this role in suggesting that step–pool sequences not only increase flow resistance but maximize it (Whittaker and Jaeggi, 1982; Abrahams *et al*., 1995), thereby conforming to the extremal condition proposed by Davies and Sutherland (1980). Their innovative experiments and field observations led Abrahams *et al*. (1995) to conclude that step–pool structures evolve towards a state of maximum resistance because that implies maximum stability, and that such a state is achieved when the approximation in equation (5.38) is satisfied. Thus an explanation of why step–pool systems develop and why they have a particular morphology can be couched in terms of their effect on energy dissipation.

Step–pool formation requires a widely graded bed load which includes large boulders that are immobile except during extreme flows, a low sediment supply, and a channel with a low width : depth ratio (Grant *et al*., 1990). During extreme floods – and a recurrence interval of 50 years or more is generally agreed for mobilization of the keystones (Hayward, 1980; Chin, 1989; Grant *et al*., 1990) – a step forms where large particles come to rest and trap smaller particles to create a small dam or bar across the stream, downstream of which a pool is scoured. Judd and Peterson (1969) attributed the formation of regularly spaced step–pools to oscillatory flow triggered by an initial boulder-anchored step, while Ashida *et al*. (1984) argued that they originate as antidunes in phase with standing waves on the water surface, large particles coming to rest under the crests of the waves. In the flume experiments of Whittaker and Jaeggi (1982) the flow initially formed regularly spaced antidunes, but, as the bed degraded, a self-armouring process stabilized the bed forms, causing an increase in resistance and a decrease in the Froude number to an extent that the step–pools could no longer be considered analogous to antidunes. On the basis of their experiments, Abrahams *et al*. (1995) have similarly argued that step–pools may develop at Froude numbers below the range normally associated with antidunes.

As in the case of the riffle–pool sequence, kinematic wave theory has been invoked as a possible mechanism of step–pool formation. The computer simulations of Naden and Brayshaw (1987), in which the probabilities of deposition (i.e. joining the particle queue) and erosion (i.e. leaving the particle queue) were linked to physically based criteria, generated bed forms having a wavelength similar to that of step–pool systems (or antidunes). Flume experiments have demonstrated that, when bed load is widely graded, longitudinal sorting of material into alternating 'smooth' and 'congested' zones can occur as a result of interparticle reactions alone (Iseya and Ikeda, 1987). Smooth zones of low gradient have high concentrations of fines, whereas congested zones of steep gradient consist of closely packed coarse particles which increase friction and cause transport to slow.

The tight interlocking of particles in steps gives them an inherent stability which only extreme floods are likely to disturb. That stability suggests that step–pools are a valid equilibrium form, especially when coupled with their apparent regularity of form (e.g. equations (5.37), (5.38)) and their role in satisfying the extremal condition of resistance maximization. Step–pools influence not only flow resistance but also sediment transport, which in flume experiments occurs as a series of waves linked to the underlying bed morphology (Whittaker, 1987). Their role as energy dissipaters can be impaired when pools become filled with sediment (Whittaker and Jaeggi, 1982), for then there is an increase in velocity and erosive capability, reactions which are opposite to the original causation.

Synthesis

Uniform beds are unstable and deform to give a wide range of configurations adjustable over various time periods. At one extreme ripples and dunes can develop during the passage of an individual flood of moderate magnitude and frequency, while at the other step–pools are significantly adjusted only during extreme floods having a recurrence interval of 50 years or more. One step–pool bed obliterated by a such a large flood was re-established within two years (Gintz *et al.*, 1996). Although sand-bed, gravel-bed and boulder-bed forms have here been treated separately, they have important elements in common. They represent sources of flow resistance which influence the nature of energy loss. They are intimately linked to the mechanics of sediment transport, reflecting organized erosions and depositions at the local scale. They are scaled to the dimensions of the channel and have strong relationships with other form elements, notably channel pattern. Dunes, riffles and steps represent concentrations of particles akin to kinematic waves in traffic flow. Indeed, kinematic wave theory may serve as a unifying principle, especially when coupled with the extremal condition of resistance maximization.

The development of a particular form depends on local flow and sediment conditions. Given the tendency for bed material size and slope to decrease and for discharge to increase downstream, systematic changes in bed configuration may also be expected in that direction, with step–pools in headwater reaches giving way, first to riffle–pool sequences where the material is gravelly, then to a mixture of gravel-bed and sand-bed forms, and finally to ripples and dunes as the sand fraction becomes dominant. Such zonation is an idealization but it emphasizes the influence of boundary composition on this element of channel form adjustment.

Channel Pattern

Channel pattern represents a mode of channel form adjustment in the horizontal plane which is additional to but nevertheless linked with transverse and lengthwise modes. It influences resistance to flow and can be regarded as an alternative to slope adjustment when valley slope is treated as constant at the short and medium timescales – an unsurprising association in view of the fact that channel pattern and slope occupy the same plane (Fig. 5.2). The effect of a meander, for example, is to increase resistance and reduce channel gradient relative to a straight reach between the same fixed points.

The conventional classification of Leopold and Wolman (1957) into straight, meandering and braided patterns is unsatisfactory. A much broader range is now recognized, which includes different types of the original three, several transitional forms, and anastomosing rivers. Schumm (1981, 1985) identified 14 patterns (Fig. 5.16A). The type of load moved through the channels is regarded as a basic criterion, with the patterns organized into bed-load (1–5), mixed-load (6–10) and suspended-load (11–14) forms. Carson (1984a) specified two types of 'wandering' gravel-bed river, the first of which is characterized by very rapid bend migration with fre-

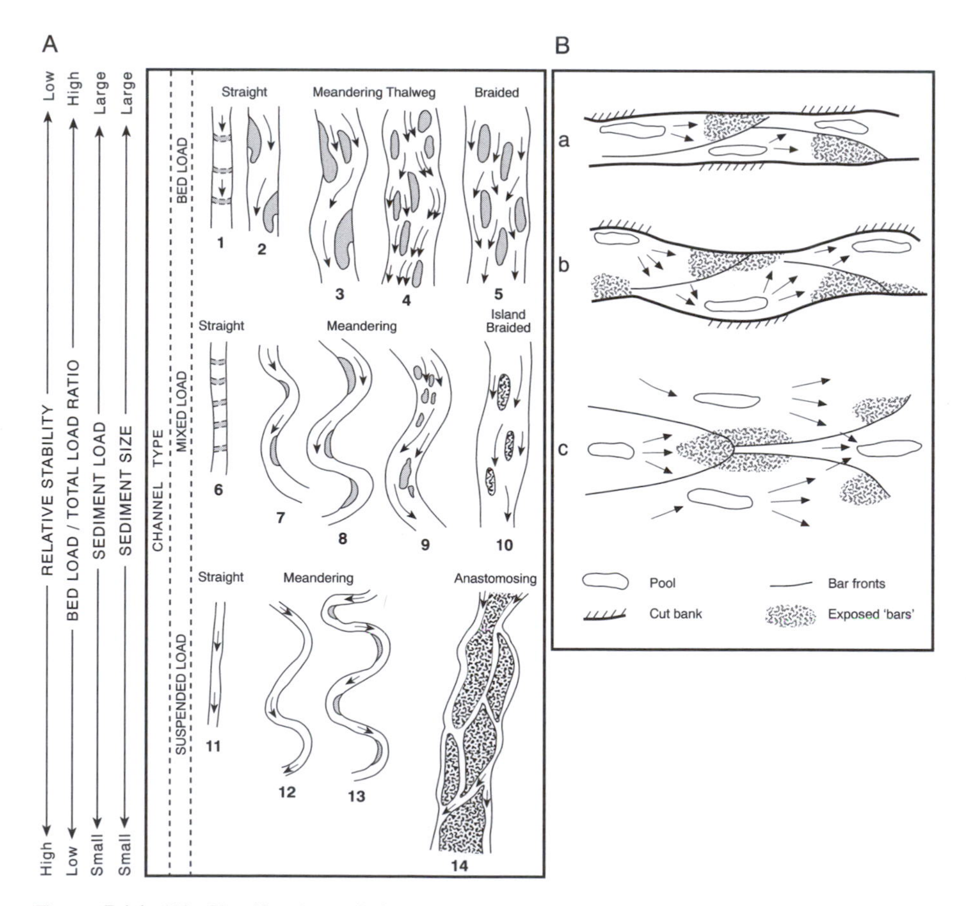

Figure 5.16 (A) Classification of channel pattern (after Schumm, 1981, 1985). (B) Overlapping pool–bar units in gravel-bed rivers of different channel pattern: (a) straight with single row of alternate bars; (b) meandering with point bars; (c) braided with medial bars composed of a back-to-back double row of alternate bars (after Ferguson, 1993).

quent dissection of point bars (akin to 3 in Fig. 5.16A) and is regarded as transitional between meandering and braided forms. The second type resembles the anastomosing pattern (14 in Fig. 5.16A) in that most of the channels are separated by vegetated islands and change is brought about largely by avulsion, but Carson maintains its separateness because of the supposed association between anastomosing and fine-grained suspended sediment transport. However, anastomosing channels carrying a relatively large bed load are not unknown (Knighton and Nanson, 1993).

Channel pattern can be regarded as a continuum with a primary subdivision into single-channel and multi-channel forms. Sinuosity defined by

$$S = \frac{\text{channel length}}{\text{straight-line valley length}} \tag{5.39}$$

provides a secondary basis for classification. Thus single-channel patterns include straight channels with (1, 6 in Fig. 5.16A) or without (11) mobile bed forms, straight channels with alternating bars and a meandering thalweg (2) grading into slightly sinuous channels, and various forms of meandering (8, 13). As regards the last of these, the degree of regularity has been used as an additional criterion, with Kellerhals *et al.* (1976) recognizing three categories: irregular meanders with only a vaguely repeated pattern; regular meanders with a clearly repeated pattern and a maximum deviation angle (between the channel and downvalley axis) of $<90°$; and tortuous meanders with a more or less repeated pattern and a maximum deviation angle of $>90°$.

Multi-channel patterns are diverse despite being grouped into the single category of 'braided' by Leopold and Wolman (1957). Braided rivers consist of flow separated by bars within a defined channel (5, 10 in Fig. 5.16A), bars which may be inundated at higher discharges to give the appearance of a single channel close to bankfull. The degree of bar development can vary considerably both horizontally and vertically, with, at one extreme, occasional, low-amplitude bars and, at the other, intense bar formation almost to the level of the surrounding floodplain. Anastomosing rivers (14 in Fig. 5.16A), the other main multi-channel form, consist of multiple channels separated by vegetated or otherwise stable islands which are usually excised from the continuous floodplain and which are large relative to the width of the channels (Fig. 5.24, p. 238; Plate 5.5, p. 239). They can be regarded as a subcategory of a diverse group of relatively uncommon anabranching rivers which are generally associated with flood-dominated regimes, banks that are resistant to erosion, and conditions that induce channel avulsion (Nanson and Knighton, 1996).

The Continuum Concept

Despite the variety of channel patterns now recognized, the original subdivision into straight, meandering and braided forms remains relevant, largely because of its association with the continuum concept introduced by

Leopold and Wolman (1957). Indeed, it is argued that, in gravel-bed rivers at least, all three patterns are composed of the same basic morphological unit, namely the 'pool–bar unit', with a different spatial arrangement for each pattern (Bridge, 1993; Ferguson, 1993; Fig. 5.16B). Assuming that channel pattern is controlled by interactions among a set of continuous variables and that all channel patterns intergrade, Leopold and Wolman argued that a continuum of channel patterns should exist, each pattern being associated with a particular combination of those variables. The problem is to identify the significant variables and determine how they control channel pattern.

Initially, emphasis was placed on discharge (Q) and slope (s) as determinants of channel pattern, with the equation:

$$s = 0.012\ Q_b^{-0.44} \tag{5.40}$$

acting as a threshold separating braided channels above the line from meandering channels below (Fig. 5.17A). Later laboratory evidence (Ackers and Charlton, 1970b; Schumm and Khan, 1972; Edgar, 1984) suggested that a similar but lower threshold exists for straight and meandering channels. Notwithstanding problems of operational definition (Ferguson, 1987) and inherent weaknesses with equations such as (5.40) to act as discriminators (Carson, 1984b), the belief that slope–discharge relationships can provide a means of explaining variations in channel pattern has become firmly established. Since discharge and slope are the two variable quantities in stream power ($= \gamma Qs$), these and subsequent results imply that the sequence of straight–meandering–braided can be associated with increasing stream power or, more appropriately, specific stream power ($= \gamma Qs/w$, where w is channel width). Chang (1979b) has introduced an additional refinement in showing analytically via his extremal hypothesis that within braided channels the number of braids also tends to increase with stream power. Most anastomosing channels plot well below the meandering–braided threshold defined by equation (5.40), and indeed below many of the meandering channels (Fig. 5.17A), suggesting that they, like straight channels, are situated towards the low end of the flow strength continuum, although their development does require periods of aggressive floodplain flow (Knighton and Nanson, 1993). In their classification of floodplains, Nanson and Croke (1992) specified specific stream power ranges of <10, 10–60 and 50–300 W m^{-2} respectively for laterally stable (including straight and anastomosing), actively meandering and braided patterns, but opinions vary as to lower and upper bounds of these classes.

Channel pattern depends not only on hydraulic factors but also on sedimentary ones, although they are less easily quantified. In an early attempt to circumvent the problem, Schumm (1963b) showed that sinuosity (S) increases with the silt–clay content of the channel boundary (M):

$$S = 0.94\ M^{0.25} \tag{5.41}$$

hinting at the possible effect of bank erodibility on channel pattern. Certainly the ability of a stream to shift laterally depends on the resistivity of the banks (Hickin and Nanson, 1984). Channels remain straight if little or

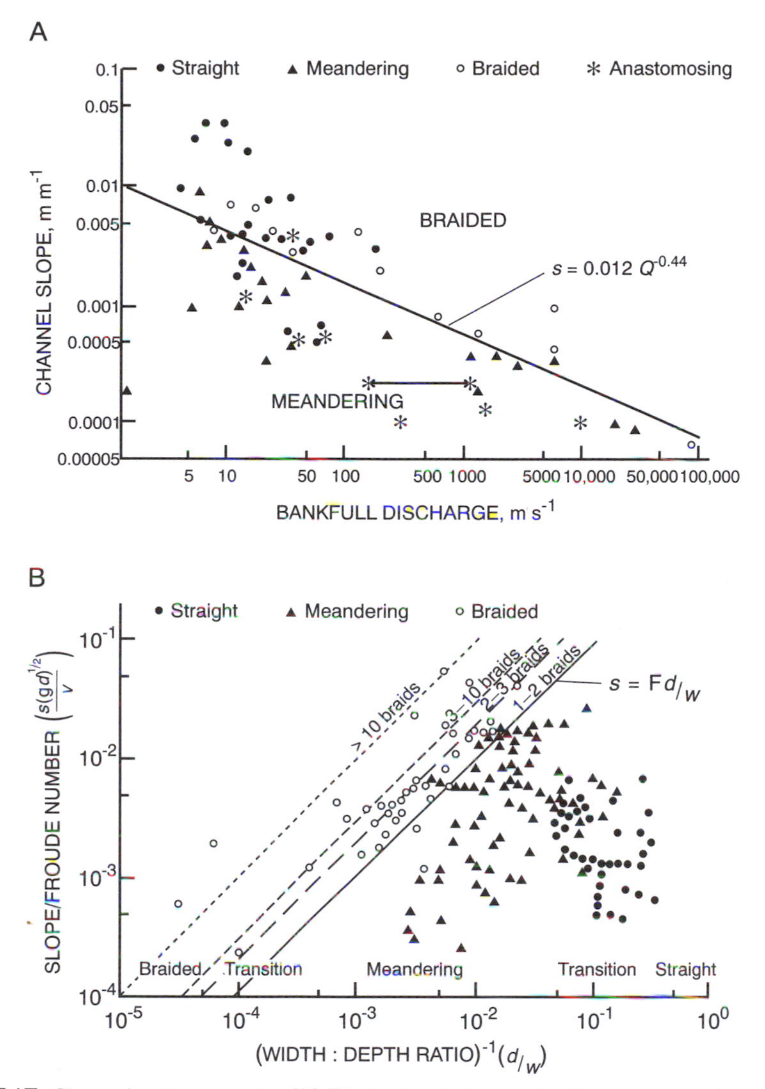

Figure 5.17 Channel pattern graphs. (A) Distinction between braided and meandering channels on the basis of a slope–discharge relationship (after Leopold and Wolman, 1957); straight and anastomosing channels are also indicated. (B) Straight, meandering and braided streams, and the degree of braiding, defined in terms of slope : Froude number and depth : width ratios (after Parker, 1976). (C) (overleaf) Gravel braided, sand braided and anastomosing channels differentiated on a slope–discharge plot, with median grain size (D_{50}) of gravel braided rivers shown (after Ferguson, 1987, and Knighton and Nanson, 1993). (D) Channel pattern in relation to grain size and unit stream power (after van den Berg, 1995).

no erosion occurs, whereas meandering requires localized bank erosion, and braiding involves extensive bank retreat so that the channel is widened and bends destroyed. Anastomosing rivers are generally agreed to have stable banks with individual channels showing little tendency to migrate (Knighton and Nanson, 1993). Supporting the effect of bank erodibility,

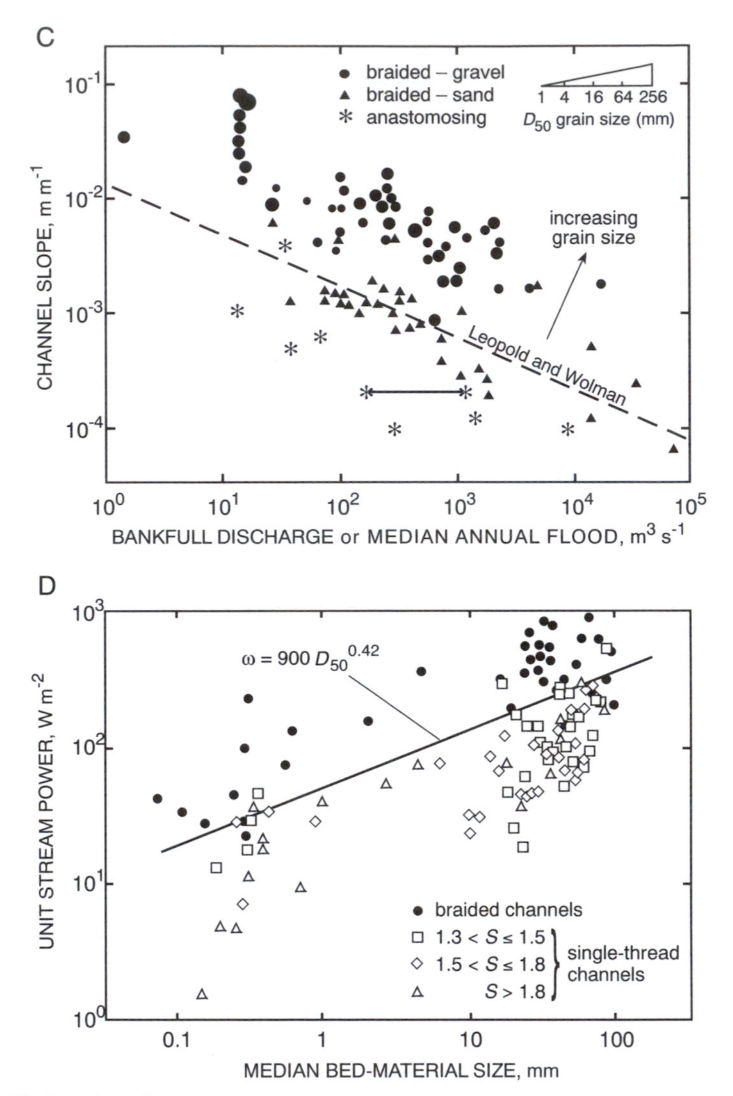

Figure 5.17 (continued)

Engelund and Skovgaard (1973) analytically defined a threshold value of width below which a river meanders and above which it braids. Parker (1976) used a similar approach to produce a regime diagram (Fig. 5.17B) which neatly illustrates the continuum idea in terms of two ratios, depth : width (d/w) and slope : Froude number (s/F), and defines the braiding criterion (solid line) as

$$s = \mathrm{F}d/w \tag{5.42}$$

Thus slopes which are steeper than those predicted by equation (5.42) will induce braiding. Combined with equation (5.22), this criterion can be manipulated to give (Ferguson, 1987)

$$s = 0.0013\ Q^{-0.24}\ B^{1.00} \tag{5.43}$$

where B is the silt–clay content of the banks. Clearly the threshold slope for braiding is sensitive to B and, at a given discharge, increases as the bank material becomes more cohesive. Bank resistance depends not only on material characteristics but also on bank vegetation, a fact well appreciated by New Zealand engineers who plant willows along banks in order to control heavily braided rivers.

Further empirical analysis has revealed that the threshold slope for braiding also depends on bed material size, with gravelly braided rivers having steeper slopes than do sandy ones at the same discharge (Fig. 5.17C). Van den Berg (1995) differentiated between braided channels and single-thread sinuous channels ($S > 1.3$) in a plot of specific stream power against median grain size (Fig. 5.17D). Thus, as boundary resistance increases through either more cohesive banks or coarser bed material, a greater stream power is required for the onset of braiding.

Consideration of bed material leads naturally to questions about the effect of sediment load on channel pattern. Brotherton (1979) argued that the relative ease of eroding and transporting bank material is a key element in differentiating between channel patterns. If erosion is more difficult than onward transport, the channel will remain straight; if, on the other hand, erosion is easier than transport, the resultant shoaling will lead to braiding. Meandering is assumed to occupy an intermediate position where erodibility and transportability are in approximate balance. Certainly Carson (1984b) regarded local shoaling of the thalweg as a prerequisite for the development of braiding, and attempted to classify pattern on the basis of the rate at which material is supplied from bank and upstream sources relative to the rate it is transported downstream. Flume experiments indicate that the sequence straight–meandering–braided can be associated with increasing bed load as well as increasing stream power (Schumm and Khan, 1972; Edgar, 1984), although supporting field data are rather sparse. That a large bed-material load promotes braiding can be ascribed to two mechanisms (Ferguson, 1987): the growth of medial bars as a result of local incompetence; and the development of large, low-angle point bars which are prone to chute dissection. Both are encouraged by the wide, shallow cross-sections characteristic of streams transporting a large bed load. Shoaling of the bed can induce avulsion over channel banks, a process associated with the development of anastomosing rivers. They, however, have a limited ability to expand laterally and a low transport capacity which, coupled with a relatively high rate of sediment supply, principally as suspended load, induces the deposition that locally forces flow out on to the floodplain and ultimately leads to the cutting of new channels (Knighton and Nanson, 1993).

Bettess and White (1983) have assessed the implications for channel pattern of the extremal condition that sediment transport is maximized (White *et al.*, 1982), by comparing the 'regime' slope (s_r) required for the transportation of a given sediment load by a known discharge with the available valley slope (s_v). Valley slope is regarded as imposed. Three possible states are identified:

(a) $s_r = s_v$ straight channel in equilibrium

(b) $s_r < s_v$ equilibrium can be achieved by either –

   (i) meandering, which reduces the (available) slope along the channel to s_r; or

   (ii) braiding when $s_r \ll s_v$, which increases the (regime) slope by sharing the total discharge and sediment load between n smaller channels

(c) $s_r > s_v$ the river cannot achieve the required equilibrium and adjustments must occur, by aggradation for example, to bring about a different equilibrium condition.

To the extent that anastomosing rivers are characterized by low gradients and aggradation, they could be considered to fall in the last category, but this would imply a state of disequilibrium and the anastomosed style can be maintained for long periods (Knighton and Nanson, 1993). This transport-based explanation of why different patterns develop is attractive, for channel pattern is generally adjustable over shorter timescales than is channel slope (Fig. 5.3), but it does not explain how they develop.

The commonest part of the pattern continuum proceeds from channels with no lateral migration or switching (straight and inactive sinuous ones) through actively meandering channels to braided channels (Ferguson, 1987). This sequence is associated with:

(a) increasing stream power, which implies increasing discharge at constant slope, increasing slope at constant discharge, or a combination of the two;
(b) increasing width : depth ratio, which is generally associated with increasing bank erodibility and increasing bed-load transport;
(c) increasing amount and calibre of bed load.

The relevant slope is the available valley slope, which is regarded as an imposed constraint adjustable only over relatively long time periods. Straight reaches are possibly restricted to low-energy environments where the available stream power is not sufficient for bank erosion at formative discharges and cross-channel currents are relatively weak. At the other extreme lies the braided pattern with its characteristically wide, shallow channel in which any flow discontinuity tends to create localized circulations conducive to sub-channel development (Chitale, 1973). Between these extremes is the range of meandering channels in which there is sufficient energy for bank erosion but sufficient bank resistance to maintain a sinuous course.

An alternative perspective defines the classic forms of channel pattern as three-dimensional vectors whose axes are respectively flow strength, bank erodibility and relative sediment supply (Knighton and Nanson, 1993):

Straight ~ (L, L, L)
Meandering ~ (M, L/M, L/M)
Braided ~ (H, H, M/H)

No more than an ordinal scaling is implied (where L is low, M is moderate and H is high), and other factors, such as bed material size, will influence the position of a given form within the space. Thus, for example, the flow strength required for braiding is higher in gravel-bed than sand-bed streams and may also increase with gravel diameter (Ferguson, 1987; Fig. 5.17C). Anastomosed rivers do not fit neatly into the sequences outlined in the previous paragraph ((a) – (c)) but they can be represented in this three-dimensional space:

Anastomosed ~ (L, L, M/H)

A similar approach has been used by Nanson and Knighton (1996) to differentiate between types of anabranching river, with seven rather than three factors involved. That no more than an ordinal scaling is applied reflects the inadequacy of the data base, particularly with respect to sediment-related factors.

The tripartite classification on which the continuum concept was originally based oversimplifies the range of channel patterns found in natural streams. However, the underlying idea of a continuum, of pattern and related variables, remains valid. The problem is to link the two in such a way that the channel pattern can be predicted for given conditions of the significant controlling variables, both hydraulic and sedimentary. Such a link has implications for predicting how a river will adjust its channel pattern in response to longitudinal variation in valley slope, tributary inflows and environmental change, whether natural or artificial. Channels that lie close to a pattern threshold are particularly susceptible to small changes in the controlling variables. A major alteration of channel pattern, for example by artificial straightening, can create problems worse than those that the alteration was designed to solve (Schumm, 1985).

Meanders

The absence of long straight reaches and the presence of sinuous flow in straight reaches is regarded as evidence of an inherent tendency in natural streams to meander, irrespective of scale or boundary material. A wide-ranging survey of river valleys in the northwestern United States revealed that 90 per cent of total valley length has meandering stream channels (Leopold, 1994). However, the definition of a meander remains somewhat arbitrary, even though a sinuosity (equation (5.39)) in excess of 1.5 is often used. There is no guarantee that the feature will be particularly regular (Fig. 5.18A) or that regularity will be maintained over a long river distance, even though definition sketches often show symmetrical bends. Indeed, asymmetry may be an inherent characteristic of meander geometry (Carson and Lapointe, 1983) and compound forms are not uncommon. In reality meanders are neither completely regular nor purely random and can be

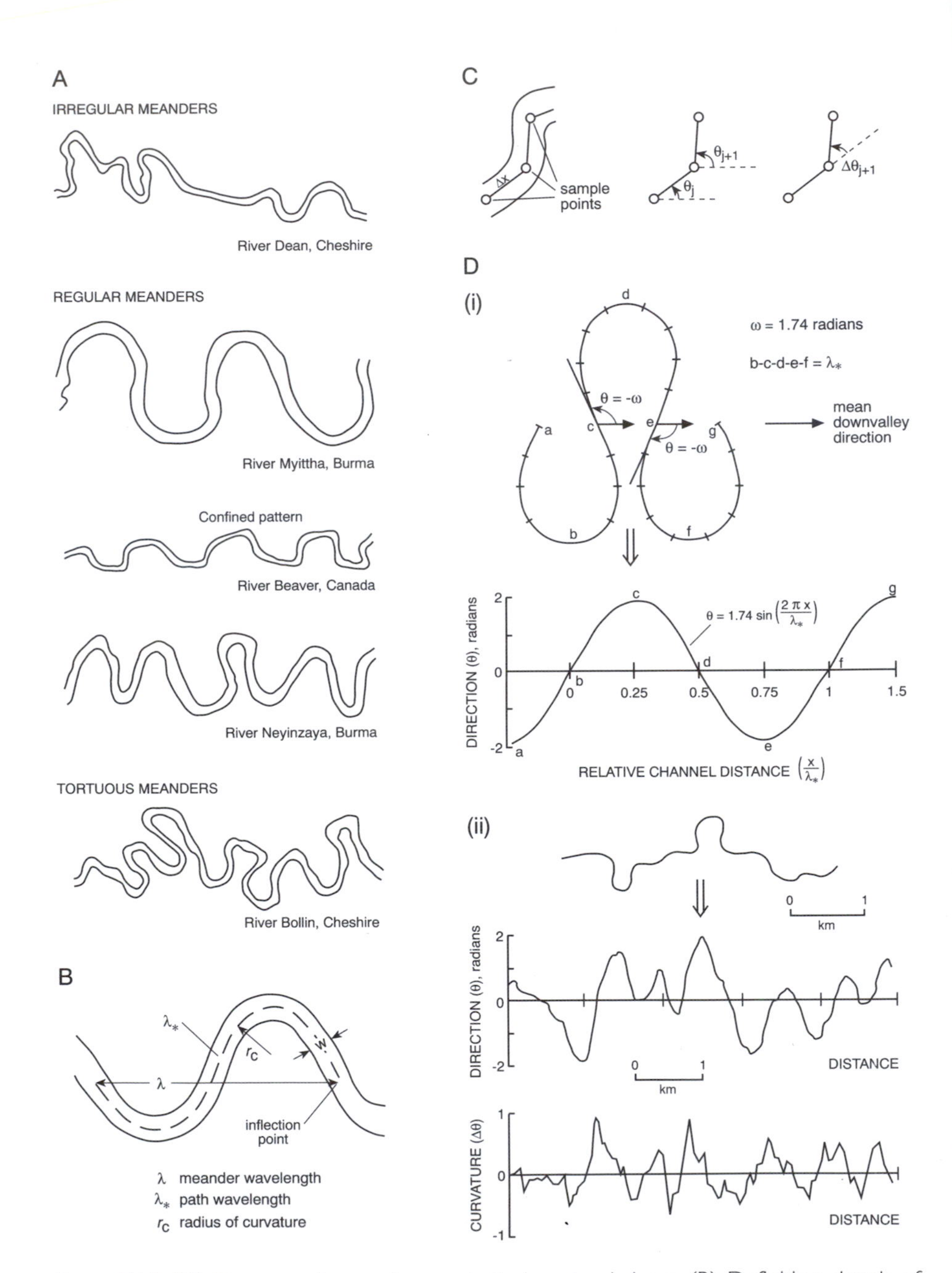

Figure 5.18 (A) Degrees of meander regularity in natural rivers. (B) Definition sketch of standard meander bend parameters. (C) Definition of path direction (θ) and change of direction or curvature ($\Delta\theta$) for series analysis. (D) Stream traces and their corresponding plots of direction and curvature against distance: (i) a hypothetical regular meander with a maximum deviation angle of 1.74 radians (110°) and the corresponding sine-generated curve (after Langbein and Leopold, 1966); (ii) irregular meander pattern of the River Trent and the corresponding plots (after Ferguson, 1979).

regarded as a compromise between the two (Ferguson, 1979). If the meandering process is assumed to be deterministic and capable of producing quasi-regular forms, then the random element is provided by the environmental conditions in which that process operates.

Meander geometry has been analysed using two main approaches: the traditional parametric approach which deals with individual bend statistics such as meander wavelength (λ) and radius of curvature (r_c) averaged over a series of bends (Fig. 5.18B); and the series approach, which spans sequences of bends and treats the stream trace as a spatial series of direction (θ) or direction change ($\Delta\theta$) in terms of distance (x) along the path (Fig. 5.18C and D). The latter is more objective and offers greater flexibility both for analysing meander traces of varying regularity and for developing theoretical models. Ferguson (1975, 1979) considered that meandering in a broad sense can be characterized by three planimetric properties: a scale variable such as wavelength (λ or λ_*), sinuosity or wiggliness, and degree of irregularity. Direction or curvature (direction-change) series can be used to estimate all three, as well as a suite of other morphometric variables (Howard and Hemberger, 1991). In addition, meander traces can be treated as fractal structures (Snow, 1989; Nikora, 1991). Stølum (1996) described river meandering as a self-organizing process which oscillates in space and time between an ordered planform and a chaotic one.

A simple model of meander geometry is provided by the equation for a sine-generated curve:

$$\theta = \omega \sin k x \tag{5.44}$$

in which channel direction (θ) is expressed as a sinusoidal function of distance (x), with parameters ω (the maximum angle between a channel segment and the mean downvalley axis) and k ($= 2\pi/\lambda_*$) (Fig. 5.18D(i)). Introduced by Langbein and Leopold (1966), this equation has an intriguing minimum variance property and describes the form of symmetrical meander paths reasonably well (Williams, 1986). However, it is much less applicable to non-regular bends or to lengthy meander traces in which a string of identical bends (constant ω and k) is unlikely (Fig. 5.18D(ii)).

It has long been recognized that consistent relationships exist between meander parameters and channel width (w), where the latter operates as a scale variable of the channel system (Fig. 1.2C, p. 7). In particular, results from a variety of fluvial environments suggest that wavelength and radius of curvature are respectively about 10–14 and 2–3 times channel width. Although Hey (1976) found no constancy in bend tightness (r_c/w) for two British rivers, the large data-set assembled by Williams (1986) has 42 per cent of r_c/w values falling in the range of 2 to 3, with a corresponding mean of 2.43. About a third of those values are also less than 2, when major changes may occur in the patterns of flow and sediment movement through bends (Hickin and Nanson, 1975, 1984).

Since width is approximately proportional to the square root of discharge, it is not unreasonable to expect that meander wavelength will also

Table 5.9 Meander wavelength–discharge relationships

Source	Relationship	Comments
Dury (1964)	$\lambda = 54\, Q_{ma}^{0.5}$	Q_{ma} = mean annual flood
Carlston (1965)	$\lambda = 166\, Q_{m}^{0.46}$	Q_{m} = mean annual discharge
	$\lambda = 126\, Q_{mm}^{0.46}$	Q_{mm} = mean monthly maximum discharge
Ackers and Charlton (1970b)	$\lambda = 62\, Q_{b}^{0.47}$	Q_{b} = constant 'bankfull' discharge in laboratory streams
Ferguson (1975)	$\lambda_s = 57\, Q_{d}^{0.58}$ $\lambda_a = 36\, Q_{d}^{0.63}$	Q_{d} = 'dominant' discharge of 1% duration; λ_s, λ_a are wavelengths estimated respectively from direction-change spectra and autocorrelograms rather than direct measurement
Dury (1977)	$\lambda_s = 33\, Q_{1.58}^{0.55}$	$Q_{1.58}$ = most probable annual flood

vary as $Q^{0.5}$. Although this approximate relationship is now well established (Table 5.9), controversy has existed as to which discharge is physically the most significant in shaping meanders. This issue, which is related to the dominant discharge concept, is reflected in the wide range of discharge indices used in the relationships (Table 5.9). The argument has centred on whether bankfull discharge or a more frequent range of flows is more important (Carlston, 1965), being based more on the strength of statistical relationships than physical reasoning. Meander geometry is probably related not to a single dominant discharge but to a range of discharges whose competence varies with the materials in which the channel is cut. However the issue is resolved, meander wavelength is clearly scaled according to the dimensions of the channel (w) and flow (Q) systems.

To reflect the influence of boundary composition, Schumm (1967) obtained multiple regression equations for a sample of non-gravelly channels

$$\lambda = 1935\, Q_{m}^{0.34} M^{-0.74} \tag{5.45}$$

$$\lambda = 618\, Q_{b}^{0.43} M^{-0.74} \tag{5.46}$$

$$\lambda = 394\, Q_{ma}^{0.48} M^{-0.74} \tag{5.47}$$

which show that, for a given discharge, meander wavelength decreases as the boundary and particularly the channel banks become more cohesive (increasing M). Combining these equations with Schumm's (1963b) sinuosity relation (equation (5.41)) and equations (5.23)–(5.25), the implication is that λ is influenced by material properties through both w and S, varying directly with the first and inversely with the second. Channels with more cohesive materials will tend to be relatively narrow, deep and sinuous and have smaller wavelengths, at least for a range of materials up to medium sand. To the extent that M reflects the type of sediment load (p. 151), larger wavelengths are associated with a higher proportion of bed-load transport, but information on the relationship between sediment load and meander form is meagre.

These various relationships indicate a self-similarity of meander geometry over a wide range of scales and environmental conditions. However, the regularity which they imply is not everywhere apparent, and the use of single parameters provides only a partial and often subjective characterization of meander form. The more objective series approach from which direction or direction-change spectra are obtained is better suited to the analysis of an irregular planform. A perfectly regular meander train should have a spectrum with a single sharp peak. Speight (1967) found that Australasian rivers seldom have a single dominant wavelength, a characteristic associated with compound forms in which short meanders are superimposed on larger ones. In contrast, most of the spectra obtained by Ferguson (1975) for 19 British rivers did have a single peak, which is strongly correlated with dominant discharge (Table 5.9), leading him to conclude that environmental irregularities distort an underlying regular process to give an observed geometry which combines random and deterministic components in varying proportions.

One important element of that process is the **flow pattern** through meander bends. The main features of that pattern are:

(a) superelevation of the water surface against the concave (outer) bank;
(b) a transverse current directed towards the outer bank at the surface and towards the inner bank at the bed to give a secondary circulation additional to the main downstream flow; and
(c) a maximum-velocity current which moves from near the inner bank at the bend entrance to near the outer bank at the bend exit, crossing the channel through the zone of greatest curvature.

These characteristics reflect the interaction between a centrifugal force which acts outwardly on the water as it flows round the bend, and an inward-acting pressure gradient force which is driven by the cross-stream tilting of the water surface. A combination of the transverse current with the main downstream component gives a spiral or helicoidal motion to the flow. Curvature induces not only a secondary circulation but also large cross-stream variations in the boundary shear stress and velocity fields (Dietrich, 1987). In particular, the superelevation of the water surface against the concave bank of a bend gives rise to a locally steep downstream energy gradient, and hence a zone of maximum boundary shear stress close to the outer bank beyond the bend apex (Fig. 5.19A). The bar–pool topography and cross-sectional asymmetry which characterize meander bends cause the maximum shear-stress zone to shift outwards further upstream than would occur as a result of curvature alone, emphasizing how important are the interactions between the flow and bed topography as water follows a curved path.

Secondary currents are usually weaker than primary ones but are nevertheless significant in that they influence the distributions of velocity and boundary shear stress. When considering the pattern of secondary flow at a bend, Markham and Thorne (1992) divided the cross-section into three regions (Fig. 5.19B):

(a) a *mid-channel region* through which about 90 per cent of the flow passes: here, the classic helicoidal motion is well established;
(b) an *outer bank region* in which a cell of opposite circulation develops: the strength of this cell increases with discharge, the steepness of the bank and the acuteness of the bend; and
(c) an *inner bank region* where, it is argued (Dietrich and Smith, 1983; Dietrich, 1987), shoaling over the point bar induces a net outward flow, forcing the core of maximum velocity more rapidly towards the outer bank; an increase in stage will tend to reduce this shoaling effect and allow an inward component of near-bed flow over the bar top.

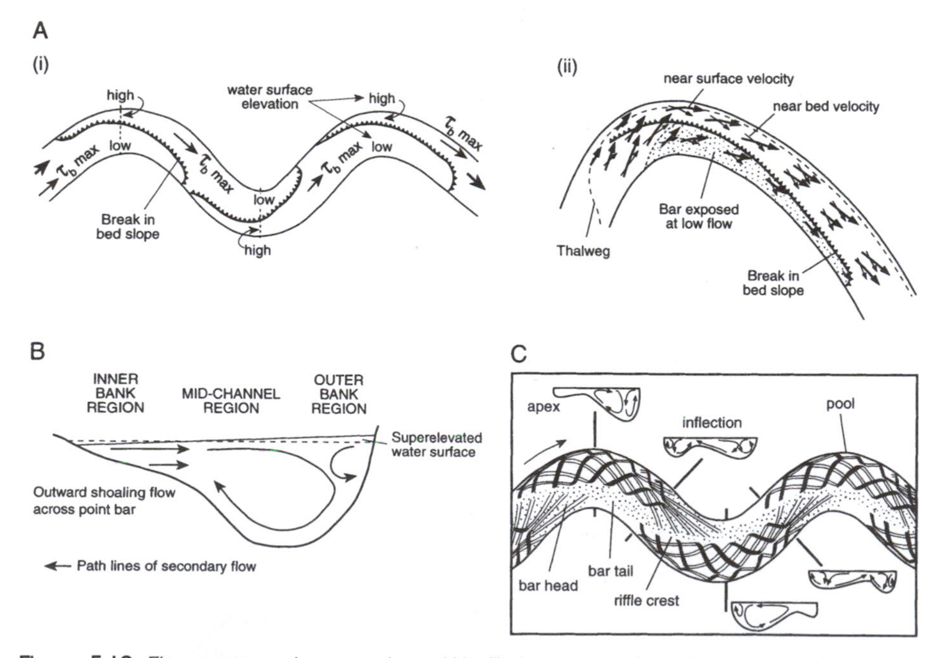

Figure 5.19 Flow pattern in meanders. (A) (i) Location of maximum boundary shear stress (τ_b), and (ii) flow field in a bend with a well-developed bar (after Dietrich, 1987). (B) Secondary flow at a bend apex showing the outer bank cell and the shoaling-induced outward flow over the point bar (after Markham and Thorne, 1992). (C) Model of the flow structure in meandering channels (after Thompson, 1986). Black lines indicate surface currents and white lines near-bed currents.

The pattern of flow is neither temporally nor spatially constant, varying with discharge, bend tightness and cross-sectional form. Bathurst *et al.* (1979) found that at intermediate discharges secondary currents are relatively strong, but at high discharges primary currents become dominant as the main flow follows a straighter path. In tighter bends (i.e. lower r_c/w), the degree of superelevation and the strength of the secondary circulation increase. Where r_c/w falls below about 2, the water encounters the outer bank at a very abrupt angle and flow separation may occur, generating a strong back-eddy adjacent to the bank near the bend apex (Hickin and Nanson, 1975, 1984). As regards cross-sectional form, the width : depth ratio appears to exert an important influence on the pattern of flow (Markham and Thorne, 1992). Where the ratio is relatively large, point-bar development is more extensive and the shoaling effect noted in (c) above directs the inner-bank flow radially outwards; but in narrow, deep channels, and especially where $w/d < 10$, bars are less likely to form and the shoaling effect is much reduced, allowing an inward movement of near-bed flow.

The pattern of primary and secondary currents in meanders influences the distribution of erosion and deposition. In broad outline, erosion in the bend tends to be concentrated against the outer bank downstream of the apex where currents are strongest, while in a parallel position against the opposite bank, point-bar building predominates with material supplied by both longitudinal and transverse currents. This pattern gives a largely downvalley component to meander migration. However, more complex planform migrations are not uncommon, with appropriate modifications to the basic flow pattern. Whiting and Dietrich (1993a) have carried out laboratory experiments with large-amplitude meanders in which multiple pools develop spaced at 3–4 times channel width. Since the banks near pools are sites of accelerated retreat, the shear-stress fields associated with these multiple forms may produce differential movement around the bend, leading to planimetric distortion.

As regards the link between flow and form, most progress has been made with the radius of curvature–width relationship, notably in the context of flow separation and migration. Bagnold (1960) showed that at $r_c/w \sim 2$, water filaments begin to separate from the inner bank. Resistance to flow through the bend is at a minimum immediately prior to breakaway but increases rapidly at $r_c/w < 2$ because the separation zone grows and turbulent activity increases. Evidence suggests that there is a corresponding reduction in the rate of erosion at the concave bank (Hickin and Nanson, 1975), possibly because the force applied to the bank also decreases rapidly as r_c/w falls below about 2 (Begin, 1981). Then also significant flow separation may occur at the outer bank, leading to the deposition of a concave-bank bench in the zone of reversed flow (Nanson and Page, 1983), or to localized bank retreat and the development of a double-headed bend if the current is strong enough at the point of separation (Markham and Thorne, 1992). Clearly adjustments to channel curvature (r_c/w) can alter the pattern of flow and the associated distribution of erosion and deposition. Through its relationship to the shear stress exerted on the outer bank, channel curvature has a major influence on the rate and character of meander migration (Fig. 5.21B, p. 226).

The **initiation of meanders** requires localized bank retreat which, if a series of bends is to develop, must alternate from one side of a channel to another in a more or less regular fashion (Plate 5.3). Periodic deformation of the channel bed and the development of a sinuous thalweg are regarded by many as necessary precursors to erosion of the channel banks (Callander, 1978). However, there is as yet no completely satisfactory explanation of how or why meanders develop. The ubiquity and underlying regularity of natural river bends suggests that they are not simply the result of random disturbances. Most explanations fall into two broad categories, regarding the oscillation either as an inherent property of turbulent flow or as a result of interaction between the flow and a mobile channel bed, an interaction in which sediment transport is an essential element.

Periodically-reversing helicoidal flow has long been recognized as an important influence on the pattern of erosion and deposition through meander bends. It is a natural consequence of flow in a curved channel. In addition, it has been argued that this type of secondary flow develops spontaneously in straight channels as a result of vortices generated at the boundary walls (Einstein and Shen, 1964; Shen and Komura, 1968). If vortices develop along both banks, a pair of surface-convergent helical cells will form. Einstein and Shen (1964) argued qualitatively that inequalities in bank roughness induce asymmetry in these cells and periodic reversal of the dominant cell, resulting in the formation of a meandering thalweg and alternating bars (Fig. 5.15A). A somewhat different pattern of secondary

Plate 5.3 (A) and (B) (opposite) Meander development over a period of 20 years, River Bollin. Localized bank retreat is compensated by point-bar construction against the opposite bank. Note the positions of the small trees and telegraph pole for comparison.

flow was envisaged by Thompson (1986), in which a streamwise alternation of convergent and divergent currents develops as the helical cells experience repeated decay and regeneration in association with an evolving bed topography (Fig. 5.15B). If erosion is concentrated on the concave bank adjacent to each scour pool, a natural progression from straight channels with riffle–pool units to meandering channels is established for the pattern of flow (Fig. 5.19C). Certainly, alternating convergent and divergent flow is a key element in Keller's (1972) model (Fig. 5.20B), but information on secondary currents in straight channels is rather sparse (Rhoads and Welford, 1991). Whether those currents would be strong enough to effect major changes in the geometry of natural channels remains a matter of debate, although they can develop meanders in laboratory flumes (Einstein and Shen, 1964).

Yalin's (1971, 1992) theory of macroturbulent flow and the bursting process discussed previously in the context of riffles and pools again emphasizes the importance of flow properties in meander development. Horizontal bursts produced by eddies at a scale proportional to the channel width deform the banks in much the same way that vertical bursts (scaled according to the depth) deform the bed. Because the turbulent structure is autocorrelated, similar perturbations to the flow field are located at distances of approximately $2\pi w$ along the channel, leading to alternating regions of high-speed and low-speed flow (Fig. 5.12A(i)) and to alternating erosions and accretions of the same periodicity. Under suitable circumstances alternate bars will develop (i.e. at distances of about $6w$), but they are not regarded as the cause of meandering, merely as catalysts which aid

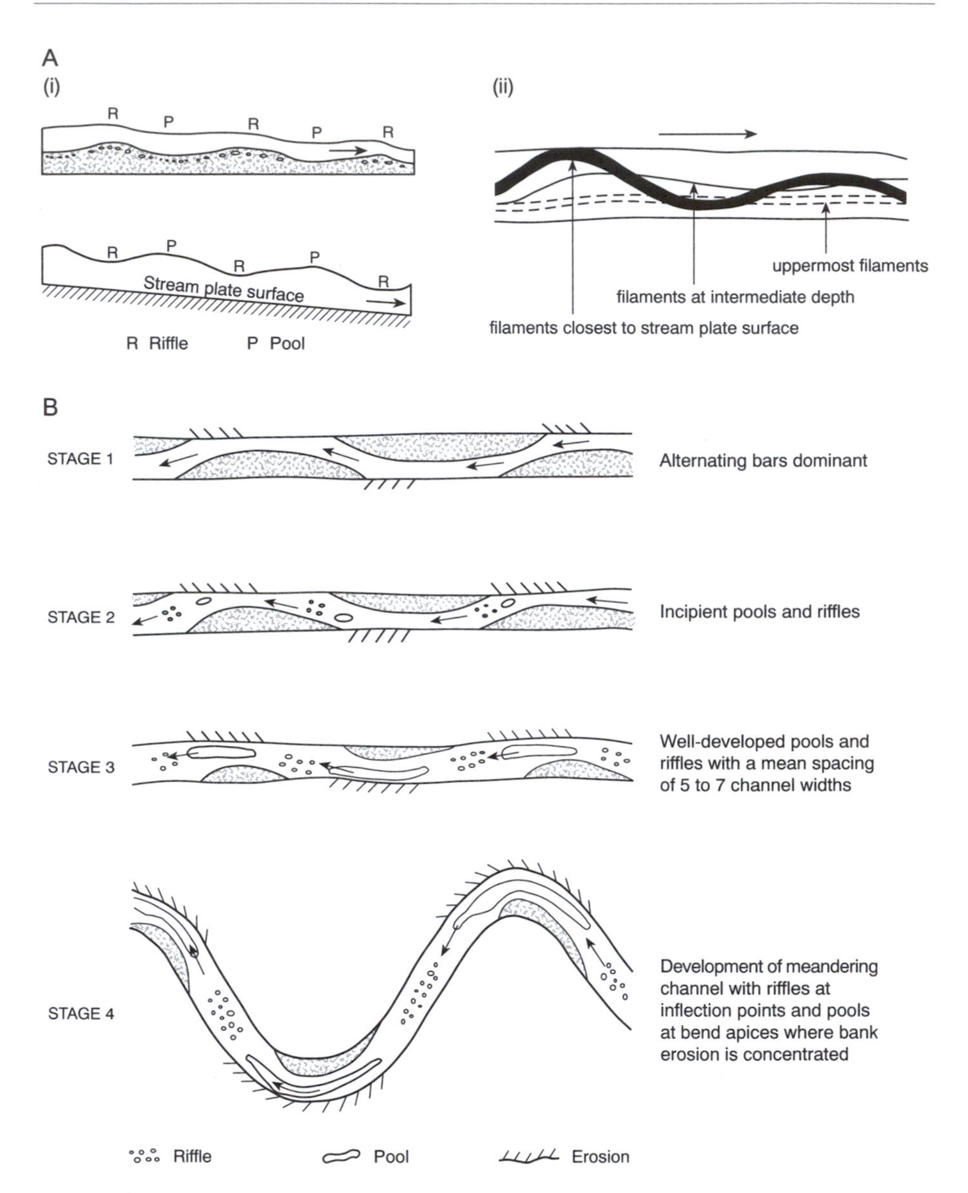

Figure 5.20 (A) Stream plate experiment (after Gorycki, 1973): (i) water surface undulations (lower diagram) akin to riffles and pools in natural streams (upper diagram); (ii) vertical variation in the extent of filament sinuosity with maximum sinuosity in those filaments closest to the plate surface where hydraulic drag is at a maximum. (B) Transformation of a straight channel with a riffle–pool bed into a meandering channel (after Keller, 1972).

or accelerate the meandering process. Whatever the merits of Yalin's argument, and field tests are sadly lacking, the main point is that given a tendency for a regular pattern of alternating regions of high velocity to develop, the resulting secondary flow is sufficient to explain meander initiation in any erodible material. This tendency is regarded as an inherent property of turbulent flow and not dependent on the presence of a deformable boundary. A deformable boundary merely allows the underlying wave-like structure in the flow to be observed. The reason for meanders is thus a horizontal version of the reason for regular alternations in bed topography.

Stream-plate experiments in the absence of sediment support the helicoidal mechanism as a cause of meandering (Tanner, 1960; Gorycki, 1973). Gorycki's results are particularly interesting. In an initially straight stream, a side view reveals upper-surface undulations akin to pools and riffles in natural streams (Fig. 5.20A(i)). Also, a vertical structuring is apparent in which the flow becomes more sinuous towards the plate surface where the frictional resistance is at a maximum (Fig. 5.20A(ii)). As the velocity and discharge are increased, progressively more of the flow depth is involved in sinuous flow until the stream itself becomes distorted and develops meanders with a characteristic geometry. Meanders thus evolve from the effect of hydraulic drag on the flow, requiring the attainment of a critical velocity or discharge. Although stream-plate experiments have produced realistic meandering tendencies, there is a large difference in scale between surface-tension streams and natural rivers. Direct evidence of oscillatory helical flow in straight channels is lacking, and detailed field measurements of flow structure are required if many of the assumptions associated with flow initiation theories are to be rigorously tested (Rhoads and Welford, 1991).

A fundamentally different approach regards meandering as the result of an inherent instability between the flow and a mobile channel boundary rather than an inherent property of the flow itself. It has two main strands: bar theory and bend theory. *Bar theory* proposes that periodic deformation of the bed is the fundamental cause of meandering (Callander, 1978), leading to the development of alternate bars which initiate curvature by deflecting flow against the opposite banks. Relationships established by Nelson and Smith (1989) suggest that such development occurs best in wide, shallow channels consisting of relatively coarse bed material. The stability analyses which underlie bar theory have become more refined over the past 20 years, but they continue to assume implicitly that the 'alternate bar' mode of instability in straight channels leads naturally to incipient meandering without attempting to analyse the occurrence of bank erosion (Seminara and Tubino, 1989). Both linear and non-linear bar theories predict that the fastest-growing perturbations, and therefore those that are likely to develop into alternate bars, have rather large migration rates. The problem is to explain how rapidly migrating bars could induce the slow bank erosion at discrete locations along a channel which is required for meander initiation.

Bend theory developed in the early 1980s in response to these shortcomings. Ikeda *et al.* (1981) found that bar and bend theories predict meander wavelengths of approximately the same order of magnitude, lending support to the notion that the bank erosion associated with alternate bars does trigger planimetric instability. A unified *bar–bend theory* has developed in which a 'resonance' mechanism operates (Blondeaux and Seminara, 1985;

Plate 5.4 Meanders in a supraglacial stream, Austre Okstindbre, Norway. The ice axe is 72 cm long.

Parker and Johannesson, 1989; Seminara and Tubino, 1989). The unified theory draws a distinction between 'free' alternate bars which migrate downstream in straight channels, and 'forced' point bars which are a consequence of curvature and remain relatively stable in position. Basically, an initial instability in the channel bed results in the formation of migrating alternate bars. As incipient sinuosity develops, the alternating flow pattern reinforces the wavelengths of those perturbations with small migration rates via the resonance mechanism. With a further increase in curvature, free bars are suppressed and forced (point) bars develop, inducing the necessary bank erosion. Unequivocal evidence of resonance has yet to emerge (Rhoads and Welford, 1991), but a similar sequence has been observed on a gravel-bed river in which a meandering pattern developed rapidly from an artificially straightened channel (Lewin, 1976). Experimental work also indicates that migrating bars tend to stall at a critical curvature and be replaced by forced, stationary bars which enhance rates of bank erosion at fixed locations (Whiting and Dietrich, 1993b). Bar–bend theory provides a physically based explanation of the initiation process, but details remain to be elucidated, particularly as regards the second phase when planimetric instability ensues (Seminara and Tubino, 1989).

The standard evidence against hypotheses that require sediment transport is the presence of meanders in environments where there is no material to form bars (supraglacial streams, Plate 5.4) and no confining banks (Gulf Stream currents, stream-plate experiments). Parker (1976) attempted to explain each of these three situations by invoking special conditions. Although not explicitly stated, the implication is that meanders are not necessarily the outcome of a single cause. Indeed, similarity of form is no guarantee of similarity of process. However, cursory observations of meander initiation in alluvial streams do suggest that the development of alternate bars precedes the initiation of channel curvature (Keller, 1972; Lewin, 1976; Fig. 5.20B).

Other explanations have been less concerned with specific processes and rather more with *why* streams meander in the broader context of channel form adjustment. In line with his minimum stream power concept, Chang (1988) argued that meander development is explained by a river's tendency to establish a minimum channel slope for given input conditions, underlining the relationship between meandering (or, more correctly, sinuosity) and slope adjustment. Along a stretch of river with essentially uniform water and sediment discharges but variable valley slope, a river can maintain a relatively constant gradient by lengthening its course where the valley slope is locally steep. Whether that gradient is minimized and whether a lengthening of course implies meandering are matters of conjecture. Langbein and Leopold (1966) postulated that meandering is the most probable channel form in that it satisfies as closely as possible the condition of minimum variance, achieved through adjustment of both the plan geometry and hydraulic variables. With respect to the latter, Langbein and Leopold argued that, whereas energy loss is concentrated at riffles in a straight reach, it is more uniformly distributed in meanders because of the additional resistance provided by curvature at pools, support for which has come from comparisons of flow conditions in neighbouring curved and straight reaches (Dozier, 1976; Richards, 1976b). Although these various explanations provide predictions that correspond with observed forms, they are essentially teleological and difficult to evaluate scientifically (Rhoads and Welford, 1991).

There is no general agreement as to how or why streams meander. Nevertheless, deformation of the channel bed appears to be an important prerequisite which modifies the pattern of flow prior to meandering. That deformation may take the form of a more or less regular variation in the degree of cross-sectional asymmetry (Fig. 5.9A; Knighton, 1982). Many of the theories contain untested assumptions or uncertainties. There is a need for accurate field measurements of primary and secondary flow components in straight streams, and for detailed observations of the interactions between flow structure, bed topography and bankline configuration as an initially straight stream evolves into a sinuous one.

That bar–bend theory regards meandering as the outcome of a double instability, affecting firstly the bed and then the planimetric form, raises a question as to the overall **stability** of the meandering planform. The under-

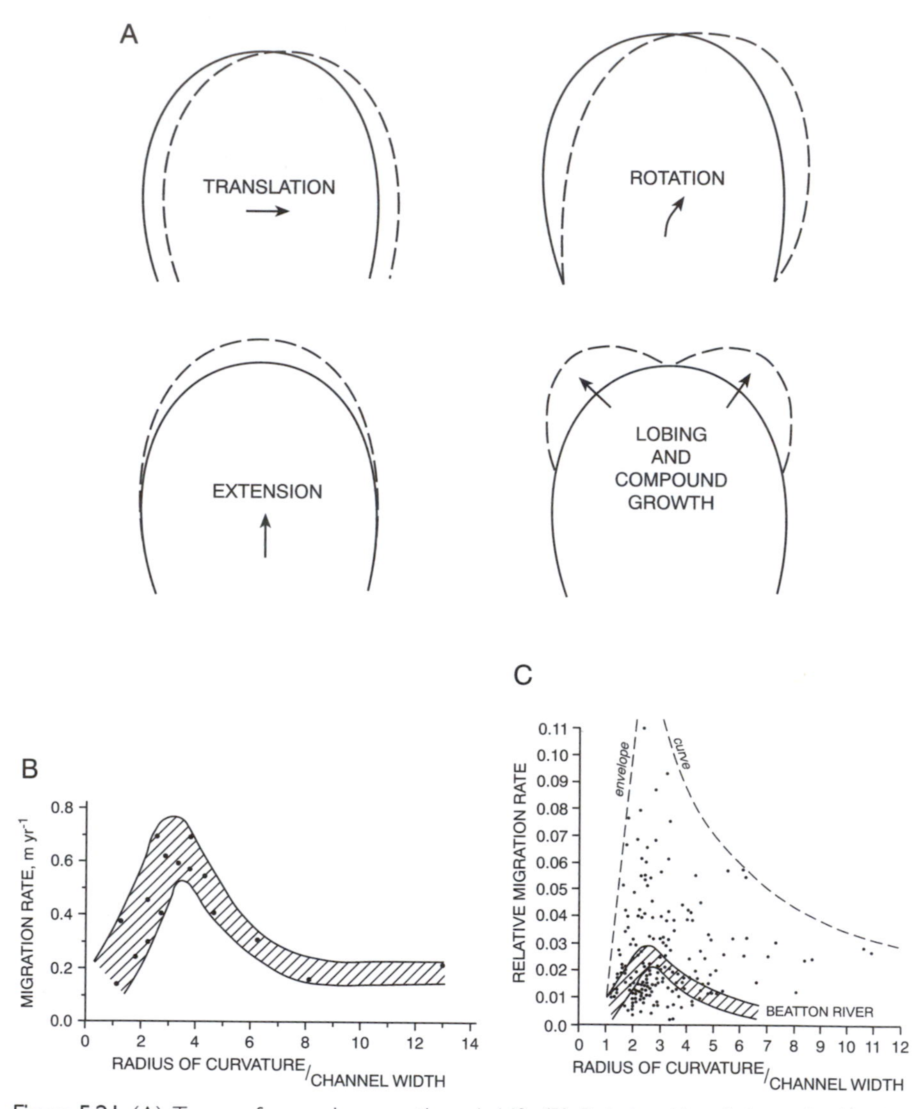

Figure 5.21 (A) Types of meander growth and shift. (B) Relationship of channel migration rate (M) to bend curvature (r_c/w), Beatton River, British Columbia (after Nanson and Hickin, 1983). (C) Relationship of relative migration rate (M/w) to bend curvature for 18 rivers (after Nanson and Hickin, 1986).

lying regularity of meander geometry implied by morphometric relationships (Fig. 1.2C, p. 7) suggests that meanders are a valid equilibrium form. However, they do have a tendency to migrate. Point-bar deposition on the convex bank complements erosion of the concave bank (Plate 5.3) which, being concentrated in the downstream limb beyond the bend apex, gives a predominantly downvalley component to migration. On average, deposition keeps pace with erosion to maintain an approximately constant width, but local imbalances are possible to give complex patterns of change. Kinoshita (1961) identified two basic types of river meander: meanders of low amplitude which migrate consistently downstream and maintain their symmetric form; and meanders of large amplitude in which differential shifting at points around the bends leads to complex planform growth.

Migration can involve various types of movement (Fig. 5.21A): *translation,* when the bend shifts in position downstream without altering its basic shape; *extension,* when the bend moves predominantly in the lateral direction, increasing its amplitude and path length; *rotation,* when the bend axis changes in orientation; and *lobing and compound growth.* The frequency and distribution of these movement types remain uncertain in a general sense, even though many meander change studies have now been carried out. In an analysis of nearly 100 bends on the highly mobile River Dane in Cheshire, Hooke (1984) showed that 25 per cent translated downstream and 15 per cent extended laterally, while only 24 per cent remained stable over the 145-year study period. Now, the River Dane flows across a piedmont zone, which in England and Wales as a whole appears to be where migration is most active (Hooke and Redmond, 1992). Meandering channels in lowland areas tend to be more stable. However, it is very difficult to relate channel behaviour to physical determinants. Translating bends tend to be of low curvature and extending ones of high curvature (Hooke, 1984), suggesting that channel curvature, via its influence on flow pattern, can affect the distribution of erosion and hence the type of movement.

Simulation models indicate that an initially regular pattern can deform into a variety of bend shapes through different combinations of these types of movement (Ferguson, 1984; Howard, 1992). Bends are seldom translated downstream without some deformation. The development of asymmetry is not uncommon, a tendency attributed by Ferguson (1984) to spatial lag, with the migration rate at one point being related to curvature at a previous upstream point. Certainly the boundary stress fields and related patterns of erosion and deposition in neighbouring bends are interdependent, so that the migration of an individual bend can only be understood properly in the context of its position within a meander train (Furbish, 1991). Indeed, certain key bends may influence much of the movement (Hooke, 1987). The meandering process is sensitive to initial bend geometry and entrance flow conditions, ensuring that diverse bend forms can arise along freely migrating rivers independently of such factors as the unsteadiness of the flow or the non-uniformity of the banks (Furbish, 1991). Given a tendency for migration and the complex patterns of growth that can result therefrom, the question arises as to whether meanders can evolve to a single characteristic form, a possibility denied by Furbish (1991).

The rate of migration is controlled to a large extent by bend geometry, and in particular by channel curvature (r_c/w). From a detailed study of scroll bars developed over 250 years along the Beatton River in British Columbia, Hickin and Nanson (1975, 1984) demonstrated that the rate of migration reaches a maximum when $2 < r_c/w < 3$, decreasing rapidly on either side of this range (Fig. 5.21B). The decrease at the lower end of the range (i.e. as bends become tighter) may be attributable to the large increase in resistance (Bagnold, 1960) or a decrease in outer-bank radial force (Begin, 1981, 1986) as r_c/w falls below about 2. On the basis of studies in a meandering flume, Hooke (1975) has argued that a uniform downvalley migration and therefore a stable meander geometry require a curvature in the range of 2 to 3. Too shallow a curve (r_c is too large) gives rise to a shear stress distribution which induces a faster rate of migration in the upstream than in the downstream limb, thereby leading to an increase in curvature. Conversely, if r_c is too small, the downstream limb migrates more rapidly and the bend becomes less arcuate. When combined, these separate studies suggest that a natural stream can control both the rate of migration and the pattern of erosion and deposition in meander bends by adjusting channel curvature.

Further work has tended to confirm the general form of the curvature–migration rate relationship but the amount of scatter can be considerable (Fig. 5.21C; Hickin and Nanson, 1984; Hooke, 1987). Part of that scatter may be explained by variations in such additional factors as stream power and outer-bank sediment size, which together account for almost 70 per cent of the non-curvature-related variance in the data of Fig. 5.21C (Nanson and Hickin, 1986). Part may be related to other aspects of bend geometry. Furbish (1988, 1991) showed that migration rate varies with bend length as well as curvature in such a way that it would increase in a bend of given curvature if the bend was sufficiently long. Indeed, he maintained, contrary to Fig. 5.21B, that average migration rate increases monotonically with average curvature if differences in bend length are taken into account. On the basis of computer simulations, Sun *et al.* (1996) showed that meander amplitude can influence migration rate. Downvalley migration rate varies inversely with amplitude, implying that smaller loops move faster, while lateral migration rate varies directly with amplitude, implying the reverse.

Curvature is but one mode of adjustment available to a natural stream. Wide, shallow channels tend to have a lower sinuosity (flatter bends) than do narrow, deep ones (equations (5.25) and (5.48)), and Chitale (1973) used this tendency to show how the cross-sectional shape (width : depth ratio) of a channel could affect the distribution of erosion in meander bends. Certainly the width : depth ratio appears to exert an important influence over the pattern of flow through bends (Markham and Thorne, 1992). Where it is small, separation at the inner and outer banks is a significant flow process, especially in bends of low r_c/w. If the migratory habit and the distribution of erosion and deposition in bends are to be better understood, both planform and other elements of channel adjustment need to be linked effectively with flow behaviour.

Not all meandering rivers migrate. The meander pattern of the lower Ohio River has changed little over the past 1000 years (Alexander and Nunnally, 1972). However, where flow and bank material conditions are conducive to a continued increase in the amplitude and tightness of bends, a threshold sinuosity may be reached which the river can no longer main-

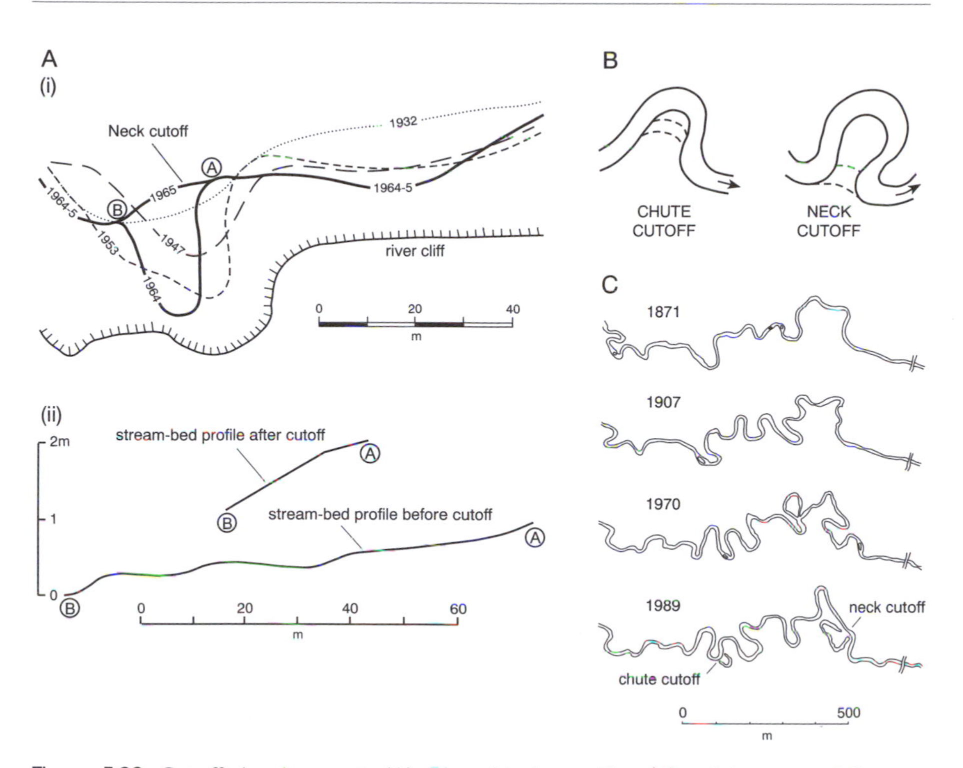

Figure 5.22 Cutoff development. (A) River Irk, Lancashire (after Johnson and Paynter, 1967): (i) course changes prior to the cutoff; (ii) increase in channel gradient resulting from cutoff development. (B) Types of cutoff. (C) Sequence of changes on the River Bollin (after Hooke and Redmond, 1992).

tain and a *cutoff* develops (Fig. 5.22A). Two main types are recognized, neck cutoffs and chute cutoffs (Fig. 5.22B), the latter of which are probably more common since they generally occur across point-bar or floodplain deposits of low elevation. Both types are present in a meandering reach of the River Bollin, which has recently entered a phase of decreasing sinuosity (Hooke and Redmond, 1992; Fig. 5.22C). Cutoffs can be regarded as a response to excessive sinuosity which so lowers the channel gradient that the stream cannot transport the load supplied. Evidence from Welsh rivers indicates that cutoff frequency is at a maximum in very tight bends where r_c/w lies in the range 1–2 (Lewis and Lewin, 1983).

The net effect of cutoff development is to increase channel gradient (Fig. 5.22A) and therefore local transporting ability. Four cutoffs on the rivers Bollin and Dane increased gradient by an average of 2.7 times (Hooke, 1995). Selective artificial cutoffs designed to improve navigation depths and reduce flood heights along the lower Mississippi increased the gradient by 12 per cent overall, but local increases were as high as 2000 per cent (Winkley, 1982). The net result has been increasing instability along the river, with the steeper slopes inducing greater bed-material mobility and the development of migrating bars, so that navigation depths are now less than they were 90 years ago. This example illustrates the problems that can result from channel changes which run counter to a river's natural tendencies, and underlines the need to understand the meandering process if costly or potentially dangerous outcomes are to be avoided. Meandering is one means whereby a river can adjust its rate of energy expenditure and ability to transport sediment. In both respects a meandering channel may be more efficient than a straight one (Onishi *et al.*, 1976).

Braided Channels

Braided reaches are multi-channel forms in which the channels are separated by bars or islands (Plate 4.4, p. 145). Bars tend to be more transient, unvegetated and submerged at bankfull stage, whereas islands are often quite stable, vegetated and emergent at bankfull stage. However, this distinction may not always be sharp, for one grades into the other as bars accrete vertically and become vegetated. The characteristic feature of the braided pattern is the repeated division and joining of channels, and the associated divergence and convergence of flow, which contributes to a high rate of fluvial activity relative to other river types. Hierarchies of channels have been recognized in braided rivers (e.g. Bristow, 1987), reflecting variations in the level of channel dominance, but the proposed ordering schemes may be difficult to apply consistently. Distinctive topographic levels can often be identified across the channelled area, ranging from the most active channels to elevated, abandoned areas. Former channels may be reoccupied and enlarged during flood discharges when rapid shifts in channel position are not uncommon.

In comparison with meanders, the geometrical properties of braided channels have received little attention. Consequently, few relationships have been elucidated between braided channel geometry and variables of any causative significance. Part of the problem stems from the fact that the degree of braiding is not necessarily constant in the short term, tending to decrease at high stages as channel bars are drowned out. Measures of the degree of braiding generally fall into two categories, involving either the mean number of active channels across the channel belt, or the ratio of the sum of channel lengths in a reach to the reach length (Bridge, 1993). The index devised by Friend and Sinha (1993):

$$B_r = \frac{\text{sum of mid-channel lengths of all primary channels in reach}}{\text{mid-channel length of widest channel in reach}} \tag{5.48}$$

exemplifies the latter category, with B_r varying from 1 for a single-thread channel to >5 for an intensively braided one, although values in excess of 3.5 are rare. Theoretical analyses (Fig. 5.17B; Parker, 1976; Chang, 1979b) predict that the degree of braiding expressed in terms of the number of channels increases with slope. Parker's analysis emphasizes how adjustments to braided channel geometry can alter resistance properties within a reach and that, if a number of multi-channel forms are possible, those with the least excess energy are the most likely to occur.

Although not as frequent as single-thread channels, braided channels occur in a wide range of environments, from proglacial to semi-arid, and at a large range of scales, from the small streams on sandy beaches to the largest continental rivers. Various conditions have been suggested as conducive for the development of this planimetric form:

(a) *An abundant bed load.* The availability of large amounts of sediment supplied either from upstream or locally (notably the channel banks) is regarded as a necessary condition for braiding. Braiding may result from either a lack of capacity to transport the amount of bed material supplied, or a lack of competence to remove the size of sediment supplied. Where the load contains a wide range of size fractions, local incompetence to transport the larger sizes may induce the initial deposition from which mid-channel bars evolve. Concentrated deposition in the form of bars diverts the flow against channel banks and thus contributes to the bank erosion needed for the development of the wide, shallow channel commonly associated with bed-load transport. Changes in the channel pattern of the Rangitata River in New Zealand illustrate the importance of sediment supply (Schumm, 1980). The river descends from the mountains to the Canterbury Plain through a narrow bedrock gorge, above which there is a plentiful supply of sediment and a braided pattern. In common with the other rivers which cross the Canterbury Plain, it seems that the Rangitata should also be braided below the gorge, but the reduced sediment supply has caused it to adopt a meandering planform. However, further downstream the river has cut into poorly consolidated materials and rapidly converts from a meandering to a braided form which then persists to the sea. If a high supply rate of bed material is the key cause of braiding, the question arises as to whether a braided river, for a given discharge, slope and grain size, has a greater transporting capacity than a single-thread river (Carson and Griffiths, 1987), a convincing answer to which is still awaited.

(b) *Erodible banks.* Banks composed of readily erodible material are an important source of sediment as well as being necessary for the channel-widening characteristic of braided reaches. In a wider channel, maintenance of laterally uniform flow and transport conditions becomes less likely, and the probability increases of the river cross-section breaking down into one or more zones of deeper flow. Without erodible banks, any incipient bar deposits would tend to be destroyed rather than added to. Theoretical analyses (Engelund and Skovgaard, 1973; Fredsøe, 1978; Fukuoka, 1989) indicate that the main control on braiding is the width : depth ratio, being >50 for braiding to occur. Results from simulation experiments suggest that the only factors essential for braiding are bed-load transport and laterally unconstrained flow so that the river can freely adjust its width (Murray and Paola, 1994). Examples exist to show that rivers with resistant banks meander rather than braid, even though bed-load transport may be dominant (Miall, 1977). Mackin (1956) attributed a sequence of meandering–braided–meandering along the Wood River in Idaho to changes in bank resistance as the river passes through a corresponding sequence of forest–prairie–forest environments. The Turandui River in New Zealand has changed over a period of years from a braided to a meandering pattern as a result of the planting of willow shrubs at appropriate places (Nevins, 1969). Thus bank erodibility depends not only on flow and sediment conditions but also on the stabilizing effect of vegetation.

(c) *A highly variable discharge*. Rapid fluctuations in discharge are often associated with high rates of sediment supply, notably in proglacial areas. They also contribute to the bank erosion and irregular bed-load movement which are important to braided channel formation. At a longer timescale, the increases in discharge, sediment transport and width : depth ratio associated with extreme floods may precipitate braiding which persists for many years thereafter (Schumm and Lichty, 1963; Burkham, 1972). However, the fact that braided reaches can be interspersed with meandering ones along rivers with a given flow regime, and that braiding can be produced in the laboratory under steady flow conditions (Leopold and Wolman, 1957; Hong and Davies, 1979; Ashmore, 1991a; Germanoski and Schumm, 1993), suggests that discharge variability is not a primary factor in braided channel development.

(d) *Steep valley slopes*. Empirical and theoretical evidence indicates that braiding develops when the slope is above some threshold value (Leopold and Wolman, 1957; Schumm and Khan, 1972; Parker, 1976). Brierley and Hickin (1991) described a downstream sequence of braided–wandering–meandering along 20 km of the Squamish River where valley slope decreases from 0.0058 to 0.0015 and 0.0013 m m^{-1} respectively. Theory also suggests that the degree of braiding increases as slope steepens (Parker, 1976; Chang, 1979b). However, the critical factor is possibly a high stream power (γQs) or high specific stream power ($\gamma Qs/w$) rather than simply a steep slope, since braiding can persist at low slopes in very large rivers. Also, any threshold slope for braiding is dependent on channel boundary resistance, increasing with more cohesive banks or coarser bed material (Ferguson, 1987). A river must be sufficiently powerful to erode its banks and achieve the high bed mobility which is a crucial requirement for braiding.

None of these conditions appears to be sufficient on its own to produce braiding, although an abundant bed load, erodible banks and a relatively high stream power are probably necessary. Where these factors occur in conjunction, as in proglacial areas, braiding tends to be most prevalent.

Braiding can be initiated by either depositional or erosional means, although the latter depends to some extent on a preceding depositional phase. On the basis of on laboratory flume experiments, Ashmore (1991a) identified two depositional (central bar deposition, transverse bar conversion) and two erosional (chute cutoff, multiple bar dissection) mechanisms.

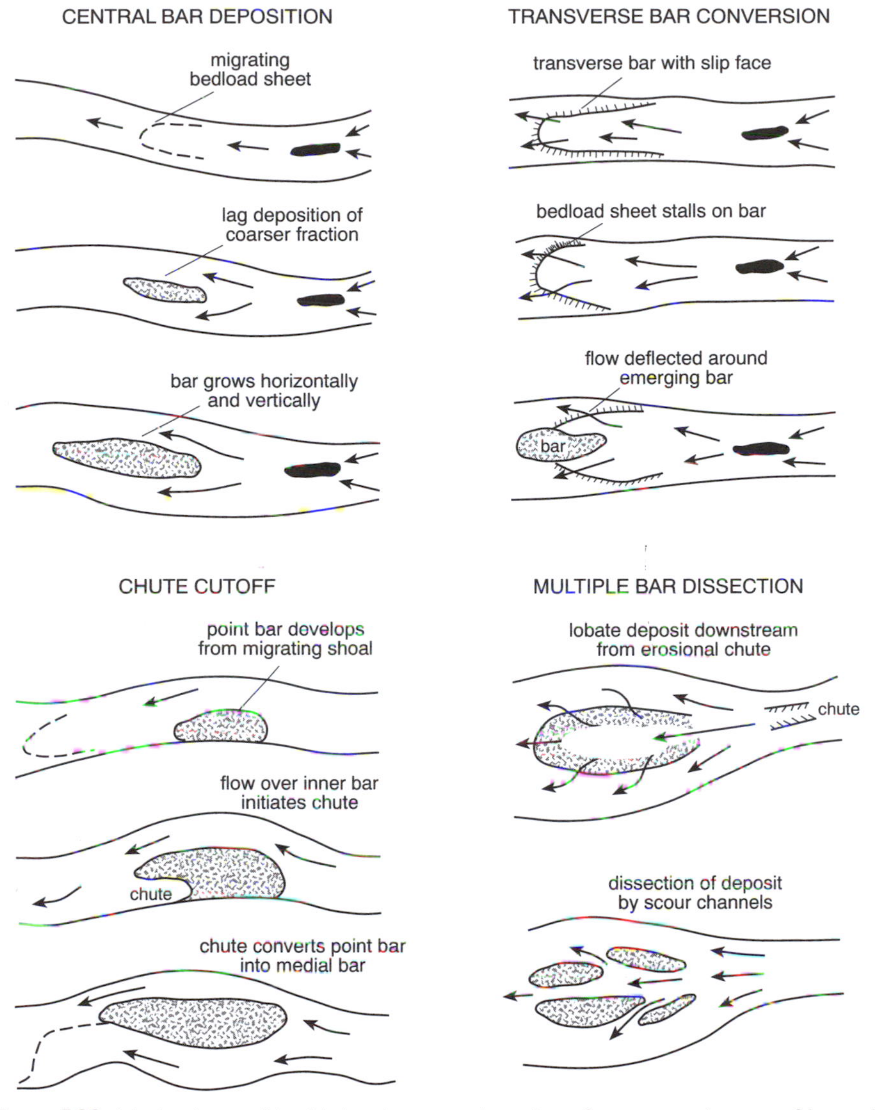

Figure 5.23 Mechanisms of braid development, based on flume experiments of Leopold and Wolman (1957) and Ashmore (1991a), and cartoons from Ferguson (1993).

The classic *central bar mechanism* of Leopold and Wolman (1957) involves the lag deposition of coarser material which the stream is locally incompetent to transport. This provides a locus for further deposition so that the incipient bar grows surfaceward and downstream by successive addition (Fig. 5.23). The bar becomes more sharply defined once the main flow is sufficiently concentrated in the flanking channels to cause significant bank erosion and bed scour. A symmetrical deposit in the form of a transverse bar is the initial stage of *transverse bar conversion*. It forms downstream of a pool where flow convergence scours the bed and provides sufficient material for substantial deposition where the flow diverges (Fig. 5.23). Ashmore (1991a) observed the stalling of thin bed-load sheets as they crossed the bar surface, contributing to vertical accretion and leading eventually to flow deflection around an elevated central lobe.

An erosional origin for braiding has been recognized for some time (e.g. Carson, 1984b) but has until recently been accorded a subordinate role. However, in the experiments of Germanoski and Schumm (1993) the initial development of braid bars was linked almost exclusively to the dissection of stalled linguoid bars (actively migrating, lobate bed forms). Ashmore (1991a) identified *chute cutoff* as the dominant process in his experiments, which involves headwards incision by flow as it follows a more direct route across an alternate or point bar, capture of progressively larger volumes of water, and rapid enlargement of a cutoff channel, the original bar surviving in modified form as a medial bar (Fig. 5.23). The cutoff process was often triggered by local thalweg shoaling related to the arrival of bed-load pulses. *Multiple bar dissection* occurs in wider channels where flow becomes concentrated into multiple chutes which dissect a lobate bar constructed in a previous phase of deposition at a flow expansion (Fig. 5.23). Although of limited occurrence in Ashmore's experiments, this process may assume greater significance in natural rivers and appears to be the characteristic braiding mechanism in New Zealand rivers (Rundle, 1985). To these erosional processes can be added *avulsion*, when larger-scale switching of the main flow occurs and new channels are cut or older ones reoccupied.

The underlying cause of braiding in all of these processes is essentially the same: local aggradation (often linked to the stalling of bed-load sheets or migrating bars) and loss of competence at a developing flow expansion (Ashmore, 1991a). Which process is prevalent depends on the particular channel morphology and sediment mobility conditions. Ashmore associated the central bar mechanism with low excess shear stress ($\tau_0 - \tau_c$) when small differences in depth or grain size can produce relatively large local variations in competence, and multiple bar dissection with channels of very high width : depth ratio. The distinction between transverse bar conversion and chute cutoff is less clear, but the former appears to require symmetrical and the latter asymmetrical flow configurations in the entrance reach. Once braiding has been established, all of these processes may recur to promote further braiding within a given stream, provided short-term transience in bed-load transport is maintained.

That transience is linked to the pronounced convergence and divergence of flow at junctions and bifurcations. Confluence zones play a key role in controlling the movement of bed material through braided reaches in that they act as funnels for the conveyance of material supplied from upstream, are themselves a source of sediment as a result of bed scour, and often have bars deposited immediately downstream from which renewed braiding can develop (Davoren and Mosley, 1986; Ferguson *et al.*, 1992). In flume experiments, Ashworth (1996) observed mid-channel bar growth after the stalling of coarse sediment downstream from a confluence scour. The flow changed from a convergent to a divergent state when the bar had reached a height of approximately 55 per cent of thalweg depth.

Confluence zones share many of the features associated with fixed tributary junctions (Fig. 2.17, p. 59), where relative discharge, junction angle and the orientation of the joining channels influence confluence geometry (Mosley, 1976; Best, 1986, 1988). In particular, the maximum depth of scour tends to increase as junction angle increases and as the discharges in the two channels approach equality. However, confluence zones are much more dynamic than tributary junctions, undergoing frequent adjustments in size, orientation and position, which has implications for depositional patterns downstream (Ashmore, 1993). Sedimentation downstream from a confluence is often the prelude to the development of a bifurcation by one of the mechanisms described above: central bar deposition or transverse bar conversion at symmetrical confluences, or chute cutoff at asymmetrical ones. Many of the changes in braided river morphology occur episodically, even at constant discharge, as a result of short-term fluctuations in the transport rate, related in particular to bed-load pulses (Ashmore, 1991b; Hoey and Sutherland, 1991). Increasing bed-load input to a reach causes local aggradation and channel multiplication, while reducing input leads to local degradation and channel simplification as the number of braid bars decreases through coalescence (Germanoski and Schumm, 1993).

Braided streams are characterized by frequent shifts in channel position. Between 1736 and 1964 the Kosi River in India migrated westward over a distance of 112 km, reworking its own deposits and maintaining a braided appearance in the process (Gole and Chitale, 1966). Shifts of up to 90–120 m per day have been reported along the lower Yellow River in China, where the amount of lateral movement is controlled by the spacing of valley-side constrictions (Chien, 1961). Major floods may be responsible for the onset of braiding as well as triggering large shifts in channel position. At a smaller scale, the formation and destruction of individual bars and channels can be observed over the few hours or days in which discharge fluctuates in a proglacial stream (Fahnestock, 1963). Where juxtaposed channels differ in flow competence, rapid changes are not unexpected. The subsidiary channel may become choked with coarser material at its upstream end so that the flow becomes concentrated in the other, which scours its bed and erodes its banks preparatory to a new phase of braiding.

Individual channels in a braided river are seldom in equilibrium and may be very unstable (Hoey and Sutherland, 1991; Ferguson, 1993). Given the high stream power, the large fluctuations in bed-load transport rate associated with successive convergence and divergence, and the changeable sharing of discharge between channels, such instability is not unexpected. As a result the braided pattern has been regarded as a disequilibrium, aggradational response to an increased load which a river cannot totally transport. However, many consider that braiding can be a valid equilibrium form given the right combination of discharge and slope, even though individual channels might be transient. Leopold and Wolman (1957) found braided reaches showing little change over several decades, while Klaassen and Vermeer (1988) regard the braided Brahmaputra River as stable in both the horizontal and the vertical directions despite a major avulsion more than two centuries ago. The branching channels of 16 out of 24 braided reaches along the middle and lower Yangtze River have lasted for more than a century, aided by moderate amounts of silt and clay in the banks which give the islands some degree of stability (You Lianyuan, 1987).

Leopold and Wolman (1957) argued that braiding is the type of adjustment made by rivers with erodible banks and a debris load too large to be carried by a single channel, representing a particular combination of the adjustable variables which, once established, can be maintained thereafter with only slight modification. There is some theoretical justification for this opinion (Parker, 1976; Chang, 1979b), although Chang differentiates between braiding due to overloading and braiding due to steep slopes, only the latter of which is deemed capable of maintaining a quasi-equilibrium between discharge, sediment inflow and transport capacity. This has echoes in the explanation proposed by Bettess and White (1983), who claimed that braiding would develop where the valley slope was much steeper than the regime slope required for the transportation of a given load. Bagnold (1977, 1980) has shown that bed-load transport is directly related to stream power and that it is more efficient in relatively wide, shallow channels (equation (4.13, p. 130)). To the extent that braiding is favoured where stream power and width : depth ratios are sufficiently high, it would seem to fulfil the requirements needed for transportation of a large bed load. In common with meandering, braiding involves changes in form elements other than strictly planimetric ones and represents a mode of adjustment that can modify the way in which a stream consumes its energy. Such modification has implications for transporting ability.

Anabranching Channels

A braided river consists of multi-channel flow separated by bars and contained within a dominant pair of floodplain banks, whereas an anabranching river consists of multiple channels separated by stable islands which are large relative to the size of the channels and which divide the flow at discharges up to and including bankfull. Consequently the flow patterns in adjacent channel segments are essentially independent of one another, unlike those in braided channels (Bridge, 1993). In anabranching rivers the islands usually persist for decades or centuries, support well-established vegetation and are at approximately the same elevation as the surrounding floodplain. Indeed, they may have been excised by avulsion from that floodplain during periods of aggressive overbank flow. Individual channels may meander, braid or remain relatively straight (Brice, 1984; Schumm, 1985), and eventually rejoin, in contrast to distributary systems where channels divide but do not rejoin.

Anabranching rivers are uncommon but are found in a wide range of environments, from subarctic to tropical, from monsoonal to semi-arid, implying that climate is not of itself a determining factor in their formation

Table 5.10 Distinguishing characteristics of channel patterns (after Smith and Putnam, 1980)

Characteristic	Anastomosing	Meandering	Braided
Gradient	Low	Moderate	High
Sinuosity	Variable	High	Low
Width : depth ratio	Low	Moderate	High
Percentage silt–clay in banks	High	Moderate	Low
Lateral migration	Rare	Common	Common

(Knighton and Nanson, 1993). On the basis of various hydraulic and physiographic conditions, Nanson and Knighton (1996) have proposed a classification which recognizes six main types, the first being the most extensive.

Type 1: Cohesive-sediment anabranching rivers (anastomosing rivers)
Anastomosing rivers are characterized by low gradients, very small stream powers (usually ≤ 8 W m^{-2}), and cohesive banks which produce laterally stable channels of low width : depth ratio. Table 5.10 lists some of the characteristics used to distinguish anastomosing from other channel patterns. Anastomosed channel development has been regarded as symptomatic of an aggradational regime which is imposed on the river through either basin subsidence or a rising base-level downstream (Smith, 1983), but this condition, although conducive, is probably not sufficient by itself or even universal. However, when combined with limited flow strength, either to transport a relatively large sediment input or erode resistant banks, aggradation can lead to shoaling and reduced channel capacity, which locally forces flow out of single-thread pattern on to the floodplain where new channels are cut. As Smith and Smith (1980) have shown in three Canadian rivers, anastomosing reaches can grade gradually into entirely braided ones but are distinguished from them by greater bankline stability. Channel shifting or relocation occurs on a larger (floodplain) scale and possibly over longer time periods, with avulsion, rather than the deposition or dissection of within-channel bars, as the dominant process. Smith *et al.* (1989) describe how a major avulsion in 1873 diverted most of the lower Saskatchewan River into a previously existing wetland where an anastomosing channel system developed as the floodplain aggraded. In the virtual absence of lateral migration, vertical accretion is the primary mechanism of floodplain construction. A system of anastomosing channels provides a means not only of effecting the widespread distribution of sediment over a floodplain, but also of accommodating infrequent but high-magnitude floods where channel capacity is constrained (Schumann, 1989).

Nanson and Knighton (1996) recognized three subtypes differentiated on the basis of environment and associated sediment conditions: organic systems, of which the Okavango megafan in Botswana is the sole representative (McCarthy *et al.*, 1992); organo-clastic systems, such as those found in Canada (Smith and Smith, 1980; Smith, 1983; Smith *et al.*, 1989); and mud-dominated systems, of which those in the Red Desert of Wyoming

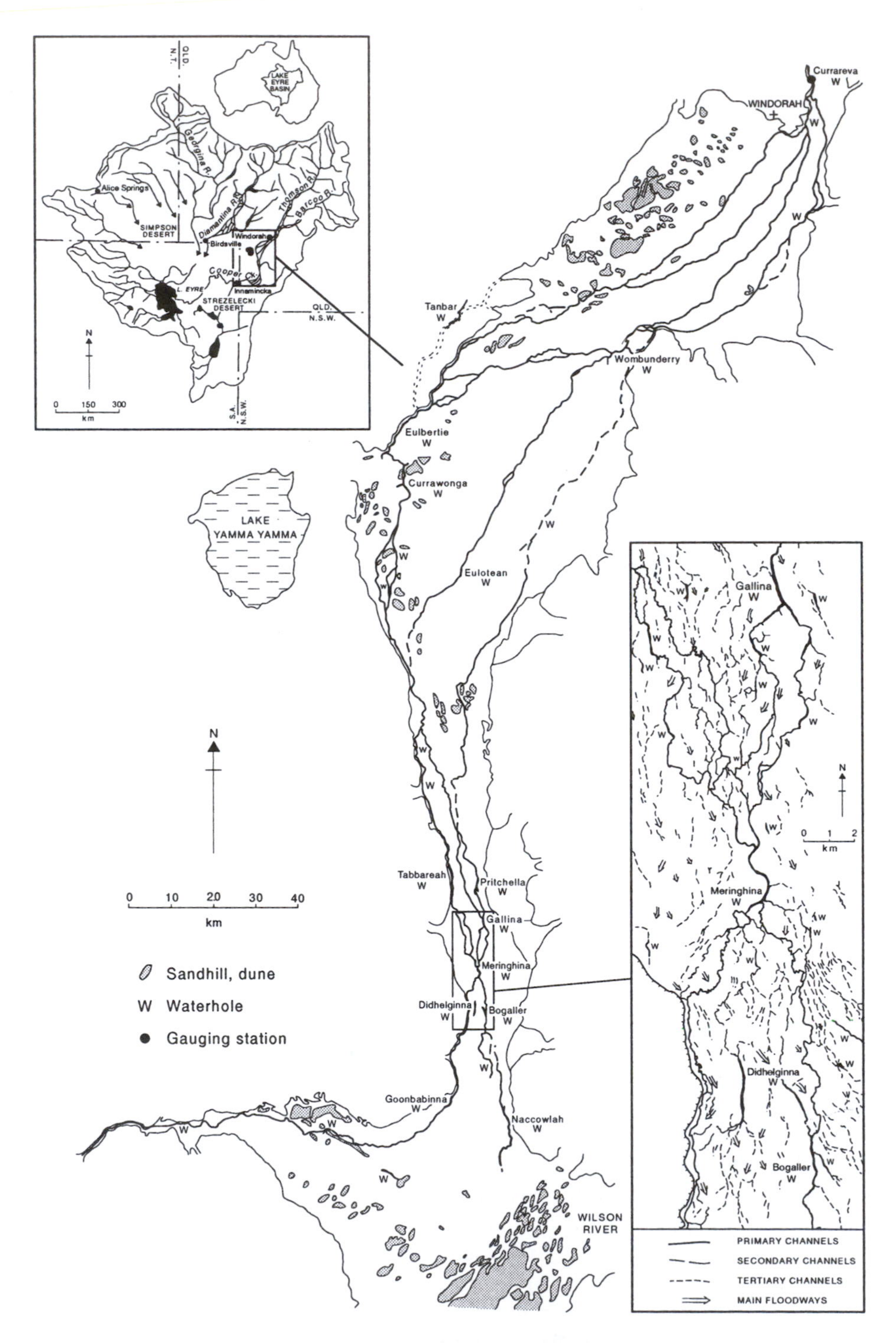

Figure 5.24 The anastomosing channel system of Cooper Creek, Australia. The inset map shows the position of Cooper Creek in the Lake Eyre Basin.

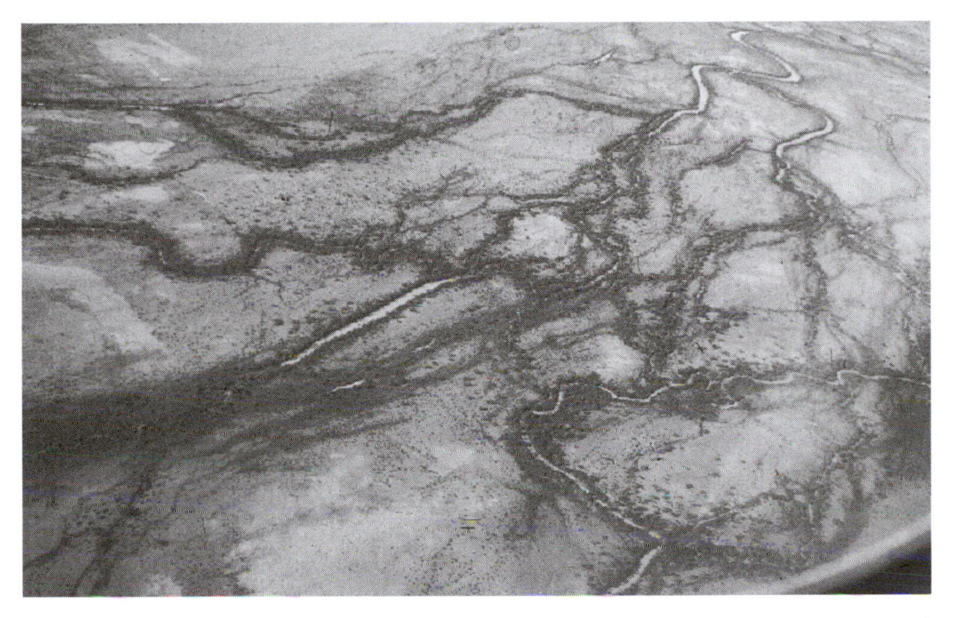

Plate 5.5 Anastomosing channel system of Cooper Creek, with the upstream end of Meringhina Waterhole in the top right of the photograph.

(Schumann, 1989) and the semi-arid Channel Country of east-central Australia (Nanson *et al.*, 1988) are prime examples. As a representative of the last, Cooper Creek maintains an anastomosed pattern for long river distances (>400 km) despite transmission losses of over 75 per cent (Knighton and Nanson, 1994a). A well-integrated primary system of one to four anastomosing channels is supplemented by secondary and tertiary channels which become active at different stages during the filling phase of a flood (Fig. 5.24). A distinctive feature of the Cooper system is the preponderance of waterholes, which represent lines of preferential scour and bank erosion along a floodplain whose surface sediments are largely clay-rich mud and therefore relatively resistant (Knighton and Nanson, 1994b; Plate 5.5).

Type 2: Sand-dominated, island-forming anabranching rivers
Similar in other respects to Type 1 rivers, these have less cohesive boundary sediments. In order to ensure anabranch stability in sandy alluvium, low stream powers and stabilizing bank vegetation are required.

Type 3: Mixed-load, laterally active anabranching rivers
These multi-channel rivers have well-established anabranches which meander and migrate laterally across part of their floodplain. They are similar to organo-clastic Type 1 rivers but are laterally more active.

Type 4: Sand-dominated, ridge-forming anabranching rivers
Observed only in parts of arid central and northern Australia, these rivers have subparallel channels separated by narrow, steep-sided sandy ridges

which are topped by lines of stabilizing trees. The ridges form as a result of either within-channel tree growth or excision from the floodplain. They effectively compartmentalize the flow, thereby enhancing the ability of the rivers to transport a sandy bed load along low-gradient valleys.

Type 5: Gravel-dominated, laterally active anabranching rivers

Common in the Cordilleran region of western Canada, these 'wandering' gravel-bed rivers have been described as transitional between meandering and braided (Church, 1983; Desloges and Church, 1989). They have specific stream powers ranging from 30 to 100 W m^{-2}, sufficient to promote vigorous lateral activity. They appear to develop at locations where sediment deposition is favoured, referred to as 'sedimentation zones' by Church (1983), but, despite the large volumes of gravel in storage, bed-load transport rates are relatively modest and less than those associated with braided rivers. In the last century, substantial reaches of the Bella Coola River in British Columbia have changed from a laterally unstable anabranching river with large forested islands to a more stable, single-channel form, probably as a result of

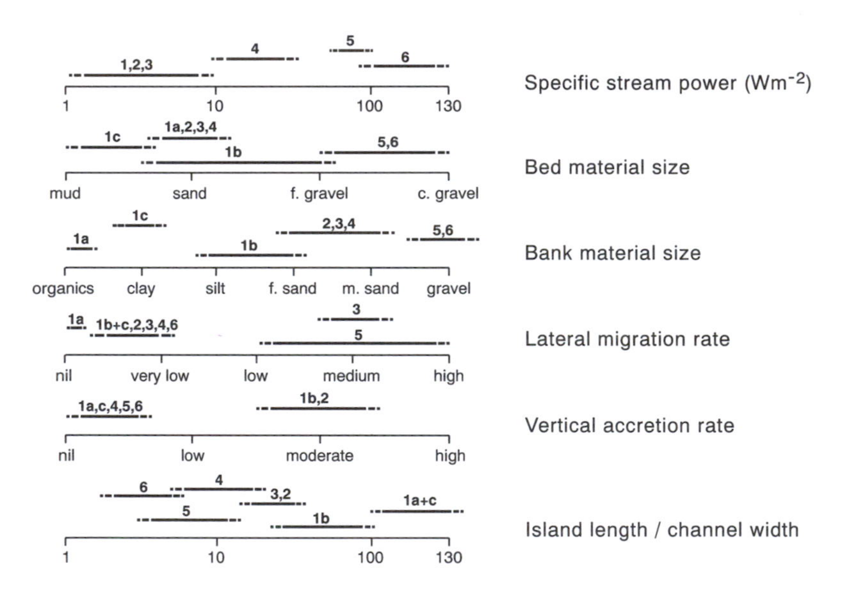

1 Cohesive sediment anabranching rivers (anastomosing):
  a Organic systems
  b Organo-clastic systems
  c Mud-dominated systems

2 Sand-dominated, island-forming anabranching rivers

3 Mixed-load, laterally active anabranching rivers

4 Sand-dominated, ridge-forming anabranching rivers

5 Gravel-dominated, laterally active anabranching rivers

6 Gravel-dominated, stable anabranching rivers

Figure 5.25 Differentiation of anabranching river types on the basis of flow, sediment, form and behavioural characteristics (after Nanson and Knighton, 1996).

a reduction in bed-material supply (Church, 1983). Anabranching behaviour largely stems from avulsion channels incising into the existing floodplain, but in some cases islands grow vertically to floodplain height from large bars stabilized by vegetation within the channel.

Type 6: Gravel-dominated, stable anabranching rivers
A small group of steep-gradient streams that respond rapidly to rainfall events can exhibit anabranching characteristics with well-vegetated gravel or boulder islands. Anabranches form in a similar way to Type 5, log jams and sediment accumulation being primary causes.

The six types are arranged approximately in order of increasing specific stream power ($\gamma Qs/w$) but can be differentiated further on the basis of sediment texture and morphology (Fig. 5.25). Two factors appear to be particularly important for their formation: a highly variable, flood-prone flow regime; and banks that are erosionally resistant relative to ambient stream power. A highly seasonal or extremely episodic flow regime is characteristic of all the anabranching rivers considered by Nanson and Knighton (1996), giving rise to concentrated flooding of relatively large magnitude. Combined with resistant banks which limit channel capacity, this increases the frequency of overbank flow and enhances the development of floodplain channels by avulsion processes. Other mechanisms may be instrumental in displacing flow from a single channel, including localized sedimentation, the formation of log or ice jams, and ponding in very low-gradient streams. It appears that the major advantage of an anabranching system is to divide and concentrate flow into several narrower, deeper and unobstructed channels, thereby reducing the total bed width and enhancing the ability of a river to maintain its water and sediment flux without increasing channel gradient. Channel banks must be sufficiently resistant to withstand the increased stress.

A distinction can be drawn between 'river pattern', which describes the planform geometry within a reach of river, and 'channel pattern', which refers to the planform geometry within an individual channel. This distinction is useful where rivers anabranch but contain a variety of channel patterns, and serves as a basis for a revised classification of single-channel and multiple-channel forms. Types 1 and 2 form a group recognized as anastomosing by Knighton and Nanson (1993). Types 3 and 4 occupy the mixed-load meandering and sand-load braided range, while Types 5 and 6 are equivalent to transitional braided gravel rivers where braiding is partially suppressed by bank resistance or low bed-sediment concentrations. Thus anabranching rivers are not distinguishable from laterally inactive, meandering or braided rivers in terms of the simple slope–discharge plots commonly used to discriminate between river types (Fig. 5.17), but represent a diverse group generally associated with flood-dominated flow regimes, banks that are resistant to erosion, and mechanisms that constrict channels and induce avulsion. Anabranching does not preclude meandering or braiding, although it is commonly associated with laterally stable channels.

Channel Gradient and the Longitudinal Profile

River profiles are an important element of drainage basin geomorphology in that, together with the channel network, they fix the boundary conditions for hillslope processes. The longitudinal profile is simply a plot of height (H) against distance downstream (L) expressed formally by

$$H = \mathrm{f}(L) \tag{5.49}$$

the tangent to which indicates channel gradient at the reach scale. Regarding the profile as an ideal, graded form, the initial problem is to determine the possible form that this function might take.

Hack (1957) argued that slope (s) and distance (L) are related by a power function,

$$s = k\mathrm{L}^{n} \tag{5.50}$$

where the exponent n is an index of profile concavity. Since $s = -\mathrm{d}H/\mathrm{d}L$ (the minus sign denotes decreasing gradient downstream), equation (5.50) can be integrated to give the two profile equations:

$$H = H_1 - k \ln L \qquad \text{where } n = -1 \tag{5.51}$$

$$H = H_0 - \frac{k}{n+1} L^{n+1} \qquad \text{where } n \neq -1 \tag{5.52}$$

where H_0 and H_1 are the heights at $L = 0$ and $L = 1$ respectively. For gravel-bed and boulder-bed streams in Virginia and Maryland, Hack believed that the two equations with their respective logarithmic and power functional forms apply to different conditions, the first where grain size remains constant along a stream, and the second where grain size changes systematically downstream. However, the association of a logarithmic profile (equation (5.51)) with constant grain size is unlikely to apply generally. Indeed, the logarithmic function performs well where gravels decrease sharply in size downstream (Snow and Slingerland, 1987). Such is the case along the River Bollin (Knighton, 1975b; Fig. 5.26C).

Assuming that slope is proportional to height,

$$-\frac{\mathrm{d}H}{\mathrm{d}L} = k_1 H \tag{5.53}$$

a simple model of profile form can be derived in which height decreases exponentially downstream:

Figure 5.26 (opposite) Longitudinal stream profiles. (A) The River Amazon, illustrating how Langbein's (1964b) index of concavity is calculated where H is total fall and A is the height difference between the profile at mid-distance and a straight line joining the end points of the profile. (B) The River Rhine. (C) The River Bollin (dashed line) and a logarithmic profile (solid line) calculated on the basis of equation (5.51). (D) Nigel Creek, Canadian Rockies. (E) The River Towy and its main tributaries, with profiles (dashed lines) fitted to indicate previous base-levels (after Jones, 1924). (F) Random-walk profiles generated without (i) and with (ii) length constraints (after Leopold and Langbein, 1962).

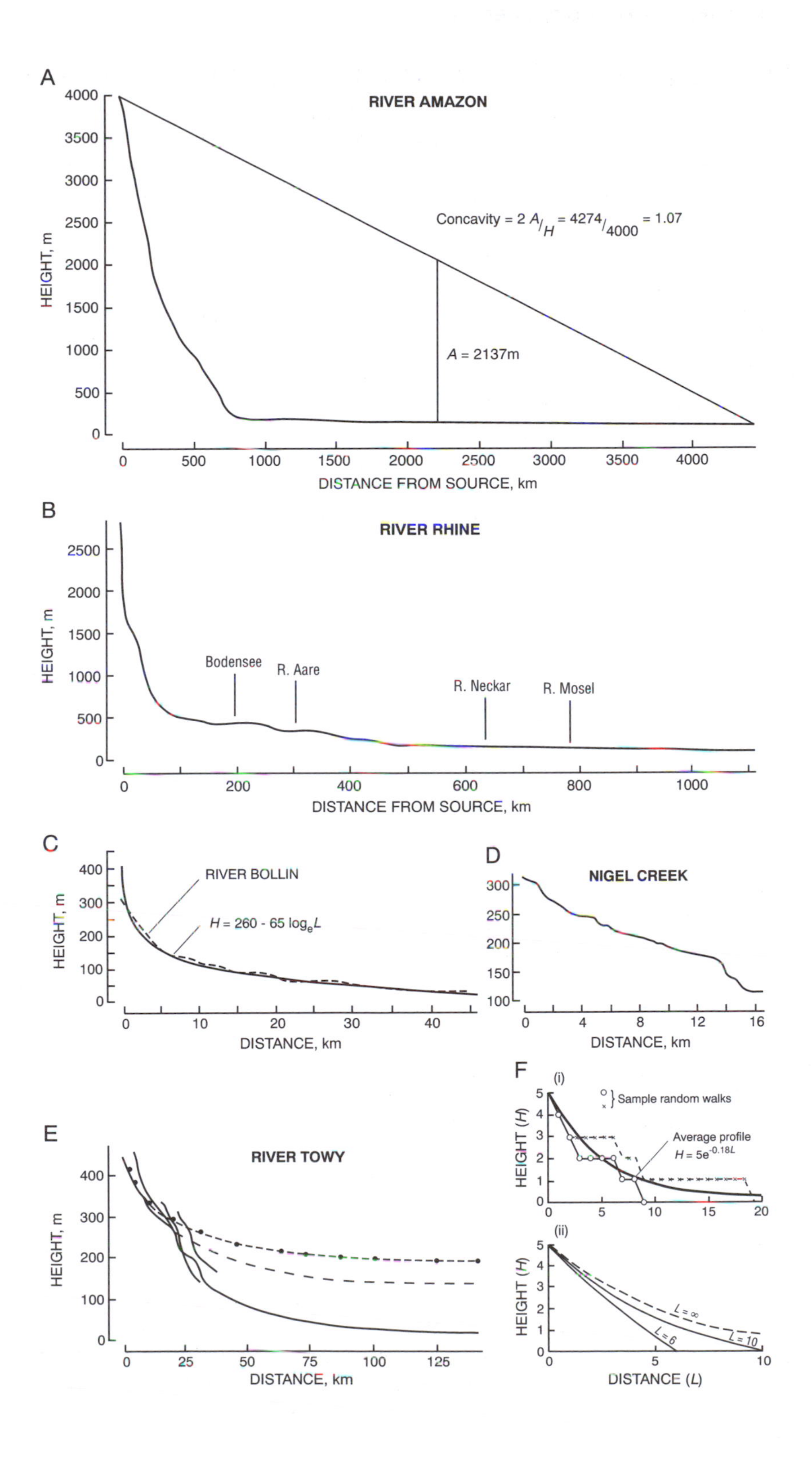

A
RIVER AMAZON
HEIGHT, m
Concavity = 2 A/H = 4274/4000 = 1.07
A = 2137m
DISTANCE FROM SOURCE, km
B
RIVER RHINE
HEIGHT, m
Bodensee
R. Aare
R. Neckar
R. Mosel
DISTANCE FROM SOURCE, km
C
RIVER BOLLIN
H = 260 - 65 log$_e$L
HEIGHT, m
DISTANCE, km
D
NIGEL CREEK
DISTANCE, km
E
RIVER TOWY
HEIGHT, m
DISTANCE, km
F
(i)
Sample random walks
Average profile
H = 5e$^{-0.18L}$
HEIGHT (H)
(ii)
L = ∞
L = 10
L = 6
HEIGHT (H)
DISTANCE (L)

$$H = H_0 \, e^{-k_1 L} \tag{5.54}$$

Since the particle size model (equation (4.14) p. 136) has a similar form, a close link between profile form and the rate of change of grain size downstream is implied. Leopold and Langbein (1962) attempted to justify the general applicability of equation (5.54) by demonstrating with a random-walk model that the most probable profile should have an exponential form (Fig. 5.26F(i)). Morris and Williams (1997) have presented a theoretical model which shows that streams with small sediment loads and limited lateral inflows have exponential profiles if the bed-load sediment undergoes either abrasion or hydraulic sorting. If both processes are significant, only short profiles are exponential.

From the results of numerical experiments, Snow and Slingerland (1987) showed that any of the three functions – exponential, logarithmic, power – can provide a reasonable fit to stream profiles, and that which function performs best may depend on which of the controlling variables is/are having a dominant influence on profile form. Stream systems characterized largely by downstream increases in water and sediment discharge tend to develop profiles which are best described by a power function. If bed material size decreases rapidly downstream, a logarithmic curve gives the best approximation, a somewhat surprising result in view of equation (4.14). Profiles approach an exponential form most closely when grain size decreases downstream under conditions of low sediment transport and near constancy of other major controls.

All three functional forms give rise to smooth, concave-upward curves. The longitudinal profiles of natural rivers tend to be concave upward (Fig. 5.26) but they are rarely smooth. In particular, they often contain convexities. Such local steepening of channel gradient can result from one of several causes: (a) more resistant bedrock strata; (b) the introduction of a coarser or larger load; (c) tectonic activity (e.g. Gregory and Schumm, 1987); or (d) the effect of past events, notably a fall in base-level. Based on the assumption that convexities in compound profiles indicate the positions of rejuvenation heads, attempts have been made to estimate former sea-levels by downvalley extrapolation of curves fitted to profiles and terrace remnants (e.g. Jones, 1924; Fig. 5.26E). However, whether knickpoints can migrate far upstream before becoming indistinct is debatable and, in any case, convexities can be produced other than by headward recession following a fall in base-level. The profile of Nigel Creek in the Canadian Rockies provides an extreme example (Fig. 5.26D). The large step towards the basin mouth reflects the discordant junction of two pre-existing glaciers, while the smaller convexities upstream are related to resistant bedrock outcrops which provide local base-level control. Interestingly, well-defined knickpoints did develop in small channels tributary to Nigel Creek in response to downcutting by the main stream as it passed through braided reaches, although they had not progressed far upchannel (Knighton, 1976). These various explanations imply that convexities are in some way abnormal and that, in line with Davis (1902), a smooth concave-upward profile is diagnostic of the graded or equilibrium state. That view has variously been challenged and a more dynamic approach developed, which focuses more on what controls channel gradient and ultimately the form of the longitudinal profile.

Controls on Channel Gradient

Rubey's (1952) equation for 'large adjusted streams'

$$s^3 X = k \frac{Q_s^{\,2} M}{Q^2} \tag{5.55}$$

summarizes the main factors controlling channel gradient (s). For a constant cross-sectional shape X ($= d/w$), slope varies directly with sediment load (Q_s) and the size of material in the load (M), and inversely with discharge (Q). Since sediment load data are difficult to obtain, empirical analyses have tended to focus on the effects of M and Q, with M being characterized by bed material size (D) and drainage area (A_d) often being used as a surrogate for Q.

Slope is in general negatively correlated with discharge but slope–discharge correlations for natural rivers tend to be rather poor (e.g. Bray, 1982; Hey and Thorne, 1986). Also, exponent (z) and coefficient (g) values in $s = gQ^z$ tend to vary over a wide range. For four major rivers in the US, z ranges from –0.50 to –0.93 with a mean of –0.65 (Carlston, 1968), while Osterkamp (1978) obtained an average value of –0.25 for 76 Kansas channels. Of particular interest in the latter case is the dependence of g on bed material properties, hinting at a dual control. Despite the similarity of exponents in the gravel- and sand-bed bivariate relationships listed in Table 5.11, the degree of variation in z and g values is wide enough to suggest that any slope–discharge relationship is principally a function of the particular physiographic province in which measurements are made. Even a good correlation does not necessarily mean that discharge is a major determinant of slope (Prestegaard, 1983b). However, downstream changes in discharge influence the ability of a river to transport sediment, upon which the adjustment of slope ultimately depends.

Bed material size and degree of sorting at any point along a stream are determined by the initial conditions of supply and the subsequent action of sorting and abrasion processes during downstream transport (Fig. 4.6, p.119). Both initial size and resistance to wear are influenced by bedrock lithology. Working in an area of diverse lithology, Hack (1957) found no simple relationship between channel slope and median bed material size (D_{50}). Only when sites of equivalent drainage area (approximately equal discharge) were compared was a significant relationship obtained:

$$s = 0.006 \left(\frac{D_{50}}{A_d} \right)^{0.6} \tag{5.56}$$

The implication is that slope depends not on grain size alone, but on the combined effect of grain size and discharge. The influence of lithology was reflected in the progressively steeper slopes and coarser debris of similar-sized streams flowing on limestone, shale and sandstone bedrock.

Table 5.11 Channel slope relationships

Source	Location/ applicable conditions	Type of equation: Bivariate	Multivariate
(1) *Threshold channels*			
Kellerhals (1967)	Western Canada		$s = 0.086\, Q_3^{-0.40}\, D_{90}^{0.92}$
Li *et al.* (1976)	Threshold theory	$s = g\, Q_b^{-0.4}$	
(2) *Gravel-bed channels*			
Charlton et *al.* (1978)	Britain		$s = 0.40\, Q_b^{-0.42}\, D_{65}^{1.38}\, D_{90}^{-0.24}$
Bray (1982)	Alberta	$s = 0.011\, Q_2^{-0.34}$	$s = 0.060\, Q_2^{-0.33}\, D_{50}^{0.59}$
Hey and Thorne (1986)	Britain		$s = 0.096\, Q_b^{-0.31}\, D_{50}^{0.71}$
			$s = 0.038\, Q_b^{-0.28}\, D_{84}^{0.62}$
			$s = 0.087\, Q_b^{-0.43}\, D_{50}^{-0.09}\, D_{84}^{0.84}\, Q_{sb}^{0.10}$
Parker (1979)	Theoretical – momentum diffusion		$s = 0.223\, \widetilde{Q}_b^{-0.41}$ ($\approx 0.395 Q_b^{-0.41}\, D_{50}^{1.02}$)
(3) *Sand-bed channels*			
Lacey (1929)	Canals – Punjab		$s = 0.211\, Q^{-0.17}\, D_{50}^{0.83}$
Simons and Albertson (1963)	Indian and US canals:		
	sandy banks	$s = 0.00007\, Q^{-0.30}$	
	cohesive banks	$s = 0.0026\, Q^{-0.30}$	
(4) *Undifferentiated*			
Rundquist (1975)	Gravel- and sand-bed channels	$s = 0.0032\, Q_b^{-0.30}$	$s = 0.002\, Q_b^{-0.25}\, D_{50}^{0.36}$

Symbols: s, channel slope; Q, discharge (bankfull (Q_b), with a recurrence interval of 2 (Q_2) or 3 (Q_3) years, dimensionless ($\widetilde{Q}$)); D, bed material size (median (D_{50}), that size at which x% (D_x) is finer); Q_{sb}, bed-load discharge.

Subsequent work has tended to confirm the basic message contained within equation (5.56). Positive relationships between slope and average bed material size remain rather weak unless some index of discharge is also included so that sites of similar flow conditions are being compared (Cherkauer, 1972; Penning-Rowsell and Townshend, 1978; Schröder, 1991). Kirkby (1993) has derived a similar expression from two different standpoints: a dynamic one involving a combination of resistance and hydraulic geometry equations, and an evolutionary one which implies that humid rivers progressively develop towards a form suggested by equation (5.56). In the multivariate relationships of Table 5.11 the bed material exponent always exceeds the discharge exponent in absolute value, suggesting that bed material size has the greater influence. Prestegaard's (1983b) results for gravel-bed streams support that point, but Penning-Rowsell and Townshend (1978) found that bed material was the more important factor only at the local scale.

Given that bed material size influences particle mobility and channel roughness, both of which affect the adjustment of channel gradient, the question arises as to the relative significance of different grain-size parameters. For British gravel-bed rivers Charlton *et al.* (1978) suggested that slope is better related to D_{65} (the threshold grain diameter at bankfull flow) than to D_{90} (the diameter used to represent the size of roughness elements), which emphasizes the transport rather than the resistance component of bed material. However, the resistance effect on slope through bed material size has also been invoked (Leopold and Bull, 1979; Prestegaard, 1983b), and in that respect the larger diameters play the dominant role, at least in coarse-bed streams. Along the River Hodder, Wilcock (1967) found that slope correlated poorly with D_{50} but significantly with 'residual' bed-material size (defined as that fraction which is immobile at present bankfull flow), implying that slope is largely controlled by the bigger sizes. Hey and Thorne's results (1986; Table 5.11) carry a similar message.

Following a similar line, Howard (1980, 1987) argued that a small range of grain sizes coarser than the mean is critical in determining channel gradient. Using 15 different transport formulae, he has calculated the gradient required to transport the load supplied for a given specific discharge. When required gradient is plotted against grain size, the various formulae give the same basic pattern (Fig. 5.27A): a maximum in the range of 0.1–2 mm, a minimum near 10 mm, and an indefinitely increasing required gradient as grain size increases beyond 10 mm. In effect the minimum near 10 mm separates gravel-bed from sand-bed streams. In the former the required gradient is determined by the supply of coarse grains. If, for some critical grain size, the required gradient is greater than the fine-grained maximum (A in Fig. 5.27A), the bed will be dominated by that coarse fraction. On the other hand, if the supply of coarse grains is so restricted that the required gradient for the critical size is less than the fine-grained maximum (B in Fig. 5.27A), the bed will be dominated by fine grain sizes because of differential transport. Thus, in the coarse size range the required gradient is controlled by the threshold of motion criterion and is largely independent of the quantity of sediment, whereas in the fine size range the required gradient is determined by the quantity as well as the grain size of the load.

By keeping other factors constant, Howard (1980, 1987) went on to show how the required gradient varies directly with sediment concentration (or sediment load with Q fixed) and indirectly with unit discharge $q = Q/w$ (Fig. 5.27B, C and D). Thus alluvial streams with a high imposed load or low unit discharge are more likely to be fine bed. Increasing the range of grain sizes in transport also alters the gradient curves (Fig. 5.27E). In particular it broadens the curve of the fine-grained maximum while simultaneously reducing the depth of the valley at intermediate sizes, and therefore the distinction between coarse and fine beds. Streams receiving a wide range of sediment sizes are therefore more likely to be coarse bed.

Only one of the multivariate relationships in Table 5.11 includes sediment load as a controlling variable and even then the exponent has a low

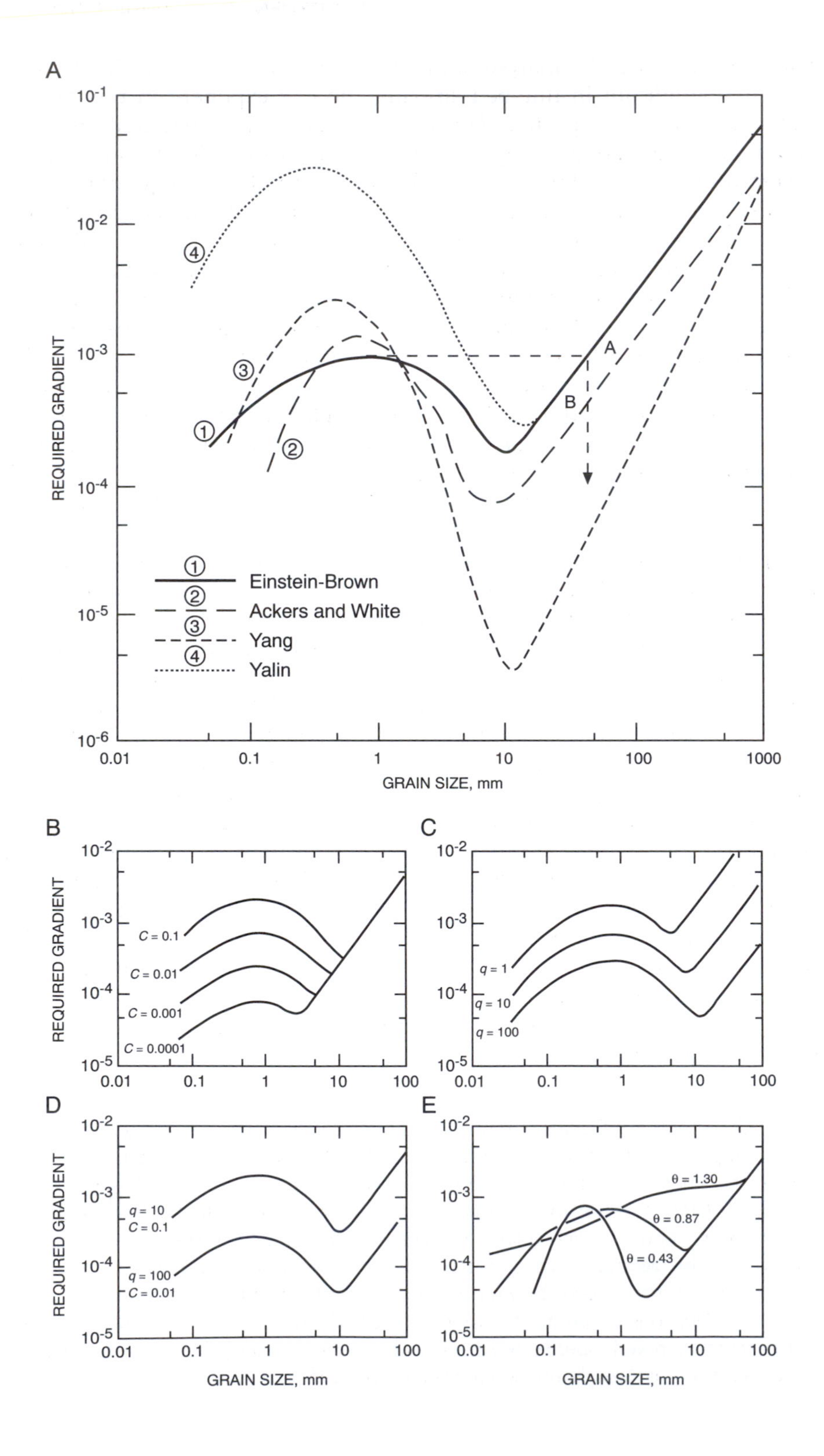
A
REQUIRED GRADIENT
GRAINSIZE, mm
①
②
③
④
A
B
Einstein-Brown
Ackers and White
Yang
Yalin
B
C = 0.1
C = 0.01
C = 0.001
C = 0.0001
C
q = 1
q = 10
q = 100
D
q = 10
C = 0.1
q = 100
C = 0.01
E
θ = 1.30
θ = 0.87
θ = 0.43
GRAIN SIZE, mm

value. Since Howard's analysis suggests that the quantity of sediment only becomes significant in fine-bed streams, the low exponent in this gravel-bed case is not unexpected. The transport of a larger bed-material load could be accommodated by an increase in width : depth ratio or an increase in channel slope or both. Rubey's formulation (equation (5.55)) implies that gradient and channel form are mutually adjusted but that, contrary to expectation, the depth : width ratio varies directly with sediment load and grain size. In an attempt to explain this anomaly, Miller (1991b) investigated the mutual adjustability of gradient and cross-sectional variables using Missouri River data, and concluded that the anomaly could be attributed to the inverse effect of grain size on the depth : width ratio being outweighed by the direct effect of grain size on gradient, to which the depth : width ratio is positively related. However, cross-sectional variables are more dependent on the mutual adjustment mechanism than is channel gradient. The cross-sectional dimension is more likely to absorb change than is the longitudinal dimension, particularly if slope adjustment is constrained.

Application of the principle of sediment transport maximization suggests that slope is strongly dependent on sediment transport rate, especially in sand-bed channels (White *et al.*, 1982). An equivalent extremal hypothesis, that of stream power minimization (Chang, 1980), has been used to show a similar dependence. Chang (1985) demonstrated that, for a given water and sediment discharge, the gradient required for equilibrium is at a minimum at a particular channel width, that minimum satisfying the requirement of stream power minimization (Fig. 5.28; see also Fig. 5.7B). Where two local minima in required gradient occur (at higher sediment transport rates), the one of lower gradient and larger width in the lower flow regime (Min 1) would be more stable. Despite misgivings about their physical justification (Griffiths, 1984; Ferguson, 1986), these extremal hypotheses do indicate how slope can vary with transport rate and how other elements of channel form, notably width, can be involved in the adjustment process. It should be remembered, however, that stream power, a product of discharge and slope (equation (4.7, p. 106)), influences the sediment transport rate (equation (4.13, p. 130)), emphasizing that the relationship of slope and transporting capacity is not simply one-way.

Figure 5.27 (opposite) Curves of required gradient against grain size calculated from sediment transport equations (after Howard, 1980). (A) The curves for four equations using nominal parameter values: Q = 1000 m^3 s^{-1}, sediment concentration (C) = 0.01, w = 100 m, n = 0.02, D_{50} = 0.3 mm, variance of grain size distribution (θ) = 0.87. The dashed line emanating from the fine-grained maximum of the Einstein–Brown curve separates regions A and B described in the text. (B)–(E) The effects of changing input parameters on the shape of the Einstein–Brown curve. (B) Variations in sediment concentration ($C = Q_s/Q$). (C) Variations in specific discharge ($q = Q/w$). (D) Variations in specific discharge and concentration with specific sediment discharge ($q_s = C.Q/w$) fixed. (E) Variations in the variance of grain size distribution (or degree of sorting).

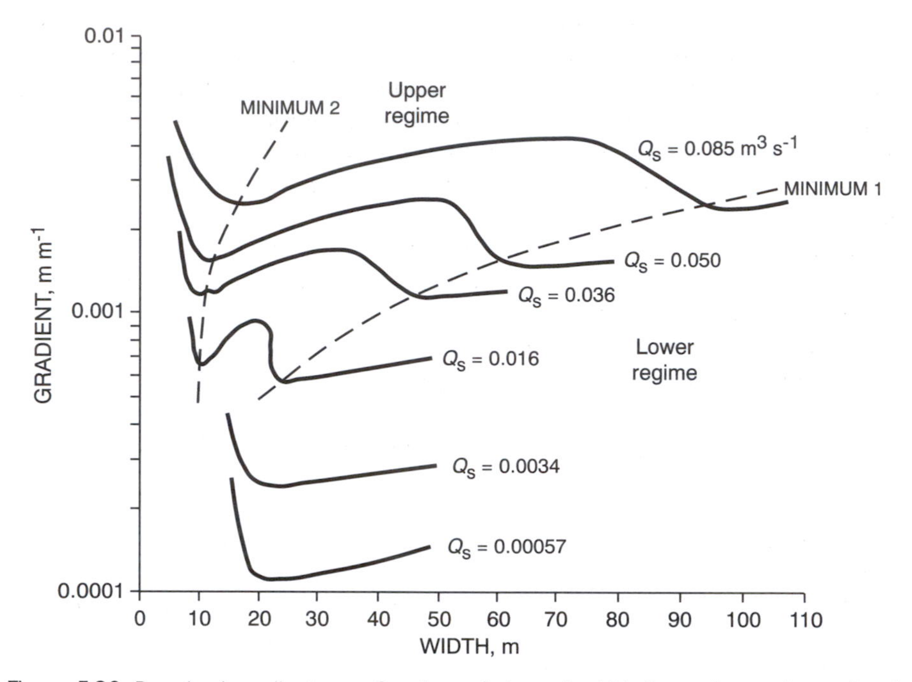

Figure 5.28 Required gradient as a function of channel width for various values of sediment discharge (Q_s), with discharge and grain size fixed (after Chang, 1985).

All of the variables indicated in equation (5.55) exert some control over channel gradient, although their separate effects are difficult to isolate, particularly in the case of sediment load where data are sparse, and their relative influence may vary with the type of channel and scale of analysis. Grain size is probably more influential in coarse-bed than in fine-bed streams, while the opposite may be the case with the quantity of sediment load. Penning-Rowsell and Townshend (1978) found that drainage area (their index of discharge) was more significant at the reach scale, but of less importance relative to bed-material size at the local scale when discharge is essentially uniform. If real, their results emphasize the need to choose an appropriate spatial scale when relating form elements to control variables. Many of the empirical studies considered above deal with variation in channel gradient over regions of diverse geology and hydrology, but truly consistent relationships may be expected only along single streams where the conditions of sediment supply and discharge addition remain relatively similar.

Causes of Profile Concavity

Initially at least, the location and height of the end points of a longitudinal profile and therefore its average gradient are determined by tectonic and other geohistorical events. From an analysis of 115 British rivers, Wheeler

(1979) found a positive correlation between profile concavity and total fall, indicating that a more concave profile is associated with greater relative relief. On the basis of a random-walk model, Leopold and Langbein (1962) suggested that a constraint on stream length reduces concavity, so that shorter streams should have straighter profiles (Fig. 5.26F(ii)). However, profile form is only partly dependent on incident relief and available length. Since the slope at any point is related to bed material size, discharge and sediment load, it is logical to explain variations in profile concavity (or the downstream rate of change of channel gradient) in terms of the downstream behaviour of those variables.

Downstream changes in sediment characteristics reflect a complex history of transport through the fluvial system. To the extent that a negative exponential function can describe downstream change in both grain size (equation (4.14), p. 136) and channel gradient (Rana *et al.*, 1973)

$$s = s_0 \, e^{-\delta L} \tag{5.57}$$

their respective rates of change (α and δ) ought to be related. Empirical data (Hack, 1957; Ikeda, 1970; Cherkauer, 1972) and numerical modelling (Snow and Slingerland, 1987) confirm that profiles are more concave where bed material size decreases more rapidly and, in addition, suggest that profiles have little concavity if particle size remains constant or increases downstream. Strong profile concavity may indeed force rapid downstream fining (Hoey and Ferguson, 1994). In so far as bedrock type influences initial size and its downstream rate of reduction, different lithologies could give rise to profiles of different concavity. Although based on data from several streams, the slope–length relationships obtained by Brush (1961) provide some support for this argument:

Shale $$s = 0.050 \, L^{-0.81} \tag{5.58}$$

Limestone $$s = 0.027 \, L^{-0.71} \tag{5.59}$$

Sandstone $$s = 0.063 \, L^{-0.67} \tag{5.60}$$

With the exponent as an index of profile concavity, the relationships imply that streams flowing over shale have more concave profiles than those flowing over limestone or sandstone. Thus the distribution of bedrock types along a stream could influence profile form overall, and the concavity of individual segments in compound profiles in particular.

Further evidence of profile–particle size association comes from the study of profiles having a marked discontinuity. In an analysis of nine Japanese rivers, Yatsu (1955) identified a distinct break of slope in the longitudinal profiles which corresponded to a rapid transition from gravel- to sand-bed conditions. Both slope and grain size decreased more rapidly above the discontinuity than below. The deficiency of intermediate grain sizes (2–10 mm) at the break can be explained as an effect of differential sorting (Howard, 1980), although the rapid mechanical breakdown of gravel into sand may also have been a contributory factor. Profile and grain size discontinuities do not always correspond (Ikeda, 1970), but a significant threshold related to the mechanics of sediment transport seems to exist between gravel-bed and sand-bed channels. Translating his results into a downstream context, Howard (1980, 1987) found that the variation of slope with length and grain size was similar within each type, but differed considerably between the two types of channel in parallel with Yatsu's results, with slope decreasing more rapidly in gravel-bed streams. Along the River Bollin-Dean, slope decreases markedly as bed material is reduced in size from 64 to 4 mm but changes only slowly once the bed becomes sandy (Knighton, 1975b). Although it is easy to appreciate the need for steeper slopes in the transport of coarser debris, especially in the gravel range, the underlying relationships are unlikely to be simple given the complex way in which bed material of varying size and sorting interacts with transport and resistance conditions.

In common with bed material size, the downstream rate of change of discharge is not necessarily constant from one river to another. The sequence of tributary inflows controls the pattern of discharge addition (Fig. 2.18B, p. 62) and is a major source of discontinuity in the fluvial system (Knighton, 1987b). Although rivers with relatively uniform discharge can have concave profiles provided they are long enough, increasing discharge is a major cause of concavity (Langbein, 1964b). Profiles tend to be more concave where discharge increases more rapidly (Wheeler, 1979; Snow and Slingerland, 1987; Xu Jiongxin, 1991), while arid-zone streams in which discharge decreases downstream because of transmission losses have profiles of reduced concavity or convex-upward form.

In broad terms, increasing discharge implies an ability to transport the same bed-material load over progressively lower slopes. However, a river can accommodate increasing discharge in various ways and different combinations of the adjustable variables may be associated with different profile forms. Following Leopold *et al.* (1964), simple deductions about the effects of different combinations can be made in terms of downstream hydraulic geometry by holding certain factors constant and then successively relaxing them. Although downstream relationships are not as straightforward as originally supposed, an assumption that width varies as the square root of discharge seems reasonable (Table 5.4). For simplicity, a smaller z in $s = gQ^z$ is equated with greater concavity, although the actual profile associated with a given slope–discharge relationship depends on the rate at which discharge increases with stream length.

State 1. Without any reference to sediment load and assuming constant resistance, increasing profile concavity requires that depth accommodates progressively more of the downstream change in discharge, and velocity correspondingly less (Table 5.12). A small increase in velocity is common along rivers (Table 5.4), associated here with a profile of moderate concavity. The requirement of constant resistance can apparently be satisfied by realistic adjustments of the dependent variables and still be reflected in a concave profile. Since such a requirement could be associated with constant

Table 5.12 Downstream changes in channel slope deduced for various states

	Assumptions	Implications	Results
State 1	$w \propto Q^{0.5}$	$f = m = 0.5$ (from continuity equation)	
	constant resistance ($y = 0$)	$z = 2m - 1.33f$ (from Manning equation)	$z = -0.33 \Leftrightarrow$ **$f = 0.40$, $m = 0.10$**
			$z = -0.50 \Leftrightarrow f = 0.45, m = 0.05$
			$z = -0.67 \Leftrightarrow f = 0.50, m = 0$
State 2	$w \propto Q^{0.5}$	$f + m = 0.5$	
	(i) $y = -0.15$	$z = 2m - 1.33f - 0.3$	$z = -0.33 \Leftrightarrow$ **$f = 0.31$, $m = 0.19$**
			$z = -0.50 \Leftrightarrow$ **$f = 0.36$, $m = 0.14$**
			$z = -0.67 \Leftrightarrow f = 0.41, m = 0.09$
	(ii) $y = -0.30$	$z = 2m - 1.33f - 0.6$	$z = -0.33 \Leftrightarrow f = 0.22, m = 0.28$
			$z = -0.50 \Leftrightarrow f = 0.27, m = 0.23$
			$z = -0.67 \Leftrightarrow$ **$f = 0.32$, $m = 0.18$**
State 3	$w \propto Q^{0.5}$	$f + m = 0.5$	
	constant bedload discharge	$z = 0.5\alpha' - 0.17 - f$ (from equation (5.61))	
	(i) $y = 0, \alpha' = 0$	$z = 2m - 1.33f$	$z = -0.67 \Leftrightarrow f = 0.50, m = 0$
	$\alpha' = \alpha_1 < 0$		$z = -0.76 \Leftrightarrow f = 0.53, m = 0.03$
	(ii) $y = -0.15, \alpha' = \alpha_1 < 0$	$z = 2m - 1.33f - 0.3$	$z = -0.63 \Leftrightarrow$ **$f = 0.40$, $m = 0.10$**
	$\alpha' = 2\alpha_1$		$z = -0.73 \Leftrightarrow f = 0.43, m = 0.07$
	(iii) $y = -0.30, \alpha' = \alpha_1 < 0$	$z = 2m - 1.33f - 0.6$	$z = -0.52 \Leftrightarrow f = 0.27, m = 0.23$
	$\alpha' = 2\alpha_1$		$z = -0.60 \Leftrightarrow$ **$f = 0.30$, $m = 0.20$**

Symbols: f, m, y and z are respectively the downstream rates of change of depth, velocity, Manning's resistance coefficient and slope with discharge (see equations (5.2)–(5.5)); α' is a nominal exponent intended to express the downstream rate of change of bed material size, but given a particular value in *State 3* to enable comparisons to be made. The more plausible f and m values (see Table 5.4) are shown in bold type.

grain size, or even increasing grain size if the change in depth is sufficient to render relative roughness (*d*/*D*) constant, decreasing bed material size does not appear to be a necessary condition for profiles to be concave.

State 2. There is evidence to suggest that, on average, resistance decreases slightly downstream. However, relationships are never based on the direct measurement of resistance, and the separate effects of different resistance components remain obscure. Grain roughness should decrease downstream because of decreasing grain size and increasing depth, its rate of change being largely determined by the behaviour of those two variables. As regards form roughness the situation is less clear. In moving from a gravel-bed to a sand-bed channel, the resistance due to bed forms may assume a greater relative significance through the development of dunes in particular. Also, the resistance offered by channel bends may increase downstream if the channel becomes more sinuous as a larger proportion of the load is carried in suspension (Schumm, 1963b). Consequently, resistance may change less rapidly than expected if grain roughness was the sole component. Although hydraulic efficiency generally improves downstream, the behaviour of total resistance could vary considerably within and between rivers, depending on how each component changes longitudinally.

A more rapid rate of resistance decrease leads to a larger m/f ratio for a given profile form (constant z) or greater profile concavity for a given m (Table 5.12). Large decreases in resistance are perhaps most likely in small basins or the headwaters of large streams, where also the flattening of channel gradient is usually most marked. A slightly greater increase in velocity can be maintained along more concave profiles than was the case for constant resistance (*State 1*).

State 3. The inclusion of sediment load as a variable markedly increases the complexity of the analysis. Yet both the competence and capacity for transport may be expected to vary longitudinally. It has been suggested, for example, that unit stream power ($\gamma Qs/w$), a variable which profoundly influences the sediment transport rate (equation (4.13, p. 130)), may reach a peak in the middle reaches of rivers (Lawler, 1992), so that longitudinal variation in transporting ability may be far from straightforward.

Leopold and Maddock (1953) sought to explain the concavity of the longitudinal profile in terms of the downstream decrease of sediment load relative to discharge. As this ratio decreases, transport capacity can be lowered and the channel gradient become flatter. Numerical experiments support this view and show also that profiles have reduced concavity where sediment load increases downstream relative to discharge (Snow and Slingerland, 1987). A recent analytical model suggests that at relatively high sediment transport rates, a decrease in relative bed-material load is a necessary condition for concavity, while at lower sediment transport rates a concave profile can be maintained even if relative bed-material load increases downstream, albeit at a mild rate (Sinha and Parker, 1996). Rarely are sediment load data available as a downstream series, but along the Huang Ho and Huaihe Rivers in China there is evidence for a downstream decrease in relative sediment discharge (suspended load) (Xu Jiongxin, 1991). There, profile concavity increases as the load : discharge ratio decreases.

The adjustment of slope to sediment load is closely bound up with the concept of a stream at grade or in equilibrium. Mackin (1948) believed that graded alluvial streams with a stable flow regime have slopes which are just sufficient to transport the load supplied. Aggradation increases a slope that is too small, while degradation decreases one that is too large. Evidence for this comes largely from laboratory experiments conducted under steady flow conditions, where gradients are adjusted to equalize the input and output of sediment load over short distances. However, the situation in natural streams with variable discharge and fluctuating sediment supply is infinitely more complex. Bogardi (1974) emphasized the difficulty of obtaining generalized trends, even for a single river system, since sediment transport rates can vary considerably over short river distances (Andrews, 1982). Setting aside the possibility that any imbalance in sediment transport can be accommodated by other adjustments, an assumption of continuity in bed-material transport enables some deductions to be made regarding channel slope.

For this purpose a simplified version of the Einstein–Brown bed-load function is used:

$$Q_{sb} = K\frac{w(d.\,s)^3}{D^{1.5}} \tag{5.61}$$

To maintain a constant bed-load discharge (Q_{sb}) where w and d increase and D decreases downstream, slope must decrease. The profile of a graded river must therefore be concave. The degree of concavity will depend in particular on the rates at which w, d and D change in the downstream direction. Osterkamp *et al.* (1983) have demonstrated the extent to which b, f and z can covary. Since downstream hydraulic geometry requires that all variables are related to discharge, further analysis is limited. By assigning nominal values to the rate of change of bed material size, it can be shown that a more rapid decrease in grain size is associated not only with a more concave profile, but also with a larger rate of change of depth relative to velocity for a given value of y (Table 5.12). Greater profile concavity is suggested for this state than for states 1 and 2 if the more plausible values for f and m are alone taken into account.

The simplified nature of this analysis cannot be overemphasized. However, it should have illustrated that increasing discharge and decreasing bed material size provide only a general explanation of profile concavity. Underlying those changes are adjustments to many variables whose interaction in different combinations can lead to a wide variety of profile forms. The definition of a graded stream needs to recognize that adjustments to form elements other than slope, notably cross-sectional form, can be made in order to maintain the continuity of sediment transport.

Adjustability of Channel Gradient

Slope adjustment can be brought about by aggradation, degradation or changes to channel sinuosity, occurring singly or in combination. Channel

Table 5.13 Main causes of change in stream bed elevation

Type	Degradation	Aggradation
Downstream progressing:		
Change in water discharge	Increase	Decrease
Change in sediment supply	Decrease	Increase
Upstream progressing:		
Change in base-level	Fall	Rise

gradient decreases where a river adopts a more sinuous course. The processes of aggradation and degradation respectively increase and decrease stream bed elevation, usually over long river distances, and operate most often in response to changes in watershed controls (notably flow regime and sediment supply) or changes in base-level (Table 5.13). The spatial distribution of aggradation and degradation is frequently more complex than Table 5.13 first suggests. Largely as a result of a major flood in 1964, the Eel River in northern California degraded its bed in the upper and middle basin but aggraded in downstream reaches in response to the increased flux of sediment, giving rise to a substantially modified longitudinal profile (Patrick *et al.*, 1982; Fig. 5.29A).

Modifications to bed elevation and slope result from changes in watershed conditions which upset the continuity of sediment transport, degradation reflecting bed-load starvation and aggradation excessive bed-load input. Sediment loading as determined by climate has been the dominant factor controlling gradient changes along rivers which cross the Canterbury Plains in New Zealand (Wilson, 1985). High sediment loads during glacial periods produced progressive downstream aggradation as the rivers emerged from the Southern Alps, while reduced river loads during interglacials instigated trench-cutting at the inner margin of the plains, which facilitated degradation along the rivers. In postglacial times the Waimakariri River has progressively shallowed its gradient across the plains as a result of upstream degradation and downstream aggradation, the intersection point between the two regimes migrating downstream more than 35 km in the past few thousand years so that it now stands some 19 km from the coast (Fig. 5.29B). A pattern of upstream incision and downstream sedimentation also characterizes the recent history of Walker Creek in the Coast Range of California (Haible, 1980). In response to fluctuating conditions over the past 5000 years, the valley has experienced one episode of infilling followed by two episodes of incision, the most recent beginning about 100 years ago (Fig. 5.29C). Despite the removal of large quantities of material during these episodes, profile concavity has remained essentially constant. Gradient adjustments of 2 to 4 per cent have occurred but are insignificant when compared with cross-sectional changes. This example underlines the potential for alternative forms of adjustment as a river seeks to accommodate a change in flow or sediment regime.

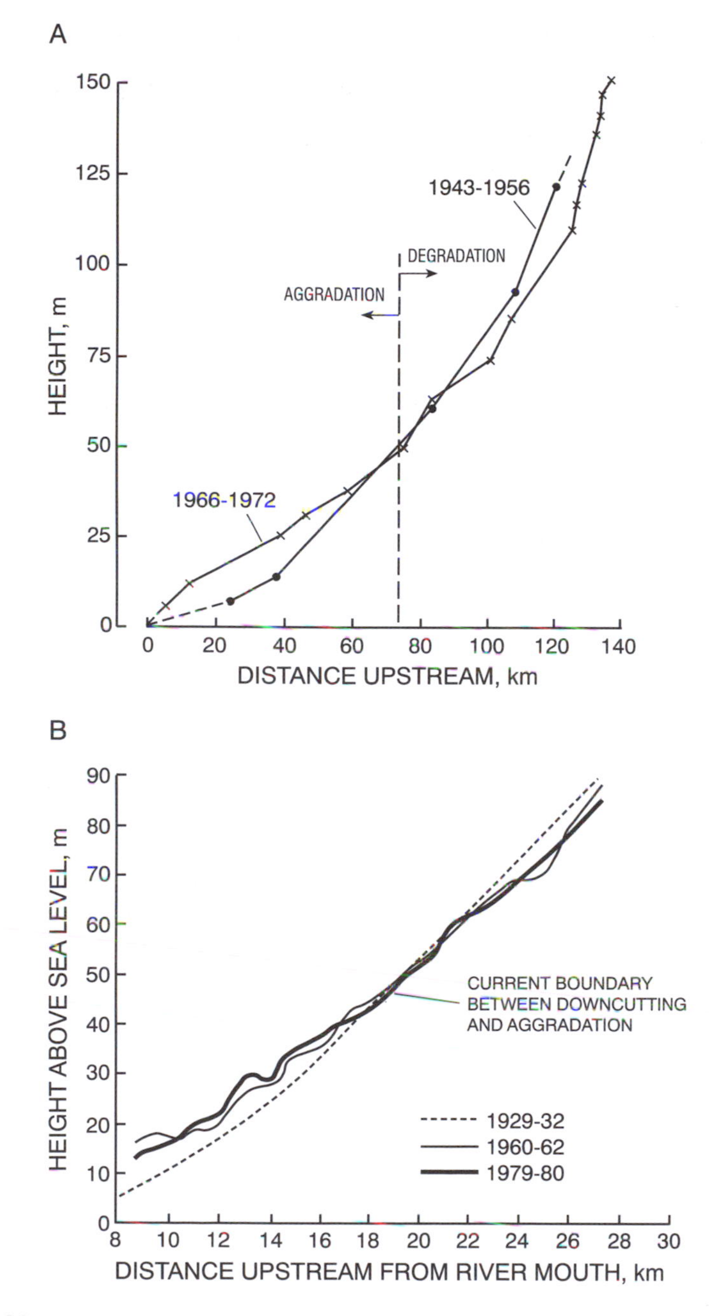

Figure 5.29 (A) Profile adjustment following a large flood, Eel River, California (after Patrick *et al.*, 1982). (B) Changes in bed height, Waimakariri River, New Zealand (after Wilson, 1985). (C) (overleaf) Stream-bed and terrace profiles, Walker Creek, California (after Haible, 1980). (D) Profile development after base-level lowering in laboratory models, without (i) and with (ii) channel bed armouring. Numbers denote time, in minutes, from the start of the run (after Begin *et al.*, 1981).

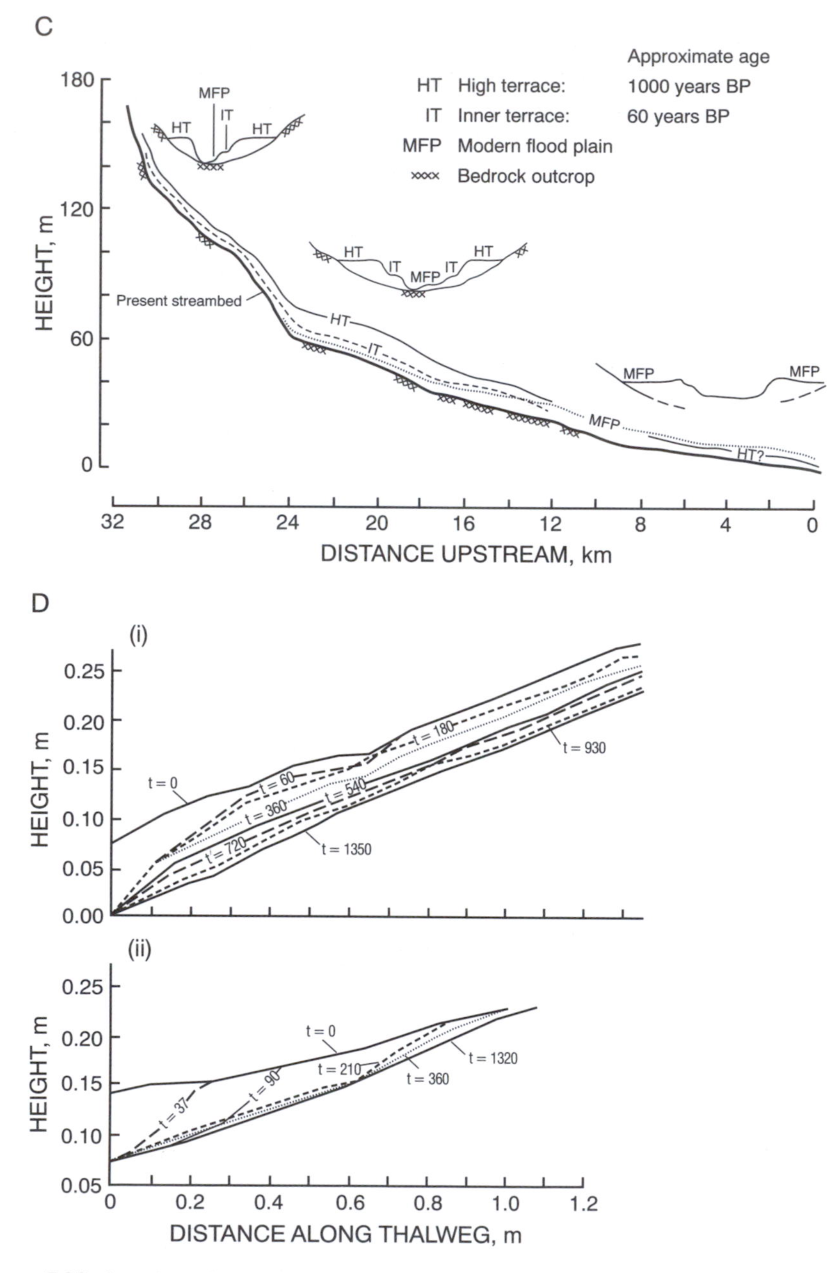

Figure 5.29 (continued)

Artificial modifications to the fluvial system can have a more immediate impact on the continuity of sediment transport, the most widely reported being associated with dam construction. The release of sediment-free water below dams commonly results in downstream-progressing degradation and a flatter slope. Although initially localized close to the dam, the effect can be transmitted rapidly over long river distances. Below the Hoover Dam on the Colorado River, degradation extended 130 km in nine years, reaching a maximum of 7.1 m at a downstream distance of 12.4 km (Galay, 1983). However, the rate, depth and extent of degradation are highly variable, so that each dam is unique as regards profile adjustment (Williams and Wolman, 1984). The degradation process becomes more complex where there is interaction between a main river and its tributaries, with downstream-progressing degradation along the former inducing upstream-progressing degradation within the latter, and thereby increasing tributary sediment contributions (Germanoski and Ritter, 1988).

Upstream-progressing degradation is a characteristic response to a fall in base-level. On the basis of a diffusion model and laboratory experimentation, Begin *et al.* (1981) demonstrated the main effects of base-level lowering (Fig. 5.29D). Local steepening at the distal end increases transporting ability, and over time the abrupt break of slope migrates headward, flattening as it does so. The rate of degradation reaches a peak relatively quickly but then declines slowly over time and also decreases with distance from the outlet. If stream-bed lowering leads to bank heights which are unstable, excessive bank erosion may occur and supply the channel with sediments that are temporarily stored, preventing further degradation until they have been flushed out. The results show that the ultimate effect of base-level lowering by a given amount is degradation along the entire length of a channel by the same amount, leading to a new profile with essentially the same form as the original (Fig. 5.29D(i)). The final bed gradient may be steeper if the bed material is heterogeneous and armouring of the bed accompanies the incision (Fig. 5.29D(ii)). However, the extent to which laboratory experiments can simulate the response of natural rivers to a fall in base-level is strictly limited because of scale and material differences, even though a diffusion model can be applied to field situations with reasonable accuracy (Begin, 1988). There are many ways in which this simple picture can be undermined.

A rise in base-level, whether eustatic or local, may be similar in effect to the construction of a barrier across a river. From repeated surveys above artificial barriers, it seems that the aggradation that occurs as a result of reduced competence and capacity proceeds rapidly at first but does not continue far upstream thereafter, the profile remaining largely unaffected beyond the head of the depositional wedge (Leopold and Bull, 1979). Van Haveren *et al.* (1987) show that over a period of 25 years a depositional wedge extended 1.5 km upstream and reached a height of 18 m above the dam, so that a slope which had been <30 per cent of the original during the early part of the period had reached 83 per cent of the original towards the end. In other rivers alternative adjustments such as increased channel width or decreased hydraulic roughness may compensate for the reduced gradient in order to maintain the continuity of sediment transport, for it appears that the effect of a rise in base-level on the upstream profile is generally rather limited in extent.

The impact of a change in base-level, whether positive or negative, is likely to be very variable, dependent on three main groups of factors (Schumm, 1993): base-level controls, geological controls and geomorphic controls. As regards the first, the direction, magnitude, rate and duration of the change are all expected to determine the degree of impact, with magnitude probably being the most important. Where the change is large, the potential for alternative forms of adjustment may be restricted and the effect could be significant, a lowering of base-level causing river incision to extend far into the upstream network. In contrast, bedrock or structural controls may prevent or delay a fall in base-level from moving upstream. Overall, Schumm (1993) agreed with Leopold and Bull (1979) that the effects of base-level change will tend not be propagated over long distances upstream, particularly in large alluvial rivers which have more time to absorb the change by adjusting their sinuosity, channel dimensions and roughness characteristics. The impact will probably be greatest where the base-level falls and does so by a large amount, where the resultant incision is particularly rapid, and where the river is confined. Channel confinement by dint of valley morphology or cohesive alluvium may inhibit gradient adjustment through a change in sinuosity, so that vertical adjustment becomes the only option.

To reflect the opinion dominant before 1950, that gradient is the principal means of adjustment whereby a river attains a state of grade or equilibrium, channel gradient has been treated as a response variable in preceding sections. There are undoubtedly instances where a sudden change in conditions has caused a river to adjust its gradient and profile form but that adjustment may be severely constrained. Because successive channel segments are interdependent, major changes in gradient require the redistribution of very large quantities of material. Also, valley slope, an inherited characteristic reflecting a history of past discharges and sediment loads, and non-alluvial controls such as bedrock outcrops, impose limits on the possible range of gradient adjustment. Even when conditions allow, a river may not alter its gradient to a marked extent (Fig. 5.29C; Haible, 1980). Leopold and Bull (1979) concluded that changes in channel gradient account for only a small part of the adjustment required to achieve a balance between the input and output of sediment load, most of the adjustment being accomplished by changes in other components of flow and channel geometry which respond more rapidly. This assumes that those components are not themselves constrained in their range of adjustment, which is not always the case. Channel slope can be regarded either as an imposed variable or, with valley slope imposed, as a variable adjustable only over a limited range (Church, 1980). At least part of that limited adjustment can be accommodated by changes in channel sinuosity. Consequently Mackin's (1948) definition of a graded stream (see p. 159) might be rephrased as 'one in which, over a period of years, slope, velocity, depth, width, roughness, pattern and channel morphology . . . mutually adjust to provide the power and the efficiency necessary to transport the load supplied from the drainage basin without aggradation or degradation of the channels' (Leopold and Bull, 1979, p. 195).

6

CHANNEL CHANGES THROUGH TIME

The fluvial system can be regarded as either a physical system or an historical system (Schumm, 1977). At one time or another both viewpoints have held sway in geomorphological approaches to the study of river channels. The first focuses on how the system functions, usually over short timespans, while the second deals more directly with evolutionary tendencies or changes to the system over various timescales.

In reality, the fluvial system is a physical system with a history (Schumm, 1977). Present form is influenced by both past and present conditions, where 'present' in this context could be defined as that time period over which inputs to the fluvial system remain reasonably constant on average and the action of constant processes tends to produce characteristic forms. There is no guarantee that input conditions will remain stable long enough for an equilibrium state to be attained, as in the coastal rivers of New South Wales where the flow regime oscillates between flood- and drought-dominated every 30–50 years, giving rise to a kind of cyclic disequilibrium (Warner, 1987, 1994). At a longer timescale, many rivers were directly affected by events during and immediately after the last glaciation, both in unglaciated and glaciated areas, and may well have retained influences from that time. The channel patterns and deposits of Amazon Basin rivers draining Andean source areas reflect the effect of Pleistocene glacial episodes when the uplands supplied large quantities of coarse bed load, which only the major rivers are now able to rework (Baker, 1978). Church and Slaymaker (1989) suggest that fluvial adjustment to post-glacial conditions is incomplete in British Columbian rivers because of the constraints imposed by relict boundary sediments. The fluvial system thus has a memory.

In broad terms the length of that memory depends on the ability of the system (or any component of the system) to absorb or respond to change, which is in turn dependent on the sequence of past events and on the inherent resistance of the system. As suggested in the previous chapter (see Fig. 5.3, p. 158), different components of channel morphology adjust at different rates and presumably reflect the influence of past conditions to different extents. Where the response time is short, the historical legacy becomes proportionately less. Stevens *et al.* (1975) maintained that the influence of past flood events on river channel form varies both with their absolute magnitude (more extreme floods having a longer-lasting effect) and with their relative magnitude expressed by the ratio of individual flood peaks to the mean annual flood. Where the ratio is large, the river channel should exhibit non-equilibrium tendencies, at least over the timespans of up to 10^2 years which they considered. Recognition of an historical dimension switches attention from those treatments which assume quasi-steady and quasi-uniform conditions to those which emphasize the unsteady nature of river channel adjustment.

The ability to interpret the past is partly dependent on the level of understanding of the present. Brunsden and Thornes (1979) maintained that geomorphologists are now in a better position to extrapolate the short-term record to timespans of 10^2–10^4 years and longer because of the knowledge of process action gained over the last few decades and the wider range of techniques now available. Such extrapolation is not without its problems. Only rarely can initial and boundary conditions be accurately specified. Continuous or even irregular studies of present-day process activity are usually limited to one or two decades, so that average rates may be unrepresentative of longer-term trends. Even when available, average rates of erosion and deposition may reveal little about the complexity of landform change. Events are unlikely to be synchronous over large areas because the threshold conditions that partly determine the magnitude and direction of change are influenced by factors which are themselves spatially variable. Even in a single area there is no guarantee that two drainage basins and their constituent channels will respond in the same way to the same external stimulus.

Further problems are created by the changing status of environmental variables over time (Table 6.1). Geology and climate retain an independent status at all timespans but, whereas climate can be regarded as fixed in the short term, it ceases to be invariant over longer time periods. Sympathetic changes are produced in those environmental factors which, together with geology and climate, control the main variables (notably discharge and sediment load) influencing channel form. Not only does the number of dependent factors increase with time but so does the level of indeterminacy in the system, so that specific details become increasingly obscure. Whatever significance is attached to the arbitrary division of time in Table 6.1, the key insight of Schumm and Lichty (1965) remains – that relationships between independent and dependent variables (input and output) change with the timescale considered. A wide range of timescales is probably appropriate for the study of river channel change, since some features retain or are influenced by inherited characteristics to a greater extent than others.

River channels are one of the most sensitive components of the physical landscape, with an ability to respond rapidly to disturbance. Sensitivity to change is an important basic idea but it is spatially very variable and has not

Table 6.1 Changing status of river variables with time (modified from Schumm and Lichty, 1965)

Variables	Long timescale $>10^5$ years	Medium timescale 10^3–10^4 years	Short timescale 10^1–10^2 years	Instantaneous time $<10^{-1}$ years
Geology	E	E	E	E
Climate	E	E	E	E
Regional relief	D	E	E	E
Slope morphology	D	D	E	E
Soil properties	D	D	E	E
Vegetation properties	D	D	E	E
Mean water and sediment discharge	*I*	D	E	E
Channel morphology	*I*	D	D	E
Instantaneous flow characteristics	*I*	*I*	*I*	D

E, environmental or independent variable; D, dependent variable; *I*, indeterminate or irrelevant variable.

yet been defined in an unequivocal way. Recognizing its relevance to river channel management, Downs and Gregory (1993) exemplify several definitions, including the ratio of disturbing to resisting forces, the propensity for change as indicated by a system's proximity to a significant threshold, and the ability of a system to recover from change, but advocate a fourth which integrates all these elements. Distinctions need also to be made between types of disturbance. In the long term the main changes are brought about by climatic, tectonic and base-level effects, while in the medium (largely post-glacial) term, human activities become an increasingly important influence in conjunction with climatic fluctuations and a rising sea-level. In the short term, human modification of the fluvial environment becomes the dominant form of disturbance, with individual extreme events as a significant secondary cause of change. The ability to reconstruct the past at whatever timescale, and thereby assess the relative contributions of past and present processes to present form, depends on how well the historical record can be read.

Evidence of Change

The primary evidence for change in the fluvial system comes from morphologic remnants such as relict or buried channels and from sediment sequences which reveal a complex and usually incomplete record of changing conditions. Geomorphologists in the past have relied heavily on longitudinal profile form and associated terrace remnants in reconstructing river histories, but purely morphologic evidence needs to be supported by other information if ambiguous interpretations are to be avoided. Table 6.2 lists some of the methods available, wherein a distinction can be drawn between those methods which provide more direct evidence of change and those which relate to the dating of past events.

Direct observations can be used to monitor change at specific locations over periods of hours (Fahnestock, 1963), days (Lane *et al.*, 1996), months (Knighton, 1972) or years (Madej and Ozaki, 1996), and are particularly valuable when tied to measurements of streamflow and sediment properties. Lane *et al.* (1996) demonstrated how photogrammetric and tacheometric survey techniques can be combined to obtain very detailed information on the response of a dynamic alluvial channel to short-term fluctuations in discharge and sediment supply. Direct observation of river channel adjustment to events of varying frequency, including extreme floods, enables a better assessment of such events in an historical context (McEwen and Werritty, 1988; Gintz *et al.* 1996).

Repeated field surveys play an important role in monitoring the variable location, timing and magnitude of river channel change, especially that induced by human activities. Twenty years of detailed cross-sectional data enabled the downstream transmission of a 'sediment wave' to be traced following a series of large floods and widespread timber harvesting (Madej and Ozaki, 1996). However, direct observations are seldom maintained for more than a decade or two. They may fail to include a wide enough range of erosional and depositional events to be representative of longer-term change. Few directly useful morphologic variables are regularly measured, although several important parameters (notably precipitation and discharge) are continuously monitored, and discharge records from the early nineteenth century do exist for a select group of rivers (Probst, 1989).

Table 6.2 Sources of evidence for river channel change

Direct observations	Instrument records (rarely continuous) Photographic records Ground surveys
Historial records	Maps and photographs from different dates Historical documents
Sedimentary evidence	Surface forms and palaeochannels Texture, structure and architecture of deposits
Dating techniques	(1) Relative methods Relative height Organic remains, including pollen Artefacts (2) Absolute methods Radioactive isotopes Dendrochronology Thermoluminescence dating

Sediment load data are particularly elusive, and the bed-load yields measured over more than 10 years from Plynlimon catchments represent an unusually long record for Britain (Moore and Newson, 1986). The main value of short-term observations lies in establishing relationships between applied stresses and stream channel response, and in evaluating the significance of individual (extreme) events.

Historical records, including documents, survey notes, maps and photographs, can provide valuable information about channel and related changes, particularly but not exclusively over the last hundred years. Systematic discharge records may be extended by reference to high-water marks on buildings, newspaper reports or personal recollections in written or verbal form. Old mine records have been used as a basis for reconstructing the sediment transport history of a river severely disturbed by the input of mining waste over more than a century (Knighton, 1989, 1991). Of greater vintage, land rent records for the Josterdalsbre area in Norway enabled Grove (1988) to demonstrate how the incidence of mass movements and floods has varied since the sixteenth century, climaxing during the Little Ice Age (Fig. 6.1A).

Maps and photographs of different dates have been increasingly used to identify changes in the planimetric properties of river channels. In many parts of Britain at least five separate surveys have been undertaken over the last 150 years, and this type of historical map information has been extensively used by Hooke (1977, 1984) to chart the distribution of course changes, to estimate rates of bank retreat and to identify types of channel migration (Fig. 6.1B). Hooke (1977) found that up to 40 per cent of the length of a Devon river was laterally mobile, with movements of up to 100 m in one 50-year period.

Direct comparisons of channels at different times are readily made using photographs, although most early photographs were not taken expressly for that purpose and it may be difficult to ensure exactly the same field of view. Photographs taken of the Platte River in Nebraska show the spectacular reductions in channel width which have resulted from river regulation schemes since 1900 (Williams, 1978c). Using evidence from both aerial and ground photographs taken between 1946 and 1977, Anderson and Calver (1977, 1980) identified how a natural channel has adjusted to a major flood (in 1952) and during the post-flood period of lower discharges. The recent advent of satellite photography raises the possibility of repeated, large-scale observations of river channels but as yet applications are relatively few.

The range of historical sources is considerable and has yet to be fully exploited. However, care needs to be exercised in using this type of information, particularly the earlier material which may be of doubtful accuracy. Since information is usually available only for specific dates, it has to be assumed that change was uniform between each pair of surveys unless evidence from other sources can be obtained. Without data for intervening periods when the channel may have been significantly altered, changes can seldom be related to the events which produced them. Generally, historical

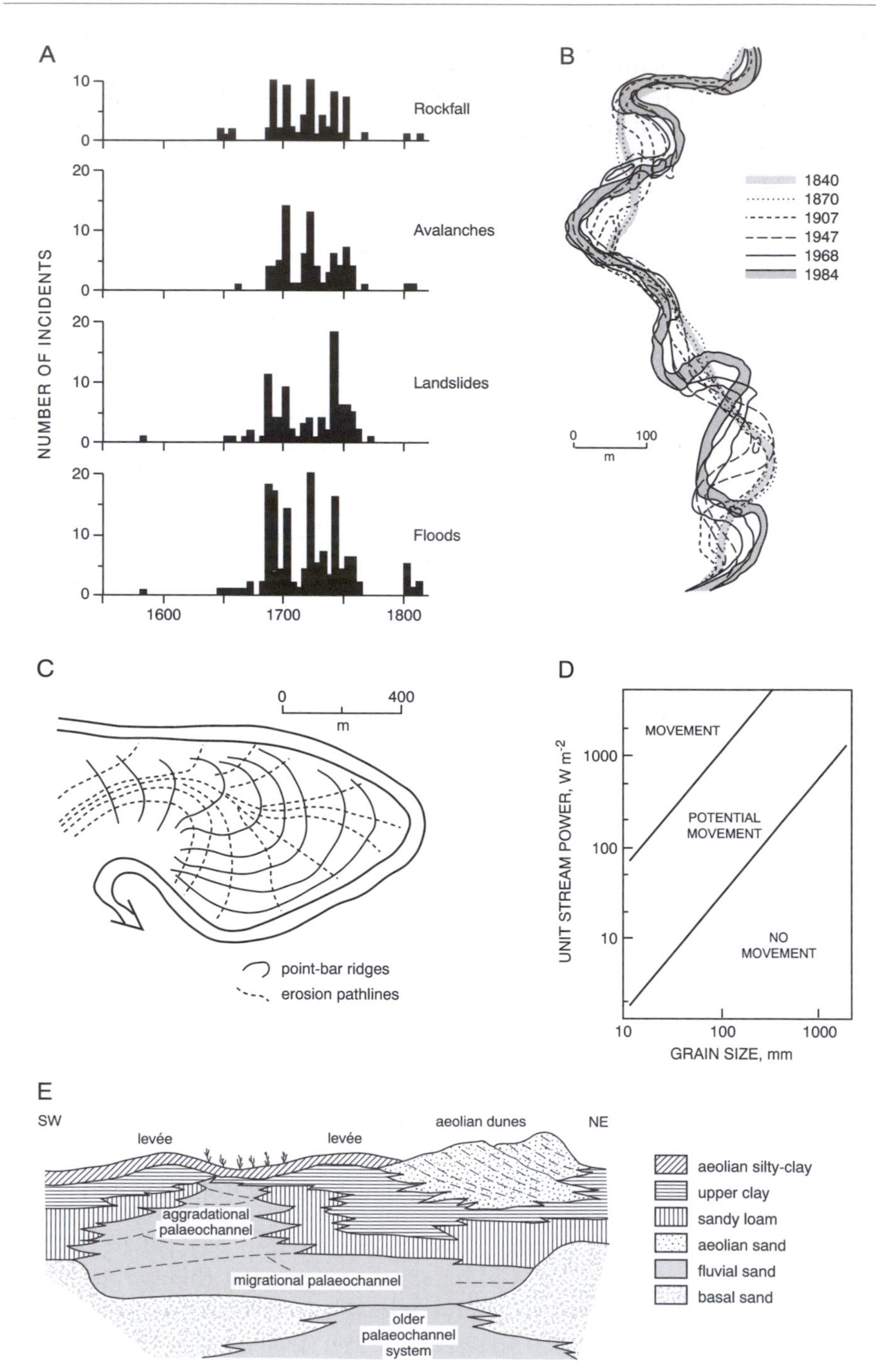

methods indicate only the overall direction and rate of channel change, and then only for planimetric properties if maps are the primary source.

Since rivers must erode and deposit sediment in changing their elevation, course or channel form, **sedimentary evidence** is a major source of information. It includes the forms developed in or on sedimentary surfaces as well as the internal structure and character of the deposits themselves. Thus Hickin (1974) has traced the pattern of meander development along the Beatton River in British Columbia over 250 years by mapping point-bar ridges formed during bend migration and defining erosional pathlines as orthogonals through the ridge systems (Fig. 6.1C). Since the ridges could be dated dendrochronologically, it was possible to show how channel curvature (r_c/w) has influenced the rate of migration (Hickin and Nanson, 1975, 1984; Fig. 5.21B, p.226).

Palaeochannels are most commonly preserved as exposed cross-sections, abandoned surface forms, or buried channels. On the basis of empirical relationships established for present-day rivers, Williams (1988) shows how the dimensions of former channels, particularly those with a meandering habit, can variously be used to estimate other palaeochannel features and former flow conditions. Thus Dury (1977) estimated past bankfull discharges (Q_b) from measurements of the width (w) and meander wavelength (λ) of former channels using contemporary relationships wherein the dependency has been reversed:

$$Q_b = (w/2.99)^{1.81} \tag{6.1}$$

$$Q_b = (\lambda/32.86)^{1.81} \tag{6.2}$$

Such relationships can be converted to a form where past (upper case) and present (lower case) values of the variables are expressed as ratios:

$$Q/q = (W/w)^{1.81} \tag{6.3}$$

thereby providing a more direct indication of change. Retrodiction from wavelength is possibly less accurate than that from width or from some combination of cross-sectional area (A_b) and slope (s) (Dury, 1985), as exemplified by the relationship (Williams, 1978a):

$$Q_b = 4.0\, A_b^{1.21}\, s^{0.28} \tag{6.4}$$

However, former width and cross-sectional area can be difficult to determine to an acceptable level of precision if there are insufficient well-preserved sections, and palaeoslope is one of the hardest variables to estimate if it cannot be measured directly (Williams, 1988). A wide range of equations has now been used in palaeohydrologic reconstruction, including bivariate form–discharge relationships of the type illustrated by equation (6.1), and more hydraulically based equations such as that developed by Rotnicki (1991):

Figure 6.1 (opposite) Evidence of change. (A) Incidence of mass movements and floods determined from land rent records, Josterdalsbre, Norway (after Grove, 1988). (B) Course changes along the River Dane, Cheshire, England, determined from maps and aerial photography of different dates (after Hooke and Redmond, 1989). (C) Erosion pathlines defined as orthogonals to point-bar ridges, River Beatton, British Columbia (after Hickin, 1974). (D) Zones of no movement, potential movement and movement in the relationship between grain size and unit stream power (after Williams, 1983). (E) Stratigraphic model of Riverine Plain palaeochannels showing sequential development of a migrational mixed-load system and an aggradational bed-load system (after Page and Nanson, 1996).

$$Q_b = (0.921/n)\, A_b\, R^{0.67}\, s^{0.5} + 2.362 \quad (6.5)$$

where R is hydraulic radius and n is Manning's resistance coefficient, a particularly difficult variable to estimate. All are subject to error, because the original relationship may have been based on limited data, because it may have been applied beyond its environmental range, or because the palaeochannel data may be unreliable (Williams, 1988). Only in exceptional circumstances are palaeochannels perfectly preserved, so former channel boundaries may be difficult to define. The most easily measured palaeo-dimension is the wavelength of valley meanders in underfit streams, but it is not necessarily the most accurate for retrodiction purposes (Dury, 1985).

Fluctuations in runoff and sediment supply are reflected in the character of *valley fill sediments*, whose interpretation can indicate the nature of channel change over longer timespans than purely morphologic evidence. The relevant sediment characteristics can be arranged in a hierarchical sequence, ranging from individual particles (sand-sized and larger), through aggregates of particles preserved as bed forms (see Table 5.8, pp. 188–9), to the architecture of large-scale sedimentary bodies. Smaller-scale features have been widely used as a means of interpreting palaeoflow hydraulics, while larger-scale ones have principally been used as indicators of former facies conditions, notably in relation to channel pattern.

On the basis of plots between unit stream power, bed shear stress and mean flow velocity as dependent variables, and particle diameter as an independent variable (see Figs 4.5C, p. 110, and 6.1D), which on the basis of published data were subdivided into zones of no movement, potential movement and movement, Williams (1983) used measurements of particle size to estimate the minimum palaeoflows that could have transported ancient fluvial deposits in Sweden. Because the estimates represent only the minimum values, as indicated by the lower boundary line, the actual transporting flows could have been much higher. Komar (1988, 1989) has developed a similar methodology based on flow-competence relations which take account not only of individual clast size but of clast size relative to the median diameter of the deposit as a whole, while pointing out that attempts to estimate flow hydraulics from maximum sizes are fraught with difficulty. Many transport equations apply with questionable validity to flood flows.

When preserved in fluvial deposits as well-defined structures, the sequence of bed forms associated with increasing flow intensity (see Fig. 4.3A, p. 104) can also provide a means of identifying former hydraulic conditions. Dune cross-bedding is probably the most reliable indicator. However, the associations are largely based on a narrow range of experimental conditions, so that precise estimates of palaeoflow parameters are unlikely to be valid for many field situations (Gregory and Maizels, 1991).

Large-scale sedimentary structures are most often interpreted in terms of channel pattern. Successful facies models have been developed for meandering river deposits but, although numerous models have been proposed for braided and anastomosing systems, the characteristic sedimentary sequences are less clearly defined (Gregory and Maizels, 1991). Nevertheless, distinct changes in facies type have been recognized, as in Central European river valleys for which Starkel (1983) proposed a pseudo-cyclic variation between braided and meandering patterns for the last 15 000 years. Provided there are sufficient subsurface data to indicate both lateral and vertical shifts in facies type, models of fluvial architecture can be constructed which characterize longer-term and larger-scale variability in fluvial activity. Such evidence enabled Page and Nanson (1996) to produce a revised fluvial history for the Riverine Plain of southeastern Australia covering more than 100 000 years. A simpler model presented by Schumm (1968), in which bed-load 'prior streams' evolved firstly into suspended-load 'ancestral rivers' and then into the present Murrumbidgee river system, is replaced by one which recognizes four distinct phases of palaeochannel activity (Table 6.8, p. 304). The complex stratigraphy extending over many hundreds of kilometres shows that the rivers on the Plain have experienced alternating episodes of channel activity, with repeated shifts from a laterally migrating mixed-load mode (migrational channels) to an aggrading bed-load mode (aggradational channels) (Fig. 6.1E).

Other forms of deposit provide evidence of former flood heights, the most extensively used being *slackwater deposits* (Kochel and Baker, 1988). They consist largely of fine-grained sediments which during floods settle from suspension in areas of reduced flow velocity, being best preserved in protected sites such as bedrock caves in canyoned valleys. Using the highest elevation of each deposit as a minimum estimate for the peak stage of the flood that emplaced it, palaeoflood magnitude can be quantified by the step-backwater method. In this manner did O'Connor *et al.* (1994) identify 15 major floods affecting the Colorado River over the last 4500 years, helping to set the gauged record of twentieth-century floods in a broader historical context.

Clearly many types of sedimentary evidence are available for reconstructing the morphologic and flow characteristics of past conditions, particularly but by no means exclusively during post-glacial times. They vary considerably in the level and accuracy of quantitative information provided. A common assumption has been that relationships established for present-day rivers can be readily applied to former environments, but such extrapolation may not be strictly valid. Post-depositional modifications may have produced an incomplete or inconsistent record, especially where periodic erosion has occurred. Although floodplain accumulations can remain stable for relatively long periods, reworking of part or all of the depositional record could lead to misinterpretation. Detailed local studies need to be integrated if intra- and inter-regional correlations of sedimentary sequences are to be developed and greater reliability achieved. Invariably the interpretation of complex fluvial histories is improved when a chronological framework can be established.

A wide range of **dating techniques** is now available, only a sample of which is given in Table 6.2. Many depend on finding suitable remains which have been fortuitously preserved without contamination. Relative methods attempt to place events in their correct temporal sequence, but require calibration against samples of known age before absolute dates can be estimated. Absolute methods partly overcome this shortcoming in enabling a numerical date to be attached to particular events.

Relative height has been one of the methods most often used to determine the relative position in time of morphological features, notably river terrace sequences. In that case height is usually taken relative to the present channel bed with the underlying assumption that the highest terrace is the oldest. Only isolated fragments remain in many instances and, without continuity of terrace surfaces along a valley, correlation on the basis of height alone is liable to large errors.

In addition to the relative chronology provided by sediments themselves, *organic and artefactual remains* contained therein can indicate relative positions in time. Floral and faunal evidence has been primarily used to establish past environmental conditions, but if a particular assemblage can be identified as time-specific, its presence may be taken as indicating sediments of that age. Pollen data have been particularly useful for reconstructing the post-glacial sequence of climatic changes. On the basis of modern analogues, Guiot *et al.* (1993) inferred continental-scale patterns of mean annual temperature and precipitation at 3000-year intervals for Europe.

Since archaeological periods can be characterized by specific assemblages of objects, artefacts such as pottery and tools also provide a basis for dating fluvial sediments. By reference to the pottery chronology of the Rio Grande valley as a whole, Miller and Wendorf (1958) were able not only to date periods of deposition and erosion in a tributary valley, but also to suggest that rates of sedimentation in the area have remained relatively uniform over the past 2000 years. Since artefacts are commonly dated by reference to their stratigraphic position, there is a need for independent dating so that diagnostic types can be identified and the dangers of circular argument avoided. This type of information is likely to have an irregular distribution, as illustrated by the fortuitous recovery of car licence plates which enabled Costa (1975) to estimate the rate of overbank deposition in a Piedmont stream since 1924.

Absolute methods tend to be more specialized and more costly to use. Major advances in absolute dating have come from the development of various *radiometric techniques* based on the known rates of decay of radioactive isotopes, radiocarbon dating being the geochronologic method of choice (100–40 000 years). Most are applicable to timespans of 10^3–10^8 years and are particularly relevant to Pleistocene and early Holocene studies, but some are suitable for shorter periods, notably lead-210 and caesium-137 (Wise, 1980). Provided suitable materials can be found, they are often used in combination with stratigraphic methods to date particular levels in deposits and thereby provide reference horizons. Both lead-210 and caesium-137 have been used to date horizons in lake sediments from which estimates of basin sediment yield have been made (Dearing, 1992; Foster and Walling, 1994). Radiocarbon dating of charcoal and other organic fragments provided the basis for correlating isolated slackwater deposits in the reconstruction of the flood history of the Colorado River (O'Connor *et al.*, 1994).

Dendrochronology, the study of tree rings to indicate time intervals or past variations in climate, is a very accurate method for dating events over the last 2000 years. The dating of floodplain trees has been used as a basis for determining the form and rate of channel migration (Everitt, 1968; Hickin, 1974) and the effects of river regulation on the downstream channel (Petts, 1977). By relating tree age to floodplain elevation along the Little Missouri River in North Dakota, Everitt was able to plot 25-year isochrones delimiting the dates of floodplain reworking. Annual ring thickness varies with climate, so it may be possible to reconstruct past climatic or streamflow conditions from variations in ring width. The reliability of tree-ring data, especially in a climatic context, depends on an understanding of the factors which determine the thickness and growth rate of rings in individual tree species.

The range of techniques considered above is far from exhaustive. Thus, for example, thermoluminescence dating of fluvial deposits is becoming more widely applied (Nanson *et al.*, 1991). Although they are discussed separately, it must be emphasized that techniques need to be used in combination if the possibility of misinterpretation is to be minimized. Dating methods are not of themselves direct evidence of change, but as the timescale lengthens and the resolution with which an event sequence can be reconstructed tends to decrease, the greater is the need for a chronological framework, especially when correlating events in different areas. Even then successive phases of development may be obscure if the information available allows only dated reconstructions at widely separated times. Sequences are invariably complex, involving events of varying frequency, magnitude and duration.

Causes of Change

Various types of change can be envisaged. During the long-continued evolution of a drainage basin and its constituent channel network, the potential energy status of the landscape varies as landscape mass is removed and uplift occurs. Davis (1899), in his cycle of erosion, assumed that landforms develop sequentially through time as relief is gradually reduced. However, since landforms are the net effect of many controlling factors, some of which change with time while others do not, reduction of relief does not of necessity lead to significant and progressive change in landscape geometry. In addition, this type of time-directed change needs to be assessed against the background of the major alterations to environmental conditions which have occurred over the past millennia, either naturally or as a result of human activities.

In so far as the fluvial system can be represented as a set of outputs (specifically, channel form elements) related to a set of inputs or controls (notably, discharge and sediment load), changes in channel form can result from a change in either input conditions or input–output relationships. As regards the former, Brunsden and Thornes (1979) distinguished between two types of external disturbance, pulsed and ramp, which may have different response characteristics. The *pulsed type* of disturbance is associated with episodic events of low frequency and high magnitude, such as extreme floods, whose effects may be spatially and temporally limited. An approximate return to initial conditions may follow the imposed disturbance, but extreme events can produce a lasting effect, particularly if some significant threshold is crossed. Schick (1974) has attributed the formation of desert stream terraces to extreme floods which incise channels in the alluvium previously built up by more moderate flows – a mode of terrace formation which does not rely on classical concepts of rejuvenation or climatic change.

In the *ramp type* of disturbance, controlling variables are sustained at a new level as a result of more permanent shifts in input conditions. Such changes are likely to apply over much larger areas, although the response may not be spatially uniform. Much of the unsteadiness in the fluvial system is caused by fluctuations in climate, which, through its influence on vegetation patterns, catchment water balances and process activity, affects the flow regime of rivers and the production, supply and transportation of material. On the basis of a wide range of evidence, an impressive picture has been constructed of climatic fluctuations over the last million years (Fig. 6.2). For time periods exceeding 10^4 years the scale of climatic fluctuation increases dramatically, but even over the past 10^3 years distinct variations have occurred, leading to unsteadiness in input conditions and geomorphic activity (Fig. 6.1A). Most is known about palaeotemperatures. Geomorphologically more relevant information on past precipitation and runoff conditions is more elusive, but Lamb (1977) has shown how broad averages can be estimated (Table 6.3), and the techniques of palaeohydrologic reconstruction outlined in the previous section are now being widely applied in the estimation of past discharges, particularly during the post-glacial period (Starkel *et al.*, 1991; Gregory *et al.*, 1995).

Although the timing, level and direction of climatic change may have varied from one area to another, all parts of the Earth's surface have been affected to some extent by events during the Quaternary. Climatic changes over timescales of $>10^5$ years are most spectacularly manifest by the repeated expansion and contraction of the great continental ice-sheets in mid- and high-latitude regions, and by the related cyclical variations in precipitation regime in lower-latitude areas. Contrasts between full glacial and interglacial conditions have been pronounced, involving changes in annual temperature of over 15°C and marked fluctuations in precipitation (Lamb, 1977). Glacial periods are characteristically marked by abrupt terminations, with transitions from maximum glacial to maximum interglacial in as little as 7000 years (Broecker, 1984).

Of greater relevance to the present fluvial environment are the changes that have occurred over the last 20 000 years. Knox (1995) recognized three

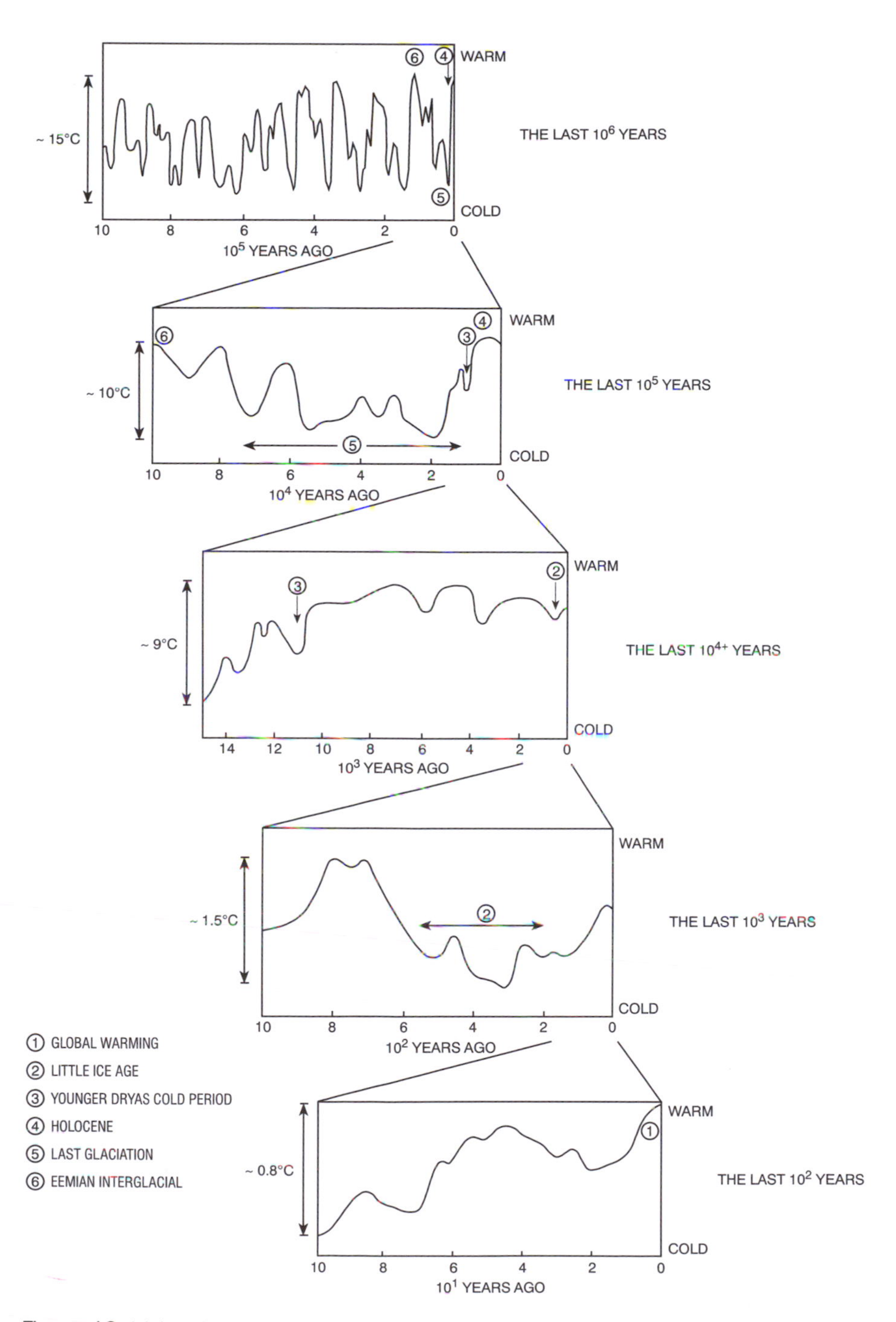

Figure 6.2 Main trends in temperature over different timescales in northwest Europe (after Lamb, 1977, and other sources).

Table 6.3 Estimates of average rainfall (R), evaporation (E) and surplus water (D) in different epochs in England and Wales, expressed as percentages of 1916–1950 averages (after Lamb, 1977)

Approximatae dates	Epoch	Estimate of average annual temperature (°C)	R	E	D
			(% of 1916–1950 averages)		
7000 BC	Pre-Boreal	9.3	92–95	94–98	89–94
4500 BC	Atlantic	10.7	110–115	108–114	112–116
2500 BC	Sub-Boreal	9.7	100–105	102–104	100–106
900–450 BC	Sub-Atlantic (onset)	9.3	103–105	97	110–115
AD 1150–1300	Little Optimum	10.2	103	104	102
AD 1550–1700	Little Ice Age	8.8	93	94	92
AD 1900–1950	20th century warm decades	9.4	(932 mm)	(497 mm)	(435 mm)

distinctive periods of fluvial activity: 20–14 ka BP, 14–9 ka BP and 9–0 ka BP. The first was characterized in temperate latitudes by major continental ice sheets and mountain glaciers which had reached their maximum extent by 18 ka BP. They contributed large volumes of glacial meltwater and sediment load to proglacial areas where extensive braided river systems developed. Active aggradation occurred along the braided upper Mississippi and its tributaries at this time, producing a longitudinal profile which was considerably steeper than that of the modern river, so much so that in headwater reaches the surface of the former river now stands as a terrace 45 m above the present floodplain.

The second period was one of major climatic transition from glacial to post-glacial conditions. During deglaciation, rates of sediment supply were initially very high but subsided as lakes in front of retreating glaciers became effective sediment traps, and as former periglacial areas were invaded by forest vegetation. Meltwater sources and increased precipitation maintained high river discharges, sometimes of catastrophic proportions when the drainage outlets of glacial lakes failed. Applying equation (6.5) to palaeomeander dimensions in the Prosna valley of central Poland, Rotnicki (1991) estimated that mean annual discharges in the Lateglacial were up to 6.5 times as large as they are today (Fig. 6.3). A reduced sediment load coupled with relatively high discharges resulted in a change from an aggradational to an erosional regime, and a transition from braiding to meandering in many temperate river valleys. In addition, the widespread development of large relict meanders (Dury, 1977) is generally associated with this period. Increased precipitation may also have characterized subtropical and tropical

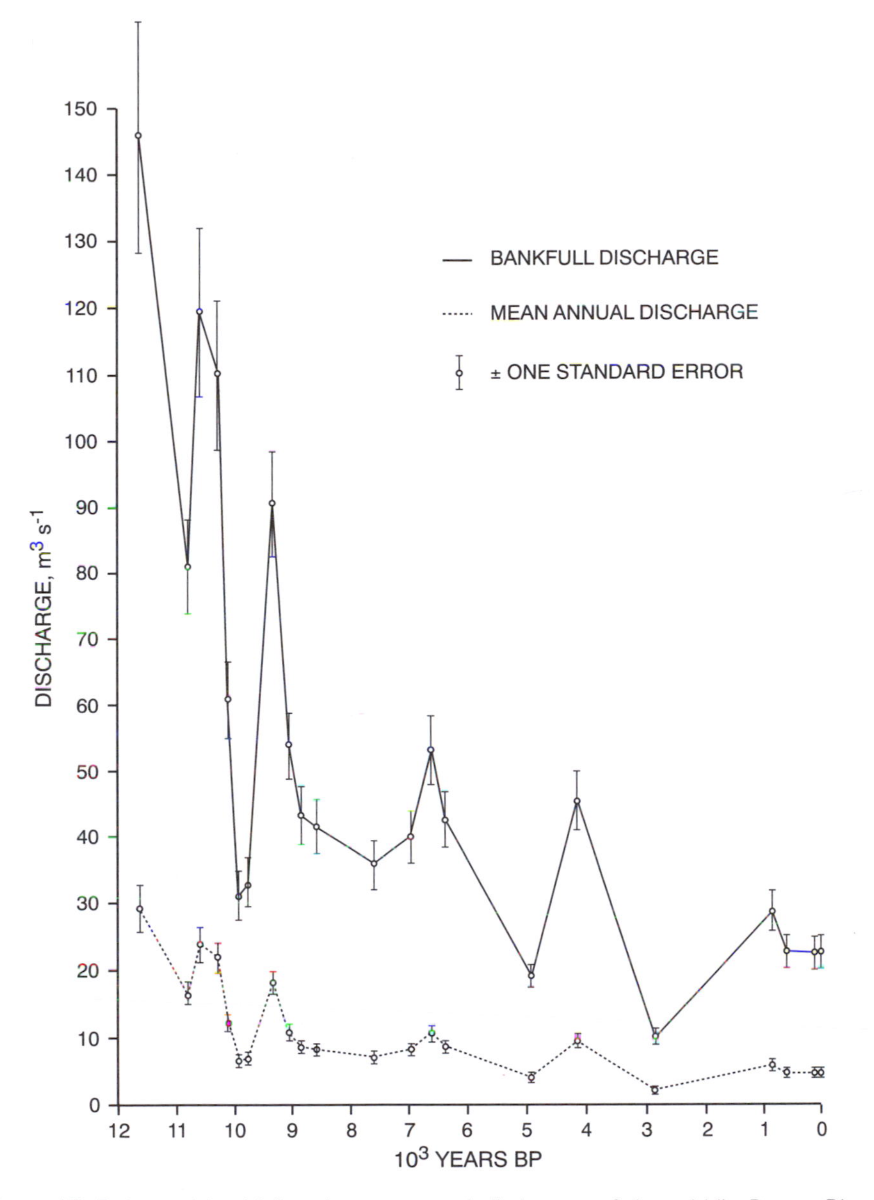

Figure 6.3 Estimated bankfull and mean annual discharges of the middle Prosna River, central Poland, over the last 12 000 years (after Rotnicki, 1991).

latitudes, leading to the higher discharges which are believed to have caused 30 m of degradation along the Nubian Nile between 12 ka and 10 ka BP (Said, 1993).

The global warming which ushered in the Holocene around 10 ka BP marked the start of a more climatically stable period (Fig. 6.2). However, that does not deny the continuation of climatic fluctuations, albeit of a lesser scale. The Climatic Optimum of 9–4 ka BP had temperatures that were higher than those of today across much of the northern temperate zone, to be followed by an episode of climatic deterioration which became most marked at 3–2.5 ka BP. Two prolonged climatic fluctuations characterized the Christian era (Table 6.3): the Little Optimum centred on about AD 1200, and the Little Ice Age which affected the sixteenth to nineteenth centuries and during which the incidence of flooding increased markedly, at least in certain areas (Fig. 6.1A). Indeed, cooler episodes during the Holocene tend to be associated with more frequent large floods in mid-latitude regions (Knox, 1995), serving to underline the point made by Knox (1993) that large changes in the magnitude–frequency characteristics of floods, and therefore in the level of fluvial activity, can be produced by modest changes in climate, of only 1–2°C in mean annual temperature and $\leq$ 10–20 per cent in mean annual precipitation. Thus major discontinuities in Holocene alluvial chronologies could be attributable to climatically induced shifts in the flood regime of rivers (Knox, 1995). Starkel (1991b) found good regional correlation between the alluvial chronologies of central European river valleys, with intensified fluvial activity at 8500–7700, 6500–6000, 5000–4500, about 2800, 2200–1800 and 400–100 years BP. These more active phases were associated with periods of glacial advance and tree-line lowering in the Alps, and with increased frequency of landslides and solifluction activity in European mountains. However, it becomes increasingly difficult during the Holocene to isolate the effects of climatic change on fluvial system behaviour, largely because of the accelerating impact of human activity, most notably through forest clearance. Deforestation increased runoff potential and particularly the supply of sediment to streams, as illustrated by the Huang Ho River in China where clearances in the loess plateau region between 200 BC and AD 600 increased sediment load by an order of magnitude (Milliman *et al.*, 1987).

The lengthiest streamflow records are for the River Nile in Egypt and extend back to AD 622, although there are notable gaps. They indicate high flood levels in 630, 850–930, 1100–1150, 1350–1470, 1737–1770 and 1850–1900. From miscellaneous reports of floods and droughts, it seems that many parts of Europe experienced a marked period of flooding from AD 1150 to 1500, with peaks at about 1310 and 1450 (Lamb, 1977). Large floods were also characteristic of upper Mississippi tributaries between AD 1300 and 1500 during the transition to the relatively cool years of the Little Ice Age (Knox, 1993). More recent records for European rivers reveal quite large short-period variations in discharge: a dry period (1863–1875), a humid period (1876–1882), a long dry phase (1883–1910), a very long humid period (1911–1941), the European drought of 1942–1956, a moderate humid period (1957–1968), and the dry period of the 1970s (Probst and Tardy, 1987; Probst, 1989). Although they are broadly synchronous, there are regional differences, and contrasts in runoff variations become even more apparent when analysis is carried out at the continental scale (Probst and Tardy, 1987). A more reliable picture of previous discharges is being built up not only for the period of historical records but also for the post-glacial period as a whole, although the problem of differentiating between the effects of climatic oscillation and human activities remains, especially over more recent times. In mid-Wales, for example, recent increases in flooding have been triggered by a greater incidence of intense storm events, but concomitant land-use changes have accentuated the tendency (Howe *et al.*, 1967).

The accumulated evidence points to the dynamic quality of climate. Major and minor shifts of climate and related vegetation changes have significantly altered input conditions and levels of fluvial activity over a range of timescales. In addition, important changes have occurred in the relative heights of land and sea, caused principally by eustatic fluctuations in sea-level and tectonic activity. Sea-level has risen by about 130 m since the last glacial maximum, thereby altering the potential energy status of the landscape and the position of the base-level to which streams may adjust. During glacial periods of lower sea-level, river courses would have lengthened and downcutting would have affected the lower reaches of most large rivers draining directly into the oceans. The Holocene marine transgression was largely accomplished by 6000 years BP, with relatively minor oscillations in sea-level thereafter, producing coastline recession and the submergence of river valleys. The effect of a change in sea-level will be felt first near the mouth of a river but may be transmitted far enough upstream to alter base-level conditions throughout a network. However, Leopold and Bull (1979) have maintained that base-level has only a limited upstream influence, and that the vertical position of a river is more the effect of aggradation and degradation caused by varying inputs of upstream discharge and sediment load than a base-level far removed in space.

Tectonic activity can produce significant modifications to river channels, but it tends to be rather localized and have a relatively minor influence overall. The primary effect is a local change in valley slope or cross-valley tilting, which induces a secondary response within the river via the processes of aggradation and degradation. Thus, domal uplift is commonly associated with the development of a convexity in the longitudinal profile, reducing slope in the upstream reaches and possibly initiating a change in channel pattern in accordance with the continuum concept (see Fig. 5.17, pp. 209–10). On the basis of laboratory experiments, Ouchi (1985) has observed that a meandering river subject to slow uplift will increase its sinuosity on the downstream side of the uplift in adjusting to the steeper valley floor, while it will tend to straighten or anastomose if the upstream slope becomes appreciably flatter. Adjustment at a larger spatial scale may be expected in those areas which have experienced considerable isostatic rebound since deglaciation. The Lower Attawapiskat River in Canada has developed an anastomosing habit as it crosses the low gradient of the Hudson Bay Lowland, which is still rising at the rate of about 0.7 m per hundred years (King and Martini, 1984).

Human activities have had a fundamental and increasing impact on fluvial system behaviour from as early as 9000 years BP when deforestation began in Europe. Via such processes as land-use change and direct

Table 6.4 Human-induced river channel change (after Gregory, 1995)

Cause of change	Channel variable	Change ratio (after disturbance/before disturbance)		
		Average	**Minimum**	**Maximum**
Land-use change	Width	1.41	0.96	1.88
	Depth	2.31*	1.58*	3.97*
	Capacity	2.15*	1.53*	4.11*
Reservoir construction	Width	0.85	0.29	1.49
	Depth	0.92	0.34	1.62
	Capacity	1.25	0.29	2.29
Channelization	Width	1.33	1.00	2.02
	Depth	1.06	0.59	1.67
	Capacity	1.39	1.00	2.53

Data are for the UK unless otherwise indicated (*).

modification of the channel, they have significantly altered the runoff regime of many rivers and the pattern of sediment supply, with consequences for channel morphology (Table 6.4). Starkel (1991a) has identified two main phases of human-induced valley aggradation in central Europe, the first during Roman times and the second during the Medieval period. Elsewhere, significant disturbance has been longer delayed, as in the US where the introduction of European agricultural practices began to have a major impact from AD 1700 onwards. Land-use activities involving forest clearance and cultivation have been the dominant form of human disturbance during the Holocene, increasing the sensitivity of catchment runoff and sediment yield to climatic events. Direct modification of the channel is largely of more recent vintage, but its consequences can nevertheless be far-reaching both spatially and temporally.

Disturbances to external factors such as climate, vegetation, land use and base-level are undoubtedly a major cause of instability in the fluvial system. However, change can also be initiated internally during the continued operation of prevailing input levels, a type of unsteady behaviour commonly associated with the exceedence of critical levels termed *intrinsic thresholds* by Schumm (1973). Without a change in external factors, a landform may through time reach a condition of incipient instability when a critical threshold may be crossed and significant change initiated. The development of meander cutoffs is a simple example of such instability. Although meanders are regarded as a stable form, they may attain a sinuosity and amplitude which cannot be maintained under existing conditions of flow and sediment transport. By increasing channel gradient, selective cutoffs may return a river to a more stable sinuosity in which there is a better balance between transporting ability and gradient. At a broader scale, Patton and Schumm (1981) suggested that, in regions with high sediment yields and high ratios of sediment discharge to water discharge, episodes of channel cutting and filling can occur under relatively stable climatic conditions because of the control exerted by geomorphic parameters that are intrinsic to the system, notably those that are slope-related. Nanson (1986) has used a similar argument to explain the evolution of New South Wales floodplains in confined valleys through a combination of episodic stripping and vertical accretion. After a period of gradual floodplain reconstruction, erosional thresholds are eventually exceeded both in the channel and on the floodplain, leading to instability and wholesale erosion by the next major flood. Response to internal change may not always be distinguishable from that due to external change, but its recognition adds a further dimension to understanding the course of fluvial history.

Adjustment to changing external or internal conditions contrasts sharply with the type of progressive change conceptualized in the cycle of erosion. Regarding that adjustment, three possibilities can be envisaged: response to an impulse may be simply damped out and the previous state restored; it may be sustained at a new level of activity; or it may be reinforced by positive feedback in which one change leads autocatalytically to another (Brunsden and Thornes, 1979). Additionally, response may be spatially uniform or spatially variable. Thus, for example, the level of climatic change may be relatively uniform over large areas but all parts of an area need not be equally affected. Erosion in one part of a stream system can be associated with deposition in another to give alluvial chronologies which are temporally and spatially out of phase. On the other hand, highly localized changes to input conditions may have effects which are propagated throughout an entire system. These characteristics of adjustment make it difficult to interpret the historical record, to effect regional correlations, and to attach specific causes to specific events.

Philosophies of Change

Two philosophies have dominated opinion regarding the course of Earth history: *catastrophism* and *uniformitarianism*. As the name suggests, the first assumes that much of the erosional and depositional work carried out at the Earth's surface is attributable to sudden and often violent events of large magnitude. This philosophy held sway until the early 1800s, with Noah's flood regarded as the event of most significance.

Largely through the work of Hutton, Playfair and Lyell, the tenets of the original catastrophe theory were undermined and replaced by the doctrine of uniformitarianism with its precepts:

(a) the basic laws of nature are time-invariant;
(b) similar processes and rates prevailed in the past as at present;
(c) change takes place gradually and progressively rather than suddenly.

Although the second proposition is now refuted by the available evidence, the main themes of uniformity of laws and gradual change (*gradualism*) have remained. That philosophy permeates Davis's model of landform development with its emphasis on the action of gradual subaerial processes, and forms the basis of climatic geomorphology as practised on the continent of Europe. The more recent assertion of Wolman and Miller (1960) that events of moderate frequency and magnitude perform the bulk of fluvial work is in substantial agreement with the uniformitarian doctrine. The main themes can also be seen in the concept of dynamic equilibrium initially proposed by Gilbert (1877) and articulated further by Hack (1960), in which small-scale adjustments are continuously being made in order to maintain an approximate state of balance between processes and forms. For given environmental conditions, there is a tendency over time to produce a set of characteristic forms.

Largely as a result of work carried out by palaeontologists in the second half of the twentieth century, a modern form of catastrophism (*neocatastrophism*) has emerged but without the intellectual baggage of the original: a short geological record (the Creation had originally been dated at 4004 BC) and an appeal to Biblical events. In geomorphology the idea of long-continued, low-amplitude change has been challenged on the grounds that controls may not remain constant for a long enough period to allow the development of characteristic forms, and that a few extreme events may produce substantial change of lasting effect. In particular, the role played by both ancient and modern floods in shaping the fluvial landscape has been emphasized. Thus, the sudden release of meltwater from Pleistocene Lake Missoula has been invoked to explain valley morphology in the Channeled Scablands of the western United States, with discharges estimated as high as 10×10^6 m^3 s^{-1} (Benito, 1997). The movement and emplacement of large clasts by extreme floods may so alter the character of a valley's morphology that subsequent floods of more moderate magnitude are unable to effect significant modification. However, to accept the importance of catastrophes in Earth history is not to deny the efficacy of more gradual processes. In any case the impact of major floods on the fluvial landscape is spatially very variable. There is a need to combine slow and gradual change with sudden and violent change, for they are both part of a continuum.

Whether or not a major event has a large or lasting effect depends partly on whether or not a significant threshold is exceeded (Fig. 5.4, p. 159). Thresholds separate different system regimes, each of which may have its own characteristic geometry, so that recovery may be difficult once a threshold has been crossed. Schumm (1973) has recognized two types – extrinsic and intrinsic – where the first is associated with change in an external factor such as climate, and the second reflects an inherent property of geomorphic systems to evolve to a critical state when adjustment or failure occurs. The cusp catastrophe model of Graf (1988b) illustrates in terms of braided and meandering channel patterns how the operation of gradual processes can lead to sudden change when the system is forced across an extrinsic threshold. On the one hand a gradual transformation can be achieved when the control variables (stream power and bank resistance) change in such a way that the system follows a smooth transitional path (path 1 in Fig. 6.4A), but

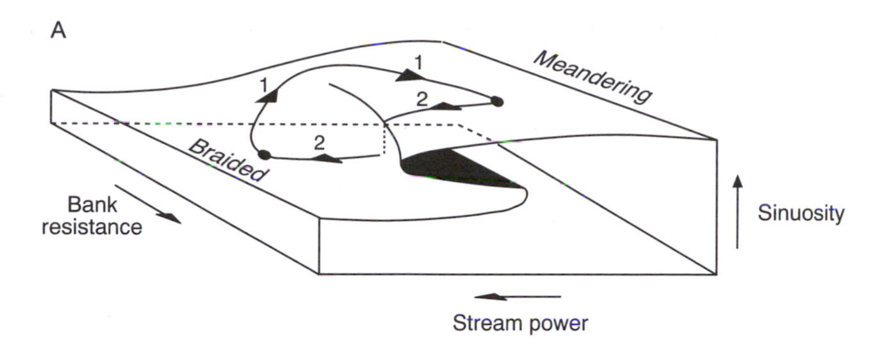

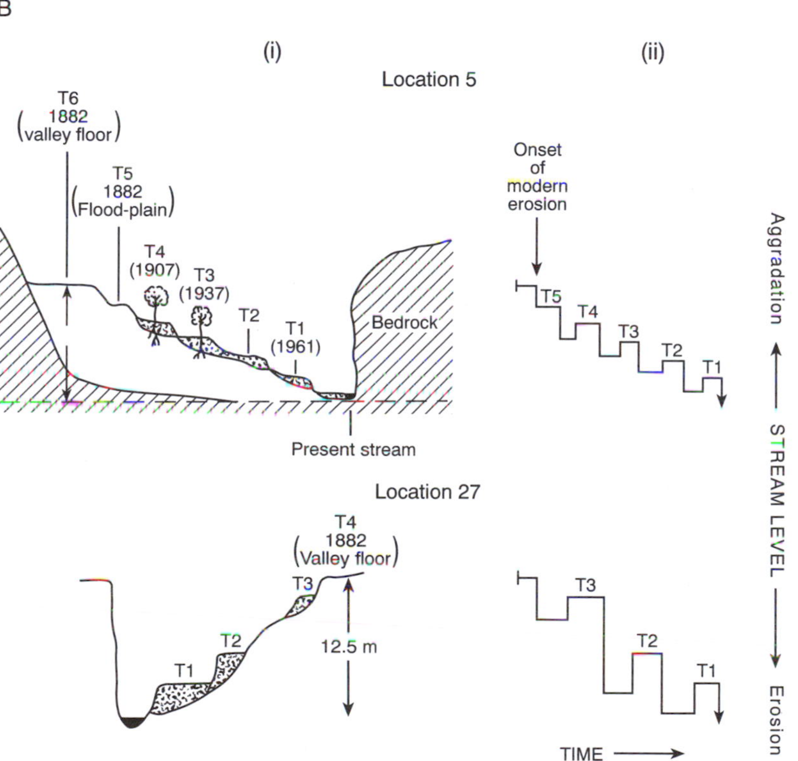

Figure 6.4 (A) A cusp catastrophe surface relating channel sinuosity to the control variables of stream power and bank resistance. Path 1 shows smooth and gradual change from a braided channel to a meandering channel due to a reduction in both stream power and bank resistance. Path 2 shows an abrupt, discontinuous switch from a meandering channel to a braided channel due to an increase in stream power at constant bank resistance (after Graf, 1988b). (B) An episodic model exemplified by terrace development along Douglas Creek (after Womack and Schumm, 1977): (i) terraces are numbered from lowest to highest, so that the 1882 valley floor may be assigned a different number depending on the number of terraces present at each location; (ii) schematic diagrams representing the episodic deposition and erosion at each location. Vertical segments indicate incision or deposition, and horizontal segments periods of relative stability. At location 5 the surfaces were dated using tree-ring and historical data.

on the other an abrupt change occurs if the control variables force the system across the fold, which is an extrinsic threshold (path 2 in Fig. 6.4A).

Change related to the crossing of an intrinsic rather than an extrinsic threshold is exemplified by the development of terrace sequences along Douglas Creek in Colorado (Womack and Schumm, 1977). Modern incision of the valley fill began in about 1882, possibly as a result of overgrazing, and led to downcutting by tributary streams. Large quantities of sediment were delivered from the steep tributaries, causing localized deposition and oversteepening of the valley floor at the new level. Repeated entrenchments in these deposits produced flights of terraces at various locations along the main stream (Fig. 6.4B). At one section six terraces have developed, but terrace surfaces vary in number and continuity downvalley, indicating spatial and presumably temporal differences in the erosional and depositional history. Nevertheless, the overall trend is episodic, with incision to a new level being followed by deposition, the valley floor steepening until a critical slope (intrinsic threshold) is reached, and renewed downcutting (Fig. 6.4B(ii)). Although the episodic model as exemplified by Douglas Creek may be difficult to apply to longer temporal and larger spatial scales because of the high rates of sediment supply required, it does emphasize that change need not be gradual and progressive. To a certain extent the threshold concept reconciles the conflicting philosophies of catastrophism and uniformitarianism, in that the associated model envisages periods of rapid adjustment separated by periods of more gradual change or relative stability when characteristic forms may develop.

Arguing along similar lines, Huggett (1990) envisages a unification of catastrophism and uniformitarianism (gradualism) via the relatively new field of *non-linear dynamics*, a view echoed by Phillips (1992). It is concerned with dissipative systems or structures in which energy is dissipated to maintain order in states removed from equilibrium. The dynamics of dissipative systems typically involve self-organizing processes which produce an orderly sequence of system configurations, such as that associated with transient bed forms as flow strength varies (see Fig. 4.3A, p. 104). Non-linear dynamical systems often exhibit discontinuities during their evolution, called bifurcations, which are directly analogous to thresholds in geomorphic systems. The system will change slowly between bifurcations but rapidly at them, so that both gradual and sudden changes are included. The discontinuities may represent a transition from one equilibrium state to another, or from regular behaviour to chaotic behaviour. Chaos theory, which deals with the irregular behaviour of systems in which order is hidden within apparent disorder, charts the paths along which non-linear systems will progress in converging towards multiple solutions, called 'strange attractors'. Although geomorphic applications of these various ideas are as yet rather limited, Phillips (1992) argues that the potential for chaotic behaviour exists in many geomorphic systems and that many fundamental concepts in geomorphology can be interpreted in terms of non-linear dynamical systems theory. Certainly its concern with the evolution or trajectory of systems subject to either internally or externally generated fluctuation matches the way in which a river will respond to the types of change considered in the preceding section.

Reaction and Relaxation Times

In the language of non-linear dynamics, systems move away from 'repellors' and towards 'attractors', defined respectively as asymptotically unstable and stable equilibrium states (Phillips, 1992). If perturbed, a system near stable equilibrium may be able to accommodate the disturbance unless it reaches a critical level, whereas one in unstable equilibrium will not return to its pre-disturbance state, remaining in disequilibrium or moving to a new, different equilibrium state. Thus, after perturbation a system may be fluctuating about a stable equilibrium, out of equilibrium (where the dynamics of non-linear systems are most relevant), or approaching a new equilibrium, depending on its initial status and the size of the disturbance.

A pervasive tendency of natural physical systems is for response to lag behind changes in process intensity. Conceptually at least, and at a simpler level than non-linear dynamical systems theory, geomorphic time can be divided into (Brunsden, 1980):

(a) the time taken for the system to react to a change in conditions (*reaction time*);
(b) the time taken for the system to attain a characteristic (equilibrium) state (*relaxation time*); and
(c) the time over which that state is expected to persist (*characteristic form time*).

Considering the complexity of the fluvial system, in which a hierarchy of form elements is related to a set of controls having space-dependent and time-dependent properties, response curves could show a wide range of forms with variable reaction and relaxation times.

Figure 6.5A shows the basic elements of response to a step-like change in a single control variable, assuming that the change is sustained at a new level. The step function, which implies a rapid transition in input conditions from one regime to another and therefore resembles a bifurcation, is one of the simplest models that can be devised, but it does seem to have merit for describing changes in climatic conditions in particular. Evidence from Greenland ice cores suggests that at the end of the last cold stage, temperatures in southern Greenland rose 7°C in 50 years (Dansgaard *et al.*, 1989). Major discontinuities in the botanical and cultural record of the Holocene, identified from an analysis of radiocarbon dates for the northern hemisphere, support an abrupt rather than smooth transition of climate, with vegetation lagging climate by about a hundred years (Wendland and Bryson, 1974).

Proceeding further to evaluate the potential response in the fluvial system, two main phases are involved: changes in the external environment need to be related firstly to changes in the flow and sediment conditions in the river, which must then be related to the morphological response of the

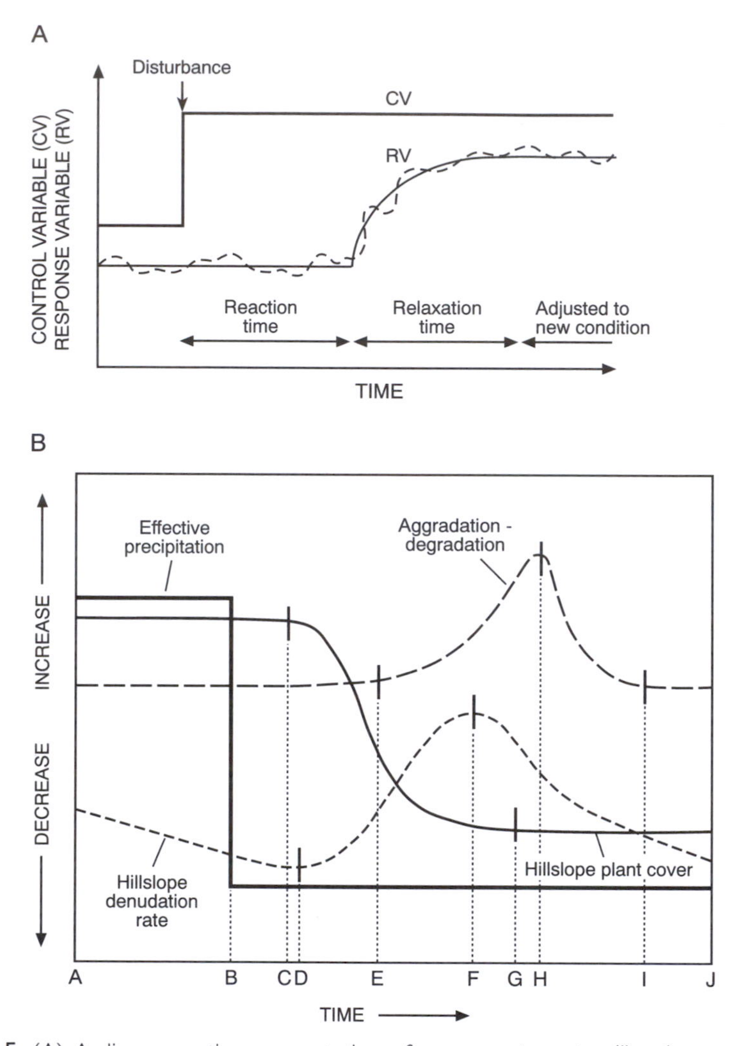

Figure 6.5 (A) A diagrammatic representation of response to a step-like change in a control variable, where the solid line of the response curve indicates the mean condition about which fluctuations occur (dashed line). (B) Hypothetical response to a decrease in effective precipitation from semi-arid to arid conditions, illustrating different response times of subsystems (after Bull, 1991).

channel. Both phases involve temporal lag. Consequently, even if climate is assumed to change in a step-like fashion, channel morphology need not behave similarly. Figure 6.5B illustrates the nature of the problem. Assuming a single abrupt decrease in effective precipitation from semi-arid (A–B) to arid (B–J) conditions, the vegetation, hillslope and valley subsystems exhibit different response times (= reaction time + relaxation time) to the climatic perturbation. Vegetation reacts quickly (B–C) but has a long relaxation time (C–G). Hillslope denudation rates react to the decrease in plant cover (C–D), then increase rapidly (D–F) before declining (F–J) as sediment sources become depleted. There is even greater delay in the reaction of the valley subsystem (E), which then responds to the variable sediment input through successive aggradation (E–H) and degradation (H–I). Even this simple diagrammatic representation of climatic change produces a complex response in conditions when relaxation effects are introduced, and changes in channel morphology are not even explicitly included.

Following the format laid out by Chorley and Kennedy (1971), reaction and relaxation times can be considered in terms of four main factors:

(a) The *resistance to change* offered by the various morphological components of the system and by the system as a whole. A hierarchy of increasing resistance and therefore response time can sensibly be constructed according to the dominant type of boundary material: from sand-bed, through gravel-bed and boulder-bed, to bedrock streams. The last are particularly slow to change, and catastrophic events may be required to effect significant response. Since change in the fluvial system requires work to be done in the form of material transfer, similar hierarchies can be envisaged for both the scale of the system (larger networks and channels adjusting more slowly) and the morphological component involved, in the order: bed configuration in sand-bed streams, channel width, channel depth, bed configuration in gravel-bed streams and planimetric geometry, channel bed slope, and the overall shape of the longitudinal profile (see Fig. 5.3, p. 158). Although the amount of work required to move from one state to another as regards these various components is not known, so that their position in the hierarchy may be debatable, their variable resistance and sensitivity to change cannot be denied. River channel width appears to be one of the most adjustable components, and available data, albeit for extreme floods, suggest relatively short relaxation times in some cases (Table 6.5).

Table 6.5 Adjustment of river channel width to extreme events

River	Frequency of event, years	Maximum widening, %	Relaxation time, years	Source
Patuxent River	200	64	15	Gupta and Fox (1974)
Baisman Run	200	160 (average of 10–20)	1–10	Costa (1974)
Appalachian rivers	100	300–400	10	Hack and Goodlett (1960)
Plum Creek	900–1600	160+	>30	Osterkamp and Costa (1987)
Gila River	200	600	45–50	Burkham (1972)
Wollombi Brook	87	100	>40	Erskine (1996)

(b) The *complexity of the system*, which includes the number of components involved and the character of their interconnections. In the fluvial system where many possible adjustments exist, response to changing discharge and load in the system as a whole may be relatively rapid, particularly if change is concentrated in more sensitive components. Schumm (1968), for example, suggested that the Murrumbidgee River responded to changes in hydrologic regime by selectively altering its channel width and sinuosity without significant aggradation or degradation. Relaxation times are dependent on the ability of a system to transmit impulses. Where drainage densities are higher and interconnections stronger, effects may be propagated away from the point of disturbance more rapidly. Howard (1982) compared the response times in a long, unbranched channel and a channel network, and found that, whereas they increased with the square of channel length in the former, the rate of increase was much lower in the latter.

(c) The *magnitude and direction of change* in input conditions. Smaller-scale changes may have noticeable effects only in more sensitive areas, or be absorbed by the system in such a way that the overall effect is reduced. Response times and the type of response depend partly on whether the changes reinforce or counteract existing tendencies. In evaluating the impact of climatic change, the initial climate is probably of critical importance. Baker and Penteado-Orellana (1977) argued that the Colorado River in Texas adjusted rapidly with the onset of more arid conditions but more gradually during arid-to-humid transitions, whereas Bull (1991) found that both reaction and relaxation times are longer in arid than in humid systems.

(d) The *energy environment* of the input. To the extent that rates of change are determined by permissible rates of material transfer, relaxation times are strongly influenced by the prevailing levels of energy available to perform work. A distinction can be drawn between headwater and downstream areas, where the former have more variable flow regimes, higher potential energy and stronger links between hillslopes and channels. They may therefore be more responsive. Howard (1982) has suggested that response time is more dependent on the magnitude of input variables than the magnitude of change and, in particular, that it decreases as water and sediment discharges increase. Consequently there may be an asymmetry of adjustment in that an increase in discharge or load causes a more rapid change than does a decrease of the same amount.

Even though the discussion above is rather general, the difficulty of defining representative reaction and relaxation times should be clear. Sensitivity to change varies not only from one climatic and physiographic environment to another, but also between the various components that characterize river channel form. Notwithstanding this complexity, it has been widely assumed that geomorphic systems respond rapidly at first after disruption, but at a steadily declining rate thereafter as a new equilibrium state is approached. In other words, relaxation paths have a basically exponential form. Expressed formally, if the rate of adjustment (dy/dt) of some channel parameter y is regarded as proportional to the difference between that parameter at time t ($y(t)$) and its new equilibrium value (y_e), then

$$\frac{dy}{dt} = \beta(y_e - y) \qquad (6.6)$$

which has the general solution,

$$\begin{aligned} y(t) &= y_e + \alpha\, e^{-\beta t} \qquad \text{when } y > y_e \\ y(t) &= y_e - \alpha\, e^{-\beta t} \qquad \text{when } y < y_e \end{aligned} \qquad (6.7)$$

where β indicates the rate at which the equilibrium state is being approached, and α is a constant equal to the difference between y and y_e at $t = 0$. Thus y_e represents the state towards which y is asymptotically trending from above ($y > y_e$) or below ($y < y_e$).

In one form or another, the exponential model has been used to describe relaxation paths in a variety of fluvial contexts. Graf (1977) investigated the successive development of gullies in the Colorado Piedmont, the initial formation of a protogully (about 1826) being followed by later dissection (about 1906) when a second gully began to develop within the first (Fig. 6.6A). Headward extension was rapid at first in both cases, so that by 18.7 and 15.4 years respectively the older and younger gullies had reached half their new equilibrium length. Thereafter the rate of headward growth steadily declined as the drainage area at the headward end (and therefore the amount of energy available for erosion) progressively decreased. In their study of channel adjustment downstream of dams, Williams and Wolman (1984) detected a similar mode of behaviour. Reaction times were generally very short (averages of 1.2 years for bed height and 0.6 years for width), while relaxation times were rather variable, although at degrading and widening sections as a whole about 50 per cent of the total change was achieved within the first 5 per cent of the adjustment period. Below the Fort Peck Dam on the Missouri River, relaxation times for channel width were mostly in the range of 20–35 years, and Fig. 6.6B illustrates how good a fit to the data can be achieved with equation (6.7). Following the eruption of Mount St Helens in 1980, the Toutle River system was catastrophically altered by a debris avalanche, with subsequent adjustment of the channel bed in the form of degradation and aggradation respectively above and below the point of maximum disturbance (Simon, 1992). Both forms of adjustment have been successfully described by an exponential function (Fig. 6.6C), which can then be used to predict future changes in bed elevation and provide a basis for assessing the suitability of various mitigation measures (Simon, 1995). Although the exponential model fits the several data-sets quite well (Fig. 6.6), it should not be assumed that this is always the case or that the universality of the underlying principle is necessarily

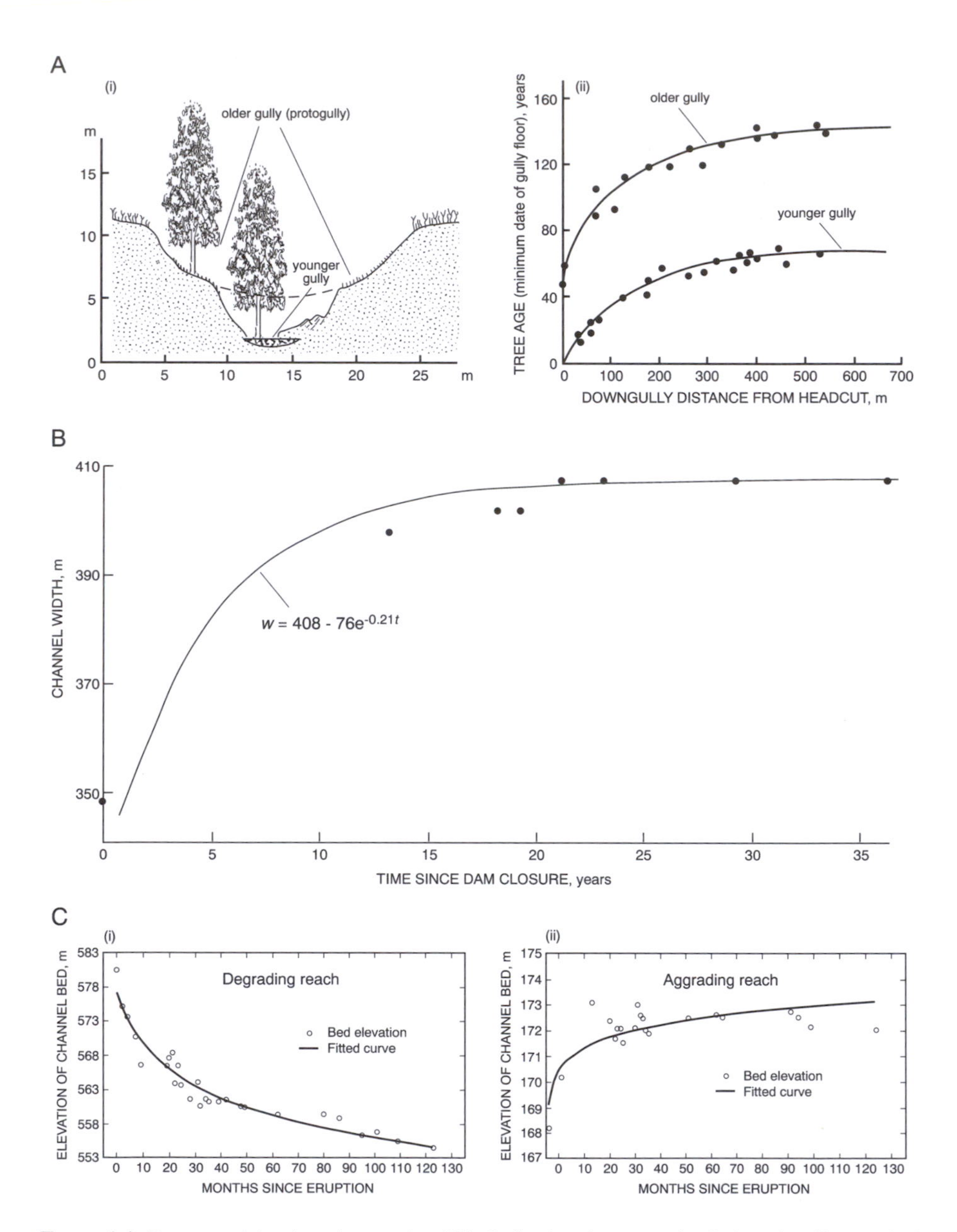

Figure 6.6 Exponential relaxation paths. (A) Gully development in Colorado: (i) a typical cross-section illustrating the two phases of gully development; (ii) dendrochronologic data obtained from trees growing on the gully floors show an exponential decline in the rate of gully extension, where younger trees grow on more recently exposed surfaces further headward (after Graf, 1977). (B) Width adjustment 9.2 km downstream from the Fort Peck Dam, Missouri River (data from Williams and Wolman, 1984). (C) Changes in channel bed height upstream (i) and downstream (ii) of the area of maximum disturbance following the eruption of Mount St Helens (after Simon, 1992).

proven. Indeed, there are sections investigated by both Williams and Wolman (1984) and Simon (1992) where adjustment follows a different course.

Relaxation paths and times have a key role to play in assessments of river channel adjustment to externally or internally generated change over a wide range of timescales. They can indicate not only the trajectory towards and potential for recovery after disturbance, but also the conditions under which system equilibrium might be attained and the extent to which the effects of past events might persist in the fluvial landscape. In terms of a transient form ratio,

$$\mathrm{TF} = \frac{\text{mean relaxation time}}{\text{mean recurrence time of events}} \tag{6.8}$$

Brunsden and Thornes (1979) defined two possible states:

(i) $\mathrm{TF} < 1$: the system has the potential to adjust to new conditions before the next major disturbance occurs so that characteristic forms will tend to prevail after the recovery period, leading to more reliable process–response relationships.
(ii) $\mathrm{TF} > 1$: because the mean recurrence time of events capable of producing change is shorter than the time taken for the system (or component of the system) to recover or equilibrate, there is likely to be a poor correspondence between process agents and resulting landforms; that is, forms will be predominantly transient.

Equation (6.7) assumes that adjustment is towards some new equilibrium state; in other words, that (i) applies. However, as in the coastal rivers of New South Wales which experience alternating drought- and flood-dominated regimes (Warner, 1987, 1994; Nanson and Erskine, 1988), there may be insufficient time for recovery between significant disturbances and the system remains in disequilibrium, with the possibility of chaotic behaviour. Neither of the times given in equation (6.8) can be specified with any assurance for the various components of river channel morphology or the fluvial system as a whole, although improved estimates are now being made. In general, river channels respond relatively rapidly except possibly in the profile dimension, but even so, scale and environmental differences produce a large range of sensitivities. Where sensitivity is high and relaxation times are short, geohistorical influences become progressively less important.

Predicting the Course of Channel Adjustment

Faced with a continuing and increasing amount of human interference to the fluvial system and the prospect of global warming, there is an even greater need for a quantitative methodology with which to predict how a river will respond to disturbance. An initial problem is to determine the susceptibility of a given river to change, which involves questions about the sensitivity of different channels to events of different magnitude. Rivers in arid and semi-arid environments are probably more susceptible to change than those in more humid ones (Burkham, 1981), and major morphologic adjustments in alluvial rivers generally do not occur unless unit stream power exceeds 300 W m^{-2} (Magilligan, 1992), but the more general problem is far from a solution. Downs and Gregory (1993) have considered several definitions of sensitivity, including equation (6.8), while Darby and Thorne (1996) have developed a numerical model which evaluates the sensitivity of channel form to changes in a range of flow, sediment and boundary conditions. In sand-bed streams with cohesive banks the model predicts that both width and depth are most sensitive to variations in discharge, with the threshold of bank material entrainment also important in the case of width. Other variables, including sediment supply, are less significant by an order of magnitude or more. Beyond the question of susceptibility, there are four main elements to the prediction problem: the direction of response, the magnitude of response, the rate of response, and the extent of response. Most progress has been made with the first.

Schumm's (1969) work on river metamorphosis provides the springboard for predicting **direction of response**. The approach is based on empirical relationships using data from largely sand-bed channels in semi-arid and sub-humid regions of the United States and southeastern Australia (Table 6.6), with the assumption that M (percentage silt–clay in channel perimeter) can be used to indicate not only the type of sediment load (p. 151) but also

Table 6.6 Empirical equations relating channel variables to flow and sediment characteristics

Channel variable	Schumm's (1969) equations	Other equations	Source
Width	$w = 44\,Q_m^{0.38} M^{-0.39}$	$w = 5.6\,Q_b^{0.51} S^{-0.30}$	Rundquist (1975)
	$w = 5.5\,Q_{ma}^{0.58} M^{-0.37}$	$^*w = 2.4\,Q_b^{0.41} s^{-0.10}$	Charlton *et al.* (1978)
Depth	$d = 0.51\,Q_m^{0.29} M^{0.34}$	$d = 0.08\,Q_b^{0.26} S^{0.50} s^{-0.22}$	Rundquist (1975)
	$d = 0.12\,Q_{ma}^{0.42} M^{0.35}$		
Width : depth ratio	$w/d = 255\,M^{-1.08}$		
	$w/d = 80\,Q_m^{0.10} M^{-0.74}$		
	$w/d = 41\,Q_{ma}^{0.18} M^{-0.74}$		
Meander wavelength	$\lambda = 1935\,Q_m^{0.34} M^{-0.74}$	$\lambda = 1.2 Q_b^{0.30} s^{-0.58}$	Derived from Larras (1968)
	$\lambda = 394\,Q_{ma}^{0.48} M^{-0.74}$		
Sinuosity	$S = 0.94\,M^{0.25}$		
Channel gradient	$s = 0.0036\,Q_m^{-0.32} M^{-0.38}$		

Symbols: Q_m, mean annual discharge; Q_{ma}, mean annual flood; Q_b, bankfull discharge; M, silt–clay in channel perimeter; w, width; d, mean depth; λ, meander wavelength; S, channel sinuosity; s, channel gradient.

* Assumed to apply to gravelly rivers with an imposed slope and appreciable sediment load.

the quantity of bed load (Q_{sb}). Thus, if M varies inversely with the quantity of bed load, then $1/Q_{sb}$ can be substituted for M in the equations. This set of relationships is one of the few to include a sediment load factor, albeit indirectly, but others (Table 6.6, equations (5.18)–(5.21)) can be used as a partial check on the consistency of predicted directions of change.

Using a + or – sign to denote an increase or decrease, the effects of a change in discharge (Q) or bed load (Q_{sb}) on channel form variables can be hypothesized:

$$Q^+ \rightarrow w^+, d^+, (w/d)^+, \lambda^+, s^- \quad (6.9)$$

$$Q^- \rightarrow w^-, d^-, (w/d)^-, \lambda^-, s^+ \quad (6.10)$$

$$Q_{sb}^+ \rightarrow w^+, d^-, (w/d)^+, \lambda^+, S^-, s^+ \quad (6.11)$$

$$Q_{sb}^- \rightarrow w^-, d^+, (w/d)^-, \lambda^-, S^+, s^- \quad (6.12)$$

However, changes in discharge or sediment load rarely occur alone because of their joint dependence on watershed variables. Four combinations are possible:

$$Q^+, Q_{sb}^+ \rightarrow w^+, d^\pm, (w/d)^+, \lambda^+, S^-, s^\pm \quad (6.13)$$

$$Q^-, Q_{sb}^- \rightarrow w^-, d^\pm, (w/d)^-, \lambda^-, S^+, s^\pm \quad (6.14)$$

$$Q^+, Q_{sb}^- \rightarrow w^\pm, d^+, (w/d)^\pm, \lambda^\pm, S^+, s^- \quad (6.15)$$

$$Q^-, Q_{sb}+ \rightarrow w^\pm, d^-, (w/d)^\pm, \lambda^\pm, S^-, s^+ \quad (6.16)$$

Thus, the net effect of an increase in both discharge and bed-material load is to produce wider, less sinuous channels with a larger meander wavelength (equation (6.13)). Changes to depth and gradient are less clearly defined, but since the width : depth ratio increases, depth may be supposed to remain constant or decrease. The predicted responses are reversed when both discharge and bed load decrease (equation (6.14)), as might occur following reservoir development.

Changes in discharge and sediment load of opposite sign could be relatively common. Equation (6.15) shows that increasing discharge and decreasing bed load tend to result in narrower, deeper channels of greater sinuosity and lower gradient. In the case of the Colorado River in central Texas, such changes have been equated with a return to more humid conditions when streamflow becomes more evenly distributed and catchment erosion is reduced (Baker and Penteado-Orellana, 1977). With the onset of a more arid climate when aggradation follows a period of limited incision, equation (6.16) is more applicable. Although runoff declines as a whole, high-magnitude flows become increasingly dominant, transporting coarser and larger sediment loads and forming broad channels of lower sinuosity.

The relationships (6.9)–(6.16) suggest no more than possible directions of change. The equations on which they are based (Table 6.6) cover a limited range of conditions and their universality is by no means proven. The use of M as an index of both the type of sediment load and the quantity of bed load is questionable and reflects an expediency caused by the lack of suitable data. No consideration is given directly to the differential rates of adjustment of the various form elements, to possible changes in channel pattern other than through sinuosity and wavelength, or to the effects produced by different amounts of relative and absolute change in the independent variables. Regarding this last point, Starkel (1983) speculated on the likely response to different levels of change in discharge and sediment load within the specific context of channel adjustment in central European rivers during the last 15 000 years. Thus, for example, a trend towards lower runoff (Q) could result in one of two channel responses depending on the relative change in sediment load:

$$Q^- < Q_{sb}^- \quad \rightarrow \quad w^-, d^+, (w/d)^-, \lambda^-, S^+, s^- \tag{6.17}$$

$$Q^- > Q_{sb}^- \quad \rightarrow \quad w^-, d^-, (w/d)^-, \lambda^-, S^+, s^- \tag{6.18}$$

where the first is associated with incision and the second with aggradation.

Several schemes similar to Schumm's have been developed subsequently (e.g. Santos-Cayade and Simons, 1973; Li and Simons, 1982), but despite an increase in the number of variables considered, predictions continue to be based on empirical relationships with limited information on the role of sediment load. Nevertheless, the approach has been widely and successfully applied, particularly in assessing the effects of human activities on river channels. However, there is no guarantee that the same change in input conditions will produce the same channel response. Dams generally decrease the discharge and sediment load in the downstream river so that equation (6.14) ought to apply, but below 17 dams in the United States, Williams and Wolman (1984) found that some sections widened (46 per cent), some narrowed (26 per cent) and some retained a constant width (22 per cent). The implication is that direction of response is not as straightforward to predict as relationships (6.9)–(6.16) would at first suggest, and that this aspect of the prediction process is still in need of improvement.

Predicting the **magnitude of response** presents a much more difficult problem. On the face of it, equations such as (5.18)–(5.21) and those in Table 6.6 could provide a basis for estimating the new dimensions of the channel for given changes in the control variables, but their general applicability is far from certain and Hey and Thorne (1986) warn against such a step. Chang's (1985, 1986) theoretical approach based on his extremal hypothesis of minimum stream power represents one of the very few systematic attempts to address the problem. The net outcome of his analysis was a pair of regime diagrams in which sand-bed rivers of distinct morphological character are classified into four regions separated by thresholds (I–IV in Fig. 6.7). These thresholds or discontinuities are largely defined in terms of slope–discharge relationships. Chang illustrated the application of the methodology with various examples, two of which are given here. The first relates to the adjustment of a semi-arid river subject to a large flood which caused the channel to move from Region 1 to Region 3 across two thresholds, increasing its width dramatically from 15 m to 75 m but changing its depth relatively little

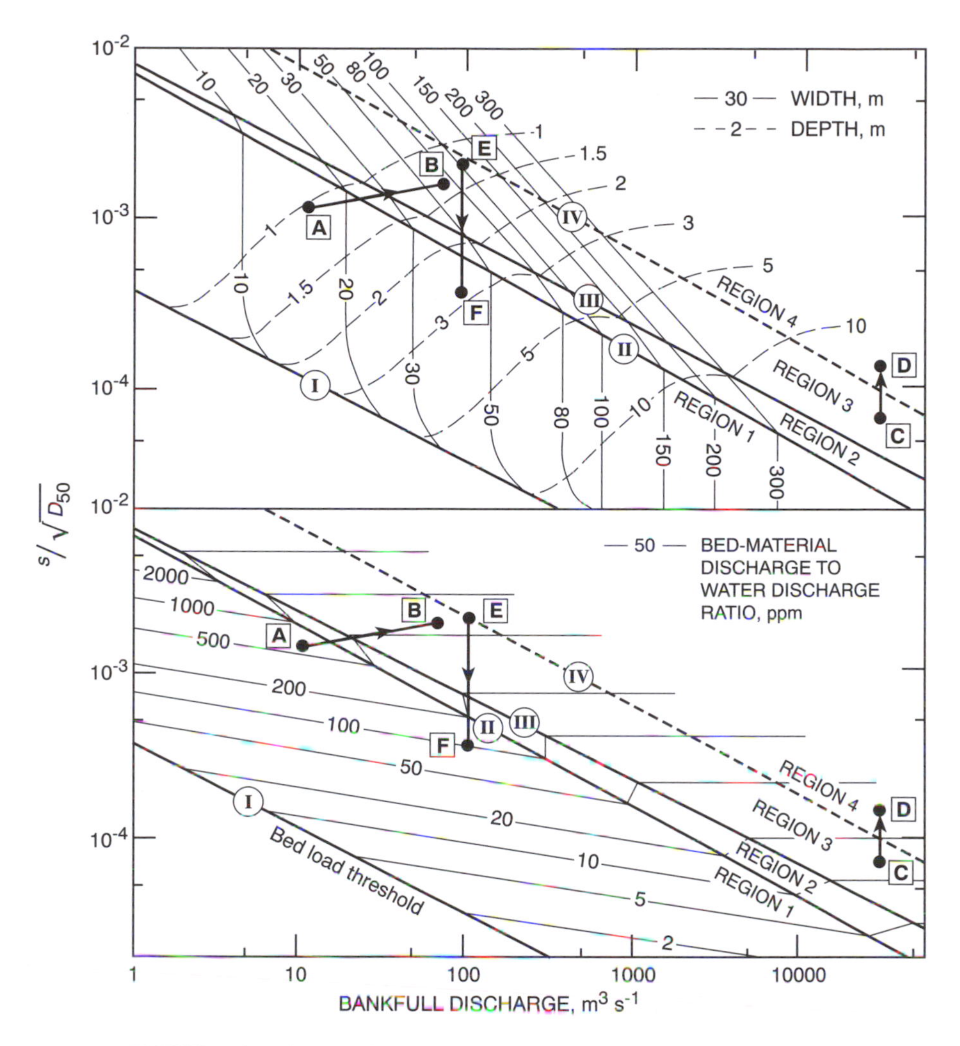

REGION 1 : low slope, relatively deep and narrow (*w*/*d* ~ 4–20), straight or mildly sinuous.

REGION 2 : flattish slope, *w*/*d* ~ 20–25, width sensitive to slope.

REGION 3 : moderately steep, *w*/*d* ~ 25–100, width sensitive to slope, sinuous on lesser slopes and braided on steeper ones.

REGION 4 : steep slopes, *w*/*d* > 100, highly braided.

Figure 6.7 Regime diagram of Chang (1986), where *s* is slope and D_{50} is median grain size. Adjustments: A→B = effect of a large flood on a semi-arid river; C→D = effect of meander cutoffs, lower Mississippi River; E→F = effect of degradation following cessation of mining, Ringarooma River, Tasmania (after Knighton, 1991).

(A → B in Fig. 6.7). The second is at a much larger scale since it concerns the lower Mississippi, where artificial cutoffs have increased channel gradient by 12 per cent overall (Winkley, 1982). Because the pre-cutoff river was in Region 3 (C in Fig. 6.7), where channel width is sensitive to slope, steepening of the channel by cutoffs (C → D in Fig. 6.7) has resulted in substantial widening and even braiding, so that an expensive levée system has had to be constructed in order to maintain the river in its new alignment.

The model predicts the magnitude of adjustment with reasonable accuracy in both of these cases, but independent tests of its validity and effectiveness are very sparse. It performed well when applied to a reach of river that was recovering from the effects of alluvial mining by degrading its bed and significantly increasing the size of its bed material (Knighton, 1991). The path of adjustment covering more than 30 years crosses two thresholds and indicates a change from a multi-thread to a single-thread pattern, with a corresponding decrease in the width : depth ratio achieved jointly by a decrease in width and an increase in depth (E → F in Fig. 6.7). On the other hand, the underlying concept of minimum stream power did not successfully predict the response of the Toutle River to the deposition of a debris avalanche, underpredicting channel widths and overpredicting channel depths (Simon and Thorne, 1996). A possible reason for this poor performance was the lack of stability in the study reaches, for Chang's approach assumes a channel in quasi-equilibrium. Whether or not this approach is ultimately vindicated, it does have predictive limitations, and there is a great deal of scope for developing methodologies that will predict the magnitude of response in a wide range of fluvial situations.

Rate of response is represented by the sum of reaction and relaxation times, which have already been shown to be highly variable and poorly predictable, even though an exponential model has been widely accepted. It is controlled by many factors, notably those which influence the inherent resistance, available energy and scale of the fluvial system. Predictability is complicated by the fact that river channels consist of various morphological components, each of which may respond to change at different rates, and by the long time-periods which may be required for complete adjustment to a new regime, particularly if the redistribution of substantial volumes of sediment is involved. Below dams, Petts (1984) has estimated that such time periods may range from 10 to more than 500 years.

Disturbance comes in many forms, ranging from the point scale (e.g. dams) to the basin scale (e.g. climatic change). Whatever the scale, effects are likely to be propagated upstream and/or downstream, maybe even reaching all parts of a drainage network, so that **extent of response** becomes a critical issue. Propagation away from the source of maximum disturbance commonly involves a change in the vertical position of the channel through the processes of aggradation and degradation, with associated modifications to channel morphology. In his stage-by-stage model of channel evolution, Simon (1995) argues that in most alluvial channels disruption generally results in a certain amount of upstream degradation and downstream aggradation, the latter following the former as degradation migrates upstream. However, the longitudinal distribution of degradation and aggradation is not necessarily so straightforward. Galay (1983) drew a distinction between downstream-progressing and upstream-progressing degradation, the first being associated with a decrease in bed material discharge or increase in water discharge, and the second with a fall in base-level. Upstream-progressing degradation generally proceeds at a much faster pace than its downstream counterpart, because the one increases and the other decreases slope as the degradation head migrates.

Initiation of the propagation process is commonly associated with the depletion or addition of bed material. Dams are a classic cause of sediment load depletion, the resultant downstream-progressing degradation usually reaching a maximum close to the dam and decreasing gradually downriver thereafter. However, some long river distances may be affected. Below the Hoover Dam on the Colorado River, degradation extended by 130 km within nine years (Galay, 1983). Large distances may be required before a river regains, by boundary erosion and tributary inputs, the same sediment load that it transported prior to dam construction. On the North Canadian River below Canton Dam, that distance may be as high as 500 km (Williams and Wolman, 1984).

The addition of sediment as 'slugs' can be attributed to a wide range of causes, with a correspondingly wide range of temporal and spatial scales of adjustment. At its simplest the downstream transmission of the increased input produces a cycle of aggradation and degradation as the slug moves through a reach, with accompanying changes to planform and cross-sectional geometry. Aggradation may involve a transition from a single-channel to a multi-channel pattern and channel widening, while degradation may result in reversion to a single, narrower channel. Typical migration rates range from 0.1–0.5 km y^{-1} for the smaller slugs to 1–5 km y^{-1} for slugs generated by the input of mining waste (Nicholas *et al.*, 1995). At an intermediate rate of 0.8–1.6 km y^{-1}, a sediment wave was propagated along Redwood Creek in California following a series of large floods and widespread timber harvesting (Madej and Ozaki, 1996). The amplitude of the wave attenuated by more than 1 m during its downstream progress, while its duration increased from eight years in upstream reaches to over 20 years closer to the mouth. Nicholas *et al.* (1995) propose a one-dimensional sediment routing procedure as a basis for modelling the propagation and attenuation of sediment slugs, but the rate, scale and extent of the propagation process in its broader context remain poorly understood. Accurate prediction of river channel adjustment in all its variety represents one of the most pressing problems in fluvial geomorphology.

The Effects of Floods

Floods generated by natural causes or dam failure represent a classic example of the pulsed type of disturbance. Their role in channel development has long been recognized but it remains controversial. Much geomorphic research has focused on channel-forming discharges within the context of equilibrium and dominant discharge concepts, promoting the belief that equilibrium channel geometry is adjusted to relatively frequent floods of moderate magnitude, but, in line with the revival of a catastrophic doctrine, there is an increasing need to evaluate the potentially destabilizing effects of extreme floods.

River channels and their associated floodplains must accommodate a wide range of flow magnitudes which can vary considerably in their frequency characteristics. Regional growth curves, in which discharge magnitude relative to the mean annual flood ($Q_{2.33}$) is plotted against return period, indicate large environmental contrasts, with monsoonal and semi-arid areas (Sri Lanka, southwestern United States) producing much larger flood magnitudes relative to $Q_{2.33}$ than do tropical ones (Congo, Guyana)

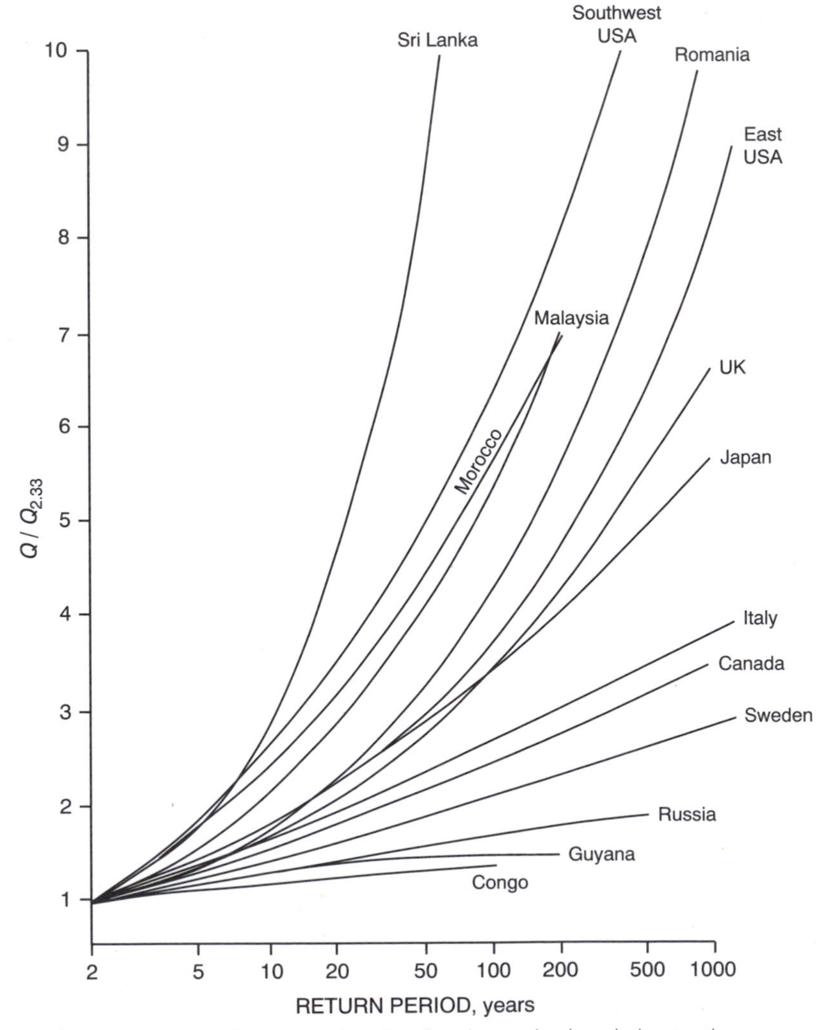

Figure 6.8 Regional growth curves, showing flood magnitude relative to the mean annual flood ($Q_{2.33}$) plotted against return period for selected countries (after Lewin, 1989).

(Fig. 6.8). In addition, relative flood magnitude tends to decrease with increasing catchment size (Lewin, 1989), even though absolute flood magnitude generally increases with distance downstream. Spatial variability in the impact of major floods can therefore be expected.

The **effectiveness** of a large flood should not be evaluated solely in terms of the amount of transportational work performed, but rather in terms of its ability to modify the fluvial landscape (Wolman and Gerson, 1978). Kochel (1988) defined a large flood as one with a recurrence interval of over 50 years, but effectiveness can vary greatly for a given recurrence interval. Catastrophic channel response to large floods probably occurs when peak velocities and depths exceed the threshold values needed for the development of macroturbulence (Kochel, 1988), with the onset of cavitation as another possible threshold (Baker and Costa, 1987). Building on the work of Baker and Costa (1987), Magilligan (1992) defined a threshold for catastrophic modification in terms of a unit stream power ($\omega = \gamma Qs/w$) of 300 W m^{-2} or shear stress ($\tau = \gamma ds$) of 100 N m^{-2}. Such values are unlikely to be attained in humid alluvial channels and probably require flow constrictions associated with localized lithologic or structural constraints. The importance of a restricted valley width can be appreciated from a comparison of the maximum stream powers attained by floods on the Amazon River and the Katherine River in Australia as it flows through a bedrock gorge, with respective values of 12 and 1000 W m^{-2} despite a contrary 50-fold difference in peak discharge (Baker and Costa, 1987). It is generally agreed that flood power varies non-linearly with distance downstream, with maximization of unit stream power or shear stress likely in moderately-sized basins of about 10^1–10^2 km^2 where slopes are steep enough and adequate flood depths can be generated.

Maximum power during a flood is not the sole determinant of effectiveness. Costa and O'Connor (1995) reported two floods of very high stream power (3300 and 2900 W m^{-2}) which, because of their short duration, made little impact on the floodplain or channel morphology. The importance of flood duration is also emphasized by Huckleberry (1994) in a comparison of two floods on the Gila River in Arizona. One flood in 1983 had minimal effect on the channel despite a larger peak discharge than a 1993 flood. The latter, because of its greater duration and total volume of flow, produced the most widening in 87 years, the width : depth ratio increasing by up to 800 per cent. Effective floods seemingly require an optimal combination of stream power, duration and total energy expenditure, but hydraulic factors are only part of the force–resistance equation (Costa and O'Connor, 1995). Effectiveness is also dependent on floodplain and channel resistance thresholds.

Channels vary considerably in their **susceptibility** to flood discharges. There is an irregular downstream pattern, related partly to the longitudinal distribution of unit stream power and partly to localized effects which increase flood depths and promote steep gradients (Magilligan, 1992). Relatively small basins of high relief tend to have flashy flow regimes, one of the main factors influencing flood impact (Kochel, 1988). Flashiness of regime gives rise to high values of Q_p/Q_m when floods are more likely to produce significant geomorphic change, and is more commonly encountered in semi-arid and arid regions where vegetation is also less dense and floodplains are therefore less resistant. Consequently streams in those regions can be particularly susceptible to large floods. Wolman and Gerson (1978) compared a change from humid to arid conditions with a decrease in basin area in a humid region as regards the potential for catastrophic response.

Influential factors having less dependence on climate include steep channel gradient, abundant coarse bed load, low bank cohesion, and channel geometry (Kochel, 1988). Bedrock channels often have deep, narrow cross-sections which encourage the high depths and velocities associated with macroturbulent flow. Indeed, because of their inherent resistance, such channels may change little except during rare catastrophic events, and yet they can develop bed configurations bearing a striking resemblance to the riffle–pool morphology more commonly associated with alluvial rivers (Baker and Pickup, 1987; Wohl, 1992). Another influential factor is the temporal sequence of events. Kochel (1988) exemplified its importance with the Pecos River in Texas, which was subject to extreme floods in 1954 and 1974, with estimated recurrence intervals of 2000 and 500–800 years respectively. The first resulted in dramatic erosion and redistribution of channel gravel, whereas the second produced only minor change. Apparently insufficient time had elapsed between the two flows for channel recovery in this arid area, so that the 1974 channel morphology was still largely adjusted to the high flow of 1954. The complex interaction of factors controlling susceptibility to large floods results in a wide variation in the scale and character of channel response.

The variation in **channel response** can be represented diagrammatically (Fig. 6.9A): from no response at one extreme to transient behaviour at the other (TF > 1, see p. 289), with various intermediate states involving different styles of recovery and non-recovery. The frequencies of these several responses cannot be determined with any certainty, but humid alluvial channels often show little change or recover rapidly (States 1 and 2). Localized channel widening occurred in response to floods having recurrence intervals exceeding 200 years in the Piedmont area of the United States, but the effects were largely temporary and in one case channel widths had almost recovered their pre-flood values within one year (Costa, 1974; Gupta and Fox, 1974).

One of the best examples of State 3 is Burkham's (1972) study of channel change in part of the Gila River, Arizona, over the period 1846–1970. Prior to 1905 the river was sinuous, relatively stable and narrow, with an average width of 90 m. Between 1905 and 1917 (possibly the wettest period since 1650) a series of large winter floods carrying low sediment loads destroyed the floodplain and widened the channel to about 600 m (Fig. 6.9B), while channel sinuosity decreased and gradient increased. From 1918 onwards the floodplain was reconstructed by smaller floods carrying large loads, so that by 1964 the stream channel had developed a meandering pattern and narrowed to an average width of 60 m, giving a relaxation time of 45–50 years.

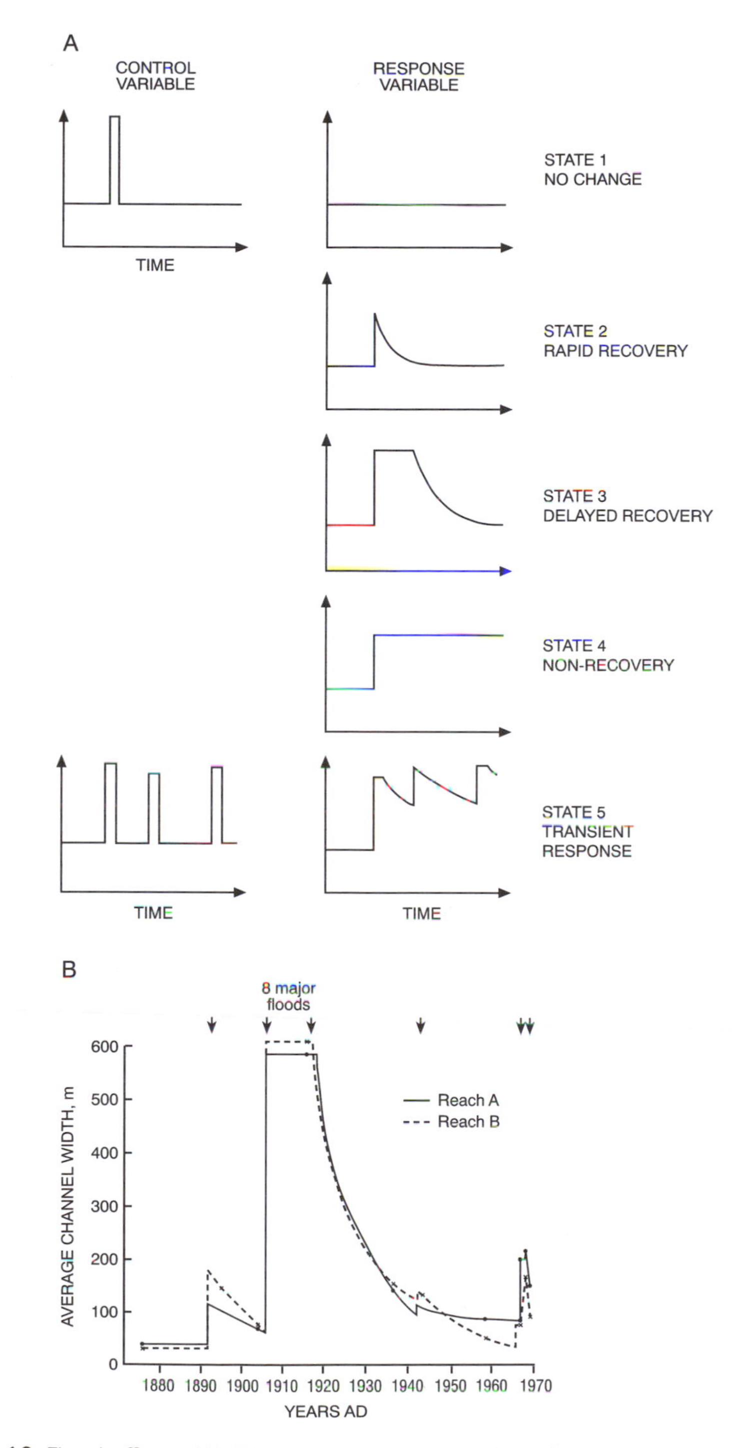

Figure 6.9 Flood effects. (A) Diagrammatic representation of potential channel response to large floods. (B) Changes in channel width, Gila River, Arizona (after Burkham, 1972). Arrows indicate major floods.

Also in a semi-arid environment, width increased greatly from 15 m to 365 m along the Cimarron River in Kansas after major floods had followed a series of dry years when riparian vegetation and bank resistance were much reduced (Schumm and Lichty, 1963). The river width remained large until a series of wet years enabled the vegetation to become re-established. These and other examples indicate that the main changes induced by major floods are the widening, straightening and steepening of channels, although such factors as antecedent conditions and valley width can modify the effects. Along Wollombi Brook in New South Wales the amount of change varied in direct proportion to the degree of channel confinement (Erskine, 1996).

The conditions that generate major floods may also trigger slope instability and lead to the input of large quantities of sediment. Consequently the discharge : load ratio of floods may be expected to vary over a wide range, with variable effects on the channel. Where input amounts are relatively large, effects are likely to persist for longer. A 1951 flood produced hillslope erosion in the upper basin of the Kowai River in New Zealand, the resultant increase in sediment supply causing aggradation and channel widening which progressed downstream at an average rate of 1 km y^{-1} over 30 years (Beschta, 1983). A series of large floods combined with widespread timber harvesting produced similar effects in Redwood Creek, with width increases of over 100 per cent and gravel berms deposited by flood flows persisting because subsequent flows were unable to erode them (Nolan and Marron, 1995). The Toutle River was completely transformed from a low-sinuosity gravel-bed stream into a straighter, braided sand-bed stream following the eruption of Mount St Helens, which introduced enormous sediment loads via a debris avalanche (Simon and Thorne, 1996), a disturbance representing one extreme of the discharge/load scale and exemplifying State 4 (or State 3 if recovery eventually occurs). Channel widening occurred at rates as high as 200 m y^{-1}, although most sites widened between 10 and 20 m y^{-1} (Simon, 1992). In the Toutle River system, channel widening in conjunction with degradation and bed-material coarsening represents an important mode of energy dissipation as the river adjusts to the huge influx of material.

Extreme floods may represent the only occasions when large boulders are mobilized in river systems. In the Herbert Gorge of northeastern Australia, boulders up to 5 m in diameter form bars which alternate with pools in a quasi-periodic way along straight reaches (Wohl, 1992). On the basis of slackwater deposit reconstructions, it has been estimated that flows with a magnitude sufficient to mobilize substantial parts of these boulder bars have occurred only six times over the last 900 years. The entire morphological character of the valley of Coffee Creek, a small upland catchment in California, was transformed by a major flood which transported boulders up to 2 m in diameter (Stewart and LaMarche, 1967). Natural levées were formed, attaining widths of 15 m and constraining future channel migration during lesser flows. The emplacement of large quantities of very coarse debris which subsequent floods have not the capacity or competence to transport is one of the primary means by which a river may fail to recover its original morphology and thus move to a new state (State 4 in Fig. 6.9A). At the extreme end of the scale, cataclysmic releases from Pleistocene Lake Missoula produced exceptionally large floods with unit stream powers of up to 100 000 W m^{-2}, completely transforming the fluvial landscape of the Columbia River (Benito, 1997).

In those cases where **recovery** does take place, the length of time involved can be very variable, depending upon the initial amount and direction of displacement, the subsequent flow regime, the supply of sufficient sediment for restoration purposes, and climatic conditions, especially as they relate to vegetation. Burkham (1972) drew a distinction between destructive and constructive flows in the history of the Gila River, the first being associated with large floods carrying small sediment loads and the second with more moderate floods having high sediment concentrations. As regards the climatic element, alluvial rivers in humid regions tend to recover quickly from the changes produced by major floods, whereas the slow rate of vegetative regeneration in more arid areas inhibits rapid recovery (Wolman and Gerson, 1978). However, even in humid catchments recovery can be long delayed where vast amounts of sediment have been introduced. Such is the case in Redwood Creek, where, even without additional inputs of sediment, at least 30 years will be required to transport the sediment introduced by the 1964 storm, with recovery taking longest in the downstream reaches (Nolan and Marron, 1995; Madej, 1995).

With widening as one of the principal impacts of floods, recovery often takes the form of floodplain reconstruction and channel narrowing. Along the Gila River five processes were involved: (a) the development of islands within the main channel and their subsequent attachment to one bank as a result of infilling of the intervening channel; (b) direct deposition against the banks of the main channel; (c) the formation of natural levées; (d) direct deposition on the floodplain; and (e) deposition on alluvial fans at the mouths of tributary streams (Burkham, 1972). All require an adequate sediment supply and are greatly aided by the growth of vegetation, which retards the flow and stabilizes deposits. Friedman *et al.* (1996) emphasize both the role of vegetation and bed-level fluctuation in the narrowing process. During periods of high flow, sand and fine gravel are delivered to downstream reaches of Plum Creek, where a 1965 flood widened the channel by an average of 160 per cent (Osterkamp and Costa, 1987). This material temporarily raises the bed level, into which subsequent flows incise a narrower channel whose upper surfaces become stabilized through vegetative colonization. Recovery, in so far as it implies the re-establishment of a pre-flood morphology, can take a variety of forms, including the return to a former bed elevation and the redevelopment of a riffle–pool bed, but each may take a different length of time to become re-established.

The overall extent to which channel morphology reflects the influence of extreme events cannot readily be assessed, but certain rivers do seem to have long memories for past floods. Bedrock gorges in arid environments are particularly distinctive in preserving the effects of rare, catastrophic events (Baker and Pickup, 1987). Where the time between major floods is shorter than the recovery time, the channel is likely to have a flood-dominated morphology and exhibit transient response characteristics (State 5 in Fig. 6.9A). Controversy remains as to the role of large, infrequent floods relative to the cumulative effects of more frequent floods of lower (near-bankfull) magnitude in controlling channel form, but the fact that good relationships can be established between form and flow variables suggests that most channels, and especially alluvial ones in humid environments, are not simply the product of the last major flood. Significant channel modification can alter the downstream transmission and magnitude–frequency characteristics of subsequent flood flows, with the consequence that channel morphology may change from a state of dependency on the flow to one of independence.

Fluvial Response to Climatic Change

Current concern with the potential impact of global warming provides one reason why interest in the reconstruction of past fluvial environments has been reawakened over the last two decades. Climate, expressed in terms of temperature and precipitation, together with the associated vegetation cover exerts a major control on the flow regime of rivers as well as the type and volume of sediment load. On the basis of curves assembled by Schumm (1968) in which mean annual runoff and sediment yield are separately related to mean annual precipitation, the possible effects of climatic change can be estimated (Table 6.7). Clearly the initial climate is an important influence on the direction of change. In a temperate climate such as Britain's, a shift to cooler and wetter conditions would lead to an increase in runoff and a decrease in sediment yield, but if the climate was originally semi-arid, a similar change would produce increases in both variables (although a denser vegetation cover may eventually become effective in reducing erosion). Because the sediment yield curve of Langbein and Schumm (1958) is at its steepest for precipitation amounts characteristic of semi-arid environments (see Fig. 3.9A, p. 91), those areas are likely to be particularly sensitive to repeated fluctuations in climate, and even small changes might induce a marked response.

This methodology indicates only broad-scale directions of change and, being based on United States data, may have limited overall applicability. In addition, mean annual precipitation and mean annual runoff have limited value as indices of climate and flow regime respectively. Given the sensitivity of alluvial channels, significant modifications to fluvial system behaviour can occur in response to changes not only in the mean climate but also in the magnitude–frequency characteristics of events under a constant mean. Indeed, relatively modest climatic changes have been held responsible for major episodes of fluvial adjustment during the Holocene (Knox, 1993, 1995). Fluvial chronologies and palaeohydrologic reconstructions also indicate that rivers can respond very rapidly to changes in climatic conditions.

Table 6.7 Possible effects of climatic change on mean annual runoff and sediment yield (+ denotes an increase, – a decrease and 0 no change)

Original climate	New climate			
	Cooler (T_m –5°C), wetter (P_m + 250 mm)	**Warmer (T_m +2.5°C), wetter (P_m + 250 mm)**	**Cooler (T_m –5°C), drier (P_m – 125 mm)**	**Warmer (T_m +2.5°C), drier (P_m – 125 mm)**
Temperate:				
T_m = 10°C	R_u^+	R_u^+	R_u^0	R_u^-
P_m = 750 mm	S_y^-	S_y^- or S_y^0	S_y^0	S_y^+
Sub-humid:				
T_m = 12.5°C	R_u^+	R_u^+	R_u^0	R_u^-
P_m = 500 mm	S_y^-	S_y^-	S_y^0	S_y^0
Semi-arid:				
T_m = 15°C	R_u^+	R_u^+	R_u^-	R_u^-
P_m = 350 mm	S_y^+	S_y^+	S_y^0	S_y^-

Symbols: T_m, mean annual temperature; P_m, mean annual precipitation; R_u, mean annual runoff; S_y, mean annual sediment yield

In line with temperature variation (Fig. 6.2), fluvial response to climatic fluctuations can be considered at various timescales. At the timescale of the last cold stage (approximately 115 to 10 ka BP), general global cooling was occasionally interrupted by short-lived warm episodes. On the basis of detailed interpretations of fluvial sediments and surface stratigraphy in the Riverine Plain of southeastern Australia (Fig. 6.1E), Page and Nanson (1996) recognized four distinct phases of palaeochannel activity between 105 and 12 ka BP, with relatively arid intervening periods. The first three phases were characterized by mixed-load, laterally migrating sinuous palaeochannels, with occasional transitions to a straighter bed-load-dominated mode. Termed 'aggradational palaeochannels', these channel systems typically terminated with a bed-load-dominated episode of vertical aggradation and reduced sinuosity in response to a change in the ratio of sediment to water delivered from the confined upstream valley. In the final phase following the Last Glacial Maximum (about 18 ka BP), mixed-load sinuous 'migrational palaeochannels' developed, but with no terminating bed-load episode. By then the approach of Holocene climatic conditions had reduced the size of flood peaks and greatly diminished the supply of bed load from the upper catchment, resulting in the establishment of highly sinuous, slowly migrating, suspended-load rivers that have continued with only slight modification until the present.

This model replaces an earlier and simpler version devised by Schumm (1968) in which bed-load 'prior streams' evolved firstly into suspended-load 'ancestral rivers' and then into the present Murrumbidgee system. Nevertheless, estimates of previous bankfull discharges based on reconstructed cross-sections are consistent with Schumm's, exceeding present values by a factor of 4 to 8 (Table 6.8). Climatic change is clearly implicated in the higher discharges and loads of the four distinct phases during the last glacial cycle, but the within-phase shifts from dominantly migrational to dominantly aggradational probably did not require major continental-scale climatic forcing. Page and Nanson (1996) suspect that shifts to bed-load aggrading episodes were initiated by large floods or sequences of floods which mobilized large quantities of coarse sediment in the upper valley. There are thus different timescales of channel adjustment represented on the Riverine Plain, within and between the phases themselves, and between phases and the intervening periods. During the phases channel widths increased by a factor of 2 to 3 in response to the higher bed loads, while modifications to channel gradient were relatively modest. In an area where the valley slope is gentle, gradient adjustments could be accommodated almost entirely by changes in channel sinuosity (Schumm, 1968).

A timescale of 10^4 years covers the post-glacial period. Some of the most spectacular transformations in the river channels of western Europe have occurred in the past 14 ka as a result of the changes accompanying deglaciation. Braided channels that characterized extra-glacial areas at the close of the last cold stage gave way to meandering morphologies as sediment loads decreased and the vegetation cover expanded, with temporary reversion to braided forms during the colder episode of the Younger Dryas (11–10 ka BP). This transition has been identified in both upland (Maizels and Aitken, 1991) and lowland (Brown, 1991) Britain, where the high runoffs associated with deglaciation represented periods of maximum instability in fluvial systems. Subsequent change has been much more muted, partly because of the constraints imposed at that time. Indeed, the lower Severn has changed little since then, unable to cross critical erosional thresholds (Brown, 1991).

Table 6.8 Estimates of palaeochannel characteristics from reconstructed cross-sections, Riverine Plain, New South Wales (after Page and Nanson, 1996)

Palaeochannel system	Date, ka BP	Slope, m m^{-1}	Width, m	Bankfull discharge, $m^3 s^{-1}$	Palaeo-discharge ratio
Phase 1	105–80	0.00026	175	1620	5.2
Phase 2	55–35	0.00026	220	2610	8.3
Phase 3	35–25	0.00018	215	1220	3.9
Phase 4	20–13	0.00010	250	1240	4.5

Note: The palaeo-discharge ratio is the ratio of the estimated palaeochannel discharge to the present bankfull discharge of the Murrumbidgee River.

The period from 14 to 9 ka BP is known not only for a general braided–meandering transition in temperate-latitude rivers but also for the widespread development of large relict meanders. Dury (1964) identified two types of underfit river: where modern meanders have a wavelength much less than that of the valley within which they are confined, termed 'manifestly underfit'; and where relatively straight channels in wider meandering valleys have a riffle–pool spacing that does not correspond to the valley meander wavelength, termed 'Osage-type underfit'. Although several causal mechanisms have been proposed, the widespread distribution of underfit rivers suggests climatic change as the main reason for channel shrinkage. Discharges representative of these relict meanders have been estimated at five to ten times those in modern rivers (Rotnicki, 1991).

The early part of the Holocene saw a marked reduction in the discharge of most temperate-zone rivers (Fig. 6.3). Nevertheless, fluvial activity has continued to vary during the Holocene. Macklin and Lewin (1993) reviewed the stratigraphic records from many sites in Britain and identified major phases of valley alluviation: 9600–8400 (lowland Britain), 4800–4200 (countrywide), 3800–3300 (largely southern Britain), 2800–2400 (largely southern Britain), 2000–1600 (countrywide), 1200–800 (countrywide), 800–400 (southern Britain), and 400–0 years BP (upland Britain), the last of which coincides with climatic deterioration during the Little Ice Age. The rarity of dated alluvial units between 8000 and 5200 years BP implies a period of channel incision, slow alluviation or stability. The good correspondence between these alluvial units and many postulated episodes of climatic change, and their approximate synchroneity with similar events in North America and Europe, suggest a climatic control of fluvial activity, despite the increasing influence of anthropogenic disturbance. Macklin and Lewin (1993) concluded that, although forest clearance and agricultural practices were undoubtedly important in the initiation of erosion, major redistribution of this material and widespread alluviation tended to occur only during relatively short periods of abrupt climatic change when the incidence of flooding was significantly different.

As pointed out in the previous section, flood discharges can play an important role in the movement and storage of basin sediment, and in the evolution of river channels and their adjoining floodplains. The general synchroneity of prominent discontinuities in Holocene alluvial chronologies suggests that major shifts in the magnitude–frequency characteristics of floods have occurred during the last 9000 years as a result of periodic changes in large-scale circulation patterns (Knox, 1995). Usually, more frequent large floods occur in the middle latitudes during cooler periods when a stronger meridional circulation allows deeper penetration of air masses. On the basis of an analysis of overbank deposits, Knox (1993) reconstructed a 7000-year history of flooding in the upper Mississippi basin (Fig. 6.10). Following a relatively warm and dry period with low flood magnitudes (5000–3300 years BP), an abrupt shift to wetter and cooler conditions increased the incidence of large floods of a size that now recur only once

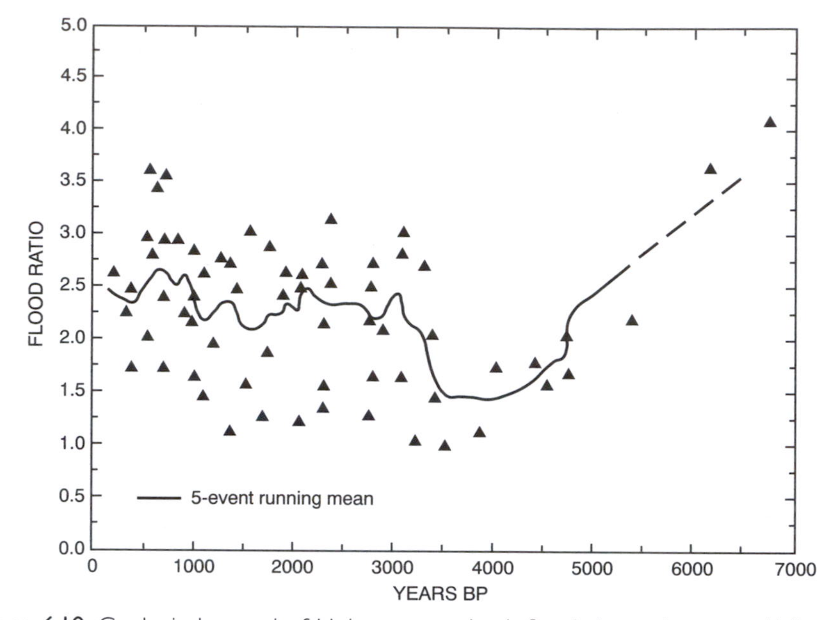

Figure 6.10 Geological record of Holocene overbank floods in southwestern Wisconsin. A flood ratio of 2 corresponds to modern floods expected about once every 30–50 years, whereas a flood ratio of 3 relates to modern floods expected about once every 500–1000 years in southwestern Wisconsin (after Knox, 1993).

every 500 years or more. Still larger floods occurred between AD 1250 and 1450 during the transition from the warm Medieval interval to the cooler Little Ice Age, a period also of anomalous wetness and large-scale flooding in many parts of northern and central Europe (Lamb, 1977). Significantly, such changes were apparently produced by relatively modest fluctuations in climate, of only 1–2°C in mean annual temperature and ≤ 10–20 per cent in mean annual precipitation.

In a small Pennine basin in northern England, 21 large flood events have been identified from flood deposits, all but one of which date from the mid-eighteenth century when the Little Ice Age was at its height (Macklin *et al.*, 1992). Over the last 1600 years rates of valley-floor incision have periodically accelerated, notably during late Roman times and in the eighteenth century when runoff and flood magnitude increased as a result of wetter and cooler conditions, with human interference as an additional factor. The more recent phase saw the replacement of a relatively stable meandering stream and a floodplain accreting fine-grained sediment by a vertically and laterally unstable, low-sinuosity stream carrying a large coarse sediment load. As much as 4 m of incision occurred, locally through bedrock. Possibly starting as early as the late seventeenth century, the incision continues to the present day but on a more limited scale.

Focusing on the period since AD 1700, Rumsby and Macklin (1994) have broadened the scope of this work and identified alternating phases of river-bed incision and stability in the Tyne basin, related to high-frequency (10–30 year) fluctuations in climate and flood regime. Episodes of widespread incision (1760–1799, 1875–1894, 1955–1969) result from a higher frequency of large floods (>20 year return period), while phases of more moderate floods (5–20 year return period) are characterized by enhanced reworking of sediments in the upper catchment and channel narrowing downstream. Short-term variation in flow regime is also a characteristic of coastal rivers in New South Wales, with a periodicity of 30–50 years (Warner, 1987, 1994). Since average climate has varied by only modest amounts in the Tyne basin since 1700 (± 0.5°C in mean annual temperature, ± 40 per cent in mean annual rainfall), the implication is clear: short-term and relatively small-scale fluctuations in climate can, through alteration of the flood regime, produce significant changes to the fluvial system. The scale of those fluctuations is within the range of global warming scenarios.

Channel Change and Human Activity

Over the last 2000 years, and especially the last 300 years, human activities have had an increasing influence on drainage basins and their constituent channels. The scale of climatic fluctuation has been much less during that period (Fig. 6.2), but human modification of the physical environment may have induced changes similar in scale to those produced by climatic change in the more distant past. Indeed, one of the main incentives for taking a longer-term view of stream channel response has been to provide a framework for estimating the effects of human activities, since they operate through the same basic mechanisms as climatic change. If dangerous or expensive outcomes are to be avoided, the consequences of interference need to be understood.

Two broad types of human-induced change can be identified (Table 6.9). The first includes those changes brought about by direct modification to the channel itself, which often takes the form of engineering works designed to alleviate the effects of flooding, erosion or deposition. Indirect changes result from activity in extra-channel areas, which modifies the discharge and/or sediment load of the stream and ultimately results in stream channel response. With both types, effects can be transmitted long distances away from the initial change.

River Regulation

Dams have been constructed for more than 5000 years, but the pace of construction has quickened dramatically in the last five decades and more than 200 large dams are now completed each year (Gregory, 1995). In North America and Africa over 20 per cent of total stream runoff is regulated, with a comparable figure for Europe and Asia of 15 per cent. The impact of a particular scheme on the fluvial environment will vary with the size and

Table 6.9 Types of change induced by human activity

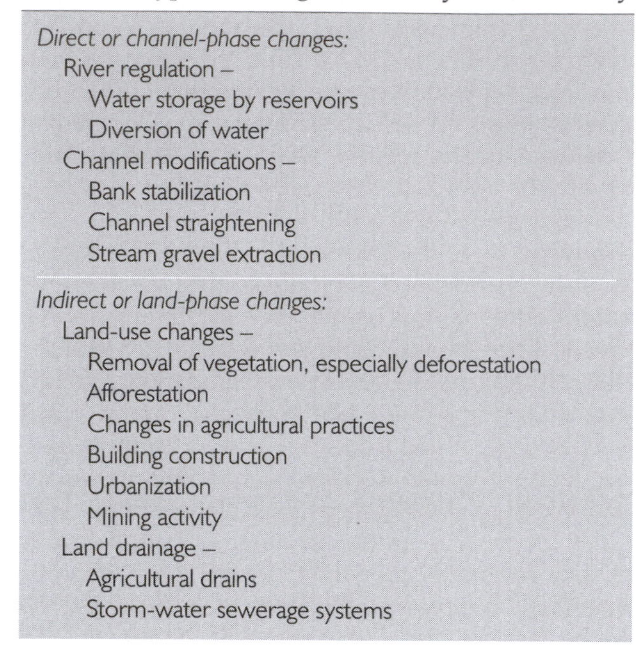

Direct or channel-phase changes:

- River regulation –
  - Water storage by reservoirs
  - Diversion of water
- Channel modifications –
  - Bank stabilization
  - Channel straightening
  - Stream gravel extraction

Indirect or land-phase changes:

- Land-use changes –
  - Removal of vegetation, especially deforestation
  - Afforestation
  - Changes in agricultural practices
  - Building construction
  - Urbanization
  - Mining activity
- Land drainage –
  - Agricultural drains
  - Storm-water sewerage systems

purpose of the dam and its physical setting, but in general both upstream and downstream adjustments can be expected.

Upstream, local base-level is raised to a position at which the water surface intersects the original bed, the maximum rise in height being determined by the crest of the dam spillway. With the reduction in transporting ability as a stream enters a reservoir, a depositional wedge is constructed and the channel gradient locally lowered. Although aggradation seems to take place rapidly at first, its upstream extent may be limited or long delayed (Leopold and Bull, 1979). Nevertheless, Van Haveren *et al.* (1987) observed the upstream progression of a depositional wedge over 25 years, by which time it had extended 1.5 km upstream and the channel gradient had reached 83 per cent of its original value. Upstream effects are rather variable, depending on sediment supply characteristics and the height of the dam relative to the pre-existing profile of the stream, but tend to be relatively insignificant.

Most work has been concerned with the downstream impacts of dam construction, of which two are widely reported: a reduction in the magnitude of flood peaks by as much as 90 per cent (Williams and Wolman, 1984); and a marked decrease in the sediment load, especially in those reaches immediately below the dam. Reservoirs are effective sediment traps, with commonly more than 90 per cent of the total load and all of the coarser fraction being retained. Depending on the amount of sediment supplied from channel erosion and tributary contributions, the downstream river may require a long channel distance to recover its load to pre-dam levels and in some cases may never do so. Since the closure of the Hoover Dam on the

Colorado River in 1935, the average annual suspended load delivered to the mouth has declined from about 130 million tonnes to under 100 000 tonnes (Meade and Parker, 1985; Fig. 6.11A). Thus, the downstream patterns of erosion, transport and deposition can be significantly altered by dam construction, with consequences for stream channel response, the particular form of which will depend on the relative changes in flow regime and sediment load. In line with the approach developed by Schumm (1969), various possibilities can be suggested (Table 6.10).

A common response to the release of sediment-free water below dams is degradation of the channel bed, typically at rates much higher than in natural rivers. Since degradation is usually at a maximum close to the dam and progressively declines in intensity downstream, channel slope tends to become flatter, although the amount of degradation can vary considerably along a river. In a study of 21 US dams, the rate at which degradation migrated downstream ranged from very little to as much as 42 km y^{-1}, varying with the pattern of flow releases and bed material characteristics (Williams and Wolman, 1984). Initial degradation rates tended to be quite high, with half the final degradation being accomplished within the first seven years, but complete adjustment could take a century or more to achieve. Somewhat exceptionally, degradation extended 130 km in nine years below the Hoover Dam, reaching a maximum of 7.1 m at a downstream distance of 12.4 km (Galay, 1983). Downstream-progressing degradation along a main river can induce upstream-progressing degradation within tributaries (Germanoski and Ritter, 1988), thereby increasing sediment supply and slowing the process.

Degradation rarely proceeds unhindered. Bedrock outcrops or the exposure of more resistant layers can prevent or slow the rate of bed erosion. In the sand-bed Hanjiang River of China the rate of downcutting slowed dramatically after 1971 when a buried gravel layer was exposed, the bed remaining relatively stable thereafter (Xu Jiongxin, 1996a; Fig. 6.11B). The stream bed may be composed of material which the regulated river is unable to move under the new regime, as is the case below the Parangana Dam in Tasmania where only one discharge has had the competence to

Table 6.10 Channel form adjustment below dams (partly after Petts, 1984)

Channel variable	$\Delta Q < \Delta L$		$\Delta Q > \Delta L$	
	$Q^{o}L^{-}$	$Q^{-}L^{--}$	$Q^{--}L^{-}$	$Q^{--}L^{o}$
Bankfull cross-sectional area	+	–	–	–
Channel width	o/+	–	–	–
Channel depth	+	o/+/–	o/–	–
Channel slope	–	–	+/–/o	+
Channel roughness	+	+	+/–	–

Symbols: ΔQ, ΔL = change in discharge (Q) or sediment load (L); +, o, – = increase, no change, decrease.

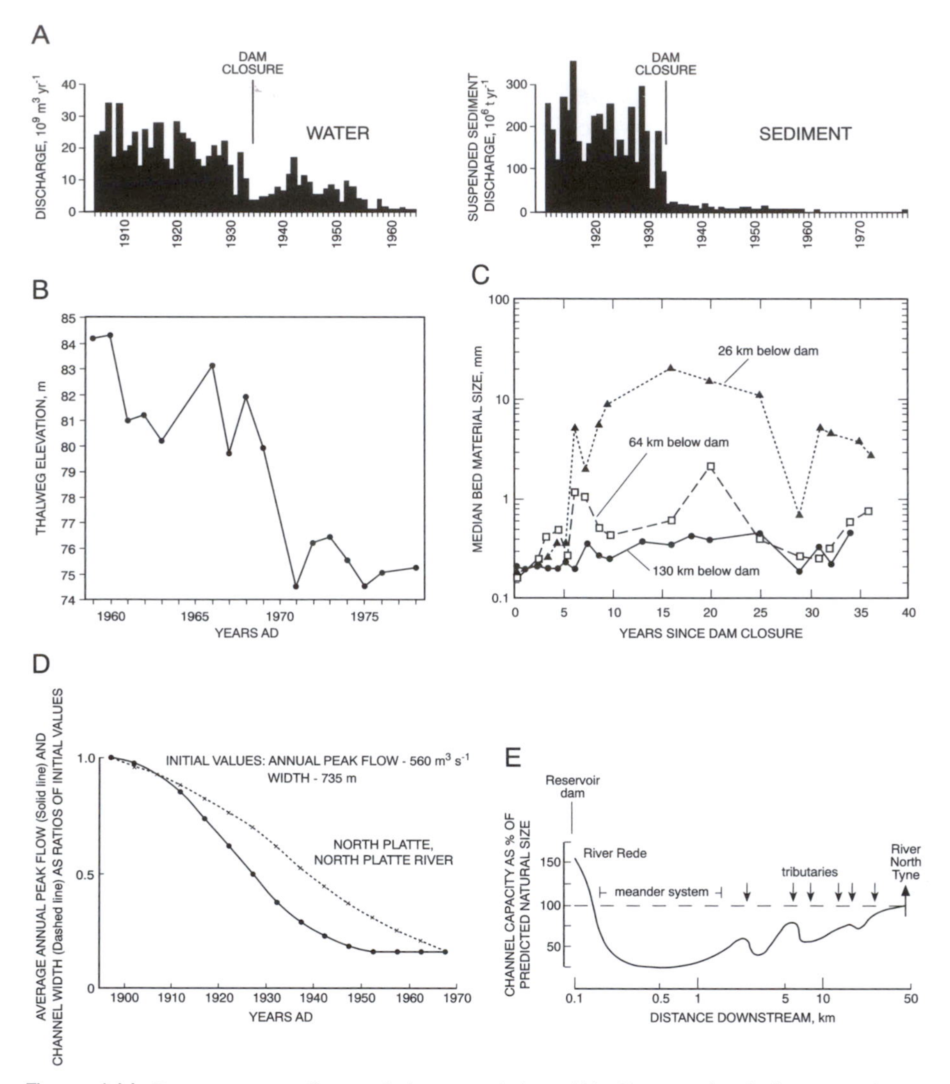

Figure 6.11 Downstream effects of river regulation. (A) Changes in discharge and suspended sediment load below Hoover Dam, Colorado River (after Meade and Parker, 1985). (B) Changes in channel bed height below Danjiangkou Reservoir, Hanjiang River, China (Xu Jiongxin, 1996a). (C) Variation in bed material size at different distances below Parker Dam, Colorado River (after Williams and Wolman, 1984). (D) Changes in annual peak discharge and channel width, North Platte River, Nebraska (after Williams, 1978c). (E) Fluctuations in channel capacity downstream of Catcleugh Reservoir, Northumberland, England (after Petts, 1979).

mobilize the coarse bed material in the post-dam period (Knighton, 1988). Where appreciable degradation does occur, it may not continue once the slope has been decreased or the channel roughness increased sufficiently to reduce local hydraulic conditions below some critical erosion threshold. If non-transportable particles are present in the bed material, selective winnowing of the smaller sizes can lead to the development of a coarse armour layer which protects the underlying material from further erosion (Fig. 6.11C). Thus external factors or internal adjustments can limit the rate, depth and extent of the highly variable degradation process.

According to Table 6.10, a primary morphological response to the change in flow regime is a decrease in channel capacity, brought about principally by a reduction in channel width. Losses of channel capacity in excess of 50 per cent are not uncommon (Petts, 1979). In the Platte River system, where peak and mean annual discharges have declined to 10–30 per cent of their pre-dam values, channel widths have decreased by equivalent amounts over 40–60 years (Williams, 1978c), the narrowing process tending to lag behind the reduction in flow by up to 15 years (Fig. 6.11D). Associated with the narrowing, the rivers have tended to become less braided and slightly more sinuous. Since width reduction is largely achieved through the formation of depositional berms and bars which require the introduction and redistribution of sediment, a favoured location for channel change is below tributary junctions where sediment is imported by non-regulated tributaries (Fig. 6.11E).

Morphological response may not, however, be so straightforward. In their comprehensive study, Williams and Wolman (1984) found a variety of width adjustments, including decreases (26 per cent), increases (46 per cent) and no change (22 per cent). On the basis of field and laboratory observations relating to Chinese rivers, Xu Jiongxin (1990, 1996b) identified different stages in the adjustment process, an initial decrease in width being followed by an increase as the bed armours and becomes relatively more resistant. Whether narrowing or widening occurs depends in part on the relative erodibility of the channel bed and banks. In situations where sediment supply from tributaries is absent and the regulated river is no longer competent to entrain and redistribute boundary sediment, the channel remains unaltered and the only adjustment occurs to the water cross-section, an extreme form of response termed 'accommodation adjustment' by Petts (1979, 1984).

Such is the diversity of release policy and catchment physiography that a general predictive model of channel response to river regulation is very difficult to develop. To all intents and purposes each reservoir–river system is unique. The direction and magnitude of channel change depend on the extent to which the flow regime and sediment load have been modified, and on the resistance properties of the boundary material in the downstream channel. The effects of river regulation tend to diminish with distance downstream as non-regulated tributaries make an increasing contribution to the flow (Fig. 6.11E), but they can persist for considerable distances. Large dams along the Missouri and Mississippi rivers have effectively reduced the annual sediment yield from 817 million tonnes to 204 million tonnes (Keown *et al.*, 1986), this loss of river sediment being a major factor in the rapid rate of shoreline recession in coastal Louisiana. Erosion of the Nile delta has been attributed in part to the construction of the Aswan High Dam nearly 1000 km upstream (Kashef, 1981). River regulation has consequences not only for the geomorphology but also for the ecology of the downstream river, and in many instances, such as the Nile (Said, 1993), river fisheries have been severely degraded. Since many dams are relatively new and since the timescales for adjustment can be very long (Petts, 1984; Williams and Wolman, 1984; Church, 1995), it may be a century or more before the effects of the twentieth-century explosion in dam construction are fully realized, especially in reaches further downstream.

River Channelization

Whereas dam construction is essentially a point type of disturbance, channelization often extends over long stretches of river. River channelization is a general term covering various forms of channel engineering used for the purposes of flood control, drainage improvement, maintenance of navigation and erosion prevention (Table 6.11). Flood control is of particular importance and may involve most of the measures listed in Table 6.11, the principal aim being to confine floodwaters within the channel and reduce or eliminate overbank flow. Channelization has a long history but has increased significantly in extent during the twentieth century. More than 8500 km of major works were undertaken in England and Wales between 1930 and 1980, with a figure for the United States of over 26 500 km for a similar period (Brookes, 1985).

Rivers respond to channelization both within and beyond the modified reach. Morphological adjustments within channelized reaches have perhaps been most dramatic where channels have been straightened (Brookes, 1985). Straightening a meandering stream increases channel gradient by providing a shorter path. Flow velocity and transport capacity are increased, leading to degradation which progresses upstream as a knickpoint. A greater load is then supplied to the downstream part of the channelized reach, which, hav-

Table 6.11 Methods of river channelization

Method	Description
Straightening	River is shortened by artificial cutoffs, thereby steepening the gradient and increasing flow velocity
Resectioning	Widening and/or deepening of the river channel to increase its conveyance capacity and thereby reduce the incidence of overbank flooding
Levée construction	Channel banks are artificially raised to confine floodwaters
Bank protection	Use of structures such as gabions and steel piles to control bank erosion
Clearing and snagging	Removal of obstructions from the watercourse, thereby decreasing resistance and increasing flow velocity

ing a flatter slope, may not be able to transport it, the excess being deposited in gradually decreasing quantities with distance downstream. Degradation within the straightened reach may also cause bank collapse, thereby increasing sediment supply still further. Thus, the modified stream attempts to establish a new equilibrium gradient through a combination of upstream-progressing degradation and downstream aggradation.

During the 1930s and 1940s the lower Mississippi was shortened by 210 km through selective cutoffs, the channel gradient increasing by 12 per cent overall but locally by as much as 2000 per cent (Winkley, 1982). The net effect has been increasing instability within the river, the steeper slopes inducing greater bed-material movement and the development of migrating bars. The channel has tended to widen and begun to braid, adjustments predicted by Chang's (1986) theoretical model based on the extremal hypothesis of stream power minimization (Fig. 6.7, C → D). Costly dredging and revetment work has been required to maintain navigation depths, which are now less than they were 90 years ago, a problem that will persist into the twenty-first century. As part of this cutoff programme the lower course of a tributary river, the Homochitto River, was realigned and severely shortened. The steeper gradient induced degradation and knickpoint migration at a rate ranging from 2.5 km y^{-1} at the lower end to less than 1 km y^{-1} at the upper end approximately 40 km upstream (Yodis and Kesel, 1993). Channel incision by up to 12.5 m has initiated bank instability and channel widening, with an appropriate delay of several years after the passage of the knickpoint. This example illustrates how modification of a main channel can have repercussions for tributary streams which themselves may or may not have undergone channelization work.

Resectioning commonly involves the widening and/or deepening of the channel in order to increase its flow-carrying capacity. Widening has the effect of reducing unit stream power and therefore sediment discharge, so that deposition may occur in the form of berms as the stream attempts to re-establish its original width (Brookes, 1988), although the magnitude of within-bank flows may increase in part-compensation. Dredging to maintain a deeper channel causes a river to degrade its bed because lowering of the bed level creates a knickpoint which then migrates upstream as material is drawn into the deepened reach. Deepening may increase the susceptibility of river banks to erosion and may trigger upstream-progressing degradation within tributaries, both of which can have consequences for the main channel through increased sediment supply.

More rarely, resectioning can involve channel narrowing. As a result of channelization work carried out along the Raba River in Poland in the 1960s and 1970s, the channel was shortened by 7 per cent but narrowed by up to 60 per cent (Wyzga, 1993, 1996). The changes produced considerable increases in mean flow velocity, particularly at the higher discharges, which were instrumental in initiating bed degradation. Up to 3 m of incision has occurred along an 8 km reach, with lesser amounts further upstream. In addition, the higher velocities have progressively washed out the finer grains from the gravelly bed material, thereby developing a surface armour of lower mobility and greater roughness. In effect, coarsening of the bed material can be regarded as a counteraction to a positive feedback mechanism initiated by channelization. Narrowing and shortening increased the erosive potential of the flow, which resulted in incision and channel enlargement; progressively larger flood flows became concentrated in the channel zone and downcutting continued, a process moderated by the coarsening bed. A new equilibrium is thus likely to be established at higher flow velocity and stream power levels than before channelization.

Channelization often involves a combination of measures which reinforce one another and whose individual effects are therefore difficult to isolate. Channel enlargement, straightening and removal of riparian vegetation along more than 250 km of West Tennessee channels in the period 1959–78 produced a complex series of morphological changes along both modified reaches and tributary streams, undoubtedly helped by the mobility of the sandy bed material (Simon, 1989, 1992, 1994). Channel lengths were shortened by up to 44 per cent and channel gradients increased by as much as 600 per cent. The character of bed-level adjustment depended on location in the network relative to the area of maximum disturbance (AMD) or the upstream terminus of channel work, the AMD acting as a fulcrum between net degradation upstream and net aggradation downstream. Adjustment took three main forms:

(a) Downstream aggradation, at a maximum rate of 0.12 m y^{-1}. Sites downstream of the AMD aggraded immediately after modification with material supplied from eroding reaches upstream, maximum aggradation occurring in the downstream-most reaches (line A in Fig. 6.12A).
(b) Migrating degradation and secondary aggradation. Amounts of degradation were at a maximum just upstream of the AMD (in response to the significant increase in stream power imposed by the work) and decreased non-linearly with distance upstream (line B in Fig. 6.12A). Degradation progressed upstream at rates as high as 1.6 km y^{-1}, lowering the bed by up to 6.1 m and reducing the channel gradient. Apparently degradation overcompensated for the increase in stream power, and a phase of secondary aggradation was initiated after 10–15 years of downcutting (C in Fig. 6.12A), increasing gradients which had become too low. If overadjustment during this secondary phase is also assumed, channel response can be envisaged as alternating and diminishing phases of aggradation and degradation during the approach to a new quasi-equilibrium (Fig. 6.12B).
(c) Pre-modified aggradation. Sites experienced mild aggradation upstream of migrating knickpoints (D in Fig. 6.12A), reflecting land-use practices and channel processes in the upper basin.

Channel width adjusted in response to these changes in bed level. As a result of bank accretion, narrowing occurred in association with aggradation, starting first in downstream reaches and progressing upstream. More

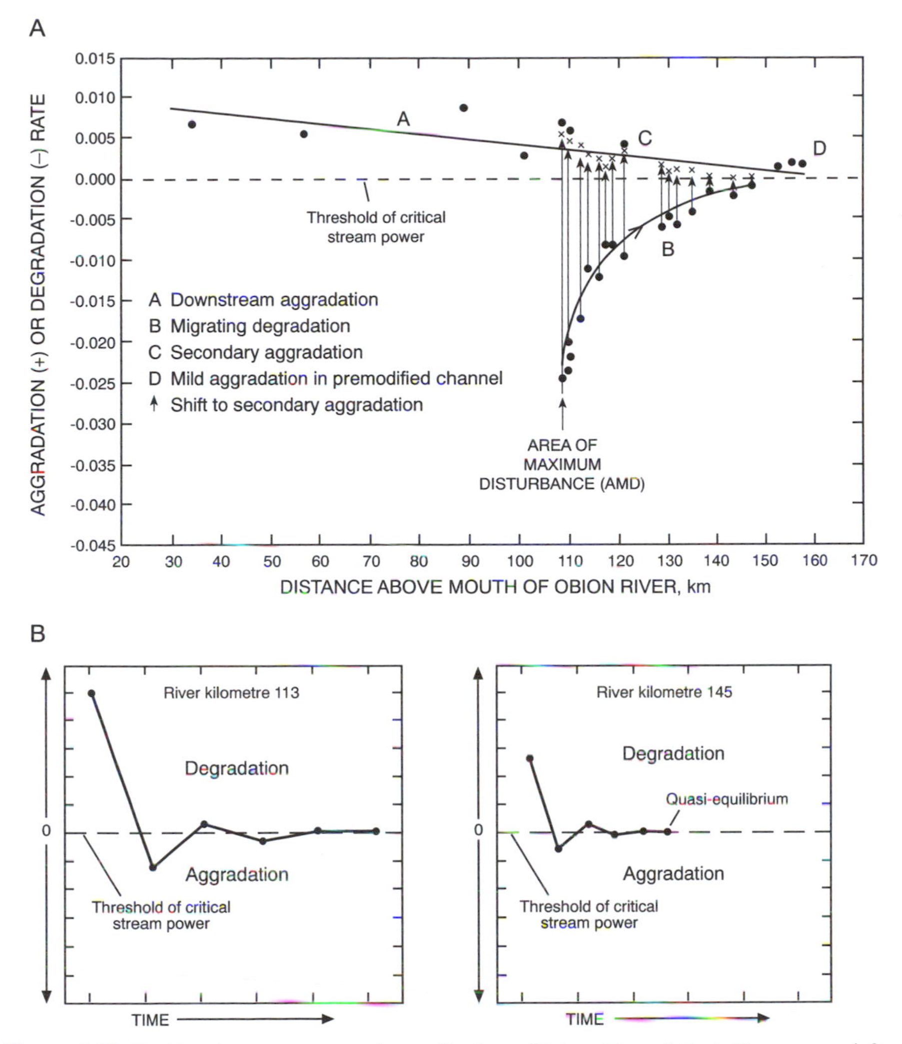

Figure 6.12 Bed-level response to channelization, Obion River, West Tennessee (after Simon, 1994). (A) Longitudinal variations in response. (B) Oscillatory behaviour during the approach to a new quasi-equilibrium.

spectacularly, degradation induced channel widening, more than doubling channel width along some rivers (Simon, 1989). Incision and basal erosion increased bank height and bank angle until a critical state was reached, estimated at 5–7 m in the case of bank height, when mass failure occurred. The effect was most potent just upstream of the AMD where degradation was at a maximum and declined in severity with distance upstream, underlining the close connection between lateral and vertical forms of adjustment. Nevertheless, vertical processes dominated in this low-energy system. Degradation in conjunction with widening is seen as an effective means of energy dissipation as the system responds to channelization work (Simon, 1992). A total of 50–100 years is deemed necessary for restabilization of the channel banks.

Connectivity in the fluvial system means that the effects of channelization can be transmitted both upstream and downstream of the modified reach, the latter being of particular importance. Aggradation will occur downstream from actively eroding reaches which supply an excess sediment load, as in the case of the West Tennessee channels, but erosive adjustments are not unknown. Below 14 channelization works located on high energy ($\omega > 25$ W m^{-2}) rivers in England and Wales, increases in channel capacity ranging from 18 to 153 per cent have been recorded, width changing preferentially to depth at the majority of sites probably because bed armouring or underlying bedrock restricted vertical erosion (Brookes, 1987). The effect persisted for relatively modest distances downstream, the maximum being 2 km along the River Caldew in Cumbria. Such channel enlargement was explained in terms of higher flood flows which increased within-stream velocities and the potential for boundary erosion. Certainly channelization can alter the magnitude and timing of downstream flood flows, for one of its main purposes is to confine floodwater within the channel as it passes through the modified reach. Along the Raba River, reduced floodplain storage and greater concentration within the channel zone have accelerated the flow to such an extent that flood waves have become progressively flashier and bankfull discharges have increased by a factor of 2.4 (Wyzga, 1996). Channelization has, however, failed to reduce the flood hazard; it has merely shifted it downstream (Wyzga, 1993).

The Raba and other rivers cited above provide an important lesson in showing how disturbance at one point can have undesirable repercussions elsewhere in the fluvial system, possibly at a magnified level. Channelization can be a costly operation requiring a long-term (>50 years) commitment of resources for maintenance, restoration and improvement (Brookes, 1988). Alternative methods of controlling rivers which work with the natural environment rather than against it have been explored in an attempt to minimize the adverse physical and ecological consequences of conventional engineering practices. Design schemes need to incorporate natural channel tendencies, such as meandering, which better preserves hydrological and biological diversity and is aesthetically more pleasing (Keller and Brookes, 1984).

Land-Use Change

When dealing with indirect effects, there is an additional dimension to the problem. Change in extra-channel areas, which can vary widely in intensity and location relative to the channel, needs to be related firstly to the flow and sediment regime of the river system before its impact on the channel itself can be properly assessed. Spatial and temporal lags of variable magnitude are commonly involved. Land-use change can take a large variety of forms but, for illustrative purposes, is considered here as a two-phase process: forest clearance for the purpose of cultivation or grazing, and urbanization. Wolman (1967) proposed a sequence for the mid-Atlantic region of the United States covering these two phases, in which changing land use is related first to sediment yield and then to general channel conditions (Fig. 6.13A).

Forest clearance began in Europe more than 5000 years ago, and England may have lost as much as half of its woodland cover by 2 ka BP. Over that timescale there is the problem of differentiating between the relative effects of climatic fluctuation and human activity as far as channel adjustment is concerned. Nevertheless, Starkel (1991a) has identified distinct phases of accelerated valley-floor sedimentation in central European valleys, during the Roman and Medieval periods, when deforestation and land-use change caused the overloading of rivers, a tendency towards braiding and the vertical accretion of floodplains being the consequences. Milliman *et al.* (1987) have estimated that the sediment load of the Huang Ho increased by an order of magnitude after the easily eroded soils of the loess plateau had been cleared of forest. However, the effects of large-scale deforestation may become clearer in areas where European-style farming practices have been introduced more recently. In the United States the farming era began at about 1700 in the Piedmont region (Wolman, 1967) and after 1820 in Wisconsin (Knox, 1977), while commercial logging in the Redwood Creek catchment of northern California did not begin until the early 1950s (Madej and Ozaki, 1996). Much of eastern Australia was relatively undisturbed before the middle of the nineteenth century.

Vegetation cover is one of the primary controls on sediment supply and catchment hydrology, and is the control most susceptible to human disturbance. A major consequence of forest clearance is accelerated soil erosion on hillslopes, associated with which are network extension through gully development and an increase in the amount of sediment supplied to streams. In both the Piedmont region and Wisconsin, changes were slow at first but intensified as the farming area expanded, particularly into the upper and steeper parts of catchments. Much of the eroded sediment is still stored, either as colluvial-sheetwash deposits on hillslopes or as alluvium in floodplains and channels (Trimble, 1974, 1983). The conversion of forest to agricultural land also affects runoff characteristics. Knox (1977, 1987) estimated that the magnitude of floods having a recurrence interval of less than five years increased by 3–5 times, with the impact being greatest in small watersheds. As well as affecting erosion rates on hillslopes, the increased runoff accelerated bank erosion along tributaries and floodplain sedimentation throughout the drainage system.

The removal of natural vegetation increases catchment sensitivity to climatic events. Large-scale timber harvesting in the Redwood Creek catchment caused storms in 1964 and 1972 (estimated as 1 in 50 year events) to have impacts disproportionate to their size (Madej, 1995; Nolan and Marron, 1995). The resultant 20 per cent increase in storm runoff triggered

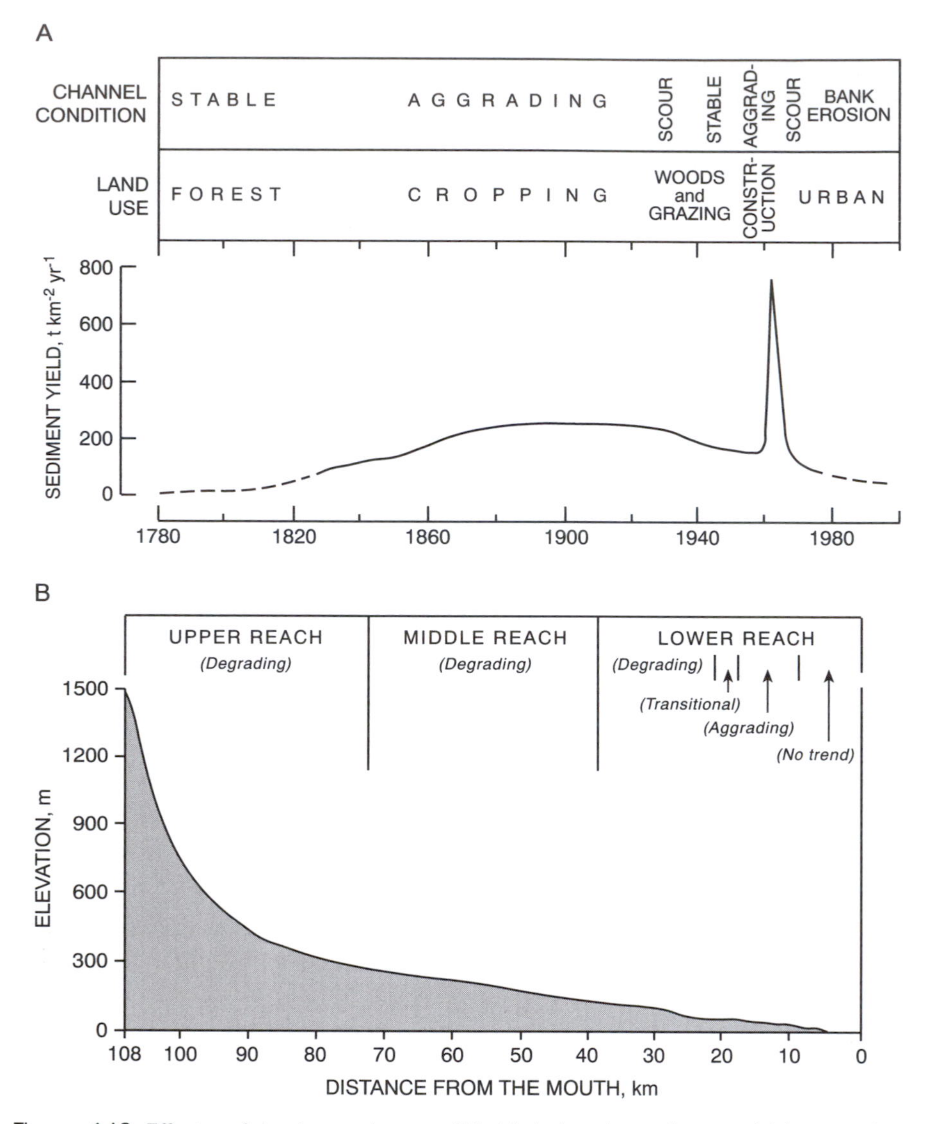

Figure 6.13 Effects of land-use change. (A) Variation in sediment yield over time, Piedmont region, USA (after Wolman, 1967). (B) Longitudinal variation in bed-level response following timber harvesting and large floods, Redwood Creek, California (after Madej and Ozaki, 1996).

extensive landsliding, particularly in the upper basin during the 1964 storm, and considerable amounts of sediment were delivered to the valley floor. The effects persist to the present day as the increased supply is transported downstream in the form of a wave whose attenuating amplitude and decreasing velocity lengthen residence times (Madej and Ozaki, 1996). Recovery will take many decades, especially if landslide scars are reactivated during subsequent storms.

The higher rates of sediment supply associated with deforestation generally produce channel-bed aggradation and accelerated deposition on floodplains, the latter being partly a consequence of the former in that bed aggradation increases the frequency of overbank flows. The most prevalent deposits in the upper reaches of Redwood Creek are gravel berms that reach heights of 9 m above the previous valley floor (Madej, 1995). In Wisconsin, rates of overbank deposition increased from 0.2 mm y^{-1} in the pre-settlement period to as much as 3–50 mm y^{-1} after forest clearance (Knox, 1987), while equivalent figures for the lower Bega River in New South Wales are 0.75 mm y^{-1} and 12.5 mm y^{-1} respectively (Brooks and Brierley, 1997).

In line with equations (6.11) and (6.13), channels have tended to become wider, shallower and less sinuous where forest clearance has significantly increased the influx of material. The incision of upland valley fills in the Bega catchment released vast amounts of sediment which transformed the lower river from a narrow, relatively deep, mixed-load system with a fine-grained floodplain, into a wide, shallow, sandy bed-load system from which sands are readily flushed on to the floodplain (Brooks and Brierley, 1997). The channel has widened by up to 340 per cent. Channel widths increased by 150–300 per cent in the upper reaches of Redwood Creek after the 1964 flood (Madej and Ozaki, 1996). In the Mangawhara catchment of New Zealand, the channel is narrow and stable in its upper 3 km where the catchment is still forested but rapidly becomes wider and more irregular beyond that point, characteristics which persist downvalley (Bennett and Selby, 1977). There also, channel length has decreased overall from 16.1 km to 13.5 km in 33 years, the decrease in sinuosity being most marked in downstream reaches.

Longitudinal variations in stream channel response have been reported elsewhere. Knox (1977) found that, as a result of increased bed-load transport and more frequent flooding, Wisconsin channels in the first 40 km have increased their bankfull widths by more than 100 per cent when compared with pre-settlement values. In the next 15 km increases in width have usually been less than 40 per cent, while in the lower reaches where bed-load transport is less important and changes in flood frequency have been less marked, the transport and deposition of finer sediments has tended to produce narrower and deeper cross-sections. Depending on how forest clearance alters the rate and type of sediment supply to streams, channel response may vary from one basin to another or between different parts of the same basin. In areas where fine gravel and sand are readily available as potential bed load, channel aggradation, widening and bar instability are common responses. Rates of lateral migration increased dramatically along a Missouri river after a post-clearance influx of gravel had greatly increased the number and size of gravel bars (Saucier, 1984).

Wolman's schematic diagram (Fig. 6.13A) identifies a phase following the farming period which can be associated with reafforestation or the introduction of conservation practices designed to alleviate the worst effects of farming. However, reafforestation need not reduce sediment yields, at least initially. The practice of digging open forest ditches has increased both bed load and suspended load yields in the upper reaches of the Severn, continuing through to the mature stage of new cover (Moore and Newson, 1986). Taking land out of cultivation is one conservation measure, the major benefit of which may be not so much a reduced sediment supply but rather lower runoff amounts which, in turn, produce smaller transport rates within the channel (Kuhnle *et al.*, 1996). Nevertheless, where sediment supply has significantly declined, a period of channel entrenchment may be initiated if flows remain effective (Trimble, 1974; Miller *et al.*, 1993). Much of the land in the Savannah River basin that was formerly under crops has reverted to less erosive uses, and the river is now actively eroding its bed, at least in those parts of the catchment where the main wave of sediment resulting from accelerated erosion has passed through (Meade, 1976). In Maryland, incision by the smaller streams has led to the formation of new floodplains within older ones (Costa, 1975).

Channels can seemingly adjust rapidly to the changed conditions, although the response is likely to be complex, with spatial and temporal lags involved. Thus, for example, if forest clearance is followed by the introduction of conservation measures or a period of lower runoff, and the wave of sediment associated with clearance has yet to complete its passage through the basin, stream channel response may vary with position depending on whether sediment supply is declining, constant or increasing. Such a variable response is evident along Redwood Creek as it adjusts to the downstream transmission of a sediment wave generated largely by the 1964 storm (Madej and Ozaki, 1996; Fig. 6.13B). Single changes seldom produce single effects.

In comparison with forest clearance, **urbanization** represents a localized form of land-use change, but its effects may nevertheless be propagated downstream. In general, the creation of impervious surfaces and the installation of more efficient drainage systems in urban areas increase the volume of runoff for a given rainfall and give rise to a flashier runoff regime with shorter lag times and higher peak discharges. The smaller, more frequent floods are much more affected than the extreme events. Hollis (1975) showed that the 1-year event is enhanced ten-fold by 20 per cent urbanization, whereas the 2-year event is only doubled or trebled. As regards sediment yield, urban development can be represented as a two-phase process, a large initial increase when soil is exposed to runoff on construction sites being followed by a sharp decline as sediment sources become more stable in the urban landscape (Fig. 6.13A). Urbanization can significantly increase the quantity of pollutants entering the river system and, if finer sediments begin to accumulate within the channels, provide a bed material that is more conducive to the adsorption of heavy metals (Thoms and Thiel, 1995).

The greater frequency and magnitude of flood discharges in urbanized catchments cause channel enlargement in line with equation (6.9). From an extensive survey of British urban sites, Roberts (1989) obtained a mean enlargement ratio (urban/rural bankfull cross-sectional area) of 1.61, with a range up to about 6. The enlargement ratio increased significantly with the degree of urbanization in the catchment, but there was no indication of any threshold level as suggested by Morisawa and LaFlure (1979). They found that significant enlargement occurred only when about 25 per cent of the basin was more than 5 per cent impervious. Average enlargement within British urban channels is strongly influenced by the flood frequency to which channels are adjusted (Roberts, 1989). Differentiating between 'less permeable' and 'more permeable' catchments, the former show approximately twice as much enlargement as the latter for the same degree of urbanization, because they are adjusted to floods with shorter return periods, which are the ones preferentially affected by urbanization (Hollis, 1975).

Hammer (1972) has evaluated the relative influence of several land-use types in promoting channel change within 78 small basins near Philadelphia (Table 6.12). The results suggest that the degree of channel enlargement is correlated with the age and type of urban development. There is a lag in channel adjustment because the effect of a land-surface change on flood peaks takes time to influence the channel itself, and because the channel has to cope with the high sediment loads characteristic of the early phase of urbanization (Fig. 6.13A). In Britain, the extent of stormwater sewerage rather than the age and type of development seems to be the dominant factor governing the amount of change (Roberts, 1989).

No distinction has yet been drawn between the relative contributions of width and depth to the total change in channel capacity. On the basis of resurveys over a 20-year period of 14 monumented cross-sections in a small Maryland catchment (A_d = 9.6 km^2), Leopold (1973) found overall that mean depth had increased by 23 per cent, largely as a result of overbank deposition rather than bed scour, but that bankfull width had decreased by 35 per cent, a response which runs counter to expectation. During the first 12 years urbanization did not markedly alter the frequency of high flows but did produce a higher rate of sediment supply, leading to within-channel

Table 6.12 Ratio of enlarged channel area to natural channel area in small basins (< 13 km^2), assuming that all the basin area is in use as specified (after Hammer, 1972)

Land use	Enlarged channel area/ natural channel area
Wooded	0.75
Previously developed land	1.08
Impervious area less than 4 years old; unsewered streets and houses	1.08
Cultivation	1.29
Houses more than 4 years old fronted by sewered streets	2.19
Sewered streets more than 4 years old	5.95
Impervious areas more than 4 years old	6.79

deposition and cross-sectional narrowing. Thereafter the number of high flows increased rapidly as the urban area expanded, and in 1967 progressive enlargement of the channel began. At two cross-sections the channel area increased by 40 and 62 per cent between 1968 and 1972. When observation ceased, the enlargement process may have been incomplete, especially as regards channel widening, although the flashy floods associated with urban catchments may not always saturate banks sufficiently to reduce their stability. Variations in cross-sectional response are influenced by the character of boundary sediment, with bed armouring restricting incision and cohesive banks inhibiting width adjustment.

Leopold's study illustrates the importance of phasing in channel adjustment. The first few years when sediment supply is higher (Fig. 6.13A) may be a period of channel contraction, and only in the post-construction phase does the increased frequency and magnitude of flood flows produce significant enlargement. Arnold *et al.* (1982) have also observed phased changes in channels affected by urbanization. Once construction was complete, sediment yields declined but discharges remained high, creating an imbalance which led to high rates of bank erosion. That erosion has not only produced channel widening but, by locally increasing sediment supply in downstream reaches, has caused pattern instability and a change from meandering to braided. Sinuosity has decreased and the slope increased through the meander cutoff process. Part of the River Bollin in Cheshire has also adjusted to a flashier regime by becoming less sinuous, a decrease in sinuosity from 2.41 to 2.34 between 1872 and 1935 being followed by a much sharper decline to 1.4 by 1973 (Mosley, 1975). Rates of change seem to have accelerated in the 1960s with an increase in annual flood maxima, related principally to suburban expansion but also to the renewal of agricultural tile drains.

Greater emphasis has been placed in the literature on the hydrologic rather than geomorphic impacts of urbanization. The two need not be simply related. Not only is there spatial and temporal lag in response, but the character of channel change varies over time with the relative extent to which discharge and load conditions are modified at different stages during the urbanization process. Many of the examples cited above refer to relatively small urban areas in basins of less than 50 km^2, but larger rivers may not show such marked changes since a smaller proportion of their catchment area is likely to be affected. Also, the extent to which enlargement is propagated downstream remains unknown, for even within urban areas themselves it can be very irregular (Roberts, 1989). Despite these limitations, it is again evident that stream channels can respond quite rapidly to external disturbance, which has implications for their design and control in the urban landscape.

Mining

Mining can generate considerable amounts of solid waste which, in the past at least, has often been disposed of in the nearest watercourse. Between 1853 and 1884 enormous quantities of material from hydraulic gold mining

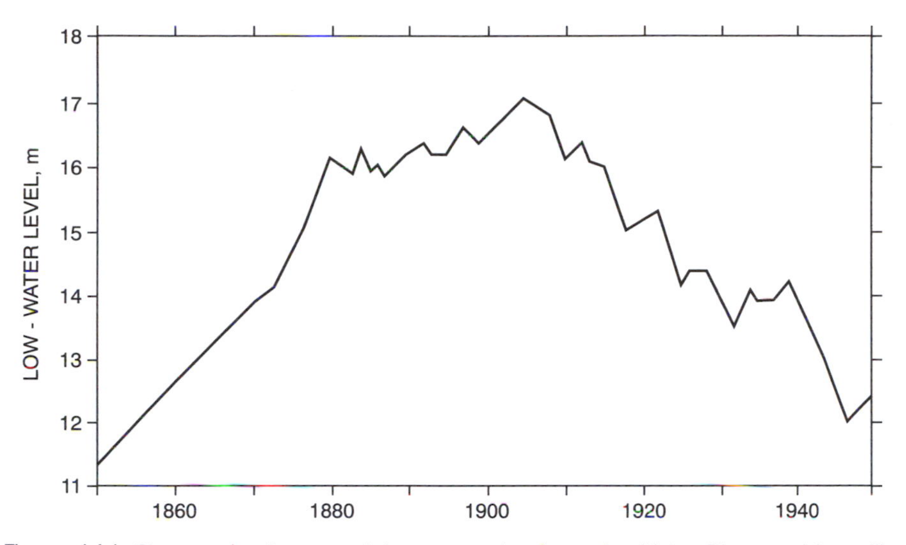

Figure 6.14 Changes in the annual low-water level on the Yuba River at Marysville, California, 1850–1950 (after Meade, 1982).

were washed into tributaries of the Sacramento River in California (Gilbert, 1917). As much as 270 million m^3 of sediment were delivered to the Bear River (James, 1991), while the Yuba River received over 15 million m^3 in a single year (Wildman, 1981). The resulting rise in bed elevation (Fig. 6.14) created such serious problems downstream (increased flooding, destruction of farmland, poorer navigation) that further dumping was prohibited in 1884. By then, however, the main damage had been done, and even now a lot of the debris remains stored in tributary valleys and on floodplains. Its removal will take much longer than the century needed for clearance of the main channels (Meade, 1982). At a more modest but nevertheless significant level, 40 million m^3 of waste from alluvial tin mining was supplied to the Ringarooma River in Tasmania between 1875 and 1982 (Knighton, 1989). The river is beginning to heal itself but the effects of the mining input will continue to be felt for at least another 50 years.

The dispersal of mining waste is physically and chemically complex. Ore particles and the residue of hydraulic mining, typically in the sand and small gravel range, are intermittently moved as bed load and have the greatest immediate impact on river channels. Hydraulic sluicing in which metal-bearing alluvium is broken down with pressurized jets of water, a technique developed in California in the 1850s, is a particularly potent means of waste generation. Since both the supply of material from mining and the occurrence of competent discharges are likely to be intermittent, the coarser waste may move downstream in the form of slugs rather than continuously (Lewin and Macklin, 1987; Nicholas *et al.*, 1995). Finer fractions move more rapidly through the river system in suspension, being deposited on the floodplain during overbank flows or, less probably, within the channel itself. However, in Cornish streams supplied with the waste products of china clay extraction, flocculated kaolinite clay has accumulated as bank accretions and caused channel narrowing (Richards, 1979). Chemical processes can also be important, particularly as far as metals are concerned. Metals may be dissolved in acid minewaters and later precipitated, possibly as adsorbed coatings on other sediment or as organic complexes (Lewin and Macklin, 1987). Although physical processes dominate dispersal, particularly during the period of mining operations, chemical dispersal processes could have a longer-lasting effect on the environment.

Mining waste introduced into a natural river can either be transported alongside the indigenous load without causing disruption, or result in the complete metamorphosis of the channel system, two ends of a spectrum of response termed respectively 'passive dispersal' and 'active transformation' by Lewin and Macklin (1987). Passive dispersal is a common condition in British catchments, where ore rather than alluvial mining has been practised and production has been relatively modest by world standards. Metal-rich sediment may accumulate temporarily within channels but with little physical effect, although the chemical composition of floodplain soils may be altered (Bradley and Cox, 1987). Where inputs have been larger, lateral accumulations of mining waste have promoted channel migration.

Active transformation is associated with a greatly increased sediment load, when equation (6.11) applies. The simplest response to the passage of an enhanced input is a cycle of aggradation and degradation. The first is frequently accompanied by channel widening and a transition from a single channel to a multi-channel pattern, while degradation induces a decline in the intensity of braiding as the channel narrows and incises, and often leads to the development of a series of terraces. In the valley of the South Tyne, where lead and zinc extraction reached a peak in 1850 and 1900 respectively, aggradation has been concentrated in discrete sedimentation zones separated by more stable reaches which have steeper gradients and are commonly associated with bedrock control (Macklin and Lewin, 1989). By 1860, aggradation involving fine metal wastes and coarser sediments unrelated to the mining had resulted in floodplain destabilization and the development of multi-thread systems up to 22 km downstream from the mining area. The first part of the twentieth century was characterized by sediment storage and lateral reworking, to be followed by incision (*c.* 1950) soon after the cessation of mining. Depths of incision have now reached 1.5–3 m. Although the river has experienced an aggradation–degradation cycle, the response has not been simple. Aggradation has been spatially discontinuous and the pattern of sediment movement is not well represented as the translation of a single downvalley wave.

The largest-scale disruptions are undoubtedly associated with hydraulic mining. In the mining districts of the upper Bear River, aggradation proceeded at rates as high as 5 m y^{-1} and had attained depths of over 40 m by 1880, but with the virtual cessation of mining in 1890, degradation was well advanced by the close of the century (James, 1993). However, immense deposits of mining sediment still remain and debris continues to be supplied to the main channels by hillslope processes, whence it is readily transported downstream by relatively frequent flow events (2-year floods being competent to move the coarsest 10 per cent of bed material). Not surprisingly, therefore, a large volume of mining material remains in the lower part of the basin, at depths of up to 5 m, and not until the 1970s did even localized incision reach pre-mining levels (James, 1991). The sustained storage, reworking and mobility of mining debris suggest that sediment transport in the Bear River has not conformed to Gilbert's (1917) symmetrical wave model originally based on observations of the neighbouring Yuba River (Fig. 6.14). A skewed wave model which recognizes the importance of long-term storage and protracted release of easily eroded deposits is more appropriate there (James, 1993). Storage was much less in the Yuba basin and, unlike the Bear River, the main channels of the South Yuba are completely stripped of mining sediment. Thus, the transport history of a large introduced load can differ even between neighbouring rivers supplied from similarly located headwater sources.

Transport history is even more complicated where supply comes not from a relatively small area in the upper basin but from a large number of sources widely distributed along the river system. Such is the case in the Ringarooma Basin (Knighton, 1989, 1991; Fig. 6.15A). The tin mining era lasted for more than 100 years and reached a peak in 1900–1920, tending to start and finish earlier in reaches further upstream. The introduced waste was much finer ($D_{50} \sim 1$ mm) than the natural bed material ($D_{50} \sim 45$ mm), and, because of the volumes involved, quickly replaced it. The pattern of sediment movement was reconstructed using a mass-conservation model and, not surprisingly in view of the longer time period and more diffuse nature of the input, does not follow the simple wave-like form proposed by Gilbert (1917). Predicted changes in bed height indicate an overall tendency for successive phases of aggradation and degradation which progress downstream, but with underlying variations dependent on position and the varying levels of input from the widely scattered sources (Fig. 6.15B). Aggradation to depths of over 10 m was most rapid in the upper reaches close to major supply points (L2 in Fig. 6.15A and B), becoming later and more gradual with distance downstream. Bridges frequently had to be replaced (Plate 6.1). Channel width increased by up to 300 per cent in suitable locations where the valley is less confined, and there also the development of a braided habit became relatively common.

While bed levels were continuing to rise in lower reaches, degradation began in upper ones (notably after 1950) and by 1984 had progressed downriver over 30 km, rates of incision reaching 0.5 m y^{-1} (Plate 6.2). In those upper reaches the bed material has again become gravelly as a result of re-exposure or the development of a surface armour, with a sharp transition downstream to a sandy bed still dominated by mining sediment (Fig. 6.15C). During the degradation process the channel has narrowed and reverted to a single-channel pattern where previously multiple channels had developed

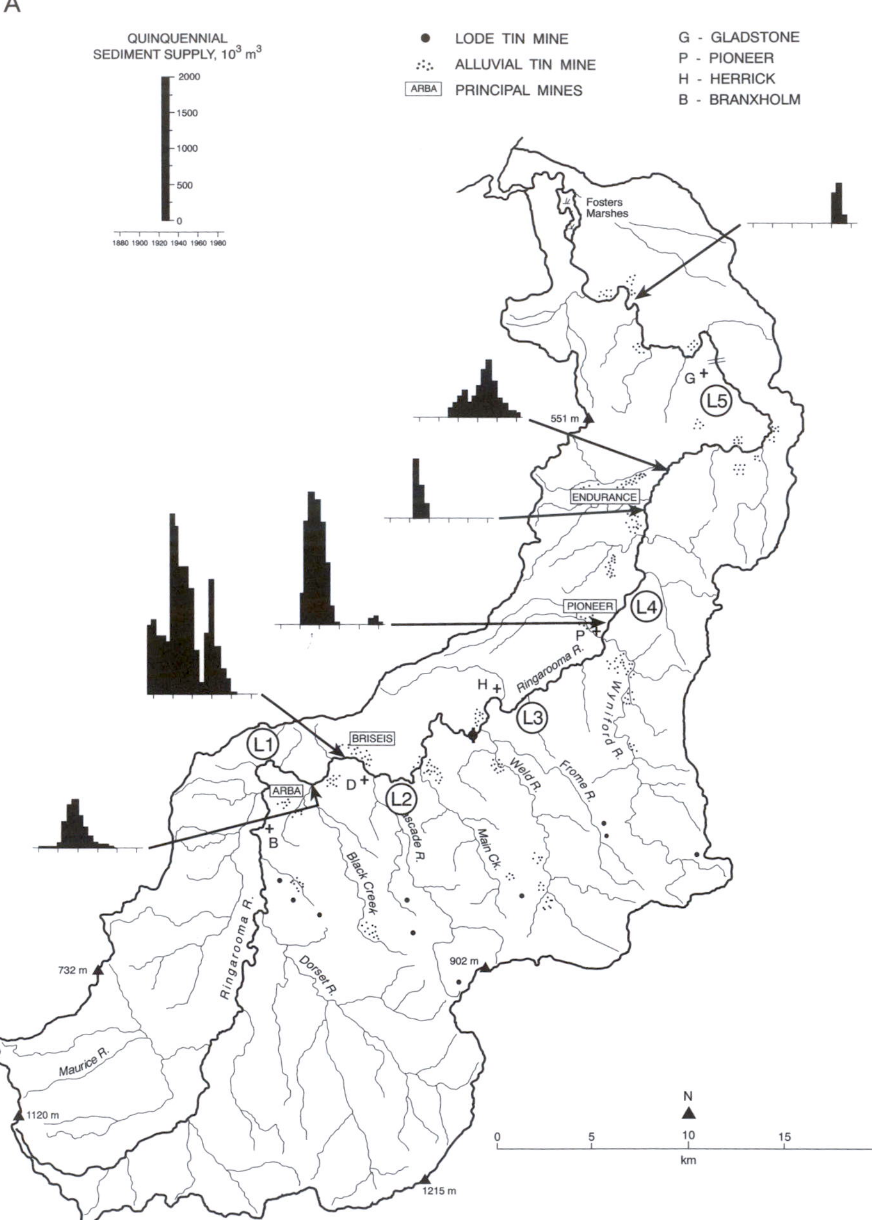

Figure 6.15 Effects of hydraulic tin mining on the Ringarooma River, Tasmania (after Knighton, 1989, 1991). (A) Fluctuations in sediment supply from the main input points. (B) (opposite) Predicted changes in bed height within selected 5 km stream lengths (see A for locations). (C) Gravel–sand transition. (D) Cross-sectional changes at Herrick during degradation.

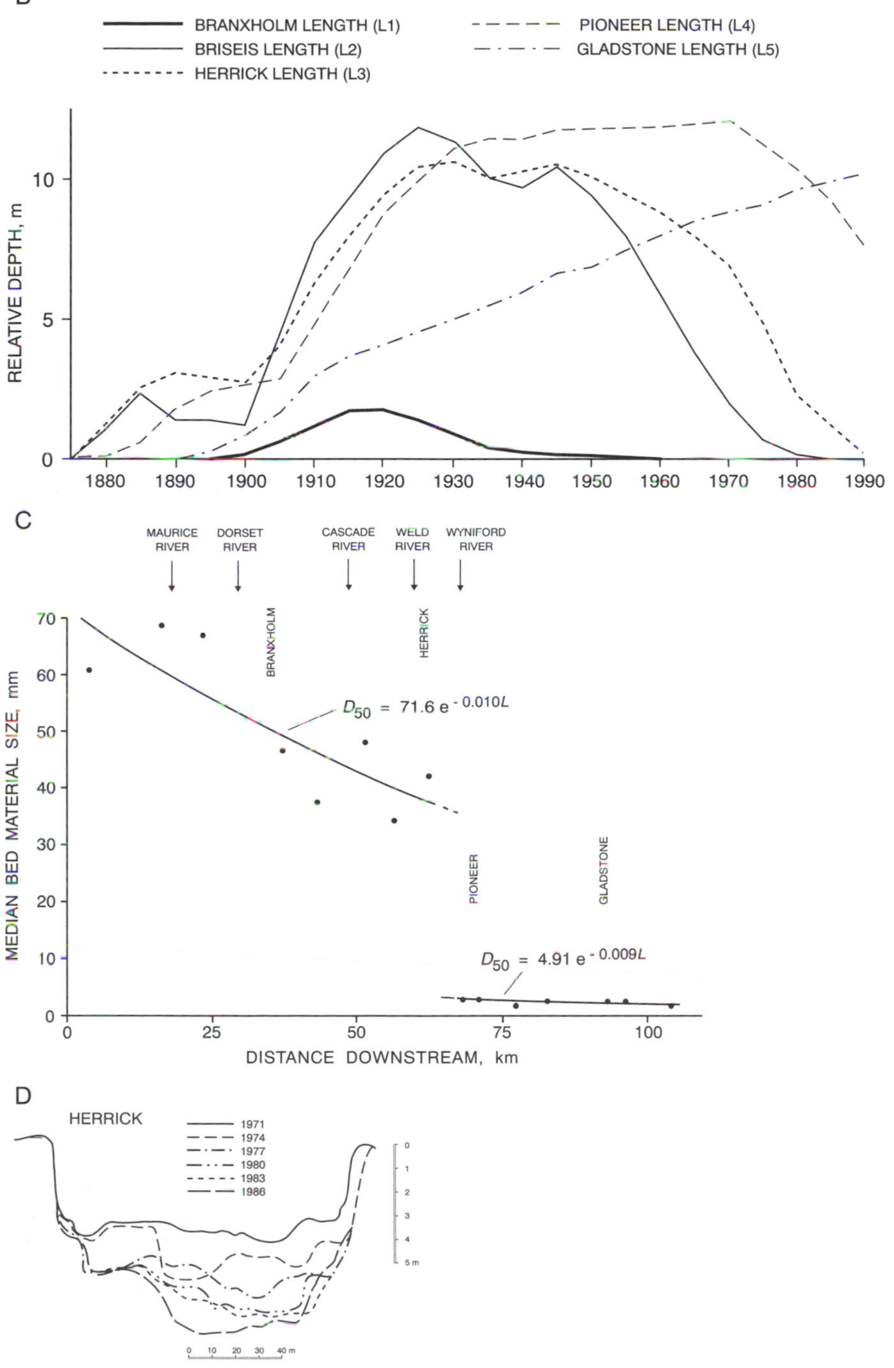

Figure 6.15 (continued)

Plate 6.1 Upstream view of the Ringarooma River near Gladstone, where the river is still aggrading. The piers of a previous (third generation) bridge are visible at the water surface.

Plate 6.2 The Ringarooma River at Herrick where degradation has exposed tree stumps previously buried beneath mining debris, and has resulted in the formation of a terrace sequence in the opposite bank.

(Fig. 6.15D; also Fig. 6.7 E → F). The Ringarooma River was in a parlous state in the 1950s. It is beginning to recover and upper reaches have regained their former bed level in about 35 years after the cessation of mining. However, recovery is increasingly sluggish with distance downstream, and at least another 50 years will be required for the river to cleanse its channel of mining debris.

Clearly mining can have serious detrimental effects on the physical and chemical status of the fluvial environment. Fluctuating bed levels and associated adjustments to the cross-sectional and planimetric form of channels constitute the most immediate effects where input amounts are large enough to cause 'active transformation', with consequences for the incidence of flooding, the safety of river crossings, and the maintenance of navigation. Where storage is a major component of transport history, effects can last much longer. The deposition of metalliferous fines and other toxic substances on floodplains can result in the long-term contamination of riparian agricultural land. Floodplain deposits are removed mainly by erosion during lateral channel migration, a process which proceeds at a much slower pace than the vertical removal of material stored in the bed of river channels. Available mining technology may have improved, but the erosion of mine tailings, the reworking of metal-contaminated alluvium and accidental releases (Graf, 1990) will continue to affect the fluvial system long after mining has ceased.

Managing the Fluvial Environment

A natural river seeks to establish a channel morphology which is adjusted to prevailing hydrologic and sedimentological conditions so that it can continue to carry a wide range of discharges and loads as efficiently as possible. Except during extreme events, the channel is essentially maintained by small-scale and short-term adjustments. However, as the preceding sections have shown, that natural equilibrium can easily be upset by human activities which either alter catchment conditions or modify the river channel itself. Too often the potential consequences of such activities have not been understood or been simply ignored. Due recognition needs to be given to:

- the sensitivity of river channels to external disturbance;
- the rapid rates at which change can be initiated and propagated;
- the extended timescales needed for complete readjustment despite the rapidity of initial reactions;
- the interrelatedness of the various channel form components, with the implication that modification of one component may initiate response within others;
- the importance of linkages – between hillslopes and channels, between contiguous parts of the stream system, between fluvial and marine environments. Change at one place can trigger change at another with an appropriate time-lag.

River channels have been adjusting to changing natural conditions since time immemorial, but human disturbance has increased the pace and scale of channel change during the course of the twentieth century. In addition, account needs to be taken of the growing demand for water and the global warming scenarios which predict an increase in global mean temperature of about 0.3°C per decade well into the twenty-first century (Houghton *et al.*, 1990). Water supply, vegetation patterns and sea-level could all be affected, with consequences for river systems especially if the incidence of flooding is significantly altered. Given the variability of natural and artificial conditions, many river channels may display symptoms of instability. Newson (1986) has compiled evidence of such instability within British upland and piedmont channels, associated with recovery from large floods, modified flow characteristics and fluctuations in sediment supply. To establish whether a given river is adjusted or adjusting and to predict the future course of channel adjustment are important considerations in the effective management of the fluvial system (Lewin *et al.*, 1988).

River management covers a wide range of objectives, sometimes complementary, sometimes conflicting, which can be broadly categorized into 'hazard' and 'resource' perspectives (Table 6.13). Many management schemes involve regulation and control, activities which are increasingly constrained by environmental concerns for aquatic habitats, fisheries and landscape aesthetics. The river engineer has largely been responsible for the design, siting and construction of control structures, using empirical experience as a primary basis for obtaining practical solutions to reach-scale problems. Geomorphological contributions to river management have been less forthcoming, but a geomorphological approach which both complements and supplements that of the river engineer is beginning to emerge (Brookes, 1995a; Sear *et al.*, 1995). That approach has various strands:

(a) Newson (1995) argues that process-based *classification* represents a significant gap in knowledge, one that can be ably filled by fluvial geomor-

Table 6.13 Perspectives and objectives of river management

River as hazard	River as resource
Bank protection	Aesthetics
Bridge stability	Conservation
Flood control – channelization, dams	Ecology – flora and fauna
Floodplain zonation	Navigation
Land drainage – agricultural drains, road drainage, urban stormwater systems	Recreation
Pollution – of water and sediment	Rivers as boundaries – international to district levels
Soil erosion and sediment transport	Sand and gravel extraction
	Water source – irrigation, industrial and municipal supplies, power generation

phologists with their fieldwork tradition. Notable contributions have been made, such as the classifications of bar forms by Church and Jones (1982), of floodplains by Nanson and Croke (1992), and of channel types by Rosgen (1994), but as yet constructive advice on how to manage particular river types has not been explicitly formulated. Classifications of river channels based on their morphology, sediments and dynamics provide common standards which can be used to evaluate problems in context and suggest remedial options (Sear *et al.*, 1995).

(b) Whereas the practical engineer has tended to focus on the reach scale, fluvial geomorphologists have adopted a broader, *catchment-scale perspective* which emphasizes the physical integrity of the drainage basin and the close links between catchment and channel dynamics. Main rivers are affected by fluctuations in the supply of sediment from hillslopes and tributaries, transport that sediment irregularly through various channel and floodplain stores, and interact with coastal processes in its eventual redistribution. Gilbert's (1917) commission on the effects of hydraulic gold-mining involved investigation of not only the movement of debris through the river system, but also its potential consequences for navigation depths in San Francisco Bay. Reservoir impoundment disrupts the continuity of sediment transport, resulting in sedimentation within the reservoir and degradation downstream from the dam. Channelization can give rise to upstream-progressing degradation and downstream aggradation, which may affect long river distances and even entire river systems. Because channel adjustments can migrate, management schemes that fail to consider spatial linkages are unlikely to be successful in the long term. Reach-based solutions tend to deal with the local symptoms of a problem rather than its more fundamental causation (Sear *et al.*, 1995; Simon, 1995).

(c) Complementary to a large spatial perspective is a long temporal one. Recognition of an *historical dimension* emphasizes the unsteady nature of channel adjustment over a variety of timescales including and extending beyond the typical engineering timescale of 10–100 years. Many drainage basins have an historical legacy, inherited from the Pleistocene glaciations, from the larger discharges which characterized the early Holocene, from the increasing clearance of natural vegetation cover after 5000 BP, from the effects of extreme events, or from the greatly enhanced human disturbance of the fluvial system over the last hundred or more years. Reconstruction of the past is seen not necessarily as an end in itself but as a means of testing models of change which, if verified, can be used to forecast future behaviour.

River engineers need information on the potential for channel destabilization, including the undesirable effects of engineering works themselves. The Afon Trannon, a gravel-bed tributary of the River Severn in Wales, has been the subject of a comprehensive flood protection scheme involving dredging, the construction of flood embankments and bank protection measures (Leeks *et al.*, 1988). The net result has been greater lateral instability which has undermined channel banks, increased the availability of sediment for transport and produced considerable problems of channel maintenance. Geomorphological evidence, in the form of historical studies (of floodplain sediments and channel change) and contemporary field surveys, has revealed the underlying reasons for those problems. If used beforehand, that evidence might have helped in the decision-making process, even if the outcome had been to abandon or completely redesign the scheme in the light of the potential instability.

(d) Given that channel adjustment through space and over time inevitably involves sediment redistribution, the *supply and movement patterns of sediment* are of primary concern to river management, and represent an area of geomorphological expertise which has expanded considerably in the last 20 years. It encompasses a wide range of processes, from the mechanics of bank erosion and the migratory habits of alluvial river channels, to the transport histories of sediment loads moving through natural and disturbed basins. The catchment-scale fluvial audit which identifies the likely sources, sinks and stores of sediment provides guidance to the river manager on the possible causes of fluvial problems and the constraints to practical solutions (Newson, 1995). The role of floodplains is particularly important since they are both the source of much of the fluvially-derived material supplied to streams and a sink for natural and contaminated sediment, with highly variable residence times.

Sear *et al.* (1995) regard sediment transport in all its variety as a common focus for engineering and fluvial geomorphology (Fig. 6.16A). With over 15 000 km of main river requiring annual maintenance in England and Wales, a distinction is drawn between fine sediment, low-energy systems in which preventative maintenance is dominant, and coarse sediment, high-energy systems where breakdown maintenance is mostly required. The main contributions from fluvial geomorphology are summarized in Fig. 6.16B, feeding through to the broader policy environment and the environmentally sympathetic actions which result. As an example of a high-energy system, Shelf Brook on the western side of the Pennines has a long history of flooding which adversely affects a small urban area. Catchment response to flooding was reconstructed from historical documents and field evidence in order to produce a fluvial audit and indicate spatial variations in geomorphological effectiveness, from which recommendations could be made as to the size and siting of remedial measures. Geomorphological guidance on active processes and cause–effect relationships at the reach and catchment scales can provide better targeting of the most appropriate engineering solutions.

(e) With increasing recognition of instability in the fluvial landscape, the regime assumptions much favoured by the river engineer need to be set in a more dynamic framework which involves the development, evaluation and application of methodologies for the *analysis of change*. A clearer understanding is required of the variable sensitivity of river

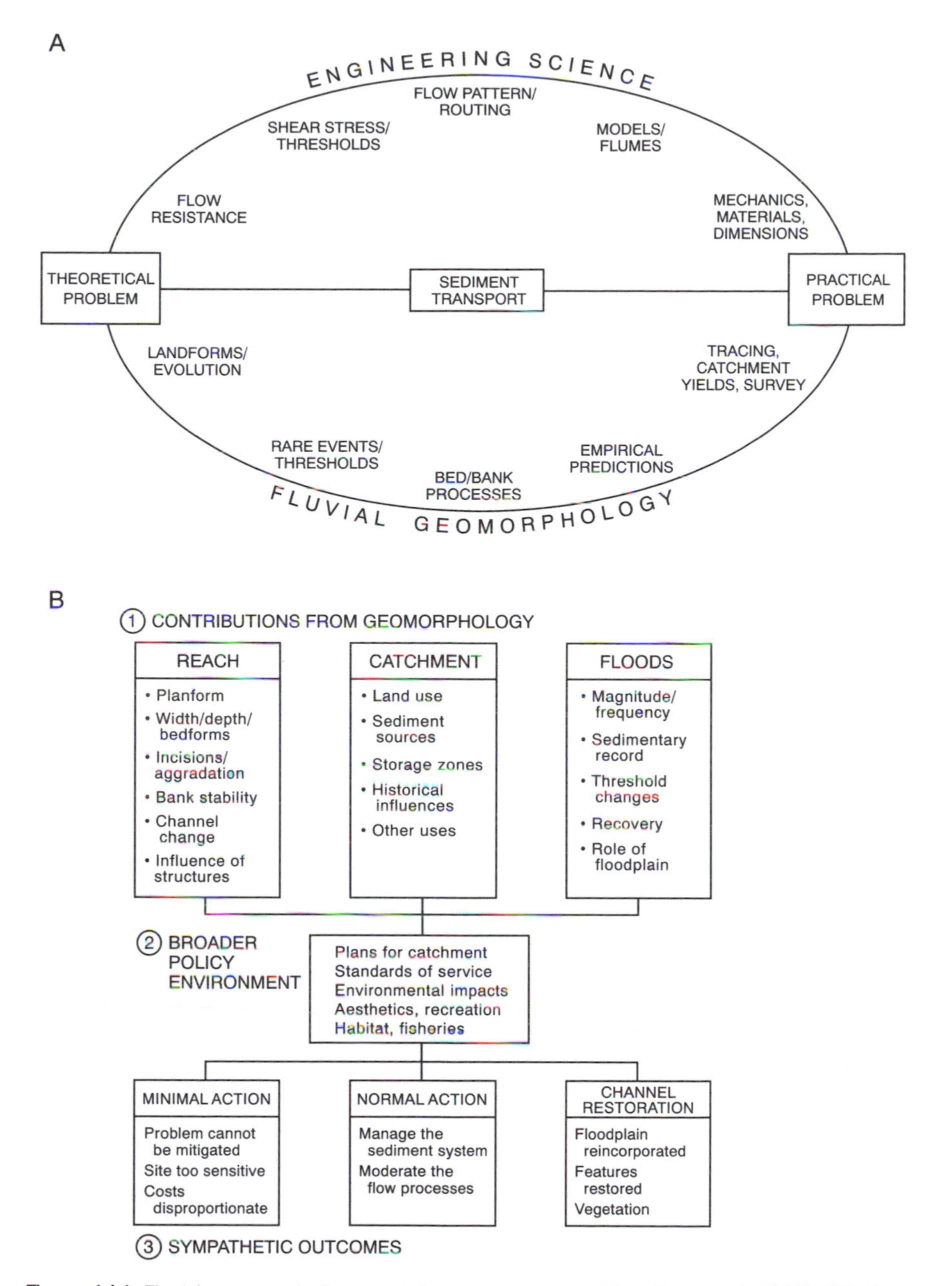

Figure 6.16 Fluvial geomorphology and river management (after Sear *et al.*, 1995). (A) Complementary research interests of fluvial geomorphology and river engineering. (B) Contributions to the broader policy environment and river engineering.

channels and of the conditions, both natural and anthropogenic, which can lead to instability. A particular geomorphological contribution is assessment of the role played by events of different magnitude and frequency. Extreme floods vary in effectiveness from one climatic environment to another, between basins of different size, and in relation to the composition of the channel boundary.

River managers need to anticipate the consequences of particular actions, both locally and systemwide. On the basis of his experience of the effects of channelization in West Tennessee, Simon (1995) regards conceptual models of channel evolution as a valuable means of interpreting past and present changes and of anticipating future trends. They can help to identify the dominant adjustment processes, to differentiate between local and more widespread instability, and to suggest the type of mitigation measures that might be required to reduce future damage. Once the dominant processes have been identified, numerical methods can be applied to quantify changes in channel form at different stages of evolution. Quantitative prediction of relaxation paths following disruption (Fig. 6.6B and C) represents the ultimate objective, but in the absence of a widely applicable methodology, qualitative models can be an important step in that direction.

(f) River management involves a range of options, including the preservation of essentially undisturbed river systems and the *restoration or rehabilitation* of systems which have experienced some degree of environmental degradation. Many lowland rivers in Europe have been directly modified through channelization: over 90 per cent in the cases of Denmark and the UK (Brookes, 1995b). Complete restoration to a predisturbance state is not usually a viable option, but partial restoration (rehabilitation) can improve the hydraulic, biological and aesthetic status of a river. The need to establish channels that are in harmony with local natural conditions is increasingly recognized as a crucial management philosophy, and geomorphology can play a large part in ensuring sustainable designs. The requirement is an ability to predict the stable form of a river, particularly in its plan and cross-sectional dimensions. Since straight channels are rare in nature, imposition of such a condition on a river is unlikely to remain successful, and yet straightening is one of the main forms of channelization (Table 6.11). The reinstatement of bends is now becoming common practice along straightened watercourses in Denmark (Brookes, 1988), along with the restoration of pools, riffles and point bars. The traditional engineering approach of imposing trapezoidal cross-sections is being replaced by one which recognizes the importance of a variable cross-sectional asymmetry, in line with the meandering theme, although this represents a more complex design and construction process.

Conservation strategies emphasize the need to maintain essential ecological processes and preserve genetic diversity. River corridors are particularly important as habitats and routeways for biological populations, and their promotion is often a major component of rehabilitation schemes. Maintenance of a riparian vegetation zone is also of considerable benefit to a river system through erosion control, as the clearances along Redwood Creek (Nolan and Marron, 1995) and the Bega River (Brooks and Brierley, 1997) ably demonstrate, and can improve the visual quality of the fluvial landscape. Environmental managers are increasingly being required to consider the aesthetic consequences of their actions, and quantitative assessment of aesthetic preferences commonly involves geomorphological variables (Leopold, 1969) as well as artistic and psychological dimensions (Gobster and Chenoweth, 1989). Restoration, whether complete or partial, is still in its infancy, but fluvial geomorphology can make a significant contribution at different stages of the restorative process (Brookes, 1995b).

These several strands suggest how the concepts, methods and results of fluvial geomorphology can be applied to management problems. However the fluvial system is managed, the consequences of particular actions need to be understood. Unfortunately, the three-dimensional adjustment of river channels to changes in external conditions defies accurate quantitative prediction. Empirical and theoretical information is needed on the relaxation paths following disturbance, with due regard to the rates of adjustment of different channel form elements. The propagation process whereby changes are transmitted away from their initial source is poorly understood. This process underlines the high degree of interaction within the fluvial system – between hillslopes and channels, between main streams and tributaries, between channels and floodplains – as water and sediment are transported from land surface to ocean.

REFERENCES

Abbott, J.E. and Francis, J.R.D. 1977: Saltation and suspension trajectories of solid grains in a water stream. *Philosophical Transactions of the Royal Society* **284A**, 225–54.

Abbott, M.B., Bathurst, J.C., Cunge, J.A., O'Connell, P.E. and Rasmussen, J. 1986: An introduction to the European Hydrological System – Système Hydrologique Européen 'SHE'. 2. Structure of the physically based, distributed modelling system. *Journal of Hydrology* **87**, 61–77.

Abrahams, A.D. 1972: Drainage densities and sediment yields in eastern Australia. *Australian Geographical Studies* **10**, 19–41.

Abrahams, A.D. 1976: Evolutionary changes in link lengths: further evidence for stream abstraction. *Transactions of the Institute of British Geographers* New Series **1**, 225–30.

Abrahams, A.D. 1977: The factor of relief in the evolution of channel networks in mature drainage basins. *American Journal of Science* **277**, 626–46.

Abrahams, A.D. 1980: Divide angles and their relation to interior link lengths in natural channel networks. *Geographical Analysis* **12**, 157–71.

Abrahams, A.D. 1984a: Channel networks: a geomorphological perspective. *Water Resources Research* **20**, 161–88.

Abrahams, A.D. 1984b: Tributary development along winding streams and valleys. *American Journal of Science* **284**, 863–92.

Abrahams, A.D. 1987: Channel network topology: regular or random? In Gardiner, V. (ed.), *International geomorphology 1986 part II*. Chichester: Wiley, 145–58.

Abrahams, A.D. and Campbell, R.N. 1976: Source and tributary-source link lengths in natural channel networks. *Bulletin of the Geological Society of America* **87**, 1016–20.

Abrahams, A.D., Li, G. and Atkinson, J.F. 1995: Step–pool streams: adjustment to maximum flow resistance. *Water Resources Research* **31**, 2593–602.

Abrahams, A.D. and Updegraph, J. 1987: Some space-filling controls on the arrangement of tributaries in dendritic channel networks. *Water Resources Research* **23**, 489–95.

Ackers, P. 1992: 1992 Gerald Lacey memorial lecture. Canal and river regime in theory and practice: 1929–92. *Proceedings of the Institution of Civil Engineers* **96**, 167–78.

Ackers, P. and Charlton, F.G. 1970a: Dimensional analysis of alluvial channels with special reference to meander length. *Journal of Hydraulics Research* **8**, 287–316.

Ackers, P. and Charlton, F.G. 1970b: The slope and resistance of small meandering channels. *Proceedings of the Institution of Civil Engineers* **47**, Supplementary Paper 7362-S, 349–70.

Al-Durrah, M. and Bradford, J.M. 1982: New methods of studying soil detachment due to raindrop impact. *Journal of the Soil Science Society of America* **45**, 949–53.

Alexander, C.S. and Nunnally, N.R. 1972: Channel stability on the lower Ohio River. *Annals of the Association of American Geographers* **62**, 411–17.

Allen, J.R.L. 1983: River bedforms: progress and problems. In Collinson, J.D. and Lewin, J. (eds), *Modern and ancient fluvial systems*. Oxford: Blackwell. Special Publication of the International Association of Sedimentologists **6**, 19–33.

American Society of Civil Engineers 1975: *Sedimentation engineering*. American Society of Civil Engineers Manuals and Reports on Engineering Practice **54**. New York: ASCE.

Anderson, M.G. and Burt, T.P. 1990: Subsurface runoff. In Anderson, M.G. and Burt, T.P. (eds), *Process studies in hillslope hydrology*. Chichester: Wiley, 365–400.

Anderson, M.G. and Calver, A. 1977: On the persistence of landscape features formed by a large flood. *Transactions of the Institute of British Geographers* New Series **2**, 243–54.

Anderson, M.G. and Calver, A. 1980: Channel plan changes following large floods. In Cullingford, R.A., Davidson, D.A. and Lewin, J. (eds), *Timescales in geomorphology*. Chichester: Wiley-Interscience, 43–52.

Andrews, E.D. 1979: Scour and fill in a stream channel, East Fork River, western Wyoming. *United States Geological Survey Professional Paper* **1117**.

Andrews, E.D. 1980: Effective and bankfull discharges of streams in the Yampa river basin, Colorado and Wyoming. *Journal of Hydrology* **46**, 311–30.

Andrews, E.D. 1982: Bank stability and channel width adjustment, East Fork River, Wyoming. *Water Resources Research* **18**, 1184–92.

Andrews, E.D. 1983: Entrainment of gravel from naturally sorted river-bed material. *Bulletin of the Geological Society of America* **94**, 1225–31.

Andrews, E.D. 1984: Bed-material entrainment and hydraulic geometry of gravel-bed rivers in Colorado. *Bulletin of the Geological Society of America* **95**, 371–8.

Andrews, E.D. and Nankervis, J.M. 1995: Effective discharge and the design of channel maintenance flows for gravel-bed rivers. In Costa, J.E., Miller, A.J., Potter, K.W. and Wilcock, P.R. (eds), *Natural and anthropogenic influences in fluvial geomorphology*. Washington, DC: American Geophysical Union. Geophysical Monograph **89**, 151–64.

Andrews, E.D. and Smith, J.D. 1992: A theoretical model for calculating marginal bed load transport rates of gravel. In Billi, P., Hey, R.D., Thorne, C.R. and Tacconi, P. (eds), *Dynamics of gravel-bed rivers*. Chichester: Wiley, 41–52.

Arnold, C.L., Boison, P.J. and Patton, P.C. 1982: Sawmill Brook: an example of rapid geomorphic change related to urbanization. *Journal of Geology* **90**, 155–66.

Ashida, K., Egashira, S. and Ando, N. 1984: Generation and geometric features of step–pool bed forms. *Disaster Prevention Research Institute, Kyoto University, Annals* **27**, 341–53.

Ashmore, P.E. 1991a: How do gravel-bed rivers braid? *Canadian Journal of Earth Sciences* **28**, 326–41.

Ashmore, P.E. 1991b: Channel morphology and bed load pulses in braided gravel-bed streams. *Geografiska Annaler* **73A**, 37–52.

Ashmore, P.E. 1993: Anabranch confluence kinetics and sedimentation processes in gravel-braided streams. In Best, J.L. and Bristow, C.S. (eds), *Braided rivers, Special Publication of the Geological Society of London* **75**, 129–46.

Ashmore, P.E. and Day, T.J. 1988: Effective discharge for suspended sediment transport in streams of the Saskatchewan River basin. *Water Resources Research* **24**, 864–70.

Ashworth, P.J. 1996: Mid-channel bar growth and its relationship to local flow strength and direction. *Earth Surface Processes and Landforms* **21**, 103–23.

Ashworth, P.J. and Ferguson, R.I. 1989: Size-selective entrainment of bed load in gravel bed streams. *Water Resources Research* **25**, 627–34.

Bagnold, R.A. 1956: The flow of cohesionless grains in fluids. *Philosophical Transactions of the Royal Society* **249A,** 235–97.

Bagnold, R.A. 1960: Some aspects of the shape of river meanders. *United States Geological Survey Professional Paper* **282E,** 135–44.

Bagnold, R.A. 1977: Bed load transport by natural rivers. *Water Resources Research* **13**, 303–12.

Bagnold, R.A. 1980: An empirical correlation of bedload transport rates in flumes and natural rivers. *Proceedings of the Royal Society* **372A,** 453–73.

Baker, V.R. 1973: Paleohydrology and sedimentology of Lake Missoula flooding in eastern Washington. *Geological Society of America Special Paper* **144**.

Baker, V.R. 1977: Stream-channel response to floods, with examples from central Texas. *Bulletin of the Geological Society of America* **88**, 1057–71.

Baker, V.R. 1978: Adjustment of fluvial systems to climate and source terrain in tropical and subtropical environments. In Miall, A.D. (ed.), *Fluvial sedimentology*. Calgary: Canadian Society of Petroleum Geologists, Memoir 5, 211–30.

Baker, V.R. and Costa, J.E. 1987: Flood power. In Mayer, L. and Nash, D. (eds), *Catastrophic flooding*. London: Allen & Unwin, 1–24.

Baker, V.R. and Penteado-Orellana, M.M. 1977: Adjustment to Quaternary climatic change by the Colorado River in central Texas. *Journal of Geology* **85**, 395–422.

Baker, V.R. and Pickup, G. 1987: Flood geomorphology of the Katherine Gorge, Northern Territory, Australia. *Bulletin of the Geological Society of America* **98**, 635–46.

Ball, J. and Trudgill, S.T. 1995: Overview of solute modelling. In Trudgill, S.T. (ed.), *Solute modelling in catchment systems*. Chichester: Wiley, 3–56.

Batalla, R.J. and Sala, M. 1995: Effective discharge for bedload transport in a subhumid Mediterranean sandy gravel-bed river (Arbúcies, north-east Spain). In Hickin, E.J. (ed.), *River geomorphology*. Chichester: Wiley, 93–103.

Bates, B.C. 1990: A statistical log piecewise linear model of at-a-station hydraulic geometry. *Water Resources Research* **26**, 109–18.

Bathurst, J.C. 1993: Flow resistance through the channel network. In Beven, K. and Kirkby, M.J. (eds), *Channel network hydrology*. Chichester: Wiley, 69–98.

Bathurst, J.C., Thorne, C.R. and Hey, R.D. 1979: Secondary flow and shear stress at bends. *Journal of the Hydraulics Division American Society of Civil Engineers* **105**, HY10, 1277–95.

Beach, T. 1994: The fate of eroded soil: sediment sinks and sediment budgets of agrarian landscapes in southern Minnesota, 1851–1988. *Annals of the Association of American Geographers* **84**, 5–28.

Beckinsale, R.P. 1969: River regimes. In Chorley, R.J. (ed.), *Introduction to physical hydrology*. London: Methuen, 176–92.

Begin, Z.B. 1981: Stream curvature and bank erosion: a model based on the momentum equation. *Journal of Geology* **89**, 497–504.

Begin, Z.B. 1986: Curvature ratio and rate of river bend migration – update. *Journal of Hydraulic Engineering* **112**, 904–8.

Begin, Z.B. 1988: Application of a diffusion–erosion model to alluvial channels which degrade due to base-level lowering. *Earth Surface Processes and Landforms* **13**, 487–500.

Begin, Z.B., Meyer, D.F. and Schumm, S.A. 1981: Development of longitudinal profiles of alluvial channels in response to base-level lowering. *Earth Surface Processes and Landforms* **6**, 49–68.

Benito, G. 1997: Energy expenditure and geomorphic work of the cataclysmic Missoula flooding in the Columbia River gorge, USA. *Earth Surface Processes and Landforms* **22**, 457–72.

Bennett, J.R. and Selby, M.J. 1977: Induced channel instability and hydraulic geometry of the Mangawhara stream, New Zealand. *Journal of Hydrology (New Zealand)* **16**, 134–47.

Berry, M., Hollingworth, T., Anderson, E.M. and Flinn, R.M. 1975: Application of network analysis to the study of branching patterns of dendritic fields. In Kreutzberg, G.W. (ed.), *Advances in neurology*. New York: Raven Press, 217–45.

Beschta, R.L. 1983: Long-term changes in channel widths of the Kowai River, Torlesse Range, New Zealand. *Journal of Hydrology (New Zealand)* **22**, 112–22.

Best, J.L. 1986: The morphology of river channel confluences. *Progress in Physical Geography* **10**, 157–74.

Best, J.L. 1988: Sediment transport and bed morphology at river channel confluences. *Sedimentology* **35**, 481–98.

Best, J.L. 1992: On the entrainment of sediment and initiation of bed defects: insights from recent developments within turbulent boundary layer research. *Sedimentology* **39**, 797–811.

Betson, R.P. 1964: What is watershed runoff? *Journal of Geophysical Research* **69**, 1541–52.

Bettess, R. and White, W.R. 1983: Meandering and braiding of alluvial channels. *Proceedings of the Institution of Civil Engineers* **75**, 525–38.

Bettess, R. and White, W.R. 1987: Extremal hypotheses applied to river regime. In Thorne, C.R., Bathurst, J.C. and Hey, R.D. (eds), *Sediment transport in gravel-bed rivers*. Chichester: Wiley, 767–78.

Bettess, R., White, W.R. and Reeve, C.E. 1988: On the width of regime channels. In White, W.R. (ed.), *International conference on river regime*. Wallingford, UK: Hydraulics Research, 149–61.

Beven, K.J. and Kirkby, M.J. 1979: A physically-based, variable contributing-area model of basin hydrology. *Hydrological Sciences Bulletin* **24**, 43–69.

Beven, K.J. and Wood, E.F. 1993: Flow routing and the hydrological response of channel networks. In Beven, K. and Kirkby, M.J. (eds), *Channel network hydrology*. Chichester: Wiley, 99–128.

Biedenharn, D.S. and Thorne, C.R. 1994: Magnitude–frequency analysis of sediment transport in the Lower Mississippi River. *Regulated Rivers: Research and Management* **9**, 237–51.

Bishop, P. 1995: Drainage rearrangement by river capture, beheading and diversion. *Progress in Physical Geography* **19**, 449–73.

Blair, T.C. and McPherson, J.G. 1994: Alluvial fan processes and forms. In Abrahams, A.D. and Parsons, A.J. (eds), *Geomorphology of desert environments*. London: Chapman & Hall, 354–402.

Blondeaux, P. and Seminara, G. 1985: A unified bar–bend theory of river meanders. *Journal of Fluid Mechanics* **157**, 449–70.

Bluck, B.J. 1987: Bed forms and clast size changes in gravel-bed rivers. In Richards, K.S. (ed.), *River channels: environment and process*. Oxford: Blackwell, 159–78.

Bogardi, J. 1974. *Sediment transport in alluvial streams*. Budapest: Akademiai Kiado.

Boon, P.J. 1992: Essential elements in the case for river conservation. In Boon, P.J., Calow, P. and Petts, G.E. (eds), *River conservation and management*. Chichester: Wiley, 11–33.

Brackenridge, G.R. 1988: River flood regime and floodplain stratigraphy. In Baker, V.R., Kochel, R.C. and Patton, P.C. (eds), *Flood geomorphology*. Chichester: Wiley, 139–57.

Bradley, J.B. and McCutcheon, S.C. 1987: Influence of large suspended-sediment concentrations in rivers. In Thorne, C.R., Bathurst, J.C. and Hey, R.D. (eds), *Sediment transport in gravel-bed rivers*. Chichester: Wiley, 645–75.

Bradley, S.B. and Cox, J.J. 1987: Heavy metals in the Hamps and Manifold valleys, north Staffordshire, UK: partitioning of metals in floodplain soils. *Science of the Total Environment* **65**, 135–53.

Bradley, W.C., Fahenstock, R.K. and Rowekamp, E.T. 1972: Coarse sediment transport by flood flows on Knik River, Alaska. *Bulletin of the Geological Society of America* **83**, 1261–84.

Brandt, C.J. and Thornes, J.B. 1987: Erosional energetics. In Gregory, K.J. (ed.) *Energetics of physical environment*. Chichester: Wiley, 51–87.

Bray, D.I. 1982: Regime equations for gravel-bed rivers. In Hey, R.D., Bathurst, J.C. and Thorne, C.R. (eds), *Gravel bed rivers*. Chichester: Wiley, 517–42.

Brayshaw, A.C. 1985: Bed microtopography and entrainment thresholds in gravel-bed rivers. *Bulletin of the Geological Society of America* **96**, 218–23.

Brebner, A. and Wilson, K.C. 1967: Determination of the regime equation from relationships for pressurised flow by use of the principle of minimum energy degradation. *Proceedings of the Institution of Civil Engineers* **36**, 47–62.

Brewer, P.A., Leeks, G.J.L. and Lewin, J. 1992: Direct measurement of in-channel abrasion processes. In *Erosion and sediment transport monitoring programmes in river basins*, Proceedings of the Oslo Symposium, August 1992. International Association of Hydrological Sciences Publication **210**, 21–9.

Brice, J.C. 1984: Planform properties of meandering rivers. In Elliott, C.M. (ed.), *River meandering*. New Orleans: American Society of Civil Engineers, 1–15.

Bridge, J.S. 1977: Flow, bed topography, grain size and sedimentary structure in open channel bends: a three-dimensional model. *Earth Surface Processes* **2**, 401–16.

Bridge, J.S. 1993: The interaction between channel geometry, water flow, sediment transport and deposition in braided rivers. In Best, J.L. and Bristow, C.S. (eds), *Braided rivers. Special Publication of the Geological Society of London* **75**, 13–71.

Bridge, J.S. and Leeder, M.R. 1979: A simulation model of alluvial stratigraphy. *Sedimentology* **26**, 617–44.

Brierley, G.J. and Hickin, E.J. 1985: The downstream gradation of particle sizes in the Squamish River, British Columbia. *Earth Surface Processes and Landforms* **10**, 597–606.

Brierley, G.J. and Hickin, E.J. 1991: Channel planform as a non-controlling factor in fluvial sedimentology: the case of the Squamish River floodplain, British Columbia. *Sedimentary Geology* **75**, 67–83.

Bristow, C.S. 1987: Brahmaputra River: channel migration and deposition. In Ethridge, F.G., Flores, R.M. and Harvey, M.D. (eds), *Recent developments in fluvial sedimentology*. SEPM Special Publication **39**, 63–74.

Bristow, C.S., Best, J.L. and Roy, A.G. 1993: Morphology and facies models of channel confluences. In Marzo, M. and Puigdefabregas, C. (eds), *Alluvial sedimentation*. Blackwell Scientific, International Association of Sedimentologists Special Publication **17**, 91–100.

Broecker, W. 1984: Terminations. In Berger, A., Imbrie, J., Hays, J., Kukla, G. and Saltzman, B. (eds), *Milankovitch and climate*. Dordrecht: Reidel, 687–98.

Brookes, A. 1985: River channelization: traditional engineering methods, physical consequences and alternative practices. *Progress in Physical Geography* **9**, 44–73.

Brookes, A. 1987: River channel adjustments downstream from channelization works in England and Wales. *Earth Surface Processes and Landforms* **12**, 337–51.

Brookes, A. 1988: *Channelized rivers. Perspectives for environmental management*. Chichester: Wiley.

Brookes, A. 1995a: Challenges and objectives for geomorphology in UK river management. *Earth Surface Processes and Landforms* **20**, 593–610.

Brookes, A. 1995b: River channel restoration: theory and practice. In Gurnell, A.M. and Petts, G.E. (eds), *Changing river channels*. Chichester: Wiley, 369–88.

Brooks, A.P. and Brierley, G.J. 1997: Geomorphic responses of lower Bega River to catchment disturbance. *Geomorphology* **18**, 291–304.

Brotherton, D.l. 1979: On the origin and characteristics of river channel patterns. *Journal of Hydrology* **44**, 211–30.

Brown, A.G. 1991: Hydrogeomorphological changes in the Severn basin during the last 15,000 years: orders of change in a maritime catchment. In Starkel, L., Gregory, K.J. and Thornes, J.B. (eds), *Temperate palaeohydrology*. Chichester: Wiley, 147–69.

Brown, A.G. and Keough, M. 1992: Holocene floodplain metamorphosis in the Midlands, United Kingdom. *Geomorphology* **4**, 433–45.

Brown, F.R. 1963: Cavitation in hydraulic structures: problems created by cavitation phenomena. *Journal of the Hydraulics Division American Society of Civil Engineers* **89**, HY1, 99–116.

Brown, L.R. 1984: Conserving soils. In Brown, L.R. (ed.), *State of the world*, 1984. New York: Norton, 53–75.

Brunsden, D. 1980: Applicable models of long term landform evolution. *Zeitschrift für Geomorphologie* Supplement 36, 16–26.

Brunsden, D. and Thornes, J.B. 1979: Landscape sensitivity and change. *Transactions of the Institute of British Geographers* New Series **4**, 463–84.

Brush, L.M. 1961: Drainage basins, channels, and flow characteristics of selected streams in central Pennsylvania. *United States Geological Survey Professional Paper* **282F**, 145–81.

Bull, W.B. 1977: The alluvial fan environment. *Progress in Physical Geography* **1**, 222–70.

Bull, W.B. 1991: *Geomorphic responses to climatic change*. New York: Oxford University Press.

Bunting, B.T. 1961: The role of seepage moisture in soil formation, slope development and stream initiation. *American Journal of Science* **259**, 503–18.

Burkham, D.E. 1972: Channel changes of the Gila River in Safford valley, Arizona, 1846–1970. *United States Geological Survey Professional Paper* **655G.**

Burkham, D.E. 1981: Uncertainties resulting from changes in river form. *Journal of the Hydraulics Division American Society of Civil Engineers* **107**, HY5, 593–610.

Burt, T.P. 1992: The hydrology of headwater catchments. In Calow, P. and Petts, G.E. (eds), *The river's handbook: hydrological and ecological principles*. Oxford: Blackwell, 3–28.

Caine, N. and Swanson, F.J. 1989: Geomorphic coupling of hillslope and channel systems in two small mountain basins. *Zeitschrift für Geomorphologie* **33**, 189–203.

Callander, R.A. 1978: River meandering. *Annual Review of Fluid Mechanics* **10**, 129–58.

Calver, A. 1978: Modelling drainage headwater development. *Earth Surface Processes* **3**, 233–41.

Cao, S. and Knight, D.W. 1996: Regime theory of alluvial channels based upon the concept of stream power and probability. *Proceedings of the Institution of Civil Engineers* **100**, 160–7.

Carling, P.A. 1983: Threshold of coarse sediment transport in broad and narrow natural streams. *Earth Surface Processes and Landforms* **8**, 1–18.

Carling, P.A. 1987: Bed stability in gravel streams, with reference to stream regulation and ecology. In Richards, K.S. (ed.), *River channels: environment and process.* Oxford: Blackwell, 321–47.

Carling, P.A. 1988: The concept of dominant discharge applied to two gravel-bed streams in relation to channel stability thresholds. *Earth Surface Processes and Landforms* **13**, 355–67.

Carling, P.A. 1991: An appraisal of the velocity-reversal hypothesis for stable pool–riffle sequences in the River Severn, England. *Earth Surface Processes and Landforms* **16**, 19–31.

Carling, P.A. and Wood, N. 1994: Simulation of flow over pool–riffle topography: a consideration of the velocity reversal hypothesis. *Earth Surface Processes and Landforms* **19**, 319–32.

Carlston, C.W. 1963: Drainage density and streamflow. *United States Geological Survey Professional Paper* **422C**.

Carlston, C.W. 1965: The relation of free meander geometry to stream discharge and its geomorphic implications. *American Journal of Science* **263**, 864–85.

Carlston, C.W. 1968: Slope–discharge relations for eight rivers in the United States. *United States Geological Survey Professional Paper* **600D**, 45–7.

Carson, M.A. 1984a: Observations on the meandering–braided transition, Canterbury Plains, New Zealand. *New Zealand Geographe*r 40, 89–99.

Carson, M.A. 1984b: The meandering–braided river threshold: a reappraisal. *Journal of Hydrology* **73**, 315–34.

Carson, M.A. 1987: Measures of flow intensity as predictors of bed load. *Journal of Hydrology* **113**, 1402–21.

Carson, M.A. and Griffiths, G.A. 1987: Influence of channel width on bed load transport capacity. *Journal of Hydraulic Engineering* **113**, 1489–509.

Carson, M.A. and Lapointe, M.F. 1983: The inherent asymmetry of river meander planform. *Journal of Geology* **91**, 41–56.

Chang, H.H. 1979a: Geometry of rivers in regime. *Journal of the Hydraulics Division American Society of Civil Engineers* **105**, HY6, 691–706.

Chang, H.H. 1979b: Minimum stream power and river channel patterns. *Journal of Hydrology* **41**, 303–27.

Chang, H.H. 1980: Geometry of gravel streams. *Journal of the Hydraulics Division American Society of Civil Engineers* **106**, HY9, 1443–56.

Chang, H.H. 1985: River morphology and thresholds. *Journal of Hydraulic Engineering* **111**, 503–19.

Chang, H.H. 1986: River channel changes: adjustments of equilibrium. *Journal of Hydraulic Engineering* **112**, 43–55.

Chang, H.H. 1988: On the cause of river meandering. In White, W.R. (ed.), *International conference on river regime.* Wallingford, UK: Hydraulics Research, 83–93.

Charlton, F.G., Brown, P.M. and Benson, R.W. 1978: The hydraulic geometry of some gravel rivers in Britain. *Hydraulics Research Station Report* IT 180.

Cherkauer, D.S. 1972: Longitudinal profiles of ephemeral streams in southeastern Arizona. *Bulletin of the Geological Society of America* **83**, 353–66.

Chien, N. 1961: The braided stream of the lower Yellow River. *Scientia Sinica* **10**, 734–54.

Chin, A. 1989: Step pools in stream channels. *Progress in Physical Geography* **13**, 391–407.

Chitale, S.V. 1973: Theories and relationships of river channel patterns. *Journal of Hydrology* **19**, 285–308.

Chorley, R.J. 1969: The drainage basin as the fundamental geomorphic unit. In Chorley, R.J. (ed.), *Introduction to physical hydrology*. London: Methuen, 37–59.

Chorley, R.J. and Kennedy, B.A. 1971: *Physical geography: a systems approach*. London: Prentice-Hall.

Church, M. 1978: Palaeohydrological reconstructions from a Holocene valley fill. In Miall, A.D. (ed.), *Fluvial sedimentology*. Calgary: Canadian Society of Petroleum Geologists, Memoir **5**, 743–72.

Church, M. 1980: *On the equations of hydraulic geometry*. Vancouver: Department of Geography, University of British Columbia.

Church, M. 1983: Pattern of instability in a wandering gravel bed channel. In Collinson, J.D. and Lewin, J. (eds), *Modern and ancient fluvial systems*. Oxford: Blackwell, Special Publication of the International Association of Sedimentologists **6**, 169–80.

Church, M. 1992: Channel morphology and typology. In Calow, P. and Petts, G.E. (eds), *The river's handbook: hydrological and ecological principles*. Oxford: Blackwell, 126–43.

Church, M. 1995: Geomorphic response to river flow regulation: case studies and time-scales. *Regulated Rivers: Research and Management* **11**, 3–22.

Church, M. and Hassan, M.A. 1992: Size and distance of travel of unconstrained clasts on a streambed. *Water Resources Research* **28**, 299–303.

Church, M. and Jones, D. 1982: Channel bars in gravel-bed rivers. In Hey, R.D., Bathurst, J.C. and Thorne, C.R. (eds), *Gravel bed rivers*. Chichester: Wiley, 291–338.

Church, M. and Mark, D.M. 1980: On size and scale in geomorphology. *Progress in Physical Geography* **4**, 342–90.

Church, M. and Slaymaker, O. 1989: Disequilibrium of Holocene sediment yield in glaciated British Columbia. *Nature* **337**, 452–54.

Clark, E.H., Haverkamp, J.A. and Chapman, W. 1985: *Eroding soils: the off-farm impacts*. Washington, DC: The Conservation Foundation.

Clifford, N.J. 1993a: Differential bed sedimentology and the maintenance of riffle–pool sequences. *Catena* **20**, 447–68.

Clifford, N.J. 1993b: Formation of riffle–pool sequences: field evidence for an autogenic process. *Sedimentary Geology* **85**, 39–51.

Clifford, N.J., Hardisty, J., French, J.R. and Hart, S. 1993: Downstream variation in bed material characteristics: a turbulence-controlled form–process feedback mechanism. In Best, J.L. and Bristow, C.S. (eds), *Braided rivers. Special Publication of the Geological Society of London* **75**, 89–104.

Clifford, N.J. and Richards, K.S. 1992: The reversal hypothesis and the maintenance of riffle–pool sequences: a review and field appraisal. In Carling, P.A. and Petts, G.E. (eds), *Lowland floodplain rivers*. Chichester: Wiley, 43–70.

Clifford, N.J., Robert, A. and Richards, K.S. 1992: Estimation of flow resistance in gravel-bedded rivers: a physical explanation of the multiplier of roughness length. *Earth Surface Processes and Landforms* **17**, 111–26.

Colby, B.R. 1963: Fluvial sediments – a summary of source, transportation, deposition, and measurement of sediment discharge. *United States Geological Survey Bulletin* **1181A**.

Colby, B.R. 1964: Sand discharge and mean-velocity relationships in sandbed streams. *United States Geological Survey Professional Paper* **462A**.

Coleman, S.E. and Melville, B.W. 1996: Initiation of bed forms on a flat sand bed. *Journal of Hydraulic Engineering* **122**, 301–10.

Cooke, R.U. and Reeves, R.W. 1976: *Arroyos and environmental change in the American South-West*. Oxford: Clarendon Press.

Costa, J.E. 1974: Response and recovery of a Piedmont watershed from tropical storm Agnes, June 1972. *Water Resources Research* **10**, 106–12.

Costa, J.E. 1975: Effects of agriculture on erosion and sedimentation in Piedmont province, Maryland. *Bulletin of the Geological Society of America* **86**, 1281–6.

Costa, J.E. and O'Connor, J.E. 1995: Geomorphologically effective floods. In Costa, J.E., Miller, A.J., Potter, K.W. and Wilcock, P.R. (eds), *Natural and anthropogenic influences in fluvial geomorphology*. Washington, DC: American Geophysical Union, Geophysical Monograph **89**, 45–56.

Cullingford, R.A., Davidson, D.A. and Lewin, J. (eds) 1980: *Timescales in geomorphology*. Chichester: Wiley-Interscience.

Cunge, J.A. 1969: On the subject of a flood propagation method. *Journal of Hydraulics Research* **7**, 205–30.

Dacey, M.F. and Krumbein, W.C. 1976: Three growth models for stream channel networks. *Journal of Geology* **84**, 153–63.

Dansgaard, W.S., White, J.W.C. and Johnsen, S.J. 1989: The abrupt termination of the Younger Dryas climatic event. *Nature* **339**, 532–4.

Darby, S.E. and Thorne, C.R. 1996: Modelling the sensitivity of channel adjustments in destabilized sand-bed rivers. *Earth Surface Processes and Landforms* **21**, 1109–25.

Davies, T.R. and Sutherland, A.J. 1980: Resistance to flow past deformable boundaries. *Earth Surface Processes* **5**, 175–9.

Davis, W.M. 1899: The geographical cycle. *Geographical Journal* **14**, 481–504.

Davis, W.M. 1902: Base-level, grade and peneplain. *Journal of Geology* **10**, 77–111.

Davoren, A. and Mosley, M.P. 1986: Observations of bedload movement, bar development and sediment supply in the braided Ohau River. *Earth Surface Processes and Landforms* **11**, 643–52.

Day, D.G. 1983: Drainage density variability and drainage basin outputs. *Journal of Hydrology (New Zealand)* **22**, 3–17.

Dearing, J.A. 1992: Sediment yields and sources in a Welsh upland lake-catchment during the past 800 years. *Earth Surface Processes and Landforms* **17**, 1–22.

Degens, E.T., Kempe, S. and Richey, J.E. 1991: Summary: Biogeochemistry of major world rivers. In Degens, E.T., Kempe, S. and Richey, J.E. (eds), *Biogeochemistry of major world rivers*, SCOPE 42. Chichester: Wiley, 323–47.

Deigaard, R. 1982: Longitudinal sorting of grain sizes in alluvial rivers. In Sumer, B.M. and Muller, A. (eds), *Mechanics of sediment transport*, Proceedings of Euromech **156**, Istanbul, July 1982, 231–6.

Desloges, J.R. and Church, M.A. 1989: Wandering gravel-bed rivers. *Canadian Geographer* **33**, 360–4.

Dietrich, W.E. 1987: Mechanics of flow and sediment transport in river bends. In Richards, K.S. (ed.), *River channels: environment and process*. Oxford: Blackwell, 179–227.

Dietrich, W.E. and Dunne, T. 1993: The channel head. In Beven, K. and Kirkby, M.J. (eds), *Channel network hydrology*. Chichester: Wiley, 175–219.

Dietrich, W.E. and Smith, J.D. 1983: Influence of the point bar on flow through curved channels. *Water Resources Research* **19**, 1173–92.

Dietrich, W.E. and Smith, J.D. 1984: Bedload transport in a river meander. *Water Resources Research* **20**, 1355–80.

Dietrich, W.E., Wilson, C.J., Montgomery, D.R. and McKean, J. 1993: Analysis of erosion thresholds, channel networks, and landscape morphology using a digital terrain model. *Journal of Geology* **101**, 259–78.

Dietrich, W.E., Wilson, C.J. and Reneau, S.L. 1986: Hollows, colluvium and landslides in soil-mantled landscapes. In Abrahams, A.D. (ed.), *Hillslope processes*. Boston: Allen & Unwin, 362–88.

Dinehart, R.L. 1989: Dune migration in a steep, coarse-bedded stream. *Water Resources Research* **25**, 911–23.

Dohrenwend, J.C., Abrahams, A.D. and Turrin, B.D. 1987: Drainage development on basaltic lava flows, Cima volcanic field, southeast California, and Lunar Crater volcanic field, south-central Nevada. *Bulletin of the Geological Society of America* **99**, 405–13.

Downs, P.W. and Gregory, K.J. 1993: The sensitivity of river channels in the landscape system. In Thomas, D.S.G. and Allison, R.J. (eds), *Landscape sensitivity*. Chichester: Wiley, 15–30.

Dozier, J. 1976: An examination of the variance minimization tendencies of a supraglacial stream. *Journal of Hydrology* **31**, 359–80.

Dunkerley, D.L. 1977: Frequency distributions of stream link lengths and the development of channel networks. *Journal of Geology* **85**, 459–69.

Dunkerley, D.L. 1990: The development of armour in the Tambo River, Victoria, Australia. *Earth Surface Processes and Landforms* **15**, 405–15.

Dunne, T. 1979: Sediment yield and land use in tropical catchments. *Journal of Hydrology* **42**, 281–300.

Dunne, T. 1980: Formation and controls of channel networks. *Progress in Physical Geography* **4**, 211–39.

Dunne, T. 1990: Hydrology, mechanics and geomorphic implications of erosion by subsurface flow. In Higgins, C.G. and Coates, D.R. (eds), *Groundwater geomorphology: the role of subsurface water in earth-surface processes and landforms*, Geological Society of America Special Paper **252**, 1–28.

Dunne, T. and Aubry, B.F. 1986: Evaluation of Horton's theory of sheetwash and rill erosion on the basis of field experiments. In Abrahams, A.D. (ed.), *Hillslope processes*. Boston: Allen & Unwin, 31–53.

Dunne, T. and Dietrich, W.E. 1980: Experimental investigation of Horton overland flow on tropical hillslopes – 2. Hydraulic characteristics and hillslope hydrographs. *Zeitschrift für Geomorphologie Supplement-Band* **35**, 60–80.

Dunne, T. and Leopold, L.B. 1978: *Water in environmental planning*. San Francisco: W. H. Freeman.

Dury, G.H. 1964: Principles of underfit streams. *United States Geological Survey Professional Paper* **452A**.

Dury, G.H. 1970: A resurvey of part of the Hawkesbury River, New South Wales, after 100 years. *Australian Geographical Studies* **8**, 121–32.

Dury, G.H. 1977: Underfit streams: retrospect, perspect, and prospect. In Gregory, K.J. (ed.), *River channel changes*. Chichester: Wiley, 281–93.

Dury, G.H. 1984: Abrupt variation in width along part of the River Severn, near Shrewsbury, Shropshire, England. *Earth Surface Processes and Landforms* **9**, 485–92.

Dury, G.H. 1985: Attainable standards of accuracy in the retrodiction of palaeodischarge from channel dimensions. *Earth Surface Processes and Landforms* **10**, 205–13.

Edgar, D.E. 1984: The role of geomorphic thresholds in determining alluvial channel morphology. In Elliott, C.M. (ed.), *River meandering*. New Orleans: American Society of Civil Engineers, 44–54.

Einstein, H.A. 1950: The bedload function for sediment transportation in open channel flows. *United States Department of Agriculture Technical Bulletin* **1026**.

Einstein, H.A. and Shen, H.W. 1964: A study of meandering in straight alluvial channels. *Journal of Geophysical Research* **69**, 5239–47.

Emmett, W.W. 1970: The hydraulics of overland flow on hillslopes. *United States Geological Survey Professional Paper* **662A**.

Emmett, W.W. 1975: The channels and waters of the Upper Salmon River area, Idaho. *United States Geological Survey Professional Paper* **870A**.

Engelund, F. and Skovgaard, O. 1973: On the origin of meandering and braiding in alluvial streams. *Journal of Fluid Mechanics* **57**, 289–302.

Erskine, W.D. 1996: Response and recovery of a sand-bed stream to a catastrophic flood. *Zeitschrift für Geomorphologie* **40**, 359–83.

Eschner, T.R. 1983: Hydraulic geometry of the Platte River near Overton, south-central Nebraska. *United States Geological Survey Professional Paper* **1277C**.

Everitt, B.L. 1968: Use of the cottonwood in an investigation of the recent history of a flood-plain. *American Journal of Science* **266**, 417–39.

Fahnestock, R.K. 1963: Morphology and hydrology of a glacial stream – White River, Mount Rainier, Washington. *United States Geological Survey Professional Paper* **422A**.

Fawcett, K.R., Anderson, M.G., Bates, P.D., Jordan, J.P. and Bathurst, J.C. 1995: The importance of internal validation in the assessment of physically based distributed models. *Transactions of the Institute of British Geographers* New Series **20**, 248–65.

Fenton, J.D. and Abbott, J.E. 1977: Initial movement of grains on a stream bed: the effect of relative protrusion. *Proceedings of the Royal Society* **352A,** 523–37.

Ferguson, R.I. 1973: Channel pattern and sediment type. *Area* **5**, 38–41.

Ferguson, R.I. 1975: Meander irregularity and wavelength estimation. *Journal of Hydrology* **26**, 315–33.

Ferguson, R.I. 1979: River meanders: regular or random? In Wrigley, N. (ed.), *Statistical applications in the spatial sciences*. London: Pion, 229–41.

Ferguson, R.I. 1984: Kinematic model of meander migration. In Elliott, C.M. (ed.), *River meandering*. New Orleans: American Society of Civil Engineers, 942–51.

Ferguson, R.I. 1986: Hydraulics and hydraulic geometry. *Progress in Physical Geography* **10**, 1–31.

Ferguson, R.I. 1987: Hydraulic and sedimentary controls of channel pattern. In Richards, K.S. (ed.), *River channels: environment and process*. Oxford: Blackwell, 129–58.

Ferguson, R.I. 1993: Understanding braiding processes in gravel-bed rivers: progress and unsolved problems. In Best, J.L. and Bristow, C.S. (eds), *Braided rivers. Special Publication of the Geological Society of London* **75**, 73–87.

Ferguson, R.I. 1994: Critical discharge for entrainment of poorly sorted gravel. *Earth Surface Processes and Landforms* **19**, 179–86.

Ferguson, R.I., Ashmore, P.E., Ashworth, P.J., Paola, C. and Prestegaard, K.L. 1992: Measurements in a braided river chute and lobe: I. Flow pattern, sediment transport and channel change. *Water Resources Research* **28**, 1877–86.

Ferguson, R.I. and Ashworth, P.J. 1991: Slope-induced changes in channel character along a gravel-bed stream: the Allt Dubhaig, Scotland. *Earth Surface Processes and Landforms* **16**, 65–82.

Flint, J.J. 1973: Development of headward growth of channel networks. *Bulletin of the Geological Society of America* **84**, 1087–94.

Flint, J.J. 1980: Tributary arrangements in fluvial systems. *American Journal of Science* **280**, 26–45.

Foley, M.G. 1980: Bed-rock incision by streams. *Bulletin of the Geological Society of America* **91**, Part 2, 2189–213.

Foster, I.D.L. and Walling, D.E. 1994: Using reservoir deposits to reconstruct changing sediment yields and sources in the catchment of the Old Mill Reservoir, south Devon, over the past 50 years. *Hydrological Sciences Journal* **39**, 347–68.

Fredsøe, J. 1978: Meandering and braiding of rivers. *Journal of Fluid Mechanics* **84**, 609–24.

Friedman, J.M., Osterkamp, W.R. and Lewis, W.M. 1996: The role of vegetation and bed-level fluctuations in the process of channel narrowing. *Geomorphology* **14**, 341–51.

Friend, P.F. and Sinha, R. 1993: Braiding and meandering parameters. In Best, J.L. and Bristow, C.S. (eds), *Braided rivers. Special Publication of the Geological Society of London* **75**, 105–11.

Fukuoka, S. 1989: Finite amplitude development of alternate bars. In Ikeda, S. and Parker, G. (eds), *River meandering*, American Geophysical Union, Water Research Monograph **12**, 237–65.

Furbish, D.J. 1988: River-bend curvature and migration: how are they related? *Geology* **16**, 752–5.

Furbish, D.J. 1991: Spatial autoregressive structure in meander evolution. *Bulletin of the Geological Society of America* **103**, 1576–89.

Galay, V.J. 1983: Causes of river bed degradation. *Water Resources Research* **19**, 1057–90.

Germann, P.F. 1990: Macropores and hydrologic hillslope processes. In Anderson, M.G. and Burt, T.P. (eds), *Process studies in hillslope hydrology*. Chichester: Wiley, 327–63.

Germanoski, D. and Ritter, D.F. 1988: Tributary response to local base level lowering below a dam. *Regulated Rivers: Research and Management* **2**, 11–24.

Germanoski, D. and Schumm, S.A. 1993: Changes in braided river morphology resulting from aggradation and degradation. *Journal of Geology* **101**, 451–66.

Gilbert, G.K. 1877: *Report on the geology of the Henry Mountains*. Washington, DC: United States Geological Survey, Rocky Mountain Region.

Gilbert, G.K. 1917: Hydraulic-mining debris in the Sierra Nevada. *United States Geological Survey Professional Paper* **105**.

Gilley, J.E., Elliot, W.J., Laflen, J.M. and Simanton, J.R. 1993: Critical shear stress and critical flow rates for initiation of rilling. *Journal of Hydrology* **142**, 251–71.

Gilman, K. and Newson, M.D. 1980: *Soil pipes and pipeflow – a hydrological study in upland Wales*. Norwich: Geobooks.

Gintz, D., Hassan, M.A. and Schmidt, K.H. 1996: Frequency and magnitude of bedload transport in a mountain river. *Earth Surface Processes and Landforms* **21**, 433–45.

Glock, W.S. 1931: The development of drainage systems: a synoptic view. *Geographical Review* **21**, 475–82.

Gobster, P.H. and Chenoweth, R.E. 1989: The dimensions of aesthetic preference: a quantitative analysis. *Journal of Environmental Management* **29**, 47–72.

Gole, C.V. and Chitale, S.V. 1966: Inland delta building activity of Kosi River. *Journal of the Hydraulics Division American Society of Civil Engineers* **92**, HY2, 111–26.

Gomez, B. 1991: Bedload transport. *Earth-Science Reviews* **31**, 89–132.

Gomez, B. and Church, M. 1989: An assessment of bed load sediment transport formulae for gravel bed rivers. *Water Resources Research* **25**, 1161–86.

Gomez, B., Naff, R.L. and Hubbell, D.W. 1989: Temporal variations in bedload transport rates associated with migration of bedforms. *Earth Surface Processes and Landforms* **14**, 135–56.

Goodchild, M.F. and Klinkenberg, B. 1993: Statistics of channel networks on fractional Brownian surfaces. In Lam, N.S.N. and De Cola, L. (eds), *Fractals in geography*. Englewood Cliffs, NJ: Prentice-Hall, 122–41.

Gorycki, M.A. 1973: Hydraulic drag: a meander-initiating mechanism. *Bulletin of the Geological Society of America* **84**, 175–86.

Govers, G. 1992: Evaluation of transporting capacity formulae for overland flow. In Parsons, A.J. and Abrahams, A.D. (eds), *Overland flow hydraulics and erosion mechanics*. London: UCL Press, 243–73.

Graf, W.L. 1977: The rate law in fluvial geomorphology. *American Journal of Science* **277**, 178–91.

Graf, W.L. 1988a: *Fluvial processes in dryland rivers*. Berlin: Springer-Verlag.

Graf, W.L. 1988b: Applications of catastrophe theory in fluvial geomorphology. In Anderson, M.G. (ed.), *Modelling geomorphological systems*. Chichester: Wiley, 33–47.

Graf, W.L. 1990: Fluvial dynamics of thorium-230 in the Church Rock Event, Puerco River, New Mexico. *Annals of the Association of American Geographers* **80**, 327–42.

Grant, G.E., Swanson, F.J. and Wolman, M.G. 1990: Pattern and origin of stepped-bed morphology in high-gradient streams, Western Cascades, Oregon. *Bulletin of the Geological Society of America* **102**, 340–52.

Gregory, D.I. and Schumm, S.A. 1987: The effect of active tectonics on alluvial river morphology. In Richards, K.S. (ed.), *River channels: environment and process*. Oxford: Blackwell, 41–68.

Gregory, K.J. 1976: Drainage networks and climate. In Derbyshire, E. (ed.), *Geomorphology and climate*. Chichester: Wiley-Interscience, 289–315.

Gregory, K.J. 1995: Human activity and palaeohydrology. In Gregory, K.J., Starkel, L. and Baker, V.R. (eds), *Global continental palaeohydrology*. Chichester: Wiley, 151–72.

Gregory, K.J. and Gardiner, V. 1975: Drainage density and climate. *Zeitschrift für Geomorphologie* **19**, 287–98.

Gregory, K.J., Gurnell, A.M., Hill, C.T. and Tooth, S. 1994: Stability of the pool–riffle sequence in changing river channels. *Regulated Rivers: Research and Management* **9**, 35–43.

Gregory, K.J. and Maizels, J.K. 1991: Morphology and sediments: typological characteristics of fluvial forms and deposits. In Starkel, L., Gregory, K.J. and Thornes, J.B. (eds), *Temperate palaeohydrology*. Chichester: Wiley, 31–59.

Gregory, K.J., Starkel, L. and Baker, V.R. (eds) 1995: *Global continental palaeohydrology*. Chichester: Wiley.

Gregory, K.J. and Walling, D.E. 1968: The variation of drainage density within a catchment. *Bulletin of the International Association of Scientific Hydrology* **13**, 61–8.

Griffiths, G.A. 1979: Recent sedimentation history of the Waimakariri River, New Zealand. *Journal of Hydrology (New Zealand)* **18**, 6–28.

Griffiths, G.A. 1984: Extremal hypotheses for river regime: an illusion of progress. *Water Resources Research* **20**, 113–18.

Griffiths, G.A. 1989: Form resistance in gravel channels with mobile beds. *Journal of Hydraulic Engineering* **115**, 340–55.

Grimshaw, D.L. and Lewin, J. 1980: Source identification for suspended sediments. *Journal of Hydrology* **47**, 151–62.

Grove, J.M. 1988: *The Little Ice Age*. London: Methuen.

Guiot, J., Harrison, S.P. and Prentice, I.C. 1993: Reconstruction of Holocene precipitation patterns in Europe using pollen and lake-level data. *Quaternary Research* **40**, 139–49.

Gupta, A. 1995: Magnitude, frequency, and special factors affecting channel form and processes in the seasonal tropics. In Costa, J.E., Miller, A.J., Potter, K.W. and Wilcock, P.R. (eds), *Natural and anthropogenic influences in fluvial geomorphology*. Washington, DC: American Geophysical Union, Geophysical Monograph **89**, 125–36.

Gupta, A. and Fox, H. 1974: Effects of high-magnitude floods on channel form: a case study in Maryland Piedmont. *Water Resources Research* **10**, 499–509.

Gupta, V.J. and Mesa, O. 1988: Runoff generation and hydrologic response via channel network geomorphology – recent progress and open problems. *Journal of Hydrology* **102**, 3–28.

Hack, J.T. 1957: Studies of longitudinal stream profiles in Virginia and Maryland. *United States Geological Survey Professional Paper* **294B**.

Hack, J.T. 1960: Interpretation of erosional topography in humid temperate regions. *American Journal of Science* **258A**, 80–97.

Hack, J.T. and Goodlett, J.C. 1960: Geomorphology and forest ecology of a mountain region in the Central Appalachians. *United States Geological Survey Professional Paper* **347**.

Haible, W.W. 1980: Holocene profile changes along a California coastal stream. *Earth Surface Processes* **5**, 249–64.

Hammer, T.R. 1972: Stream channel enlargement due to urbanization. *Water Resources Research* **8**, 1530–40.

Harrison, S.S. and Clayton, L. 1970: Effects of ground-water seepage on fluvial processes. *Bulletin of the Geological Society of America* **81**, 1217–26.

Harvey, A.M. 1969: Channel capacity and the adjustment of streams to hydrologic regime. *Journal of Hydrology* **8**, 82–98.

Harvey, A.M. 1975: Some aspects of the relations between channel characteristics and riffle spacing in meandering streams. *American Journal of Science* **275**, 470–8.

Harvey, A.M. 1990: Factors influencing Quaternary alluvial fan development in southeast Spain. In Rachocki, A.H. and Church, M. (eds), *Alluvial fans: a field approach*. Chichester: Wiley, 247–69.

Harvey, A.M. 1992: Process interactions, temporal scales and the development of hillslope gully systems: Howgill Fells, northwest England. *Geomorphology* **5**, 323–44.

Harvey, A.M. 1994: Influence of slope/stream coupling on process interactions on eroding gully slopes: Howgill Fells, northwest England. In Kirkby, M.J. (ed.), *Process models and theoretical geomorphology*. Chichester: Wiley, 247–70.

Harvey, A.M. 1997: The role of alluvial fans in arid zone fluvial systems. In Thomas, D.S.G. (ed.), *Arid zone geomorphology*. Chichester: Wiley, 231–59.

Hasegawa, K. 1989: Universal bank erosion coefficient for meandering rivers. *Journal of Hydraulic Engineering* **115**, 744–65.

Hassan, M.A. and Church, M. 1992: The movement of individual grains on the streambed. In Billi, P., Hey, R.D., Thorne, C.R. and Tacconi, P. (eds), *Dynamics of gravel-bed rivers*. Chichester: Wiley, 159–73.

Hassan, M.A. and Reid, I. 1990: The influence of microform bed roughness elements on flow and sediment transport in gravel bed rivers. *Earth Surface Processes and Landforms* **15**, 739–50.

Hattingh, J. and Illenberger, W.K. 1995: Shape sorting of flood-transported synthetic clasts in a gravel bed river. *Sedimentary Geology* **96**, 181–90.

Hayward, J.A. 1980: Hydrology and stream sediments from Torlesse stream catchment. *Tussock Grasslands and Mountain Lands Institute Special Publication* **17**, Lincoln College, New Zealand.

Heede, B.H. 1981: Dynamics of selected mountain streams in the western United States of America. *Zeitschrift für Geomorphologie* **25**, 17–32.

Helley, E.J. 1969: Field measurement of the initiation of large bed particle motion in Blue Creek near Klamath, California. *United States Geological Survey Professional Paper* **562G**.

Henderson, F.M. 1966: *Open channel flow*. New York: Macmillan.

Hewlett, J.D. 1961: *Soil moisture as a source of base flow from steep mountain watersheds*. United States Department of Agriculture Forest Service, Southeastern Forest Experiment Station, Asheville, NC, Station Paper **132**.

Hey, R.D. 1976: Geometry of river meanders. *Nature* **262**, 482–4.

Hey, R.D. 1978: Determinate hydraulic geometry of river channels. *Journal of the Hydraulics Division American Society of Civil Engineers* **104**, HY6, 869–85.

Hey, R.D. 1979: Flow resistance in gravel-bed rivers. *Journal of the Hydraulics Division American Society of Civil Engineers* **105**, HY4, 365–79.

Hey, R.D. 1988a: Bar form resistance in gravel-bed rivers. *Journal of Hydraulic Engineering* **114**, 1498–508.

Hey, R.D. 1988b: Mathematical models of channel morphology. In Anderson, M.G. (ed.), *Modelling geomorphological systems*. Chichester: Wiley, 99–125.

Hey, R.D. and Thorne, C.R. 1986: Stable channels with mobile gravel beds. *Journal of Hydraulic Engineering* **112**, 671–89.

Hickin, E.J. 1974: The development of river meanders in natural river channels. *American Journal of Science* **274**, 414–42.

Hickin, E.J. 1995: Hydraulic geometry and channel scour, Fraser River, British Columbia, Canada. In Hickin, E.J. (ed.), *River geomorphology*. Chichester: Wiley, 155–67.

Hickin, E.J. and Nanson, G.C. 1975: The character of channel migration on the Beatton River, north-east British Columbia, Canada. *Bulletin of the Geological Society of America* **86**, 487–94.

Hickin, E.J. and Nanson, G.C. 1984: Lateral migration rates of river bends. *Journal of Hydraulic Engineering* **110**, 1557–67.

Hicks, D.M. and Mason, P.D. 1991: *Roughness characteristics of New Zealand rivers*. Wellington: Water Resources Survey.

Hjulström, F. 1935: Studies of the morphological activity of rivers as illustrated by the River Fyris. *Bulletin of the Geological Institute University of Uppsala* **25**, 221–527.

Hoey, T.B. 1992: Temporal variations in bedload transport rates and sediment storage in gravel-bed rivers. *Progress in Physical Geography* **16**, 319–38.

Hoey, T.B. and Ferguson, R.I. 1994: Numerical simulation of downstream fining by selective transport in gravel bed rivers: model development and illustration. *Water Resources Research* **30**, 2251–60.

Hoey, T.B. and Sutherland, A.J. 1991: Channel morphology and bedload pulses in braided rivers: a laboratory study. *Earth Surface Processes and Landforms* **16**, 447–62.

Hollis, G.E. 1975: The effect of urbanisation on floods of different recurrence intervals. *Water Resources Research* **11**, 431–5.

Hong, L.B. and Davies, T.R.H. 1979: A study of stream braiding. *Bulletin of the Geological Society of America* **90**, Part 2, 1839–59.

Hoogmoed, W.B. and Stroosnijder, L. 1984: Crust formation on sandy soils in the Sahel. I. Rainfall and infiltration. *Soil and Tillage Research* **4**, 5–24.

Hooke, J.M. 1977: The distribution and nature of changes in river channel patterns: the example of Devon. In Gregory, K.J. (ed.), *River channel changes*. Chichester: Wiley-Interscience, 265–80.

Hooke, J.M. 1979: An analysis of the processes of river bank erosion. *Journal of Hydrology* **42**, 39–62.

Hooke, J.M. 1980: Magnitude and distribution of rates of river bank erosion. *Earth Surface Processes* **5**, 143–57.

Hooke, J.M. 1984: Changes in river meanders: a review of techniques and results of analyses. *Progress in Physical Geography* **8**, 473–508.

Hooke, J.M. 1985: The significance of mid-channel bars in an active meandering river. *Sedimentology* **33**, 839–50.

Hooke, J.M. 1987: Changes in meander morphology. In Gardiner, V. (ed.), *International geomorphology 1986 part I*. Chichester: Wiley, 591–609.

Hooke, J.M. 1995: River channel adjustment to meander cutoffs on the River Bollin and River Dane, northwest England. *Geomorphology* **14**, 235–53.

Hooke, J.M. and Redmond, C.E. 1989: Use of cartographic sources for analysing river channel change with examples from Britain. In Petts, G.E. (ed.), *Historical change of large alluvial rivers: Western Europe*. Chichester: Wiley, 79–93.

Hooke, J.M. and Redmond, C.E. 1992: Causes and nature of river planform change. In Billi, P., Hey, R.D., Thorne, C.R. and Tacconi, P. (eds), *Dynamics of gravel-bed rivers*. Chichester: Wiley, 557–71.

Hooke, R. Le B. 1975: Distribution of sediment transport and shear stress in a meander bend. *Journal of Geology* **83**, 543–66.

Hooke, R. Le B. and Rohrer, W.L. 1977: Relative erodibility of source area rock types from second order variations in alluvial fan size. *Bulletin of the Geological Society of America* **88**, 1177–82.

Horton, R.E. 1933: The role of infiltration in the hydrologic cycle. *Transactions of the American Geophysical Union* **14**, 446–60.

Horton, R.E. 1945: Erosional development of streams and their drainage basins: hydrophysical approach to quantitative morphology. *Bulletin of the Geological Society of America* **56**, 275–370.

Houghton, J.T., Jenkins, G.J. and Ephraums, J.J. (eds) 1990: *Climate change. The IPCC Scientific Assessment*. Cambridge: Cambridge University Press.

Howard, A.D. 1971a: Simulation of stream networks by headward growth and branching. *Geographical Analysis* **3**, 29–50.

Howard, A.D. 1971b: A simulation model of stream capture. *Bulletin of the Geological Society of America* **82**, 1355–76.

Howard, A.D. 1980: Thresholds in river regimes. In Coates, D.R. and Vitek, J.D. (eds) *Thresholds in geomorphology*. Boston: George Allen & Unwin, 227–58.

Howard, A.D. 1982: Equilibrium and time scales in geomorphology: application to sand-bed alluvial streams. *Earth Surface Processes and Landforms* **7**, 303–25.

Howard, A.D. 1987: Modelling fluvial systems: rock-, gravel- and sand-bed channels. In Richards, K.S. (ed.), *River channels: environment and process*. Oxford: Blackwell, 69–94.

Howard, A.D. 1988: Equilibrium models in geomorphology. In Anderson, M.G. (ed.), *Modelling geomorphological systems*. Chichester: Wiley, 49–72.

Howard, A.D. 1990: Theoretical model of optimal drainage networks. *Water Resources Research* **26**, 2107–17.

Howard, A.D. 1992: Modeling channel migration and floodplain sedimentation in meandering streams. In Carling, P.A. and Petts, G.E. (eds), *Lowland floodplain rivers*. Chichester: Wiley, 1–41.

Howard, A.D. 1994: A detachment-limited model of drainage basin evolution. *Water Resources Research* **30**, 2261–85.

Howard, A.D. and Hemberger, A.T. 1991: Multivariate characterization of meandering. *Geomorphology* **4**, 161–86.

Howe, G.M., Slaymaker, H.O. and Harding, D.M. 1967: Some aspects of the flood hydrology of the upper catchments of the Severn and Wye. *Transactions of the Institute of British Geographers* **41**, 33–58.

Huang, H.Q. and Warner, R.F. 1995: The multivariate controls of hydraulic geometry: a causal investigation in terms of boundary shear distribution. *Earth Surface Processes and Landforms* **20**, 115–30.

Hubbell, D.W. 1987: Bed load sampling and analysis. In Thorne, C.R., Bathurst, J.C. and Hey, R.D. (eds), *Sediment transport in gravel-bed rivers*. Chichester: Wiley, 89–106.

Huckleberry, G. 1994: Contrasting channel response to floods on the middle Gila River, Arizona. *Geology* **22**, 1083–6.

Huggett, R. 1990: *Catastrophism. Systems of Earth history*. London: Edward Arnold.

Ichim, I. and Radoane, M. 1990: Channel sediment variability along a river: a case study of the Siret River (Romania). *Earth Surface Processes and Landforms* **15**, 211–25.

Ichoku, C. and Chorowicz, J. 1994: A numerical approach to the analysis and classification of channel network patterns. *Water Resources Research* **30**, 161–74.

Ijjasz-Vasquez, E.J., Bras, R.L. and Rodriguez-Iturbe, I. 1993: Hack's relation and optimal channel networks: the elongation of river basins as a consequence of energy minimization. *Geophysical Research Letters* **20**, 1583–6.

Ikeda, H. 1970: On the longitudinal profiles of the Asake, Mitaki and Utsube Rivers, Mie Prefecture. *Geographical Review of Japan* **43**, 148–59.

Ikeda, S. and Izumi, N. 1990: Width and depth of self-formed straight gravel rivers with bank vegetation. *Water Resources Research* **26**, 2353–64.

Ikeda, S., Parker, G. and Kimura, Y. 1988: Stable width and depth of straight gravel rivers with heterogeneous bed materials. *Water Resources Research* **24**, 713–22.

Ikeda, S., Parker, G. and Sawai, K. 1981: Bend theory of river meanders, part 1: Linear development. *Journal of Fluid Mechanics* **112**, 363–77.

Iseya, F. and Ikeda, H. 1987: Pulsations in bedload transport rates induced by a longitudinal sediment sorting. A flume study using sand and gravel mixtures. *Geografiska Annaler* **69A**, 15–27.

Jacobson, R.B. and Coleman, D.J. 1986: Stratigraphy and recent evolution of Maryland Piedmont floodplains. *American Journal of Science* **286**, 617–37.

James, L.A. 1989: Sustained storage and transport of hydraulic gold mining sediment in the Bear River, California. *Annals of the Association of American Geographers* **79**, 570–92.

James, L.A. 1991: Incision and morphologic evolution of an alluvial channel recovering from hydraulic mining sediment. *Bulletin of the Geological Society of America* **103**, 723–36.

James, L.A. 1993: Sustained reworking of hydraulic mining sediment in California: G.K. Gilbert's sediment wave model reconsidered. *Zeitschrift für Geomorphologie Supplement-Band* **88**, 49–66.

James, W.R. and Krumbein, W.C. 1969: Frequency distributions of stream link lengths. *Journal of Geology* **77**, 544–65.

Jansson, M.B. 1988: A global survey of sediment yield. *Geografiska Annaler* **70A**, 81–98.

Jarvis, R.S. 1976: Classification of nested tributary basins in analysis of drainage basin shape. *Water Resources Research* **12**, 1151–64.

Jarvis, R.S. 1977: Drainage network analysis. *Progress in Physical Geography* **1**, 271–95.

Jarvis, R.S. and Sham, C.H. 1981: Drainage network structure and the diameter–magnitude relation. *Water Resources Research* **17**, 1019–27.

Jia, Y. 1990: Minimum Froude number and the equilibrium of alluvial sand rivers. *Earth Surface Processes and Landforms* **15**, 199–209.

Johnson, R.H. and Paynter, J. 1967: The development of a cutoff on the River Irk at Chadderton, Lancashire. *Geography* **52**, 41–9.

Jones, J.A.A. 1981: *The nature of soil piping: a review of research*. Norwich: Geobooks.

Jones, J.A.A. 1987: The initiation of natural drainage networks. *Progress in Physical Geography* **11**, 207–45.

Jones, O.T. 1924: The longitudinal profiles of the Upper Towy drainage system. *Quarterly Journal of the Geological Society of London* **80**, 568–609.

Judd, H.E. and Peterson, D.F. 1969: *Hydraulics of large bed element channels*. Utah Water Research Laboratory, Utah State University, Report PRWG 17–6.

Julien, P.Y. and Klaassen, G.J. 1995: Sand-dune geometry of large rivers during floods. *Journal of Hydraulic Engineering* **121**, 657–63.

Karlinger, M.R. and Troutman, B.M. 1989: A random spatial network model based on elementary postulates. *Water Resources Research* **25**, 793–8.

Kashef, A.-A.I. 1981: Technical and ecological impacts of the High Aswan Dam. *Journal of Hydrology* **53**, 73–84.

Kashiwaya, K. 1987: Theoretical investigation of the time variation of drainage density. *Earth Surface Processes and Landforms* **12**, 39–46.

Keller, E.A. 1971: Areal sorting of bed load material: the hypothesis of velocity reversal. *Bulletin of the Geological Society of America* **82**, 753–6.

Keller, E.A. 1972: Development of alluvial stream channels: a five-stage model. *Bulletin of the Geological Society of America* **83**, 1531–6.

Keller, E.A. and Brookes, A. 1984: Consideration of meandering in channelization projects: selected observations and judgements. In Elliott, C.M. (ed.), *River meandering*. New Orleans: American Society of Civil Engineers, 384–97.

Keller, E.A. and Florsheim, J.L. 1993: Velocity-reversal hypothesis: a model approach. *Earth Surface Processes and Landforms* **18**, 733–40.

Keller, E.A. and Melhorn, W.N. 1973: Bedforms and fluvial processes in alluvial stream channels: selected observations. In Morisawa, M. (ed.), *Fluvial geomorphology*. Binghamton, NY: New York State University Publications in Geomorphology, 253–83.

Keller, E.A. and Melhorn, W.N. 1978: Rhythmic spacing and origin of pools and riffles. *Bulletin of the Geological Society of America* **89**, 723–30.

Keller, E.A. and Swanson, F.J. 1979: Effects of large organic material on channel form and fluvial processes. *Earth Surface Processes* **4**, 361–80.

Kellerhals, R. 1967: Stable channels with gravel-paved beds. *Journal of the Waterways and Harbors Division American Society of Civil Engineers* **93**, 63–84.

Kellerhals, R., Church, M. and Bray, D.I. 1976: Classification of river processes. *Journal of the Hydraulics Division American Society of Civil Engineers* **102**, HY7, 813–29.

Kellerhals, R., Church, M. and Davies, L.B. 1979: Morphological effects of interbasin river diversions. *Canadian Journal of Civil Engineering* **6**, 18–31.

Keown, M.P., Dardeau, E.A. and Causey, E.M. 1986: Historic trends in the sediment flow regime of the Mississippi River. *Water Resources Research* **22**, 1555–64.

King, W.A. and Martini, I.P. 1984: Morphology and recent sediments of the lower anastomosing reaches of the Attawapiskat River, James Bay, Ontario, Canada. *Sedimentary Geology* **37**, 295–320.

Kinoshita, R. 1961: *Investigation of channel deformation in Ishikari River*. Report for the Bureau of Resources No. 36, Department of Science and Technology, Japan.

Kirchner, J.W. 1993: Statistical inevitability of Horton's laws and the apparent randomness of stream channel networks. *Geology* **21**, 591–4.

Kirkby, M.J. 1976: Tests of the random network model, and its application to basin hydrology. *Earth Surface Processes* **1**, 197–212.

Kirkby, M.J. 1977: Maximum sediment efficiency as a criterion for alluvial channels. In Gregory, K.J. (ed.), *River channel changes*. Chichester: Wiley-Interscience, 429–42.

Kirkby, M.J. 1987: Modelling some influences of soil erosion, landslides and valley gradient on drainage density and hollow development. In Ahnert, F. (ed.), *Geomorphological models, Catena Supplement* **10**, 1–11.

Kirkby, M.J. 1988: Hillslope runoff processes and models. *Journal of Hydrology* **100**, 315–39.

Kirkby, M.J. 1993: Long term interactions between networks and hillslopes. In Beven, K. and Kirkby, M.J. (eds), *Channel network hydrology*. Chichester: Wiley, 255–93.

Kirkby, M.J. 1994: Thresholds and instability in stream head hollows: a model of magnitude and frequency for wash processes. In Kirkby, M.J. (ed.), *Process models and theoretical geomorphology*. Chichester: Wiley, 294–314.

Kirkby, M.J. and Chorley, R.J. 1967: Throughflow, overland flow and erosion. *Bulletin of the International Association of Scientific Hydrology* **12**, 5–21.

Klaassen, G.J. and Vermeer, K. 1988: Channel characteristics of the braiding Jamuna River, Bangladesh. In White, W.R. (ed.), *International conference on river regime*. Wallingford, UK: Hydraulics Research, 173–89.

Klimek, K. 1974: The retreat of alluvial river banks in the Wisloka valley (south Poland). *Geographia Polonica* **28**, 59–75.

Knighton, A.D. 1972: Changes in a braided reach. *Bulletin of the Geological Society of America* **83**, 3813–22.

Knighton, A.D. 1973: Riverbank erosion in relation to streamflow conditions, River Bollin-Dean, Cheshire. *East Midland Geographer* **6**, 416–26.

Knighton, A.D. 1974: Variation in width-discharge relation and some implications for hydraulic geometry. *Bulletin of the Geological Society of America* **85**, 1069–76.

Knighton, A.D. 1975a: Variations in at-a-station hydraulic geometry. *American Journal of Science* **275**, 186–218.

Knighton, A.D. 1975b: Channel gradient in relation to discharge and bed material characteristics. *Catena* **2**, 263–74.

Knighton, A.D. 1976: Stream adjustment in a small Rocky Mountain basin. *Arctic and Alpine Research* **8**, 197–212.

Knighton, A.D. 1977a: Alternative derivation of the minimum variance hypothesis. *Bulletin of the Geological Society of America* **88**, 364–6.

Knighton, A.D. 1977b: Short-term changes in hydraulic geometry. In Gregory, K.J. (ed.), *River channel changes*. Chichester: Wiley-Interscience, 101–19.

Knighton, A.D. 1979: Comments on log-quadratic relations in hydraulic geometry. *Earth Surface Processes* **4**, 205–10.

Knighton, A.D. 1980a: Comment on 'Drainage network power' by K.J. Gregory. *Water Resources Research* **16**, 1128–9.

Knighton, A.D. 1980b: Longitudinal changes in size and sorting of stream-bed material in four English rivers. *Bulletin of the Geological Society of America* **91**, 55–62.

Knighton, A.D. 1981a: Asymmetry of river channel cross-sections: Part I. Quantitative indices. *Earth Surface Processes and Landforms* **6**, 581–8.

Knighton, A.D. 1981b: Local variations of cross-sectional form in a small gravel-bed stream. *Journal of Hydrology (New Zealand)* **20**, 131–46.

Knighton, A.D. 1981c: Channel form and flow characteristics of supraglacial streams, Austre Okstindbreen, Norway. *Arctic and Alpine Research* **13**, 295–306.

Knighton, A.D. 1982: Asymmetry of river channel cross-sections: Part II. Mode of development and local variation. *Earth Surface Processes and Landforms* **7**, 117–31.

Knighton, A.D. 1984: Indices of flow asymmetry in natural streams: definition and performance. *Journal of Hydrology* **73**, 1–19.

Knighton, A.D. 1987a: Estimating the mean annual flood in the Trent basin. *East Midland Geographer* **10**, 1–6.

Knighton, A.D. 1987b: River channel adjustment – the downstream dimension. In Richards, K.S. (ed.), *River channels: environment and process*. Oxford: Blackwell, 95–128.

Knighton, A.D. 1988: The impact of the Parangana Dam on the River Mersey, Tasmania. *Geomorphology* **1**, 221–37.

Knighton, A.D. 1989: River adjustment to changes in sediment load: the effects of tin mining on the Ringarooma River, Tasmania, 1875–1984. *Earth Surface Processes and Landforms* **14**, 333–59.

Knighton, A.D. 1991: Channel bed adjustment along mine-affected rivers of northeast Tasmania. *Geomorphology* **4**, 205–19.

Knighton, A.D and Nanson, G.C. 1993: Anastomosis and the continuum of channel pattern. *Earth Surface Processes and Landforms* **18**, 613–25.

Knighton, A.D and Nanson, G.C. 1994a: Flow transmission along an arid zone anastomosing river, Cooper Creek, Australia. *Hydrological Processes* **8**, 137–54.

Knighton, A.D and Nanson, G.C. 1994b: Waterholes and their significance in the anastomosing channel system of Cooper Creek, Australia. *Geomorphology* **9**, 311–24.

Knighton, A.D and Nanson, G.C. 1997: Distinctiveness, diversity and uniqueness in arid zone river systems. In Thomas, D.S.G. (ed.), *Arid zone geomorphology*. Chichester: Wiley, 185–203.

Knighton, A.D., Woodroffe, C.D. and Mills, K. 1992: The evolution of tidal creek networks, Mary River, northern Australia. *Earth Surface Processes and Landforms* **17**, 167–90.

Knox, J.C. 1977: Human impacts on Wisconsin stream channels. *Annals of the Association of American Geographers* **67**, 323–42.

Knox, J.C. 1987: Historical valley floor sedimentation in the Upper Mississippi valley. *Annals of the Association of American Geographers* **77**, 224–44.

Knox, J.C. 1993: Large increases in flood magnitude in response to modest changes in climate. *Nature* **361**, 430–2.

Knox, J.C. 1995: Fluvial systems since 20,000 years BP. In Gregory, K.J., Starkel, L. and Baker, V.R. (eds), *Global continental palaeohydrology*. Chichester: Wiley, 87–108.

Kochel, R.C. 1988: Geomorphic impact of large floods: review and new perspectives on magnitude and frequency. In Baker, V.R., Kochel, R.C. and Patton, P.C. (eds), *Flood geomorphology*. New York: Wiley-Interscience, 169–87.

Kochel, R.C. and Baker, V.R. 1988: Paleoflood analysis using slackwater deposits. In Baker, V.R., Kochel, R.C. and Patton, P.C. (eds), *Flood geomorphology*. Chichester: Wiley-Interscience, 357–76.

Kochel, R.C., Howard, A.D. and McLane, C. 1985: Channel networks developed by groundwater sapping in fine-grained sediments: analogs to some Martian valleys. In Woldenberg, M.J. (ed.), *Models in geomorphology*. Boston: Allen & Unwin, 313–41.

Kodama, Y. 1994: Downstream changes in the lithology and grain size of fluvial gravels, the Watarase River, Japan: evidence of the role of abrasion in downstream fining. *Journal of Sedimentary Research* **A64**, 68–75.

Komar, P.D. 1988: Sediment transport by floods. In Baker, V.R., Kochel, R.C. and Patton, P.C. (eds), *Flood geomorphology*. New York: Wiley-Interscience, 97–111.

Komar, P.D. 1989: Flow-competence evaluations of the hydraulic parameters of floods: an assessment of the technique. In Beven, K.J. and Carling, P.A. (eds), *Floods: hydrological, sedimentological and geomorphological implications.* Chichester: Wiley, 107–34.

Komar, P.D. and Shih, S.-M. 1992: Equal mobility versus changing bedload grain sizes in gravel-bed streams. In Billi, P., Hey, R.D., Thorne, C.R. and Tacconi, P. (eds), *Dynamics of gravel-bed rivers*. Chichester: Wiley, 73–93.

Kuenen, P.H. 1956: Experimental abrasion of pebbles. 2. Rolling by current. *Journal of Geology* **64**, 336–68.

Kuhnle, R.A., Binger, R.L., Foster, G.R. and Grissinger, E.H. 1996: Effect of land use changes on sediment transport in Goodwin Creek. *Water Resources Research* **32**, 3189–96.

La Barbera, P. and Rosso, R. 1989: On fractal dimension of stream networks. *Water Resources Research* **25**, 735–41.

Lacey, C. 1929: Stable channels in alluvium. *Proceedings of the Institution of Civil Engineers* **229**, 259–384.

Laity, J.E. and Malin, M.C. 1985: Sapping processes and the development of theater-headed valley networks in the Colorado Plateau. *Bulletin of the Geological Society of America* **96**, 203–17.

Lamb, H.H. 1977: *Climate: present, past and future.* Volume 2 *Climatic history and the future.* London: Methuen.

Lane, E.W. 1955: Design of stable channels. *Transactions of the American Society of Civil Engineers* **120**, 1234–60

Lane, L.J., Nearing, M.A., Laflen, J.M., Foster, G.R. and Nichols, M.H. 1992: Description of the US Department of Agriculture water erosion prediction project (WEPP) model. In Parsons, A.J. and Abrahams, A.D. (eds), *Overland flow hydraulics and erosion mechanics*. London: UCL Press, 377–91.

Lane, S.N. 1995: The dynamics of dynamic river channels. *Geography* **80**, 147–62.

Lane, S.N, Richards, K.S. and Chandler, J.H. 1995: Morphological estimation of the time-integrated bed load transport rate. *Water Resources Research* **31**, 761–72.

Lane, S.N, Richards, K.S. and Chandler, J.H. 1996: Discharge and sediment supply controls on erosion and deposition in a dynamic alluvial channel. *Geomorphology* **15**, 1–15.

Langbein, W.B. 1964a: Geometry of river channels. *Journal of the Hydraulics Division American Society of Civil Engineers* **90**, HY2, 301–12.

Langbein, W.B. 1964b: Profiles of rivers of uniform discharge. *United States Geological Survey Professional Paper* **501B**, 119–22.

Langbein, W.B. 1965: Geometry of river channels: closure of discussion. *Journal of the Hydraulics Division American Society of Civil Engineers* **91**, HY3, 297–313.

Langbein, W.B. and Leopold, L.B. 1964: Quasi-equilibrium states in channel morphology. *American Journal of Science* **262**, 782–94.

Langbein, W.B. and Leopold, L.B. 1966: River meanders – theory of minimum variance. *United States Geological Survey Professional Paper* **422H**.

Langbein, W.B. and Leopold, L.B. 1968: River channel bars and dunes – theory of kinematic waves. *United States Geological Survey Professional Paper* **422L**.

Langbein, W.B. and Schumm, S.A. 1958: Yield of sediment in relation to mean annual precipitation. *Transactions of the American Geophysical Union* **39**, 1076–84.

Lapointe, M. 1992: Burst-like sediment suspension events in a sand bed river. *Earth Surface Processes and Landforms* **17**, 253–70.

Laronne, J.B., Reid, I., Yitshak, Y. and Frostick, L.E. 1994: The non-layering of gravel streambeds under ephemeral flood regimes. *Journal of Hydrology* **159**, 353–63.

Larras, J. 1968: Problèmes d'hydraulique fluviale. *Annales des Ponts et Chaussées* **138**, 195–209.

Lawler, D.M. 1986: River bank erosion and the influence of frost: a statistical examination. *Transactions of the Institute of British Geographers* New Series **11**, 227–42.

Lawler, D.M. 1992: Process dominance in bank erosion systems. In Carling, P.A. and Petts, G.E. (eds), *Lowland floodplain rivers*. Chichester: Wiley, 117–43.

Lawler, D.M. 1993: Needle ice processes and sediment mobilization on river banks: the River Ilston, West Glamorgan, UK. *Journal of Hydrology* **150**, 81–114.

Leeder, M.R. 1983: On the interactions between turbulent flow, sediment transport and bedform mechanics in channelized flows. In Collinson, J.D. and Lewin, J. (eds), *Modern and ancient fluvial systems*. Oxford: Blackwell, Special Publication of the International Association of Sedimentologists **6**, 121–32.

Leeks, G.J., Lewin, J. and Newson, M.D. 1988: Channel change, fluvial geomorphology and river engineering: the case of the Afon Trannon, mid-Wales. *Earth Surface Processes and Landforms* **13**, 207–23.

Leopold, L.B. 1969: Quantitative comparison of some aesthetic factors among rivers. *United States Geological Survey Circular* **620**.

Leopold, L.B. 1973: River channel change with time: an example. *Bulletin of the Geological Society of America* **84**, 1845–60.

Leopold, L.B. 1994: *A view of the river*. Cambridge, MA: Harvard University Press.

Leopold, L.B., Bagnold, R.A., Wolman, M.G. and Brush, L.M. 1960: Flow resistance in sinuous or irregular channels. *United States Geological Survey Professional Paper* **282D**, 111–34.

Leopold, L.B. and Bull, W.B. 1979: Base level, aggradation, and grade. *Proceedings of the American Philosophical Society* **123**, 168–202.

Leopold, L.B. and Emmett, W.W. 1976: Bedload measurements, East Fork River, Wyoming. *Proceedings of the National Academy of Sciences* **73**, 1000–4.

Leopold, L.B., Emmett, W.W. and Myrick, R.M. 1966: Channel and hillslope processes in a semi-arid area, New Mexico. *United States Geological Survey Professional Paper* **352G**, 153–253.

Leopold, L.B. and Langbein, W.B. 1962: The concept of entropy in landscape evolution. *United States Geological Survey Professional Paper* **500A**.

Leopold, L.B. and Maddock, T. 1953: The hydraulic geometry of stream channels and some physiographic implications. *United States Geological Survey Professional Paper* **252**.

Leopold, L.B. and Miller, J.P. 1956: Ephemeral streams – hydraulic factors and their relation to the drainage net. *United States Geological Survey Professional Paper* **282A**.

Leopold, L.B. and Wolman, M.G. 1957: River channel patterns – braided, meandering and straight. *United States Geological Survey Professional Paper* **282B**, 39–85.

Leopold, L.B., Wolman, M.G. and Miller, J.P. 1964: *Fluvial processes in geomorphology*. San Francisco: W.H. Freeman.

Lewin, J. 1976: Initiation of bed forms and meanders in coarse-grained sediment. *Bulletin of the Geological Society of America* **87**, 281–5.

Lewin, J. 1989: Floods in fluvial geomorphology. In Beven, K.J. and Carling, P.A. (eds), *Floods: hydrological, sedimentological and geomorphological implications*. Chichester: Wiley, 265–84.

Lewin, J. and Macklin, M.G. 1987: Metal mining and floodplain sedimentation in Britain. In Gardiner, V. (ed.), *International geomorphology 1986 part I*. Chichester: Wiley, 1009–27.

Lewin, J., Macklin, M.G. and Newson, M.D. 1988: Regime theory and environmental change – irreconcilable concepts? In White, W.R. (ed.), *International conference on river regime*. Wallingford, UK: Hydraulics Research, 431–45.

Lewis, G.W. and Lewin, J. 1983: Alluvial cutoffs in Wales and the Borderlands. In Collinson, J.D. and Lewin, J. (eds), *Modern and ancient fluvial systems*. Oxford: Blackwell, Special Publication of the International Association of Sedimentologists **6**, 145–54.

Lewis, L.A. 1969: Some fluvial geomorphic characteristics of the Manati Basin, Puerto Rico. *Annals of the Association of American Geographers* **59**, 280–93.

Li, R.-M. and Simons, D.B. 1982: Geomorphological and hydraulic analysis of mountain streams. In Hey, R.D., Bathurst, J.C. and Thorne, C.R. (eds), *Gravel-bed rivers*. Chichester: Wiley, 425–40.

Li, R.-M., Simons, D.B. and Stevens, M.A. 1976: Morphology of cobble streams in small watersheds. *Journal of the Hydraulics Division American Society of Civil Engineers* **102**, HY8, 1101–17.

Li, Z. and Komar, P.D. 1986: Laboratory measurements of pivoting angles for applications to selective entrainment of gravels in a current. *Sedimentology* **33**, 413–23.

Limerinos, J.T. 1970: Determination of the Manning coefficient from measured bed roughness in natural channels. *United States Geological Survey Water-Supply Paper* **1898B**.

Lisle, T. 1979: A sorting mechanism for a riffle–pool sequence. *Bulletin of the Geological Society of America* **90**, Part 2, 1142–57.

Lisle, T.E. 1995: Particle size variations between bed load and bed material in natural gravel bed channels. *Water Resources Research* **31**, 1107–18.

Loewenherz, D.S. 1991: Stability and the initiation of channelized surface drainage: a reassessment of the short wavelength limit. *Journal of Geophysical Research* **96**, 8453–64.

Loughran, R.J. 1989: The measurement of soil erosion. *Progress in Physical Geography* **13**, 216–33.

Mackin, J.H. 1948: Concept of the graded river. *Bulletin of the Geological Society of America* **59**, 463–512.

Mackin, J.H. 1956: Cause of braiding by a graded river. *Bulletin of the Geological Society of America* **67**, 1717–18.

Macklin, M.G. and Lewin, J. 1989: Sediment transfer and transformation of an alluvial valley floor: the River South Tyne, Northumbria, UK. *Earth Surface Processes and Landforms* **14**, 233–46.

Macklin, M.G. and Lewin, J. 1993: Holocene river alluviation in Britain. *Zeitschrift für Geomorphologie Supplement-Band* **88**, 109–22.

Macklin, M.G., Rumsby, B.T. and Heap, T. 1992: Flood alluviation and entrenchment: Holocene valley-floor development and transformation in the British uplands. *Bulletin of the Geological Society of America* **104**, 631–43.

Maddock, T. 1969: The behavior of straight open channels with movable beds. *United States Geological Survey Professional Paper* **622A**.

Madduma Bandara, C.M. 1974: Drainage density and effective precipitation. *Journal of Hydrology* **21**, 187–90.

Madej, M.A. 1995: Changes in channel-stored sediment, Redwood Creek, northwestern California, 1947 to 1980. *United States Geological Survey Professional Paper* **1454O**.

Madej, M.A. and Ozaki, V. 1996: Channel response to sediment wave propagation and movement, Redwood Creek, California, USA. *Earth Surface Processes and Landforms* **21**, 911–27.

Magilligan, F.J. 1992: Thresholds and the spatial variability of flood power during extreme floods. *Geomorphology* **5**, 373–90.

Mahmood, K., Tarar, R.N. and Masood, T. 1979: *Hydraulic geometry relations for ACOP channels*. Washington, DC: Civil, Mechanical and Environmental Engineering Department, George Washington University.

Maizels, J. and Aitken, J. 1991: Palaeohydrological change during deglaciation in upland Britain: a case study from northeast Scotland. In Starkel, L., Gregory, K.J. and Thornes, J.B. (eds), *Temperate palaeohydrology*. Chichester: Wiley, 105–45.

Markham, A.J. and Thorne, C.R. 1992: Geomorphology of gravel-bed river bends. In Billi, P., Hey, R.D., Thorne, C.R. and Tacconi, P. (eds), *Dynamics of gravel-bed rivers*. Chichester: Wiley, 433–50.

Marron, D.C. 1992: Floodplain storage of mine tailings in the Belle Fourche River system: a sediment budget approach. *Earth Surface Processes and Landforms* **17**, 675–85.

Masterman, R. and Thorne, C.R. 1994: Analytical approach to flow resistance in gravel-bed channels with vegetated banks. In Kirkby, M.J. (ed.), *Process models and theoretical geomorphology*. Chichester: Wiley, 201–18.

McCarthy, T.S., Ellery, W.N. and Stanistreet, I.G. 1992: Avulsion mechanisms on the Okavango Fan, Botswana: the control of a fluvial system by vegetation. *Sedimentology* **39**, 779–95.

McEwen, L.J. and Werritty, A. 1988: The hydrology and long-term geomorphic significance of a flash flood in the Cairngorm Mountains, Scotland. *Catena* **15**, 361–77.

Meade, R.H. 1976: Sediment problems in the Savannah River basin. In Dillman, B.L. and Stepp, J.M. (eds), *The future of the Savannah River*. Clemson, SC: Water Resources Research Institute, 105–29.

Meade, R.H. 1982: Sources, sinks, and storage of river sediment in the Atlantic drainage of the United States. *Journal of Geology* **90**, 235–52.

Meade, R.H. and Parker, R.S. 1985: Sediment in rivers of the United States. In *National Water Summary 1984*, United States Geological Survey Water-Supply Paper **2275**, 49–60.

Melton, M.A. 1957: An analysis of the relations among elements of climate, surface properties, and geomorphology. *Office of Naval Research, Geography Branch, Project NR 389–042*, Technical Report **11**.

Melton, M.A. 1958: Correlation structure of morphometric properties of drainage systems and their controlling agents. *Journal of Geology* **66**, 442–60.

Mertes, L.A.K. 1994: Rates of floodplain sedimentation on the central Amazon River. *Geology* **22**, 171–4.

Mesa, O.J. and Mifflin, E.R. 1986: On the relative role of hillslope and network geometry in hydrologic response. In Gupta, V.K., Rodriguez-Iturbe, I. and Wood, E.F. (eds), *Scale problems in hydrology*. Dordrecht: Reidel, 1–17.

Meybeck, M. 1976: Total mineral dissolved transport by world major rivers. *Hydrological Sciences Bulletin, International Association of Scientific Hydrology* **21**, 265–84.

Meybeck, M. 1979: Concentrations des eaux fluviales en éléments majeurs et apports en solution aux océans. *Revue de Géologie Dynamique et de Géographie Physique* **21**, 215–46.

Meybeck, M. 1987: Global chemical weathering of surficial rocks estimated from river dissolved loads. *American Journal of Science* **287**, 401–28.

Miall, A.D. 1977: A review of the braided-river depositional environment. *Earth Science Reviews* **13**, 1–62.

Millar, R.G. and Quick, M.C. 1993: Effect of bank stability on geometry of gravel rivers. *Journal of Hydraulic Engineering* **119**, 1343–63.

Miller, J.P. 1958: High mountain streams: effects of geology on channel characteristics and bed material. *New Mexico State Bureau of Mines and Mineral Resources, Memoir* **4**.

Miller, J.P. and Wendorf, F. 1958: Alluvial chronology of the Tesuque Valley, New Mexico. *Journal of Geology* **66**, 177–94.

Miller, M.C., McCave, I.N. and Komar, P.D. 1977: Threshold of sediment motion under unidirectional currents. *Sedimentology* **24**, 507–27.

Miller, S.O., Ritter, D.F., Kochel, R.C. and Miller, J.R. 1993: Fluvial responses to land-use changes and climatic variations within the Drury Creek watershed, southern Illinois. *Geomorphology* **6**, 309–29.

Miller, T.K. 1984: A system model of stream-channel shape and size. *Bulletin of the Geological Society of America* **95**, 237–41.

Miller, T.K. 1991a: An assessment of the equable change principle in at-a-station hydraulic geometry. *Water Resources Research* **27**, 2751–8.

Miller, T.K. 1991b: A model of stream channel adjustment: assessment of Rubey's hypothesis. *Journal of Geology* **99**, 699–710.

Miller, V.C. 1975: Lateral tributary capture in the Montalban area, Spain. *The ITC-Journal* **2**, 230–5.

Milliman, J.D. and Meade, R.H. 1983: World-wide delivery of river sediment to the oceans. *Journal of Geology* **91**, 1–21.

Milliman, J.D. and Syvitski, J.P.M. 1992: Geomorphic/tectonic control of sediment discharge to the ocean: the importance of small mountainous rivers. *Journal of Geology* **100**, 525–44.

Milliman, J.D., Yun-Shan, Q., Mei-E, R. and Saito, Y. 1987: Man's influence on the erosion and transport of sediment by Asian rivers: the Yellow River (Huanghe) example. *Journal of Geology* **95**, 751–62.

Mills, H.H. 1979: Downstream rounding of pebbles – a quantitative review. *Journal of Sedimentary Petrology* **49**, 295–302.

Milne, J.A. 1982: Bed-material size and the riffle–pool sequence. *Sedimentology* **29**, 267–78.

Milne, J.A. 1983: Variation in cross-sectional asymmetry of coarse bedload river channels. *Earth Surface Processes and Landforms* **8**, 503–11.

Mock, S.J. 1971: A classification of channel links in stream networks. *Water Resources Research* **7**, 1558–66.

Mock, S.J. 1976: Topological properties of some trellis pattern channel networks. *United States Army Cold Regions Research and Engineering Laboratory, CRREL Report* **76–46**.

Moglen, G.E. and Bras, R.L. 1995: The effect of spatial heterogeneities on geomorphic expression in a model of basin evolution. *Water Resources Research* **31**, 2613–23.

Montgomery, D.R. and Dietrich, W.E. 1988: Where do channels begin? *Nature* **336**, 232–4.

Montgomery, D.R. and Dietrich, W.E. 1989: Source areas, drainage density, and channel initiation. *Water Resources Research* **25**, 1907–18.

Montgomery, D.R. and Dietrich, W.E. 1994: Landscape dissection and drainage area-slope thresholds. In Kirkby, M.J. (ed.), *Process models and theoretical geomorphology*. Chichester: Wiley, 221–46.

Moore, R.J. and Newson, M.D. 1986: Production, storage and output of coarse upland sediments: natural and artificial influences as revealed by research catchment studies. *Journal of the Geological Society of London* **143**, 921–6.

Morgan, R.P.C. 1972: Observations on factors affecting the behaviour of a first-order stream. *Transactions of the Institute of British Geographers* **56**, 171–86.

Morgan, R.P.C. 1977: Soil erosion in the United Kingdom: field studies in the Silsoe area, 1973–75. *National College of Agricultural Engineering, Silsoe, Occasional Paper* **4**.

Morgan, R.P.C. 1986: *Soil erosion and conservation*. London: Longman.

Morisawa, M.E. 1964: Development of drainage systems on an upraised lake floor. *American Journal of Science* **262**, 340–54.

Morisawa, M. and LaFlure, E. 1979: Hydraulic geometry, stream equilibrium and urbanization. In Rhodes, D.D. and Williams, G.P. (eds), *Adjustments of the fluvial system*. Dubuque, IA: Kendall-Hunt, 333–50.

Morris, P.H. and Williams, D.J. 1997: Exponential longitudinal profiles of streams. *Earth Surface Processes and Landforms* **22**, 143–63.

Mosley, M.P. 1974: Experimental study of rill erosion. *Transactions of the American Society of Agricultural Engineers* **17**, 909–13.

Mosley, M.P. 1975: Channel changes on the River Bollin, Cheshire, 1872–1973. *East Midlands Geographer* **6**, 185–99.

Mosley, M.P. 1976: An experimental study of channel confluences. *Journal of Geology* **84**, 535–62.

Mosley, M.P. 1982: The effect of a New Zealand beech forest canopy on the kinetic energy of water drops and on surface erosion. *Earth Surface Processes and Landforms* **7**, 103–7.

Mosley, M.P. 1983: Flow requirements for recreation and wildlife in New Zealand rivers – a review. *Journal of Hydrology (New Zealand)* **22**, 152–74.

Murray, A.B. and Paola, C. 1994: A cellular model of braided rivers. *Nature* **371**, 54–7.

Myers, N. 1989: *Deforestation rates in tropical forests and their climatic implications*. London: Friends of the Earth.

Naden, P.S. 1988: Models of sediment transport in natural streams. In Anderson, M.G. (ed.), *Modelling geomorphological systems*. Chichester: Wiley, 217–58.

Naden, P.S. 1992: Spatial variability in flood estimation for large catchments: the exploitation of channel network structure. *Hydrological Sciences Journal* **37**, 53–71.

Naden, P.S. 1993: A routing model for continental-scale hydrology. In *Macroscale Modelling of the Hydrosphere*, Proceedings of the International Conference, Yokohama 1993. International Association of Hydrological Sciences Publication **214**, 67–89.

Naden, P.S. and Brayshaw, A.C. 1987: Small- and medium-scale bedforms in gravel-bed rivers. In Richards, K.S. (ed.), *River channels: environment and process*. Oxford: Blackwell, 249–71.

Naden, P.S. and Polarski, M. 1990: Derivation of river network variables from digitised data and their use in flood estimation. *Report to MAFF, Institute of Hydrology*.

Nakamura, F. and Swanson, F.J. 1993: Effects of coarse woody debris on morphology and sediment storage of a mountain stream system in western Oregon. *Earth Surface Processes and Landforms* **18**, 43–61.

Nanson, G.C. 1986: Episodes of vertical accretion and catastrophic stripping: a model of disequilibrium floodplain development. *Bulletin of the Geological Society of America* **97**, 1467–75.

Nanson, G.C. and Croke, J.C. 1992: A genetic classification of floodplains. *Geomorphology* **4**, 459–86.

Nanson, G.C. and Erskine, W.D. 1988: Episodic changes of channels and floodplains on coastal rivers of New South Wales. In Warner, R.F. (ed.), *Fluvial geomorphology of Australia*. Sydney: Academic Press, 201–21.

Nanson, G.C. and Hickin, E.J. 1983: Channel migration and incision on the Beatton River. *Journal of Hydraulic Engineering* **109**, 327–37.

Nanson, G.C. and Hickin, E.J. 1986: A statistical examination of bank erosion and channel migration in western Canada. *Bulletin of the Geological Society of America* **97**, 497–504.

Nanson, G.C. and Knighton, A.D. 1996: Anabranching rivers: their cause, character and classification. *Earth Surface Processes and Landforms* **21**, 217–39.

Nanson, G.C. and Page, K.J. 1983: Lateral accretion of fine-grained concave benches on meandering rivers. In Collinson, J.D. and Lewin, J. (eds), *Modern and ancient fluvial systems*. Oxford: Blackwell, Special Publication of the International Association of Sedimentologists **6**, 133–43.

Nanson, G.C., Price, D.M., Short, S.A., Young, R.W. and Jones, B.G. 1991: Comparative uranium–thorium and thermoluminescence dating of weathered Quaternary alluvium in the tropics of northern Australia. *Quaternary Research* **35**, 347–66.

Nanson, G.C., Von Krusenstierna, A., Bryant, E.A. and Renilson, M.R. 1994: Experimental measurements of river-bank erosion caused by boat-generated waves on the Gordon River, Tasmania. *Regulated Rivers: Research and Management* **9**, 1–14.

Nanson, G.C., Young, R.W., Price, D.M. and Rust, B.R. 1988: Stratigraphy, sedimentology and Late Quaternary chronology of the Channel Country of western Queensland. In Warner, R.F. (ed.), *Fluvial geomorphology of Australia*. Sydney: Academic Press, 151–75.

Nash, D.B. 1994: Effective sediment-transporting discharge from magnitude–frequency analysis. *Journal of Geology* **102**, 79–95.

Natural Environment Research Council (NERC) 1975: *Flood studies report* (5 volumes). Wallingford, UK: Institute of Hydrology.

Nelson, J.M. and Smith, J.D. 1989: Evolution and stability of erodible channel beds. In Ikeda, S. and Parker, G. (eds), *River meandering*. American Geophysical Union, Water Research Monograph **12**, 321–77.

Nevins, T.H.F. 1969: River training – the single thread channel. *New Zealand Engineering*, December, 367–73.

Newson, M.D. 1986: River basin engineering – fluvial geomorphology. *Journal of the Institution of Water Engineers and Scientists* **40**, 307–24.

Newson, M.D. 1992: *Land, water and development*. London: Routledge.

Newson, M.D. 1995: Fluvial geomorphology and environmental design. In Gurnell, A.M. and Petts, G.E. (eds), *Changing river channels*. Chichester: Wiley, 413–32.

Nicholas, A.P., Ashworth, P.J., Kirkby, M.J., Macklin, M.G. and Murray, T. 1995: Sediment slugs: large-scale fluctuations in fluvial sediment transport rates and storage volumes. *Progress in Physical Geography* **19**, 500–19.

Nikora, V.I. 1991: Fractal structures of river plan forms. *Water Resources Research* **27**, 1327–33.

Nixon, M. 1959: A study of the bankfull discharges of rivers in England and Wales. *Proceedings of the Institution of Civil Engineers* **12**, 157–75.

Nolan, K.M. and Marron, D.C. 1995: History, causes, and significance of changes in the channel geometry of Redwood Creek, northwestern California, 1936 to 1982. *United States Geological Survey Professional Paper* **1454N**.

Nordin, C.F., Meade, R.H., Curtis, W.F., Bosio, N.J. and Landim, P.M.B. 1980: Size distribution of Amazon River bed sediment. *Nature* **286**, 52–3.

O'Connor, J.E., Ely, L.L., Wohl, E.E., Stevens, L.E., Melis, T.S., Kale, V.S. and Baker, V.R. 1994: A 4500-year record of large floods on the Colorado River in the Grand Canyon, Arizona. *Journal of Geology* **102**, 1–9.

O'Connor, J.E., Webb, R.H. and Baker, V.R. 1986: Paleohydrology of pool–riffle pattern development: Boulder Creek, Utah. *Bulletin of the Geological Society of America* **97**, 410–20.

Odgaard, J. 1987: Streambank erosion along two rivers in Iowa. *Water Resources Research* **23**, 125–36.

O'Donnell, T. 1985: A direct three-parameter Muskingum procedure incorporating lateral inflow. *Hydrological Sciences Journal* **30**, 479–96.

O'Neill, M.P. and Abrahams, A.D. 1984: Objective identification of pools and riffles. *Water Resources Research* **20**, 921–6.

Onishi, Y., Subhash, C. and Kennedy, J.F. 1976: Effects of meandering in alluvial streams. *Journal of the Hydraulics Division American Society of Civil Engineers* **102**, HY7, 899–917.

Osman, A.M. and Thorne, C.R. 1988: Riverbank stability analysis. I. Theory. *Journal of Hydraulic Engineering* **114**, 134–50.

Osterkamp, W.R. 1978: Gradient, discharge, and particle-size relations of alluvial channels in Kansas, with observations on braiding. *American Journal of Science* **278**, 1253–68.

Osterkamp, W.R. 1980: Sediment–morphology relations of alluvial channels. *Proceedings of the Symposium on Watershed Management, American Society of Civil Engineers, Boise 1980*, 188–99.

Osterkamp, W.R. and Costa, J.E. 1987: Changes accompanying an extraordinary flood on a sand-bed stream. In Mayer, L. and Nash, D. (eds), *Catastrophic flooding*. London: Allen & Unwin, 201–24.

Osterkamp, W.R. and Hedman, E.R. 1982: Perennial-streamflow characteristics related to channel geometry and sediment in Missouri River basin. *United States Geological Survey Professional Paper* **1242**.

Osterkamp, W.R., Lane, E.J. and Foster, G.R. 1983: An analytical treatment of channel-morphology relations. *United States Geological Survey Professional Paper* **1288**.

Ouchi, S. 1985: Response of alluvial rivers to slow active tectonic movement. *Bulletin of the Geological Society of America* **96**, 504–15.

Page, K.J. 1988: Bankfull discharge frequency for the Murrumbidgee River, New South Wales. In Warner, R.F. (ed.), *Fluvial geomorphology of Australia*. Sydney: Academic Press, 267–81.

Page, K.J. and Nanson, G.C. 1996: Stratigraphic architecture resulting from Late Quaternary evolution of the Riverine Plain, south-eastern Australia. *Sedimentology* **43**, 927–45.

Paola, C. and Seal, R. 1995: Grain-size patchiness as a cause of selective deposition and downstream fining. *Water Resources Research* **31**, 1395–407.

Park, C.C. 1977: World-wide variations in hydraulic geometry exponents of stream channels: an analysis and some observations. *Journal of Hydrology* **33**, 133–46.

Parker, G. 1976: On the cause and characteristic scales of meandering and braiding in rivers. *Journal of Fluid Mechanics* **76**, 457–80.

Parker, G. 1978: Self-formed straight rivers with equilibrium banks and mobile bed. 2. The gravel river. *Journal of Fluid Mechanics* **89**, 127–46.

Parker, G. 1979: Hydraulic geometry of active gravel rivers. *Journal of the Hydraulics Division American Society of Civil Engineers* **105**, HY9, 1185–201.

Parker, G. 1991: Selective sorting and abrasion of river gravel. I: Theory. *Journal of Hydraulic Engineering* **117**, 131–49.

Parker, G. and Johannesson, H. 1989: Observations on several recent theories of resonance and overdeepening in meandering channels. In Ikeda, S. and Parker, G. (eds), *River meandering*, American Geophysical Union, Water Research Monograph **12**, 379–415.

Parker, G., Klingeman, P.C. and McLean, D.L. 1982: Bedload and size distribution in paved gravel-bed streams. *Journal of Hydraulic Engineering* **108**, 544–71.

Parker, R.S. 1977: Experimental study of drainage basin evolution and its hydrologic implications. *Colorado State University, Fort Collins, Colorado, Hydrology Papers* **90**.

Patrick, D.M., Smith, L.M. and Whitten, C.B. 1982: Methods of studying accelerated fluvial change. In Hey, R.D., Bathurst, J.C. and Thorne, C.R. (eds), *Gravel bed rivers*. Chichester: Wiley, 783–812.

Patton, P.C. and Schumm, S.A. 1981: Ephemeral stream processes: implications for studies of Quaternary valley fills. *Quaternary Research* **15**, 24–43.

Pearce, A.J. 1976: Magnitude and frequency of erosion by Hortonian overland flow. *Journal of Geology* **84**, 65–80.

Penning-Rowsell, E.G. and Townshend, J.R.G. 1978: The influence of scale on the factors affecting stream channel slope. *Transactions of the Institute of British Geographers* New Series **3**, 395–415.

Petit, F. 1987: The relationship between shear stress and the shaping of the bed of a pebble-loaded river, La Rulles – Ardennes. *Catena* **14**, 453–68.

Petts, G.E. 1977: Channel response to flow regulation: the case of the River Derwent, Derbyshire. In Gregory, K.J. (ed.), *River channel changes*. Chichester: Wiley-Interscience, 145–64.

Petts, G.E. 1979: Complex response of river channel morphology to reservoir construction. *Progress in Physical Geography* **3**, 329–62.

Petts, G.E. 1984: *Impounded rivers*. Chichester: Wiley.

Phillips, J.D. 1990a: Relative importance of factors influencing fluvial soil loss at the global scale. *American Journal of Science* **290**, 547–68.

Phillips, J.D. 1990b: The instability of hydraulic geometry. *Water Resources Research* **26**, 739–44.

Phillips, J.D. 1991a: Fluvial sediment budgets in the North Carolina Piedmont. *Geomorphology* **4**, 231–41.

Phillips, J.D. 1991b: Multiple modes of adjustment in unstable river channel cross-sections. *Journal of Hydrology* **123**, 39–49.

Phillips, J.D. 1992: Nonlinear dynamical systems in geomorphology: revolution or evolution. *Geomorphology* **5**, 219–29.

Phillips, J.D. 1993: Interpreting the fractal dimension of river networks. In Lam, N.S.-N. and De Cola, L. (eds), *Fractals in geography*. Englewood Cliffs, NJ: Prentice-Hall, 142–57.

Pickup, G. 1976a: Alternative measures of river channel shape and their significance. *Journal of Hydrology (New Zealand)* **15**, 9–16.

Pickup, G. 1976b: Adjustment of stream-channel shape to hydrologic regime. *Journal of Hydrology* **30**, 365–73.

Pickup, G. and Warner, R.F. 1976: Effects of hydrologic regime on magnitude and frequency of dominant discharge. *Journal of Hydrology* **29**, 51–75.

Pizzuto, J.E. 1984a: Bank erodibility of shallow sandbed streams. *Earth Surface Processes and Landforms* **9**, 113–24.

Pizzuto, J.E. 1984b: Equilibrium bank geometry and the width of shallow sandbed streams. *Earth Surface Processes and Landforms* **9**, 199–207.

Pizzuto, J.E. 1987: Sediment diffusion during overbank flows. *Sedimentology* **34**, 301–17.

Pizzuto, J.E. 1992: The morphology of graded gravel rivers: a network perspective. *Geomorphology* **5**, 457–74.

Pizzuto, J.E. 1995: Downstream fining in a network of gravel-bedded rivers. *Water Resources Research* **31**, 753–9.

Pizzuto, J.E. and Meckelnburg, T.S. 1989: Evaluation of a linear bank erosion equation. *Water Resources Research* **25**, 1005–13.

Poesen, J.W.A. 1992: Mechanisms of overland-flow generation and sediment production on loamy and sandy soils with and without rock fragments. In Parsons, A.J. and Abrahams, A.D. (eds), *Overland flow hydraulics and erosion mechanics*. London: UCL Press, 275–305.

Poulos, S.E., Collins, M. and Evans, G. 1996: Water–sediment fluxes of Greek rivers, southeastern Alpine Europe: annual yields, seasonal variability, delta formation and human impact. *Zeitschrift für Geomorphologie* **40**, 243–61.

Prestegaard, K.L. 1983a: Bar resistance in gravel bed streams at bankfull stage. *Water Resources Research* **19**, 472–6.

Prestegaard, K.L. 1983b: Variables influencing water-surface slopes in gravel-bed streams at bankfull stage. *Bulletin of the Geological Society of America* **94**, 673–8.

Probst, J.-L. 1989: Hydroclimatic fluctuations of some European rivers since 1800. In Petts, G.E. (ed.), *Historical change of large alluvial rivers: Western Europe*. Chichester: Wiley, 41–55.

Probst, J.-L. and Tardy, Y. 1987: Long range streamflow and world continental runoff fluctuations since the beginning of this century. *Journal of Hydrology* **94**, 289–311.

Proffitt, A.P.B. and Rose, C.W. 1991: Soil erosion processes 1. The relative importance of rainfall detachment and runoff entrainment. *Australian Journal of Soil Research* **29**, 671–83.

Prosser, I.P. and Dietrich, W.E. 1995: Field experiments on erosion by overland flow and their implication for a digital terrain model of channel initiation. *Water Resources Research* **31**, 2867–76.

Ramette, M. 1979: Une approche rationelle de la morphologie fluviale. *La Houille Blanche* **8**, 491–8.

Rana, S.A., Simons, D.B. and Mahmood, K. 1973: Analysis of sediment sorting in alluvial channels. *Journal of the Hydraulics Division American Society of Civil Engineers* **99**, HY11, 1967–80.

Raudkivi, A.J. 1997: Ripples on stream bed. *Journal of Hydraulic Engineering* **123**, 58–64.

Reid, I., Frostick, L.E. and Layman, J.T. 1985: The incidence and nature of bedload transport during flood flows in coarse-grained alluvial channels. *Earth Surface Processes and Landforms* **10**, 33–44.

Reid, I. and Laronne, J.B. 1995: Bed load sediment transport in an ephemeral stream and a comparison with seasonal and perennial counterparts. *Water Resources Research* **31**, 773–81.

Reid, J.M., MacLeod, D.A. and Cresser, M.S. 1981: Factors affecting the chemistry of precipitation and river water in an upland catchment. *Journal of Hydrology* **50**, 129–45.

Reid, L.M. and Dunne, T. 1996: *Rapid evaluation of sediment budgets*. Reiskirchen: Catena Verlag.

Reinfelds, I. and Nanson, G.C. 1993: Formation of braided river floodplains, Waimakariri River, New Zealand. *Sedimentology* **40**, 1113–27.

Renwick, W.H. 1992: Equilibrium, disequilibrium, and nonequilibrium landforms in the landscape. *Geomorphology* **5**, 265–76.

Rhoads, B.L. 1987: Changes in stream channel characteristics at tributary junctions. *Physical Geography* **8**, 346–61.

Rhoads, B.L. 1992: Statistical models of fluvial systems. *Geomorphology* **5**, 433–55.

Rhoads, B.L. and Welford, M.R. 1991: Initiation of river meandering. *Progress in Physical Geography* **15**, 127–56.

Rhodes, D.D. 1977: The b–f–m diagram: graphical representation and interpretation of at-a-station hydraulic geometry. *American Journal of Science* **277**, 73–96.

Rice, S. 1994: Towards a model of changes in bed material texture at the drainage basin scale. In Kirkby, M.J. (ed.), *Process models and theoretical geomorphology*. Chichester: Wiley, 159–72.

Rice, S. and Church, M. 1996: Bed material texture in low order streams on the Queen Charlotte Islands, British Columbia. *Earth Surface Processes and Landforms* **21**, 1–18.

Richards, K.S. 1973: Hydraulic geometry and channel roughness – a non-linear system. *American Journal of Science* **273**, 877–96.

Richards, K.S. 1976a: Channel width and the riffle–pool sequence. *Bulletin of the Geological Society of America* **87**, 883–90.

Richards, K.S. 1976b: The morphology of riffle–pool sequences. *Earth Surface Processes* **1**, 71–88.

Richards, K.S. 1977a: Slope form and basal stream relationships: some further comments. *Earth Surface Processes* **2**, 87–95.

Richards, K.S. 1977b: Channel and flow geometry. *Progress in Physical Geography* **1**, 65–102.

Richards, K.S. 1979: Channel adjustment to sediment pollution by the china clay industry in Cornwall, England. In Rhodes, D.D. and Williams, G.P. (eds), *Adjustments of the fluvial system*. Dubuque, IA: Kendall-Hunt, 309–31.

Richards, K.S. 1980: A note on changes in channel geometry at tributary junctions. *Water Resources Research* **16**, 241–4.

Richards, K.S. 1982: *Rivers: form and process in alluvial channels*. London: Methuen.

Richards, K.S. 1993: Sediment delivery and the drainage network. In Beven, K. and Kirkby, M.J. (eds), *Channel network hydrology*. Chichester: Wiley, 221–54.

Richards, K.S. and Clifford, N.J. 1991: Fluvial geomorphology: structured beds in gravelly rivers. *Progress in Physical Geography* **15**, 407–22.

Ridenour, G.S. and Giardino, J.R. 1991: The statistical study of hydraulic geometry: a new direction for compositional data analysis. *Mathematical Geology* **23**, 349–66.

Rigon, R., Rodriguez-Iturbe, I., Maritan, A., Giacometti, A., Tarboton, D.G. and Rinaldo, A. 1996: On Hack's law. *Water Resources Research* **32**, 3367–74.

Rinaldo, A., Rodriguez-Iturbe, I., Rigon, R., Bras, R.L., Ijjasz-Vasquez, E.J. and Marani, A. 1992: Minimum energy and fractal structures of drainage networks. *Water Resources Research* **28**, 2183–95.

Ritter, D.F., Kinsey, W.F. and Kauffman, M.E. 1973: Overbank sedimentation in the Delaware River Valley during the last 6000 years. *Science* **179**, 374–5.

Ritter, J.B. and Gardner, T.W. 1993: Hydrologic evolution of drainage basins disturbed by surface mining, central Pennsylvania. *Bulletin of the Geological Society of America* **105**, 101–15.

Robert, A. 1990: Boundary roughness in coarse-grained channels. *Progress in Physical Geography* **14**, 42–70.

Roberts, C.R. 1989: Flood frequency and urban-induced channel change: some British examples. In Beven, K.J. and Carling, P.A. (eds), *Floods: hydrological, sedimentological and geomorphological implications*. Chichester: Wiley, 57–82.

Roberts, M.C. 1978: Drainage density variations on the morainic landscapes of northeastern Indiana. *Zeitschrift für Geomorphologie* **22**, 462–71.

Robinson, J.S., Sivapalan, M. and Snell, J.D. 1995: On the relative roles of hillslope processes, channel routing and network geomorphology in the hydrologic response of natural catchments. *Water Resources Research* **31**, 3089–101.

Rodriguez-Iturbe, I., Rinaldo, A., Rigon, R., Bras, R.L., Marani, A. and Ijjasz-Vasquez, E. 1992: Energy dissipation, runoff production, and the three-dimensional structure of river basins. *Water Resources Research* **28**, 1095–103.

Rodriguez-Iturbe, I. and Valdes, J.B. 1979: The geomorphologic structure of hydrologic response. *Water Resources Research* **15**, 1409–20.

Rosgen, D.L. 1994: A classification of natural rivers. *Catena* **22**, 169–99.

Rotnicki, A. 1991: Retrodiction of palaeodischarges of meandering and sinuous alluvial rivers and its palaeohydroclimatic implications. In Starkel, L., Gregory, K.J. and Thornes, J.B. (eds), *Temperate palaeohydrology*. Chichester: Wiley, 431–71.

Roy, A.G. 1985: Optimal models of river branching angles. In Woldenberg, M.J. (ed.), *Models in geomorphology*. London: Allen & Unwin, *Binghamton Symposia in Geomorphology* **14**, 269–85.

Roy, A.G. and Woldenberg, M.J. 1986: A model for changes in channel form at a river confluence. *Journal of Geology* **94**, 402–11.

Rubey, W.W. 1952: Geology and mineral resources of the Hardin and Brussels quadrangles (in Illinois). *United States Geological Survey Professional Paper* **218**.

Ruhe, R.V. 1952: Topographic discontinuities of the Des Moines lobe. *American Journal of Science* **250**, 46–56.

Rumsby, B.T. and Macklin, M.G. 1994: Channel and floodplain response to recent abrupt climate change: the Tyne basin, northern England. *Earth Surface Processes and Landforms* **19**, 499–515.

Rundle, A.S. 1985: Braid morphology and the formation of multiple channels: the Rakaia, New Zealand. *Zeitschrift für Geomorphologie Supplement-Band* **55**, 15–37.

Rundquist, L.A. 1975: A classification and analysis of natural rivers. Colorado State University, Fort Collins, Colorado, PhD thesis (unpublished).

Said, R. 1993: *The River Nile: geology, hydrology and utilization*. Oxford: Pergamon.

Sambrook Smith, G.H. 1996: Bimodal fluvial bed sediments: origin, spatial extent and processes. *Progress in Physical Geography* **20**, 402–17.

Sambrook Smith, G.H. and Ferguson, R.I. 1995: The gravel–sand transition along river channels. *Journal of Sedimentary Research* **A65**, 423–30.

Santos-Cayade, J. and Simons, D.B. 1973: River response. In Shen, H.W. (ed.), *Environmental impact on rivers*. Fort Collins, CO: Colorado State University, 1-1-1-25.

Saucier, R.T. 1984: Historic changes in current river meandering regime. In Elliott, C.M. (ed.), *River meandering*. New Orleans: American Society of Civil Engineers, 180–90.

Savat, J. and De Ploey, J. 1982: Sheetwash and rill development by surface flow. In Bryan, R.B. and Yair, A. (eds), *Badland geomorphology and piping*. Norwich: Geobooks, 113–26.

Schick, A.P. 1974: Formation and obliteration of desert stream terraces – a conceptual analysis. *Zeitschrift für Geomorphologie Supplement-Band* **21**, 88–105.

Schröder, R. 1991: Test of Hack's slope to bed material relationship in the southern Eifel uplands, Germany. *Earth Surface Processes and Landforms* **16**, 731–6.

Schumann, R.R. 1989: Morphology of Red Creek, Wyoming, an arid-region anastomosing channel system. *Earth Surface Processes and Landforms* **14**, 277–88.

Schumm, S.A. 1956: The evolution of drainage systems and slopes in badlands at Perth Amboy, New Jersey. *Bulletin of the Geological Society of America* **67**, 597–646.

Schumm, S.A. 1960: The shape of alluvial channels in relation to sediment type. *United States Geological Survey Professional Paper* **352B**, 17–30.

Schumm, S.A. 1963a: A tentative classification of alluvial river channels. *United States Geological Survey Circular* **477**.

Schumm, S.A. 1963b: Sinuosity of alluvial rivers on the Great Plains. *Bulletin of the Geological Society of America* **74**, 1089–100.

Schumm, S.A. 1967: Meander wavelength of alluvial rivers. *Science* **157**, 1549–50.

Schumm, S.A. 1968: River adjustment to altered hydrologic regimen – Murrumbidgee River and paleochannels, Australia. *United States Geological Survey Professional Paper* **598**.

Schumm, S.A. 1969: River metamorphosis. *Journal of the Hydraulics Division American Society of Civil Engineers* **95**, HY1, 255–73.

Schumm, S.A. 1971: Fluvial geomorphology: the historical perspective. In Shen, H.W. (ed.), *River mechanics*, volume I. Fort Collins, CO: H.W. Shen, 4-1–4-30.

Schumm, S.A. 1973: Geomorphic thresholds and complex response of drainage systems. In Morisawa, M. (ed.), *Fluvial geomorphology*. Binghamton, NY: New York State University Publications in Geomorphology, 299–309.

Schumm, S.A. 1977: *The fluvial system*. New York: Wiley-lnterscience.

Schumm, S.A. 1980: Some applications of the concept of geomorphic thresholds. In Coates, D.R. and Vitek, J.D. (eds), *Thresholds in geomorphology*. Boston: George Allen & Unwin, 473–85.

Schumm, S.A. 1981: Evolution and response of the fluvial system, sedimentologic implications. *Society of Economic Paleontologists and Mineralogists Special Publication* **31**, 19–29.

Schumm, S.A. 1985: Patterns of alluvial rivers. *Annual Review of Earth and Planetary Sciences* **13**, 5–27.

Schumm, S.A. 1993: River response to base level change: implications for sequence stratigraphy. *Journal of Geology* **101**, 279–92.

Schumm, S.A. and Hadley, R.F. 1957: Arroyos and the semi-arid cycle of erosion. *American Journal of Science* **255**, 161–74.

Schumm, S.A. and Khan, H.R. 1972: Experimental study of channel patterns. *Bulletin of the Geological Society of America* **83**, 1755–70.

Schumm, S.A. and Lichty, R.W. 1963: Channel widening and flood-plain construction along Cimarron River in south-western Kansas. *United States Geological Survey Professional Paper* **352D**, 71 -88.

Schumm, S.A. and Lichty, R.W. 1965: Time, space, and causality in geomorphology. *American Journal of Science* **263**, 110–19.

Schumm, S.A., Mosley, M.P. and Weaver, W.E. 1987: *Experimental fluvial geomorphology*. Chichester: Wiley.

Schumm, S.A. and Stevens, M.A. 1973: Abrasion in place: a mechanism for rounding and size reduction of coarse sediments in rivers. *Geology* **1**, 37–40.

Sear, D.A. 1996: Sediment transport in pool–riffle sequences. *Earth Surface Processes and Landforms* **21**, 241–62.

Sear, D.A., Newson, M.D. and Brookes, A. 1995: Sediment-related river maintenance: the role of fluvial geomorphology. *Earth Surface Processes and Landforms* **20**, 629–47.

Sedimentation Seminar, H.N. Fisk Laboratory of Sedimentology 1977: Magnitude and frequency of transport of solids by streams in the Mississippi basin. *American Journal of Science* **277**, 862–75.

Seminara, G. and Tubino, M. 1989: Alternate bars and meandering: free, forced and mixed interactions. In Ikeda, S. and Parker, G. (eds), *River meandering*, American Geophysical Union, Water Research Monograph **12**, 267–320.

Sharma, K.D. and Chatterji, P.C. 1982: Sedimentation in nadis in the Indian arid zone. *Hydrological Sciences Journal* **27**, 345–52.

Shaw, J. and Kellerhals, R. 1982: The composition of recent alluvial gravels in Alberta river beds. *Alberta Research Council Bulletin* **41**.

Shen, H.W. and Komura, S. 1968: Meandering tendencies in straight alluvial channels. *Journal of the Hydraulics Division American Society of Civil Engineers* **94**, HY4, 997–1016.

Shields, A. 1936: Anwendung der Ähnlichkeitsmechanik und der Turbulenzforschung auf die Geschiebebewegung. *Mitteilung der preussischen Versuchsanstalt für Wasserbau und Schiffbau* **26**, Berlin.

Shreve, R.L. 1966: Statistical law of stream numbers. *Journal of Geology* **74**, 17–37.

Shreve, R.L. 1967: Infinite topologically random channel networks. *Journal of Geology* **75**, 178–86.

Shreve, R.L. 1975: The probabilistic–topologic approach to drainage-basin geomorphology. *Geology* **3**, 527–9.

Simon, A. 1989: A model of channel response in disturbed alluvial channels. *Earth Surface Processes and Landforms* **14**, 11–26.

Simon, A. 1992: Energy, time, and channel evolution in catastrophically disturbed fluvial systems. *Geomorphology* **5**, 345–72.

Simon, A. 1994: Gradation processes and channel evolution in modified West Tennessee streams: process, response, and form. *United States Geological Survey Professional Paper* **1470**.

Simon, A. 1995: Adjustment and recovery of unstable alluvial channels: identification and approaches for engineering management. *Earth Surface Processes and Landforms* **20**, 611–28.

Simon, A. and Thorne, C.R. 1996: Channel adjustment of an unstable coarse-grained stream: opposing trends of boundary and critical shear stress, and the applicability of extremal hypotheses. *Earth Surface Processes and Landforms* **21**, 155–80.

Simons, D.B. and Albertson, M.L. 1963: Uniform water conveyance channels in alluvial material. *Transactions of the American Society of Civil Engineers* **128**, 65–107.

Simons, D.B., Li, R.-M., Alawady, M.A. and Andrew, J.W. 1979: *Report on: Connecticut River streambank erosion study, Massachusetts, New Hampshire and Vermont.* Colorado State University Research Institute, Fort Collins, CO, Report CSU-213.

Simons, D.B. and Richardson, E.V. 1966: Resistance to flow in alluvial channels. *United States Geological Survey Professional Paper* **422J**.

Simons, D.B., Richardson, E.V. and Haushild, W.H. 1963: Some effects of fine sediment on flow phenomena. *United States Geological Survey Water-Supply Paper* **1498G**.

Sinha, S.K. and Parker, G. 1996: Causes of concavity in longitudinal profiles of rivers. *Water Resources Research* **32**, 1417–28.

Slattery, M.C. and Bryan, R.B. 1992: Hydraulic conditions for rill incision under simulated rainfall: a laboratory experiment. *Earth Surface Processes and Landforms* **17**, 127–46.

Smart, J.S. 1968: Statistical properties of stream lengths. *Water Resources Research* **4**, 1001–14.

Smart, J.S. 1972: Channel networks. *Advances in Hydroscience* **8**, 305–46.

Smart, J.S. 1978: The analysis of drainage network composition. *Earth Surface Processes* **3**, 129–71.

Smith, D.G. 1976: Effect of vegetation on lateral migration of anastomosed channels of a glacier meltwater river. *Bulletin of the Geological Society of America* **87**, 857–60.

Smith, D.G. 1983: Anastomosed fluvial deposits: modern examples from western Canada. In Collinson, J.D. and Lewin, J. (eds), *Modern and ancient fluvial systems.* Oxford: Blackwell, Special Publication of the International Association of Sedimentologists **6**, 155–68.

Smith, D.G. and Putnam, P.E. 1980: Anastomosed river deposits: modern and ancient examples in Alberta, Canada. *Canadian Journal of Earth Sciences* **17**, 1396–406.

Smith, D.G. and Smith, N.D. 1980: Sedimentation in anastomosed river systems: examples from alluvial valleys near Banff, Alberta. *Journal of Sedimentary Petrology* **50**, 157–64.

Smith, N.D., Cross, T.A., Dufficy, J.P. and Clough, S.R. 1989: Anatomy of an avulsion. *Sedimentology* **36**, 1–23.

Smith, T.R. and Bretherton, F.P. 1972: Stability and the conservation of mass in drainage basin evolution. *Water Resources Research* **8**, 1506–29.

Snow, R.S. 1989: Fractal sinuosity of stream channels. *Pure and Applied Geophysics* **131**, 99–109.

Snow, R.S. and Slingerland, R.L. 1987: Mathematical modeling of graded river profiles. *Journal of Geology* **95**, 15–33.

Speight, J.G. 1967: Spectral analysis of meanders of some Australian rivers. In Jennings, J.N. and Mabbutt, J.A. (eds), *Landform studies from Australia and New Guinea.* Cambridge: Cambridge University Press, 48–63.

Stanley, D.J., Krinitzsky, E.L. and Compton, J.R. 1966: Mississippi river bank failure. *Bulletin of the Geological Society of America* **77**, 859–66.

Stark, C.P. 1991: An invasion percolation model of drainage network evolution. *Nature* **352**, 423–5.

Starkel, L. 1983: The reflection of hydrologic changes in the fluvial environment of the temperate zone during the last 15,000 years. In Gregory, K.J. (ed.), *Background to palaeohydrology*. Chichester: Wiley, 213–35.

Starkel, L. 1991a: The Vistula River valley: a case study for central Europe. In Starkel, L., Gregory, K.J. and Thornes, J.B. (eds), *Temperate palaeohydrology.* Chichester: Wiley, 171–88.

Starkel, L. 1991b: Long-distance correlation of fluvial events in the temperate zone. In Starkel, L., Gregory, K.J. and Thornes, J.B. (eds), *Temperate palaeohydrology.* Chichester: Wiley, 473–95.

Starkel, L., Gregory, K.J. and Thornes, J.B. (eds) 1991: *Temperate palaeohydrology.* Chichester: Wiley.

Statham, I. 1977: *Earth surface sediment transport*. Oxford: Clarendon Press.

Stevens, M.A. 1989: Width of straight alluvial channels. *Journal of Hydraulic Engineering* **115**, 309–26.

Stevens, M.A., Simons, D.B. and Richardson, E.V. 1975: Non-equilibrium river form. *Journal of the Hydraulics Division American Society of Civil Engineers* **101**, HY5, 557–66.

Stewart, J.H. and LaMarche, V.C. 1967: Erosion and deposition produced by the flood of December 1964 on Coffee Creek, Trinity County, California. *United States Geological Survey Professional Paper* **422K**.

Stoddart, D.R. 1978: Geomorphology in China. *Progress in Physical Geography* **2**, 187–236.

Stølum, H.-H. 1996: River meandering as a self-organization process. *Science* **271**, 1710–13.

Strahler, A.N. 1950: Equilibrium theory of erosional slopes approached by frequency distribution analysis. *American Journal of Science* **248**, 673–96 and 800–14.

Strahler, A.N. 1952: Hypsometric (area–altitude) analysis of erosional topography. *Bulletin of the Geological Society of America* **63**, 1117–42.

Summerfield, M.A. and Hulton, N.J. 1994: Natural controls of fluvial denudation rates in major world drainage basins. *Journal of Geophysical Research* **99**, 13871–83.

Sun, T., Meakin, P., Jøssang, T. and Schwarz, K. 1996: A simulation model for meandering rivers. *Water Resources Research* **32**, 2937–54.

Tanner, W.F. 1960: Helicoidal flow, a possible cause of meandering. *Journal of Geophysical Research* **65**, 993–5.

Tarboton, D.G., Bras, R.L. and Rodriguez-Iturbe, I. 1988: The fractal nature of river networks. *Water Resources Research* **24**, 1317–22.

Task Committee for Preparation of Sediment Manual 1971: Sediment transportation mechanics: Q. Genetic classification of valley sediment deposits. *Journal of the Hydraulics Division American Society of Civil Engineers* **97**, HY1, 43–53.

Task Force on Friction Factors in Open Channels 1963: Friction factors in open channels: progress report. *Journal of the Hydraulics Division American Society of Civil Engineers* **89**, HY2, 97–143.

Thompson, A. 1986: Secondary flows and the pool–riffle unit: a case study of the processes of meander development. *Earth Surface Processes and Landforms* **11**, 631–41.

Thoms, M. and Thiel, P. 1995: The impact of urbanisation on the bed sediments of South Creek, New South Wales. *Australian Geographical Studies* **33**, 31–43.

Thorn, C.E. and Welford, M.R. 1994: The equilibrium concept in geomorphology. *Annals of the Association of American Geographers* **84**, 666–96.

Thorne, C.R. 1982: Processes and mechanisms of river bank erosion. In Hey, R.D., Bathurst, J.C. and Thorne, C.R. (eds), *Gravel bed rivers*. Chichester: Wiley, 227–59.

Thorne, C.R., Russell, A.P.G. and Alam, M.K. 1993: Planform pattern and channel evolution of the Brahmaputra River, Bangladesh. In Best, J.L. and Bristow, C.S. (eds), *Braided rivers. Special Publication of the Geological Society of London* **75**, 257–76.

Thorne, C.R. and Tovey, N.K. 1981: Stability of composite river banks. *Earth Surface Processes and Landforms* **6**, 469–84.

Tinkler, K.J. 1970: Pools, riffles, and meanders. *Bulletin of the Geological Society of America* **81**, 547–52.

Torri, D. and Poesen, J. 1992: The effect of soil surface slope on raindrop detachment. *Catena* **19**, 561–78.

Trimble, S.W. 1974: *Man-induced soil erosion on the southern Piedmont 1700–1900*. Ankeny, IA: Soil Conservation Society of America.

Trimble, S.W. 1975: Denudation studies: can we assume stream steady state? *Science* **188**, 1207–8.

Trimble, S.W. 1983: A sediment budget for Coon Creek basin in the Driftless Area, Wisconsin, 1853–1977. *American Journal of Science* **283**, 454–74.

Trimble, S.W. 1994: Erosional effects of cattle on streambanks in Tennessee, USA. *Earth Surface Processes and Landforms* **19**, 451–64.

Troch, P.A., Smith, J.A., Wood, E.F. and De Troch, F.P. 1994: Hydrologic controls of large floods in a small basin: central Appalachian case study. *Journal of Hydrology* **156**, 285–309.

Troendle, C.A. 1985: Variable source area models. In Anderson, M.G. and Burt, T.P. (eds), *Hydrological forecasting*. New York: Wiley, 347–403.

Troutman, B.M. 1980: A stochastic model for particle sorting and related phenomena. *Water Resources Research* **16**, 65–76.

Troutman, B.M. and Karlinger, M.R. 1985: Unit hydrograph approximations assuming linear flow through topologically random channel networks. *Water Resources Research* **21**, 743–54.

van den Berg, J.H. 1995: Prediction of alluvial channel pattern of perennial rivers. *Geomorphology* **12**, 259–79.

Van Haveren, B.P., Jackson, W.L. and Lusby, G.C. 1987: Sediment deposition behind Sheep Creek Barrier Dam, southern Utah. *Journal of Hydrology (New Zealand)* **26**, 185–96.

Vanoni, V.A. and Nomicos, G.N. 1960: Resistance properties of sediment-laden streams. *Transactions of the American Society of Civil Engineers* **125**, 1140–67.

van Pelt, J., Woldenberg, M.J. and Verwer, R.W.H. 1989: Two generalized topological models of stream network growth. *Journal of Geology* **97**, 281–99.

Van Rijn, L.C. 1984: Sediment transport. Part III: Bed forms and alluvial roughness. *Journal of Hydraulic Engineering* **110**, 1733–54.

Walker, J., Arnborg, L. and Peippo, J. 1987: Riverbank erosion in the Colville Delta, Alaska. *Geografiska Annaler* **69A,** 61–70.

Walling, D.E. 1983: The sediment delivery problem. *Journal of Hydrology* **65,** 209–37.

Walling, D.E. 1987: Rainfall, runoff and erosion of the land: a global view. In Gregory, K.J. (ed.), *Energetics of physical environment*. Chichester: Wiley, 89–117.

Walling, D.E. 1988: Erosion and sediment yield research – some recent perspectives. *Journal of Hydrology* **100**, 113–41.

Walling, D.E. and Kleo, A.H.A. 1979: Sediment yields of rivers in areas of low precipitation: a global view. In *The hydrology of areas of low precipitation*, Proceedings of the Canberra Symposium, December 1979, IAHS-AISH Publication **128**, 479–93.

Walling, D.E., Quine, T.A. and He, Q. 1992: Investigating contemporary rates of floodplain sedimentation. In Carling, P.A. and Petts, G.E. (eds), *Lowland floodplain rivers*. Chichester: Wiley, 165–84.

Walling, D.E. and Webb, B.W. 1983: Patterns of sediment yield. In Gregory, K.J. (ed.), *Background to palaeohydrology*. Chichester: Wiley, 69–100.

Walling, D.E. and Webb, B.W. 1986: Solutes in river systems. In Trudgill, S.T. (ed.), *Solute processes*. Chichester: Wiley, 251–327.

Walling, D.E. and Webb, B.W. 1987a: Material transport by the world's rivers: evolving perspectives. In *Water for the future: hydrology in perspective*. Wallingford, UK: International Association of Hydrological Sciences, Proceedings of the Rome Symposium, April 1987, International Association of Hydrological Sciences Publication **164**, 313–29.

Walling, D.E. and Webb, B.W. 1987b: Suspended load in gravel-bed rivers: UK experience. In Thorne, C.R., Bathurst, J.C. and Hey, R.D. (eds), *Sediment transport in gravel-bed rivers*. Chichester: Wiley, 691–723.

Walling, D.E. and Webb, B.W. 1992: Water quality: I. Physical characteristics. In Calow, P. and Petts, G.E. (eds), *The river's handbook: hydrological and ecological principles*. Oxford: Blackwell, 48–72.

Walling, D.E., Woodward, J.C. and Nicholas, A.P. 1993: A multi-parameter approach to fingerprinting suspended-sediment sources. In *Tracers in hydrology*, Proceedings of the Yokohama Symposium, July 1993, International Association of Hydrological Sciences Publication 215, 329–38.

Walsh, R.P.D. 1985: The influence of climate, lithology and time on drainage density and relief development in the tropical terrain on the Windward Islands. In

Douglas, I. and Spencer, T. (eds), *Environmental change and tropical geomorphology*. London: Allen & Unwin, 93–122.

Warner, R.F. 1987: Spatial adjustments to temporal variations in flood regime in some Australian rivers. In Richards, K.S. (ed.), *River channels: environment and process*. Oxford: Blackwell, 14–40.

Warner, R.F. 1994: A theory of channel and floodplain responses to alternating regimes and its application to actual adjustments in the Hawkesbury River, Australia. In Kirkby, M.J. (ed.), *Process models and theoretical geomorphology*. Chichester: Wiley, 173–200.

Webb, B.W. and Walling, D.E. 1982: The magnitude and frequency characteristics of fluvial transport in a Devon drainage basin and some geomorphological implications. *Catena* **9**, 9–24.

Webb, B.W. and Walling, D.E. 1983: Stream solute behaviour in the River Exe basin, Devon, UK. In *Dissolved loads of rivers and surface water quantity/quality relationships*, Proceedings of the Hamburg Symposium, August 1983, International Association of Hydrological Sciences Publication **141**, 153–69.

Wendland, W.M. and Bryson, R.A. 1974: Dating climatic episodes of the Holocene. *Quaternary Research* **4**, 9–24.

Werritty, A. 1992: Downstream fining in a gravel-bed river in southern Poland: lithologic controls and the role of abrasion. In Billi, P., Hey, R.D., Thorne, C.R. and Tacconi, P. (eds), *Dynamics of gravel-bed rivers*. Chichester: Wiley, 333–46.

Wharton, G. 1994: Progress in the use of drainage network indices for rainfall-runoff modelling and runoff prediction. *Progress in Physical Geography* **18**, 539–57.

Wharton, G. 1995: The channel-geometry method: guidelines and applications. *Earth Surface Processes and Landforms* **20**, 649–60.

Wharton, G., Arnell, N.W., Gregory, K.J. and Gurnell, A.M. 1989: River discharge estimated from channel dimensions. *Journal of Hydrology* **106**, 365–76.

Wheeler, D.A. 1979: The overall shape of longitudinal profiles of streams. In Pitty, A.F. (ed.), *Geographical approaches to fluvial processes*. Norwich: Geobooks, 241–60.

White, W.R., Bettess, R. and Paris, E. 1982: Analytical approach to river regime. *Journal of the Hydraulics Division American Society of Civil Engineers* **108**, 1179–93.

Whiting, P.J. and Dietrich, W.E. 1993a: Experimental studies of bed topography and flow patterns in large-amplitude meanders: I. Observations. *Water Resources Research* **29**, 3605–14.

Whiting, P.J. and Dietrich, W.E. 1993b: Experimental constraints on bar migration through bends: implications for meander wavelength selection. *Water Resources Research* **29**, 1091–102.

Whiting, P.J., Dietrich, W.E., Leopold, L.B., Drake, T.G. and Shreve, R.L. 1988: Bedload sheets in heterogeneous sediment. *Geology* **16**, 105–8.

Whittaker, J.G. 1987: Sediment transport in step–pool streams. In Thorne, C.R., Bathurst, J.C. and Hey, R.D. (eds), *Sediment transport in gravel-bed rivers*. Chichester: Wiley, 545–70.

Whittaker, J.G. and Jaeggi, M.N.R. 1982: Origin of step–pool systems in mountain streams. *Journal of the Hydraulics Division American Society of Civil Engineers* **108**, HY6, 758–73.

Wilby, R.L., Dalgleish, H.Y. and Foster, I.D.L. 1997: The impact of weather patterns on historic and contemporary catchment sediment yields. *Earth Surface Processes and Landforms* **22**, 353–63.

Wilcock, D.N. 1967: Coarse bedload as a factor determining bed slope. *Publication of the International Association of Scientific Hydrology* **75**, 143–50.

Wilcock, D.N. 1971: Investigation into the relations between bedload transport and channel shape. *Bulletin of the Geological Society of America* **82**, 2159–76.

Wilcock, P.R. 1993: Critical shear stress of natural sediments. *Journal of Hydraulic Engineering* **119**, 491–505.

Wilcock, P.R. and McArdell, B.W. 1993: Surface-based fractional transport rates: mobilization thresholds and partial transport of a sand–gravel sediment. *Water Resources Research* **29**, 1297–312.

Wildman, N.A. 1981: Episodic removal of hydraulic-mining debris, Yuba and Bear River basins, California. Colorado State University, Fort Collins, CO, MSc thesis (unpublished).

Willgoose, G.R., Bras, R.L. and Rodriguez-Iturbe, I. 1991a: Results from a new model of river basin evolution. *Earth Surface Processes and Landforms* **16**, 237–54.

Willgoose, G.R., Bras, R.L. and Rodriguez-Iturbe, I. 1991b: A physically-based coupled network growth and hillslope evolution model: 1. Theory. *Water Resources Research* **27**, 1671–84.

Willgoose, G.R., Bras, R.L. and Rodriguez-Iturbe, I. 1991c: A physically-based coupled network growth and hillslope evolution model: 2. Applications. *Water Resources Research* **27**, 1685–96.

Williams, G.P. 1978a: Bankfull discharge of rivers. *Water Resources Research* **14**, 1141–58.

Williams, G.P. 1978b: Hydraulic geometry of river cross-sections – theory of minimum variance. *United States Geological Survey Professional Paper* **1029**.

Williams, G.P. 1978c: The case of the shrinking channels – the North Platte and Platte Rivers in Nebraska. *United States Geological Survey Circular* **781**.

Williams, G.P. 1983: Paleohydrological methods and some examples from Swedish fluvial environments. *Geografiska Annaler* **65A**, 227–44.

Williams, G.P. 1986: River meanders and channel size. *Journal of Hydrology* **88**, 147–64.

Williams, G.P. 1988: Paleofluvial estimates from dimensions of former channels and meanders. In Baker, V.R., Kochel, R.C. and Patton, P.C. (eds), *Flood geomorphology*. Chichester: Wiley, 321–34.

Williams, G.P. 1989: Sediment concentration versus water discharge during single hydrologic events in rivers. *Journal of Hydrology* **111**, 89–106.

Williams, G.P. and Wolman, M.G. 1984: Downstream effects of dams on alluvial rivers. *United States Geological Survey Professional Paper* **1286**.

Wilson, D.D. 1985: Erosional and depositional trends in rivers of the Canterbury Plains, New Zealand. *Journal of Hydrology (New Zealand)* **24**, 32–44.

Wilson, L. 1973: Variations in mean annual sediment yield as a function of mean annual precipitation. *American Journal of Science* **273**, 335–49.

Winkley, B.R. 1982: Response of the Lower Mississippi to river training and realignment. In Hey, R.D., Bathurst, J.C. and Thorne, C.R. (eds), *Gravel bed rivers*. Chichester: Wiley, 659–80.

Wischmeier, W.H. and Smith, D.D. 1978: *Predicting rainfall erosion losses – a guide to conservation planning*. Agriculture Handbook No. 537. Washington, DC: US Department of Agriculture.

Wise, S.M. 1980: Caesium-137 and lead-210: a review of the techniques and some applications in geomorphology. In Cullingford, R.A., Davidson, D.A. and Lewin, J. (eds), *Timescales in geomorphology*. Chichester: Wiley-Interscience, 109–27.

Wohl, E.E. 1992: Gradient irregularity in the Herbert Gorge of northeastern Australia. *Earth Surface Processes and Landforms* **17**, 69–84.

Wohl, E.E. and Grodek, T. 1994: Channel bed-steps along Nahal Yael, Negev desert, Israel. *Geomorphology* **9**, 117–26.

Wohl, E.E., Vincent, K.R. and Merritts, D.J. 1993: Pool and riffle characteristics in relation to channel gradient. *Geomorphology* **6**, 99–110.

Wolman, M.G. 1955: The natural channel of Brandywine Creek, Pennsylvania. *United States Geological Survey Professional Paper* **271**.

Wolman, M.G. 1959: Factors influencing erosion of a cohesive river bank. *American Journal of Science* **257**, 204–16.

Wolman, M.G. 1967: A cycle of sedimentation and erosion in urban river channels. *Geografiska Annaler* **49A**, 385–95.

Wolman, M.G. and Gerson, R. 1978: Relative scales of time and effectiveness of climate in watershed geomorphology. *Earth Surface Processes* **3**, 189–208.

Wolman, M.G. and Leopold, L.B. 1957: River flood plains: some observations on their formation. *United States Geological Survey Professional Paper* **282C**, 87–109.

Wolman, M.G. and Miller, J.P. 1960: Magnitude and frequency of forces in geomorphic processes. *Journal of Geology* **68**, 54–74.

Womack, W.R. and Schumm, S.A. 1977: Terraces of Douglas Creek, northwestern Colorado: an example of episodic erosion. *Geology* **5**, 72–6.

Wyzga, B. 1993: River response to channel regulation: case study of the Raba River, Carpathians, Poland. *Earth Surface Processes and Landforms* **18**, 541–56.

Wyzga, B. 1996: Changes in the magnitude and transformation of flood waves subsequent to the channelization of the Raba River, Polish Carpathians. *Earth Surface Processes and Landforms* **21**, 749–63.

Xu Jiongxin, 1990: An experimental study of complex response in river channel adjustment downstream from a reservoir. *Earth Surface Processes and Landforms* **15**, 45–53.

Xu Jiongxin, 1991: Study of overall longitudinal profile shape of alluvial rivers: an example from rivers in the Huanghuaihai Plain, China. *Zeitschrift für Geomorphologie* **35**, 479–90.

Xu Jiongxin, 1996a: Underlying gravel layers in a large sand bed river and their influence on downstream-dam channel adjustment. *Geomorphology* **17**, 351–9.

Xu Jiongxin, 1996b: Channel pattern change downstream from a reservoir: an example of wandering braided rivers. *Geomorphology* **15**, 147–58.

Yalin, M.S. 1971: On the formation of dunes and meanders. *Proceedings of the 14th International Congress of the International Association for Hydraulic Research* **3**, Paper C13, 1–8.

Yalin, M.S. 1992: *River mechanics*. Oxford: Pergamon.

Yang, C.T. 1971: Formation of riffles and pools. *Water Resources Research* **7**, 1567–74.

Yang, C.T. 1976: Minimum unit stream power and fluvial hydraulics. *Journal of the Hydraulics Division American Society of Civil Engineers* **102**, HY7, 919–34.

Yang, C.T. and Song, C.S.S. 1979: Theory of minimum rate of energy dissipation. *Journal of the Hydraulics Division American Society of Civil Engineers* **105**, HY7, 769–84.

Yang, C.T., Song, C.S.S. and Woldenberg, M.J. 1981: Hydraulic geometry and minimum rate of energy dissipation. *Water Resources Research* **17**, 1014–18.

Yatsu, E. 1955: On the longitudinal profile of the graded river. *Transactions of the American Geophysical Union* **36**, 655–63.

Yodis, E.G. and Kesel, R.H. 1993: The effects and implications of base-level changes to Mississippi River tributaries. *Zeitschrift für Geomorphologie* **37**, 385–402.

You Lianyuan, 1987: A study of the formation and evolution of braided channels with stable islands: the middle and lower reaches of the Yangtze River. In Gardiner, V. (ed.), *International geomorphology 1986 part I*. Chichester: Wiley, 649–62.

Yu, B. and Wolman, M.G. 1987: Some dynamic aspects of river geometry. *Water Resources Research* **23**, 501–9.

Zaslavsky, D. and Sinai, G. 1981: Surface hydrology: I – Explanation of phenomena; II – Distribution of raindrops; III – Causes of lateral flow; IV – Flow in sloping, layered soil; V – In-surface transient flow. *Journal of the Hydraulics Division American Society of Civil Engineers* **107**, HY1, 1–93.

Zeller, J. 1967: Meandering channels in Switzerland. *Publication of the International Association of Scientific Hydrology* **75**, 174–86.

Zernitz, E.R. 1932: Drainage patterns and their significance. *Journal of Geology* **40**, 498–521.

INDEX

Numbers in *italics* refer to figures; **bold** numbers indicate pages of particular relevance